S0-DYE-208

SEVENTH EDITION

Basic Statistical Analysis

RICHARD C. SPRINTHALL

American International College

Boston New York San Francisco
Mexico City Montreal Toronto London Madrid Munich Paris
Hong Kong Singapore Tokyo Cape Town Sydney

Series Editor: *Kelly May*
Editorial Assistant: *Marlana Voerster*
Marketing Manager: *Brad Parkins*
Editorial Production Service: *Chestnut Hill Enterprises, Inc.*
Manufacturing Buyer: *JoAnne Sweeney*
Cover Administrator: *Linda Knowles*
Electronic Composition: *Omegatype Typography, Inc.*

For related titles and support materials, visit our online catalogue at www.ablongman.com.

Between the time Website information is gathered and published, some sites may have closed. Also, the transcription of URLs can result in typographical errors. The publisher would appreciate notification where these occur so that they may be corrected in subsequent editions.

Library of Congress Cataloging-in-Publication Data

Sprinthall, Richard C.
Basic statistical analysis / Richard C. Sprinthall. — 7th ed.
p. cm.
Includes bibliographical references and index.
ISBN 0-205-36066-1 (alk. paper)
1. Social sciences—Statistical methods. 2. Statistics. I. Title.
HA29 .S658 2003
519.5'024'3—dc21 2002066543

Printed in the United States of America

10 9 8 7 6 5 4 3 2 1 07 06 05 04 03 02

Contents

Preface

As with any new edition, significant changes and additions have been incorporated into this seventh edition of *Basic Statistical Analysis*, and yet the major thrust of the coverage remains essentially the same. To that extent it may seem like deja vu, or as some naive critics of statistical analysis might say, deja voo doo. In fact, this new version may even seem, as Yogi Berra once said, like experiencing "deja vu all over again." (That's redundant, sir, isn't it, isn't it?) And if a double redundancy isn't enough, let's remember Howard Cosell, who once offered a triple-play on words when he asked his sports viewers "to reflect back nostalgically on the past." And so, as was the case with past editions, this book is still intended for students in the behavioral and health sciences who are confronting their first "stats" course. Although the book was originally written with psychology majors in mind, this edition features an infusion of new material that should be appropriate for other majors, especially criminal justice majors, education majors, and majors in the health sciences. Despite the overall continuity with previous editions, the text has changed in a number of important ways.

New features in this edition are many, but the major addition is the inclusion of the *SPSS Statistical Program*. This is probably the most popular and creative statistical program ever produced. The inclusion of this tool should add immeasurably to the increasing statistical sophistication of today's students. This edition of the text has been rewritten to fit seamlessly with SPSS and an appendix has been included that should provide students new to computing with a complete user's guide to this powerful program. For any of you who prefer to use the *AB-Stat* program that has accompanied this book in the past, it can be downloaded from the Allyn-Bacon Web site (www.ablongman.com). Other new features include:

1. More examples of statistical misdirection in Chapter 1
2. New techniques for scoring questionnaire research in Chapter 3
3. A chance to have some fun with your calculator in Chapter 4
4. A creative device for calculating various unknowns for the z-score equation in Chapter 5
5. New material for finding the *SD* or s from the *SEm*
6. A tightening of the research chapter, Chapter 9, with additional material in this chapter on the placebo effect
7. A simplified version of the Bonferroni test in Chapter 12

8. More on coefficient alpha and its relationship to the Cyril Hoyt in Chapter 17; also, new material on construct validity and test ethics
9. More coverage of confidence intervals for the single sample *t*, the two-sample independent *t* and the paired *t*
10. Many new and interesting examples from the fields of psychology, law enforcement, health sciences, education, and sociology. Also, all problems now have a research basis and include a full write-up of the conclusions.
11. And finally, we present two new members of the statistical hall of fame: Ada Lovelace, the daughter of the poet Lord Byron, who has been admitted because of her pioneering efforts in computer programming and Gertrude Cox, the founder of the first experimental statistics department in the United States.

For many students, approaching that first statistics course is a rather scary and mystifying experience. The prospect is laced with a large dose of ambivalence, like a two-edged sword that has "can't live with it" on one side and "can't live without it" on the other. The "can't live with it" feeling often owes its genesis to the memory of past traumas, perhaps a ninth-grade algebra course in which everything went smoothly "until we got to the word problems." And the "can't live without it" conviction may be even more direct and compelling, since the statistics course is probably required, or, at the very least, "strongly recommended."

To help alleviate some of the pangs of anxiety, many students adopt one of two main defenses, both of which revolve around the same core notion that this is a course that is simply not needed. First, there are those who are convinced they do not need it because the intricacies of statistical proof are irrelevant. To them reading the conclusions of a research report is enough: "If it's got numbers in it, especially significant numbers, then I believe it." Understanding whether the logic of the design warrants those significant conclusions is deemed a waste of time. Students often feel that this is like asking them to understand the theory of the internal combustion engine when all they want to do is drive the car. The attitude of these students resembles that of the person who earnestly intones that "they say to starve a cold," without having the vaguest notion as to who "they" are ("they" are always the anonymous voices of supreme and unchallenged authority). This group of students is blithely willing to accept all statistical results. Second, there are those students who are convinced that they do not need the course because statistics is all a pack of lies. To them, statistical proof is tantamount to no proof at all and so they smugly and cynically reject all statistical evidence. Their attitudes are like those of persons who say, "I don't think it works this way," and when questioned as to how it does work, they reply, "some other way."

There is of course a germ of truth in both arguments. Significant numbers do mean something, although not always what the researcher or research consumer concludes. Some statistical proofs are like a sleight-of-hand show: Now you see it, now you don't. In order to know what the significant results mean and when the statistical charlatan is operating, at least one course in statistics is essential. Newspapers, magazines, and television newscasts constantly bombard us with the results of research studies: drinking coffee causes this, drinking beer causes that, washing clothes in a certain detergent causes something else, and on and on. Determining cause-and-effect relationships should be made on the basis of fundamental and sound research practices, not by falling prey to a human need to quantify that is so strong as to embrace even dubious math as evidence of analytical rigor. Just to be an informed citizen and voter, some knowledge of statistical methdology is of enormous benefit. One must be made alert to politicized statistics, such as when a new administraiton announces in September that there are more people working than when they took over last January (ignoring the obvious seasonal variations in employment).

The avowed goal of this book is to demystify statistics—to state the case for statistical analysis and inference in clear, no-frills language. The student is told specifically what an X is and what a Y is, and whether the twain shall ever meet. As in law school, the student is presented with rules of evidence and the logic behind those rules. The focus will constantly remain on how statistical techniques can be used. It will not be a case of presenting the best method for calculating a standard deviation and then leaving it up to the student to find some use for it. Statistical concepts are embedded in the hard rock of research methodology. The student will learn at a practical level how to read and do statistical research. The student will be given a guided tour of the most important and practical exhibits in the statistician's showcase: not to create feelings of awe, but to teach for understanding.

Part of the power of being a professional has been the ability to use, take control of and protect from public scrutiny the language of the trade. The way to keep any discipline closed is to make the concepts obscure enough to only be understood by the so-called experts, thereby excluding the lay audience. Using and creating terms that have the dazzling sound of super sophistication become intimidating moats, designed to keep the great unwashed from entering the statistical fortress. This book, unlike so many others, crosses the moat.

For most students it will be easy to read and, at times, perhaps even fun to read. The book assumes little or no background in mathematics. The student will not be stunned by finding an elegant, but tricky, derivation on page 3 or by finding that the author suddenly assumes on page 5 that everyone remembers enough calculus to integrate the normal curve equation. The student does not even have to remember arithmetic, let alone calculus.

The use of this book does require, however, that the student own an electronic calculator. Although the calculator need not be expensive, it must at least have a square-root key. Pressing the square-root key is easier and more accurate than looking up and interpolating table values. Therefore, the back of this book is not cluttered with pages of square and square-root tables.

Although the text is designed primarily for a one-semester, beginning course, enough added material is presented to allow its use by students taking more advanced courses. Chapters 1 to 13 contain topics usually covered in a one-semester course and if this is what is needed, the course can end with Chapter 13. At this point, the student will have gained enough understanding of statistical reasoning and research methodology to be able to read and comprehend a large part of the research in the social-science literature. Because many students must later take courses in experimental psychology or in research methods in education and/or sociology, topics sometimes found in the more advanced courses are also included here in Chapters 14 through 19.

The book is divided into three major units: Descriptive Statistics, Inferential Statistics, and Advanced Topics in Inferential Statistics. How can a book of this size cover so much? Because some topics will not be covered here. First, little consideration will be given to grouped data problems. Finding class intervals and standard deviations from the frequency data inherently creates some error and also loses track of the individual score. When the amount of data is so large that grouped data techniques are really needed, statisticians turn to computers anyway. Second, the coverage of probability theory will be shortened. Not that probability theory is unimportant; it is absolutely crucial. But the only probability concepts found in this book are those that bear directly on practical statistical tests of significance. What is practical? How to calculate and understand the logic behind such things as z scores, the t test, ANOVA, chi square, and the Pearson r, and regression.

Special features of this book include the following:

1. *Definitions of key concept in the glossary.* Brief, but thorough, definitions are conveniently presented in the glossary. Experience has clearly shown that much of the trauma experienced by students taking their first "stats" course can be traced to confusion over terminology. Conscientious use of the glossary can alleviate most if not all of this confusion.

2. *A programmed approach to the computation of each statistical test.* Computational procedures are set forth in a step-by-step programmed format. A student who can follow a recipe or build a simple model plane can do an ANOVA.

3. *Constant stressing of the interaction between statistical tests and research methodology.* Examples are used from the literature of the social sciences to illustrate strategic methodological problems. Statistical analysis, if not carried out in the context of methodology, can degenerate into an empty and sterile pursuit. Three chapters have been specially designed to bridge the analysis-methodology gap. Chapter 9 focuses directly on the essentials of the research enterprise, Chapter 17 relates the statistical analyses to measurement theory and Chapter 19 presents 26 research simulations that are programmed in such a way as to lead directly to the appropriate statistical analysis.

4. *A literary style that is both easy to read and attention getting.* This book attempts to generate a feeling of excitement and enthusiasm, by talking directly to the students and spotlighting the student's own life space. Students obviously learn best when their interests are aroused.

5. *A large number of problems and test questions.* John Dewey's "learn by doing" axiom was never more true than in the field of statistics. Over 400 problems and test questions are placed both within and at the end of each chapter. Students need the opportunity to "try their hands" at practice problems and to get some immediate feedback as to their progress.

6. *A list of key points and names.* Each chapter contains a list of key points and names that also appears in the glossary. At the end of each chapter a convenient wrap-up summary is also provided.

7. *Computer program.* A statistical program (SPSS) is included that covers the statistical tests presented in the book. The program is totally menu driven and can be easily handled within an hour by the first-time user.

8. *Computer printouts.* A series of computer printouts, all containing errors of one kind or another, is presented in Chapter 18. Students may then use the "logic checks" found throughout the text to identify these errors. This should accomplish at least two important objectives: 1) remind us that the computer is not an infallible genius, but is instead, a fast idiot who needs a smart leader, and 2) reinforce the various statistical logic checks by making them "live" in the context of a computer printout.

9. *The binomial distribution.* A special section is included that covers the essentials of the binomial distribution and its relationship to the z distribution. This section includes methods for obtaining exact probabilities for the binomial distribution, as well as the z approximations. The coverage here also includes the z test for proportions and the t test for differences between proportions (and its relationship to chi square). Problems are worked out within the chapter, and a series of student problems presented at the end.

10. *Supplements.* A new and up-dated "Instructor's Manual" containing well over 1800 test items and problems is available. The Instructor's Manual also includes a number of suggestions for class activities, all designed to spark student interest.

Putting together a book of this type requires a lot of help. Special thanks must go to the "significant others" in my academic life, the professors who first initiated and then sustained my interest in statistical analysis: Greg Kimble, then of Brown University; the late Nathan (Mac) Maccoby, then of Boston University; and the late P. J. Rulon of Harvard University. Without them this book could never have happened. Also, I am grateful to the Literary Executor of the late Sir Ronald A. Fisher, F.R.S., to Dr. Frank Yates, F.R.S., and to the Longman Group Ltd., London, for permission to reprint Tables III, IV, and VI from their book *Statistical Tables for Biological, Agricultural, and Medical Research* (6th edition, 1974).

More recently I must thank my colleagues at American International College: Tom Fitzgerald for his creative chi square examples, Lee Sirois for his significant role in putting together Chapter 8, Gregory Schmutte for his valuable contributions to the research chapters, Gus Pesce for his "spec ed" examples, Pam Diamond for her z score matrix in Chapter 4, and Marty Lyman at the Hampden County House of Corrections for her research examples. I also wish to extend a special thanks to Barry Wadsworth at Mt. Holyoke College, Marjorie Marcotte at Springfield College and Norm Sprinthall at North Carolina State for their insightful comments on topic inclusions (and exclusions), to Barbara Anderson Lounsbury at Rhode Island College for her help and encouragement during the early stages of the manuscript's development, to both Jim Vivian at Gordon College, Norm Berkowitz at Boston College, and Chris Hakala at the Western New England College for their ideas on sampling theory, to Steve Fisk at Bowdoin College for his expertise in showing the world how to put together a computer program that is really menu driven, to Mike Plant for his superb revision of the computer program (and to Peter Flournoy at the V.A. Medical Center in Portland, Oregon, for his many suggestions for the program revision). I must also thank Ken Weaver at Emporia State University in Kansas for his careful reading of the entire manuscript. Without Ken's help, many errors of omission and commission would have been left in the text. Thanks also to the following reviewers for their helpful comments: Ken Hobby, Harding University; James Johnson, University of North Carolina, Wilmington; Neva Sanders, Canisius College; and Laura Snodgrass, Muhlberg College.

Also a special note of gratitude must go to Dianne Ratcliffe Sprinthall for her illustrations, as well as her general help in artistic design, and to both Lou Conlin and Sidney Harris for their marvelously clever cartoons.

UNIT

I

Descriptive Statistics

Chapter

1

Introduction to Statistics

During the next hour, while you are reading this chapter, 250 Americans will die, and the chances are one in a million that you'll be one of them. Don't stop reading, however, since that won't reduce the probability. In fact, putting your book down might even increase the probability, especially if you decide to leave your room and go outside to engage in some high-risk activity.

According to the TV special "Against All Odds," if you go rock climbing, the probability of your getting killed is 200 in a million; or parachuting, 250 in million; or hang gliding, 1140 in a million. So, sit safely still and read on, and while doing that let's look at some more of life's probabilities: the probability of having your car stolen this year, 1 out of 120; of a pregnant woman having twins, 1 out of 90 (or triplets, 1 out of 8000); of a young adult (18–22) being paroled from prison and then being rearrested for a serious crime, 7 out of 10; and of any single American baby becoming a genius (IQ of 135 or higher), less than 1 out of 100 (Krantz, 1992). Incidentally, by the time you finish reading Chapter 6, you'll be able to calculate that genius probability value—even if you're not a genius yourself.

As you probably know, most accidents occur at home, since typical Americans spend most of their time there. And 25% of all home accidents occur in the bathroom—falling in the tub, getting cut while shaving, and so forth. Don't assume, however, that you'll be safe if you decide to shave in the kitchen instead. Also, we can predict with a high degree of accuracy that during the next year 9000 pedestrians will be killed by a moving car. But this statistic does not tell us which 9000. Understanding probability situations is an important aspect of life, so maybe there are some good reasons for getting involved in statistical thinking.

Rather than continuing to list reasons why you should take a first course in statistics, let's assume that it is probably a required course and that you have to take it anyway. Perhaps you have put it off for quite a while, until there is no choice left but to "bite the bullet" and get it over with. This is not to say that all of you have been dragged, kicking and screaming, into this course; however, as statisticians would put it, the probability is high that this hypothesis is true for some of you.

STUMBLING BLOCKS TO STATISTICS

Let us look at some of the most common objections raised by students when confronted with this seemingly grim situation. Perhaps your feelings of intimidation arise because you know you have a math block. You're still being buffeted by lingering anxieties from some math course taken in the perhaps distant past. Or maybe it's that you have read or heard a research report and been totally confused by the seemingly endless and seemingly meaningless stream of jargon. Perhaps you're a person who simply does not trust statistical analysis. If this is the case, you're in good company. Benjamin Disraeli, Queen Victoria's prime minister, once said, "There are three kinds of liars: liars, damned liars, and statisticians." Disraeli obviously agreed with the sentiment expressed by many—that you can prove anything with statistics.

Before he died, Malcolm Forbes had been a hot-air balloon enthusiast, and one day the winds just took his balloon in so many directions that he became completely lost. Spotting what appeared to be a farmer down below tilling his field, Forbes lowered the balloon and called out to the man, "Please tell me where I am." The man called back, "You're up in a balloon, you goddamned fool." And Forbes answered, "You must be a statistician, since although your answer is complete, accurate, concise, and precise, it tells me absolutely nothing that I don't already know." And then there's the story of the three statisticians who went hunting and after a while spotted a solitary rabbit. The first statistician takes aim and overshoots. The second aims and undershoots. The third shouts out, "We must have got him."

Whatever their basis, your doubts about taking this course will probably prove unfounded. You may even, believe it or not, get to like it and voluntarily sign up for a more advanced course.

Math Block

First, although it is obvious that people do differ in math ability, a case of true math block is extremely rare and difficult to substantiate. It is true that some very fortunate people have a kind of perfect pitch for math. They take to math as gifted musicians take to harmony. (You remember the kid we all hated in high school, the one who completed calculus during his sophomore year and was angry because the school didn't offer any more advanced math courses.) At the other end of the continuum, we find those people who are definitely math phobics. To them, merely drawing a number on the chalkboard evokes strangulating feelings of sheer panic. They avoid any course or situation (even keeping a checkbook) that deals with those spine-chilling little inscriptions called numbers. If you're one of those who suffers from or borders on this condition, relax—this is not a math course. While numbers are involved and certain arithmetic procedures are required, thanks to the magic of electronics you won't have to do the arithmetic yourself.

Go to your friendly neighborhood discount store, and, for less than ten dollars (less than the price of a good slide rule), purchase a small electronic calculator. You don't need a fancy calculator with several memories, but do insist on one with a square root key. Statisticians, as you will see, love to square numbers, add them, and then extract the square root. In fact, you really must get a calculator, for this text assumes that you own one. The back of this book is not cluttered with page after page of square and square root tables. It's not only that such tables are a relatively expensive waste of space, but it is easier for you to push a button than it is to interpolate from a table. Your job is to focus on the logic of statistics. To rephrase the bus ad that says "leave the driving to us," leave the arithmetic to the calculator. While we're on the topic of arithmetic, a quick word of caution is in order. Do not scare yourself by thumbing through later chapters of the book. Until you have been provided with a contextual background, some of the procedures and equations found later on will seem absolutely harrowing to contemplate. When the time comes for you to confront and use these techniques, the mystery and fear will have long disappeared. With the background provided by the early chapters, the later chapters will, like a perfect bridge hand, play themselves. If you can follow step-by-step directions well enough to bake a cake or assemble a simple model airplane, you can do any of the procedures in this book, and, even more importantly, you can understand them. You may even come to have more appreciation for and less fear of quantitative thinking. Actually, instead of fearing numbers, students should learn to appreciate what Professor Posamentier has called the "beauty of mathematics and the charm of some numbers." For example, the years 1991 and 2002 are palindromes, numbers that read the same both forward and backwards. Experiencing two palindromic years in one lifetime won't occur again for over a thousand years (Posamentier, 2002).

Persons who can't read or understand words suffer from what we all call illiteracy, and persons who can't read or understand numbers suffer from an equally disabling, although apparently less embarrassing, condition called innumeracy. As Professor Paulos wrote, "The same people who cringe when words such as imply and infer are confused react without a trace of embarrassment to even the most egregious of numerical solecisms" (Paulos, 1988, p. 3). Professor Paulos then recounts a story of someone at a party droning on about the difference between continually and continuously. Later that evening, a TV weathercaster announced that there was a 50% chance of rain for Saturday *and* also a 50% chance for Sunday, and then erroneously concluded that there was therefore a 100% chance of rain that weekend. "The remark went right by the self-styled grammarian, and even after I explained the mistake to him, he wasn't nearly as indignant as he would have been had the weathercaster left a dangling participle."

That's like concluding that since there is a 50% chance of flipping a head with a fair coin on one toss and a 50% chance of flipping a head on a second toss, then there must be a 100% chance of flipping a head on at least one of the two tosses. Get out a coin, start flipping, and check it out.

Unlike other personal failings that tend to be kept offstage, innumeracy may even be flaunted: "I can't even balance my checkbook." "I'm a people person, not a numbers person." Or, as we have seen, the proverbial "I have a math block." Professor Paulos believes that one reason for this perverse pride in innumeracy is that its consequences are not perceived to be as damaging as those of other handicaps; but they really are. Not understanding what interest rates you're paying, or what the total cost of a loan might be, or how to tip a waiter, or what the probabilities are in a certain bet you've just made, or what a newspaper headline or story is really saying may in the long run be more personally damaging than not knowing the difference between a gerund and the subjunctive mood.

Statistical Jargon

As to the objection concerning the jargon or technical language, again, relax—it's not nearly as bad as it seems. Too often students are dismayed by the number of technical terms and the seemingly endless statistical lexicon, written in (oh no!) both Greek and English. Too often students are traumatized by the rigors of statistical analysis and its vast and mysterious body of symbols. Social scientists especially seem to make a fetish of the intricacies of significance tests and measurement theory. They seem to spend countless hours debating and fretting over statistical details that often seem trivial to the casual observer. There is also the psychology instructor who gives exam grades back in the form of standard scores. "Never mind about the standard error, or the amount of variance accounted for, did I pass?" is the oft-heard plea of many a student.

Is the researcher's use of statistical terms simply a case of sound and fury, signifying nothing? Obviously, it is not. The jargon of the trade represents an attempt to be precise in the communication of meaning. This effort is especially important in the social sciences because the concepts being considered are not always as precise as they are in the physical sciences like physics and chemistry. In short, there are some terms and symbols that must be learned. However, you can also get some help. At the end of this book, important terms, concepts, and equations are set down and defined. Faithful use of these glossary items definitely increases the retention of learned material.

Statistical Sleight of Hand

Finally, the objection that the field of statistics resembles a sleight-of-hand show (now you see it, now you don't) is valid only when the research consumer is totally naïve. The conclusions that "Figures don't lie, but liars can figure" and that one can prove anything with statistics are only true when the audience doesn't know the rules of the game. To the uninitiated, liars can indeed figure plausibly. But an audience with even a patina of statistical sophistication will not be easily misled by such artful dodgers. Unscrupulous persons will probably always employ faulty statistical interpretations to gain their ends. By the time you finish this course, however, they will not be able to lie to you.

Frankly, some statistical studies, especially correlation studies, have been grossly misinterpreted by certain researchers. This does not mean that all statisticians are charlatans. If statistics, as a result of some trickery and deceit, has a bad name, it is a result of the *misuse* and not the use of statistics. There are some booby traps lying in wait out there, but you'll be guided safely along the right path, and each potential pitfall will be carefully pointed out. It should be stressed that most statisticians *do* use the correct statistical tests, and it is almost unheard of that the data have been faked. The traps usually result from the way the conclusions are drawn from the data. Sometimes the conclusions are simply not logically derived. Let's consider a couple of examples.

An often-heard argument against capital punishment is that it simply does not have a deterrent effect. To support this, it is stated that years ago in England when convicted pickpockets were publicly hanged, other pickpockets worked the crowd that was there to witness the hanging. Can you see any problems with the logic of this example? The fallacy is that there is no comparison (or, as it will later be called, control) group. What about the number of pockets picked in crowds at a horse race or a carnival? If the frequency of pocket picking was lower at the public hangings, then perhaps capital punishment did have a deterrent effect.

Or statistics show that, say, 50% of a certain country's population are so illiterate they can't read the morning newspaper. Watch it! Perhaps 25% are under age 6, perhaps another 10% are blind, and so on.

In another example, *Parade* magazine headlined the story that "Americans are happier in marriage than their parents were." This news was based on a survey of young married couples, 70% of whom described themselves as "happily married," while only 51% could say the same about their parents. The question of course is, "Compared to whom?" Had the parents participated in the survey, which they didn't, it might be important to find out how they described their own marital satisfaction, and, more importantly, the marital success of their offspring. With those data, a meaningful comparison could at least be attempted (*Parade,* April 28, 1985).

Also in 1985, New York City released figures showing that the uniformed police made only 30% of the subway arrests, while the plainclothes police hit the 70% arrest figure. Does this mean that the plainclothes police work that much harder, or could it possibly be that the uniform is indeed a deterrent?

Or a paint company runs a TV ad showing a house that originally sold for $75,000. The new owner then repaints the house with the brand being advertised and later sells the house for $90,000. The implicit message, of course, is that the new paint job increased the value of the house by $15,000. However, in the absence of a definite time frame, there is no way to factor out the obvious effects of inflation. The house may easily have sold for $90,000, or more, given enough time—with or without its reglossed exterior.

Or in the screenplay of John Irving's *The World According to Garp,* Robin Williams knows that the probability of his home being destroyed by a plane crashing into it is extremely small, but not small enough for our cautious hero. So he deliberately buys a rebuilt house that has *already been hit* by a plane, convincing himself that the probability of being hit by two planes would be infinitesimally small.

Or consider this one. A researcher with an obvious antifeminist bias concludes that virtually all of America's problems can be traced directly to the women's liberation movement. Statistics are paraded before us to prove that as women left the kitchen and entered the fields of psychiatry, criminal justice, politics, real estate, and law, for example, the prison population tripled. The increase in the number of women entering business coincides directly with the number of small business failures. Finally, the increasing number of women's individual bank accounts correlates with the increasing number of felonies. This is the kind of misuse of statistics that causes honest, competent statisticians to blanch and also casts a pall of suspicion over the whole field. If people accept such artful juxtaposing of statistics as proof of a causal relationship, then, indeed, statistics can be used to prove anything. The point, of course, is that just because two events occur simultaneously is no reason at all to conclude that one of these events has in any way caused the other. As we shall see later, the only way we can ferret out a cause-and-effect relationship is through the use of the controlled experiment.

You must always be careful when evaluating groups of different sizes. For example, it might be argued that Alaska is a far healthier area in which to live than New York City is. Why, do you realize that in New York City last year over 70,000 people died, whereas in Ugashik, Alaska, only 22 people died? (The other 8 probably moved to Fairbanks.) Only by converting the numbers of deaths to *percentages* of the two total populations could a meaningful comparison be made in a case like this.

Unequal comparisons of this type bring to mind the fish store owner who beat all the competition by selling lobster meat for two dollars a pound less. The owner admitted that the lobster meat was stretched out some with shark meat. The best estimate of the proportions was 50–50, one shark to one lobster. Always be alert to the sizes of the groups or the things that are being compared. The death rate in the peacetime army is lower (by far) than the death rate in New York City. Therefore, New Yorkers should join the army. Right? Wrong, because again two unequal groups are being compared. The army is composed of young, largely healthy persons who are not likely to be subject to chronic diseases and certainly are not going to be ravaged by infant diseases or old age.

Also, you must be aware of the size of the sample group from which the statistical inference is being made. Recently, a well-known company reported the results of a comparative study on the effects of using its brand of toothpaste. "The study was conducted over a period of six months, and millions of brushings later," the company concluded that the other popular fluoride toothpaste was "no better than ours." This is interesting in that the company isn't even claiming victory but is apparently proud to have merely achieved a tie. Even more interesting, however, is that we are never told *how many subjects* participated in this study. Millions of brushings later could mean four people brushing every minute of every day for the full six months. If so, no wonder they don't have any cavities—they probably haven't any teeth left.

Two researchers, in an attempt to minimize the positive influence of birth fathers in the household, have cited evidence to indicate that among child abusers the absolute number of birth parents surpasses parent substitutes (Silverstein & Auerbach, 1999). Among other things, they found that genetic fathers outnumber (although only slightly) stepfathers as perpetrators of fatal baby beatings in Canada, the United Kingdom, Australia and the United States. This, of course, fails to take percentages into account. Birth parents vastly outnumber parent substitutes (including stepparents) in the countries listed. Also, as pointed out by Daly and Wilson (2000), the younger a child the less likely he or she is going to be raised by stepparents since very few babies have stepfathers. Thus, on a percentage basis, parent substitutes far exceed the rates of child abuse by genetic parents.

Thus, percentages can be grotesquely misleading if population size is left unknown. Suppose a listenership survey reports that 33% of the respondents polled said that they listen regularly to radio station WFMF. Perhaps only three people were polled, and one of those, the last one surveyed, mentioned WFMF—at which point the polling ended.

Be extremely skeptical when reading reported corporate profits, especially huge percentage increases. A company can truthfully report a 100% increase in profits and still be having a very poor year. Suppose in the previous year the company earned 1 cent for each dollar spent—a very modest return on capital. This year it earns 2 cents for each dollar. That is a bona fide 100% increase in profits, but it's hardly enough to keep a prudent stockholder happy or the company long in existence.

In his book *How to Lie with Statistics,* Darrell Huff tells us how the president of a flower growers' association loudly proclaimed that certain flowers were now 100% cheaper than they were four months ago. Huff notes that "he didn't mean that florists were now giving them away—but that's what he said" (Huff, 1954).

The National Safety Council is always telling us how dangerous it is to drive our cars near our own homes. We are warned that over 85% of all auto accidents occur within 10 miles of the driver's home. It is alleged that one person became so frightened after hearing about this high percentage of accidents occurring close to home that he moved! The other side to the story is, of course, that more than 95% of all driving is done within 10 miles of home. Where does the danger really lie? It lies in not being alert to statistical misdirection. While statistics do have an aura of precision about them, the interpretations are often anything but precise.

The Tyranny of Numbers

The authority of numbers, especially computer-generated numbers, promotes their unthinking acceptance as fact. Because the audience is more likely to believe the story, speakers like to waive computer tearsheets as they announce alleged statistics. If you had been using your computer during the presidential campaign of 2000, and if you had gone to the Bush Web site during July, you would have found a counter that said "There are 564 days 17 minutes until the end of the Clinton/Gore era." Could this be? This is easy to check. Just count the number of days from July until January's inauguration day and guess what? You would have found that the counter was off by over 200 days. But since the computer seemed to be generating the numbers, would you have checked? Now let's turn to www.algore.com for the same day. It was showing "The Bush Debate Duck Counter" which was clicking off the number of days that Bush had evaded a head-to-head debate with Al Gore. On July 11 it said 251 days and on July 12 it said 250 days. The problem here is that it was counting DOWN, as though Bush had already agreed and the debate had been arranged. It should have been counting up which would have portrayed Bush as increasing in his fear of the showdown (Fisher, 2000).

The Dirty Analogy. As we have seen, statistical thinking is logical thinking, both deductive and inductive. You must beware of the dirty analogy, the seemingly valid deductive proof that changes direction in midstream. An example is the argument that starts off "If we can put a man on the moon, then" and then fills in the rest of the sentence from a long list of non sequiturs, such as "We can end racism," or "We can put an end to the AIDS epidemic," or "We can solve the problem of the homeless." Notice how the conclusions suddenly have nothing to do with the premise. A successful space launch does, of course, take engineering skill and plenty of money, but, unlike AIDS, it doesn't require any change in attitudes or behavior. Other premises to beware of are "In a country as rich as ours, we ought to be able to," or "Since it's cheaper to send a student to Harvard than to keep a person in prison, then. . ." (Should that final conclusion be, "Therefore low-income, law-abiding citizens should be taught to commit felonies in order to get a free college education"?)

The Only Way to Travel?

Calculations trumpeting the safest way to travel have long been a favorite target of special-interest groups and a rich mother lode for unscrupulous statisticians. Perhaps the major interpretative misunderstanding has come because of the confusion between miles traveled and the number of trips involved. For example, on the miles-traveled scale, one of the safest ways to travel is to be launched into orbit aboard a space shuttle, but on a per-trip basis this is an extremely high-risk endeavor. As of 1994, the chance of being killed in a space shuttle was 1 for every 78 trips, which unless you had a death wish was a very scary prospect. Imagine if the probability of getting killed in the family sedan were 1 for every 78 times it was taken out of the garage. On that basis, most of us would think long and hard before revving up the old V6. But on a per-mile basis, the space shuttle is amazingly safe. The average NASA trip takes eight days and covers 3.36 million miles, making the risk per mile an infinitesimal 1 in 262 million. Compare that with the risk per mile of 1 in 50 million for the automobile. Incidentally, the risk of injury while driving your car varies tremendously depending on other conditions. For example, you are three times as likely to get killed driving at night, two and one-half times as likely when driving on a country road as opposed to the downtown area of a large city, and three times as likely when driving on the weekend (Berger, 1994). So, if you want to play the probabilities, drive your car on a Wednesday at high noon, dodging taxis on Manhattan's Fifth Avenue, and be sure to buckle up (which reduces your fatality risk by about another one-third). And be wary of the warnings about holiday driving. It could be that the holiday death rate *per car* is lower, since with so many cars on the road, people necessarily might have to drive at a slower speed, which itself might reduce the seriousness of accidents, if not the frequency.

It's All in the Question

When the great poet Gertrude Stein was near death, one of her friends asked her, "Gertrude, what is the answer?" and Ms. Stein whispered back, "It depends on the question." Taking Ms. Stein's advice, you must therefore be especially careful when reviewing the results of all the polling data that constantly bombard us. In the first place, the questions themselves may be loaded—phrased in such a way as to favor a specific answer. For example, a question such as "Do you prefer socialized medicine as practiced in England or free-enterprise medicine as practiced in the United States?" is obviously slanted. Patriotic respondents might opt for the American variety simply because it is American. Also the terms socialized and free-enterprise might bias the result, since some respondents might be negatively disposed to anything socialistic without always realizing what it means in context. When a *New York Times*–CBS poll asked respondents their opinion on a constitutional amendment "prohibiting abortions," the majority (67%) opposed it. But when the question was reworded, the majority (51%) favored "protecting the life of the unborn child." Also, when asked, "Are we spending too much, too little, or about the right amount on welfare?" the results seemed to paint Americans as greedy and noncaring, with only 22% saying "too little." However, when the same question was rephrased as "Are we spending too much, too little, or about the right amount on assistance to the poor?" the results cast Americans in an entirely differently light. To that question, the vast majority, 61%, said "too little" (National Opinion Research Center of the University of Chicago). The Gallup Poll asked Americans if they were for or against a "waiting period and background check before guns can be sold." The results? An impressive 91% said they were for it. However, when the Wirthin poll asked a similar question, "Are you for or against a national gun registration program costing about 20% of all dollars now spent on crime control?" the results were virtually reversed, with 61% disagreeing. A poll conducted by *Reader's Digest* in conjunction with the Roper Center for Public Opinion Research (at the University of Connecticut) asked respondents if they would be disappointed if Congress cut funding for public TV. The majority said they would be disappointed if the cuts were made. But when the *same* respondents were later asked whether cuts in funding for public TV were justified as part of an overall effort to cut federal spending, the results went in the opposite direction, with the majority now agreeing that the cuts should be made (Barnes, 1995). In another rather famous (infamous?) *New York Times*–CBS poll, people were asked if they agreed or disagreed with the statement that the "federal government ought to help people get medical care at low cost." The results indicated that three-quarters of those polled agreed, appearing to lend strong support for a government-run national health plan. But when the poll was repeated asking exactly the same question but substituting "private enterprise" for "federal government," it was discovered that, again, three-quarters of those asked voiced their agreement. At least the respondents were consistent in one way: they want low-cost medical care—whether it be by the federal government or free enterprise seems not to be of concern. Then again, perhaps there are long lists of other things and activities that people want at low cost. Even the order of questions may influence the responses. The same question presented at the beginning of the survey may get a very different answer when presented near the end (Schwarz, 1999).

Finally, poll data may even be reported in such a way as to belie the questions that were actually asked. A sample of Americans was asked, "Would you be concerned if the family unit were to disintegrate?" Of those polled, 87% said yes. The headlined conclusion in the newspaper was that "87% of Americans are concerned over the disintegration of the family unit." One might just as well ask, "Would you be concerned if the sun were to explode?" The conclusion would then be, "100% of Americans are concerned that the sun will explode."

The Numbers Game

Even the numbers themselves are sometimes only guesses, and often wild guesses at that. But after the numbers have been used for a while and have become part of the public record, they often take on a life of their own and may continue to be reported long after their shadowy origins have been exposed. Several years ago a distraught father began a campaign for finding missing children after his own son had been abducted. He announced that 50,000 children were being abducted each year, and newspapers trumpeted that figure throughout the country. The faces of missing children began appearing on milk cartons, and alarmed parents, hoping to protect their offspring, formed missing children's organizations. However, during the late 1980s the FBI announced that the number of yearly confirmed abductions was somewhat lower, a total of 57 in fact for the previous year. Also, the widely cited figure of 626,000 kidnappings committed each year by parents in custody battles may also be a tad on the high side. Since the census data for the 1990s told us that there were 6.1 million children of divorce in this country, at the rate of 600,000 per year, every one of them would have to have been kidnapped over the span of the last 10 years. Then there is Mitch Snyder, advocate for the homeless, who announced that there are 2.2 million homeless persons in the United States and that soon there would be 3 million. When confronted with scientific studies that place the number of homeless at about 250,000, Snyder told a congressional committee that his numbers were in fact "meaningless and were created to satisfy your gnawing curiosity for a number because we are Americans with Western little minds that have to quantify everything in sight, whether we can or not" (*U.S. News & World Report,* July 1986, p. 107). The trouble with inflated numbers, aside from the lie factor, is that they are so often used to guide public policy. We must constantly remind ourselves that what ought to be is not always what is. Advocacy groups typically tend to inflate the numbers, since the most accurate numbers are not always the most satisfying numbers. This can often then lead to real social problems being misdiagnosed. It is too often the case that a numerical lie gets jetted halfway across the country before the truth has had a chance to taxi out onto the runway.

Beer, Taxes, and Gonorrhea

If You Drink Don't Park: Accidents Cause People. A study released on April 27, 2000, by the U.S. Center for Disease Control, suggested that increasing taxes on beer could lower STD (sexually transmitted disease) rates, especially gonorrhea. Although about 3 million teenagers are infected with sexually transmitted diseases each year, the CDC found that there were reductions in gonorrhea rates following tax increases by the various states during the period 1981 to 1995. The conclusion: a 20-cent increase per six-pack leads to a 9 percent drop in gonorrhea. It was suggested that the increased beer price puts it over the price-range for many teenagers. Without the beer's influence on impulse control, the teenager is less likely to be sexually active.

The data did not show that teen agers were buying less beer, or having less sex. This was simply assumed. Perhaps the reduction in STD rates was a result of other factors, such as the states using the increased revenue to augment sex education programs in the schools, or more TV and radio spots extolling the virtues of "safe sex." It might have been instructive had there been some indication of whether STD rates had declined in the state's older populations, a group that would probably not be dissuaded by a few cents being added to the beer tax. We might also wish to know if consumption of other alcoholic beverages had increased as a result of the beer tax, beverages that might be more influential regarding impulse control. Also, there was no evidence put forth to indicate that the teens in those states were actually less sexually active, only that there was a small decline in STDs. Perhaps the beer influences the teen's readiness to use condoms, rather than general sexual readiness.

September 11, 2001

In 2001, almost 83% of all the homicides in New York city were committed by foreigners. Of course, almost 3000 of the total of roughly 3600 were committed on one day, September 11. Newspapers tend to meter out homicide rates evenly on a per day or per hour basis, such as someone gets killed every *X* minutes (as though homicides really occur on a regularly timed basis). Using that logic, in 2001 someone in New York got murdered every two hours. A person reading that number might check his watch and decide not to leave his hotel room for a few minutes to avoid that two-hour mark. Incidentally, if you take out September 11, the rate drops dramatically to fewer than two per 24 hours.

Can You Spot the Volvo? Do you remember the ad for Volvo that showed a lineup of cars being crushed by a giant pickup truck called the big "Bear Foot"? Big Bear Foot, with its huge tires, was shown running over the tops of the row of automobiles. The Volvo withstands the pounding while the competition's cars are flattened. The ad is titled, "Can You Spot the Volvo?" Well, you could if you were told ahead of time that the Volvo had been reinforced with extra steel columns to withstand Bear Foot's assault *and* that the structural pillars of all the other cars had been severed. In fact, *USA TODAY* (October 5, 1994) named this Volvo ad as the most effective television promotion for the entire 1991 model year. After the deception was uncovered, however, Volvo quietly dropped the ad.

A BRIEF LOOK AT THE HISTORY OF STATISTICS

The general field of statistics is of fairly recent origin, and its birth and growth were spurred on by some very practical considerations. Although some form of both mathematics and gambling has existed since the earliest days of recorded history, it wasn't until about 300 years ago that any attempt was made to bring the two together.

It is rather curious that it took the human race such a long time to gain any real understanding of the probability concept. Primitive persons used numbers, had a counting system, and were not averse to gambling. In fact, well-formed dice (which must have played true since even now they show almost no bias whatsoever) have been found dating at least as far back as 3000 B.C. Perhaps early humans were afraid to think about probability, believing that it was the sole province of the gods. Perhaps it was felt that to deny the gods' control over events would be an invitation to personal disaster, either at the hands of the gods or at the hands of the religious authorities. It was probably easier and a good deal safer to speak fatalistically in terms like "the wheel of fortune" and "the throw of the dice" rather than to dare penetrate the mysteries of the deities and thereby bring on a charge of heresy.

In Book I of *De Divinatione,* Cicero wrote 50 years before the birth of Christ, "They are entirely fortuitous you say? Come! Come! Do you really mean that? When the four dice produce the venus-throw you may talk of accident; but suppose you made a hundred casts and the venus-throw appeared a hundred times; could you call that accidental?" (David, 1962, p. 24). Implicit in his statement, of course, is that the gods

GERTRUDE COX (1900–1978)

Gertrude Cox was born in the small town of Dayton, Iowa on January 13, 1900, and graduated from Perry High School in Perry, Iowa in 1918. Her early years were filled with a strong sense of ethics and religious faith. In fact her earliest goal was to become a Methodist minister, not a typical career choice for a young girl growing up in the Midwest in the early 1900s. After attending Iowa State College in Ames for a year, however, she changed her career path and instead chose to major in mathematics, again an atypical major for a woman at that time. Like Pascal, Cox was searching for answers in both faith and probability theory, and both became strong guides throughout her remarkable life. She graduated from Iowa in 1929 with a B.S. degree in Mathematics, and then entered Iowa's graduate school, receiving her Masters degree in statistics in 1931 (which was the first Masters in statistics ever awarded at Iowa State). She also met and exchanged ideas with the great English statistician, R.A. Fisher, who spent the summer of 1931 at Iowa State. In 1933, she became the assistant director of Iowa's newly formed Statistical Laboratory. But then in 1940 she answered a call from North Carolina State's president, Frank Graham, to organize, set up, and head a new department in Experimental Statistics, the first of its kind in the country. This was also the first time that North Carolina State had ever named a woman as a full professor, (and a department head as well). By 1944, the University established what was called the "Institute of Statistics," and chose Cox as its first director. For the next several years she worked on a series of methodological and design problems that culminated in 1950, along with her coauthor William Cochran, with the publication of her first book, *Experimental Designs.* In this book, which became a bible to statisticians for many years to come, she proved to be far ahead of her time, creating tables for determining effect sizes and sample sizes for given alpha levels. She also addressed some of the thorny problems involved in setting up within-subjects designs, especially issues resulting from the residual or carryover effect from one trial to the next. She insisted that since within-subjects designs are more precise than between-subjects designs, her control techniques make ferreting out the pure effects of the IV (independent variable) more defensible. In 1949, she established the Department of Biostatistics at the University of North Carolina at Chapel Hill, and in the early 1950s, she worked on putting together the Research Triangle Institute (RTI). Then, in 1960, she took over the directorship of RTI's Statistics Research Division (Monroe & McVay, 1980). Dr. Cox was recognized around the world for her creative and profound work in statistical methods. Her contributions to our field are far too numerous to mention here, but among them were the following: the first editor of the Biometrics Journal of the American Statistical Association, a post she held from 1945 until 1956; the first woman to become elected to the International Statistical Institute in 1949; President of the American Statistical Association in 1956; President of the Biometric Society from 1968 to 1969; an honorary fellow in England's prestigious Royal Statistical Society in 1957 (Yates, 1979); an honorary doctorate for her outstanding contributions to the field of statistics from her beloved Iowa State University in 1958.

In 1960, North Carolina State erected a building in her honor, Cox Hall, a six-story structure containing the Departments of Physical and Mathematical Sciences. Then, in 1987, in her memory, North Carolina State set up the Gertrude M. Cox Fellowship fund designed to aid outstanding women who were intending to study statistics. Since 1989, four young women have been chosen each year.

Finally, in the words of Duke University's Sandra Stinnett (1990) "she was instilled with ethics, moral courage and determination. This combined with her grand dreams and the genius and tenacity to materialize them, resulted in legendary accomplishments and awed those who knew her."

must intervene to cause the occurrence of so improbable an event. In this passage, Cicero is voicing the popular view of his day, but in later writings he indicates his own mistrust of this opinion.

The advent of Christianity didn't do much to advance thinking in this area. Writing in the fifth century A.D., St. Augustine said that "nothing happened by chance, everything being minutely controlled by the will of God. If events appear to occur at random, that is because of the ignorance of man and not in the nature of the events" (David, 1962, p. 26).

Even today, many people prefer not to calculate probabilities, but instead to trust blind luck. All of us have met bold adventurers who grandly dismiss the dangers inherent in their newest sport. "After all, you can get killed in your own driveway," they intone, not caring to be bothered with the blatant probability differences between getting hurt by falling on the driveway and getting hurt while hang gliding off the top of Mount Everest.

Pascal and Gossett

During the seventeenth century, the birth of statistics finally took place. It happened one night in France. The scene was the gaming tables, and the main character was the Chevalier de Mère, a noted gambler of his time. He had been having a disastrous run of losing throws. To find out whether his losses were indeed the product of bad luck or simply of unrealistic expectations, he sought the advice of the great French mathematician and philosopher **Blaise Pascal** (1623–1662 see bio on page 116). Pascal worked out the probabilities for the various dice throws, and the Chevalier de Mère discovered that he had been making some very bad bets indeed. Thus, the father of probability theory was Pascal. His motive was to help a friend become a winner at the dice table. Although Pascal's motive may seem not to have been overly idealistic, it was extremely practical as far as the Chevalier de Mère was concerned.

Another milestone for statistics occurred in the early 1900s in Ireland at the famous Guinness brewery, now known worldwide for the record books of the same name. In 1906, to produce the best beverage possible, the Guiness Company decided to select a sample of people from Dublin to do a little beer tasting. Since there turned out to be no shortage of individuals willing to participate in this taste test, the question of just how large a sample would be required became financially crucial to the brewery. They turned the problem over to the mathematician **William Sealy Gossett** see bio on page 240. In 1908, under the pen name "Student," Gossett produced the formula for the standard error of the mean (which specified how large a sample must be, for a given degree of precision, to extrapolate accurately its results to the entire beer-drinking population).

So that's the history—craps and beer—hardly likely to strike terror in the hearts of students new to the field. The point is that the hallmark of statistics is the very practicality that gave rise to its existence in the first place. This field is not an area of mysticism or sterile speculations. It is a no-nonsense area of here-and-now pragmatism. You will not be led upstairs to a dark and dingy garret, with a taper and a crust of bread, to contemplate heavy philosophical issues. Instead, you will, with your trusty calculator in your hand or computer at your side, be brought into the well-lit arena of practicality.

BENEFITS OF A COURSE IN STATISTICS

If, as the Bible says, "the truth shall set you free," then learning to understand statistical techniques will go a long way toward providing you with intellectual freedom. Choosing to remain ignorant of statistical concepts may doom you to a life sentence of half-truths. Essentially, the benefits of a course like this are twofold. You will learn to read and understand research reports, and you will learn to produce your own research. As an intelligent research consumer, you'll be able to evaluate statistical reports read at professional conventions or printed in your field's journals. Also, as a student of the social sciences, you will probably be called on at some time to do original research work. This prospect will not seem so overwhelming after you've mastered the tools available in this book. More basic than that, you'll have a far better chance of understanding research items in newspapers or magazines, or on TV. Who should take statistics? Virtually anyone who wishes to be informed.

In fact, the argument has been made that learning the lessons of statistical methods increases a person's reasoning ability, so the rules taught in statistics courses can be generalized to situations encountered in everyday life (Lehman et al., 1988). In fact, Lehman found that large gains in the application of scientific rules of evidence to everyday life situations occurred among graduate psychology majors (who had taken research methods courses), but not among graduate students in chemistry. In another study the reasoning powers of students who had just completed an introduction to psychology course (where research methodology had been stressed) were compared with students who had had an introductory philosophy course. The reasoning power and level of critical thinking for the psychology students increased dramatically from the first to the last day of the course, whereas the philosophy students showed no such improvement on the final test (Leshowitz, 1989).

GENERAL FIELD OF STATISTICS

Statistics as a general field consists of two subdivisions: descriptive statistics and inferential, or predictive, statistics.

Descriptive Statistics

Descriptive statistics involves techniques for describing data in abbreviated, symbolic fashion. It's a sort of shorthand, a series of precise symbols for the description of what could be great quantities of data.

For example, when we are told that the average score on the verbal section of the Scholastic Assessment Test (SAT) is 500, we are being provided with a description of one characteristic of hundreds of thousands of college-bound high school students.

The descriptive tool used in this case is the arithmetic average, or the mean. To arrive at this value, the SAT verbal scores of all the high school students taking the test throughout the country were added together, and then the total was divided by the number of students involved. The resulting mean value of 500 *describes* one characteristic of this huge group of high school students.

Perhaps we would also like to know how wide the range of SAT scores was. To arrive at this value, the difference between the highest and lowest scores is calculated. In the case of the SAT distribution, where the highest score is 800 and the lowest 200, the range is found to be 600.

Knowing this value, our description of the group gains additional refinement. Other important descriptive statistics are the median, the mode, the standard deviation, and the variance. Chapters 2 and 3 will introduce these descriptive techniques.

Inferential Statistics

Inferential statistics involves making predictions of values that are not really known. Suppose we wished to estimate the height of the average American male. Since it would be impossible to line up all the men in the country and actually measure them, we would instead select a small number of men, measure their heights, and then predict the average height for the entire group. In this way, inferential statistics makes use of a small number of observations to predict to, or infer the characteristics of, an entire group.

This process of inference is obviously risky. The small group of observations from which the inference will be made must be representative of the entire group. If not, the predictions are likely to be way off target. A person who takes a small amount of blood for analysis knows that the sample is fairly representative of all the blood in the entire circulatory system. But when a researcher takes a sample of adult males, no one can be absolutely sure that true representation has been achieved. Also, the researcher seldom, if ever, gets the chance to verify the prediction against the real measure of the entire group. One exception, however, is in political forecasting. After pollsters like Gallup, Harris, Zogby, and Yankelovich make their predictions as to how the population of voters will respond, the actual results are made compellingly (and sometimes embarrassingly) clear on the first Tuesday in November.

Despite the riskiness of the endeavor, statisticians do make predictions with better than chance (actually, far better than chance) accuracy about the characteristics of an entire group, even though only a small portion of the group is actually measured. Inferential statistics is not an infallible method. It does not offer eternal truth or immutable reality carved in stone. As one statistician said, "There is no such thing as eternal truth until the last fact is in on Judgment Day." It does offer a probability model wherein predictions are made and the limits of their accuracy are *known*. As we will see, that really isn't bad.

Statistics and Fallacies

Many statistical fallacies are a result of wishful thinking. As has been sarcastically stated, statisticians are only satisfied with results that confirm something they already believe to be true. Also, many fallacies are based on a lack of understanding of independent events. Soldiers often dive into a shell hole on the assumption that a shell will never hit in exactly the same place twice. During WWII, the English chess champion, P. S. Millner-Barry, decided to refurbish and not leave his London apartment after it was bombed. Sadly, it was bombed again.

Sir Francis Galton (1822–1911) is long remembered as the father of intelligence testing and the person most responsible for developing the methods we use even today for quantifying the data of the behavioral sciences. Galton once said,

> Some people even hate the name of statistics, but I find them full of beauty and interest. Whenever they are not brutalized, but delicately handled by the higher methods, and are warily interpreted, their power of dealing with complicated phenomena is extraordinary. They are the only tools by which an opening can be cut through the formidable thicket of difficulties that bars the path of those who pursue the Science of Man (Galton, 1899).

Let us hope that throughout the semester, we will not be accused of brutalizing statistics but will try and follow Galton's advice and handle them delicately.

SUMMARY

It has often been said that one can prove anything with statistics. However, this is only true if the audience is naïve about statistical procedures and terms. The terms used by statisticians are exact and their definitions are important, since they are designed to facilitate very precise communication.

The field of statistics is of fairly recent origin. The laws of probability were not formulated systematically until the seventeenth century. Blaise Pascal is popularly credited with these first formulations. Not until the beginning of the twentieth century were the strategies devised (by W. S. Gossett in 1906) for measuring samples and then using those data to infer population characteristics.

The general field of statistics includes two subdivisions, descriptive statistics and inferential statistics.

Descriptive statistics: Those techniques used for describing data in abbreviated, symbolic form.

Inferential statistics: Those techniques used for measuring a sample (subgroup) and then generalizing these measures to the population (the entire group).

Key Terms and Names

descriptive statistics
Gossett, William Sealy
inferential statistics
Pascal, Blaise

PROBLEMS

1. During the crisis in Vietnam, the number of Americans who died there was *lower* (for the same time period) than was the number of Americans who died in the United States. Therefore, one can conclude that it is safer to participate in a war than to remain at home. Criticize this conclusion based on what you know of the two populations involved—Americans who go to war versus those who do not.
2. A certain Swedish auto manufacturer claims that 90% of the cars it has built during the past 15 years are still being driven today. This is true, so say the ads, despite the fact that the roads in Sweden are rougher than are those in the United States. What important piece of information must you have before you can evaluate the truth of this auto-longevity claim?
3. A recent TV ad tried to show the risks involved in not taking a certain antacid tablet daily. The actor, wearing the obligatory white coat and stethoscope, poured a beaker of "stomach acid" onto a napkin, immediately creating a large hole. The actor then menacingly intoned, "If acid can do that to a napkin, think what it can do to your stomach." On the basis of the "evidence" included in the commercial, what can the acid do to your stomach?
4. An oil company grandly proclaims that its profits for the fourth quarter increased by 150% over those in the same period a year ago. On the basis of this statement, and assuming that you have the money, should you immediately rush out and buy the company's stock?
5. A toothpaste company says that a large sample of individuals tested after using its brand of toothpaste had 27% fewer dental caries. Criticize the assumption that using the company's toothpaste reduces the incidence of caries.
6. A statewide analysis of speeding tickets found that state troopers in unmarked cars were giving out 37% more tickets over holiday weekends than were troopers working from cruisers that were clearly visible as police cars. Criticize the suggestion that troopers assigned to unmarked cars were obviously more vigilant and more vigorous in their pursuit of highway justice.
7. A marketing research study reported that a certain brand of dishwashing detergent was found by a test sample to be 35% more effective. What else should the consumer find out before buying the product?
8. Two major automakers, one in the United States and the other in Japan, proudly announce from Detroit that they will jointly produce a new car. The announcement further states (patriotically) that *75% of the parts for this car will be produced in the United States.* Is this necessarily good news for U.S. workers?
9. A research study reported a linkage between learning disabilities and crime. Data from the Brooklyn Family Court (1988) indicated that 40% of the juveniles who appeared in court were learning disabled. The report further suggested that these data show that juveniles with learning disabilities are likely to engage in antisocial behavior for the following reasons: (1) they are unskilled, (2) they suffer from low self-

esteem, and (3) they are easily swayed by others. Criticize these conclusions on the basis of the data being offered.

10. Two separate research studies conducted in 1994 came up with the same finding, that men in the corporate world who had working wives earn less money than the men whose wives stayed home to care for the children (as reported in *Newsweek*, October 14, 1994, page 44). In both studies the causal inference was drawn that the reason these men earned less was *because* their wives were in the workforce. It was suggested that men with working wives are therefore subject to some of the same social inequalities as are working women. Thus, when it comes to salary increases and promotions, men with working wives are treated with the same bias and disdain as are the working wives themselves. What alternative explanations might you suggest that are still consistent with these data?

11. It was reported that at a certain university in the United States only 50% of the student athletes in the major sports' programs (Division 1 football and basketball) were graduating within the standard four-year period. The announcement was greeted with great howls of anguish by both the local press and many of the university's alumni and faculty. A chorus of voices moaned that this just goes to prove that the "dumb jock" image is indeed still alive and well. Before the hand wringing gets too out of control, what might you as a fledgling researcher point out with regard to the value of comparison groups?

12. With a nice four-day holiday weekend ahead, you're cruising down the highway and listening to your favorite music station when suddenly you hear the National Safety Council's warning that over 600 persons will die on the highways this weekend. What other information should you know before you pull off the road and hide out in a motel for the entire holiday weekend?

13. In 2000, it was said that senior citizens were spending twice as much on prescription drugs as they did 8 years earlier. Does this mean that big, greedy, drug companies were charging twice as much for their drugs?

14. The heights of 2,848 students, ages 5 to 13, were recorded and it was found that at every grade level, the oldest students in the class were the shortest (Wake, Coghlan & Hesketh, 2000). The conclusion was that these students were being held back because of faulty teacher perceptions. That is, the teachers perceived the lack of height as a sign of immaturity and a lack of readiness for promotion. This was found to be especially true for boys. What other reasons might you suggest to explain this result?

15. The opening of a new Broadway musical in April of 2001, "The Producers," led news accounts to rave that box-office advance sales set an all-time popularity record, with almost $3 million worth of tickets sold the first day. What other information must we get before concluding that this actually reflected a record popularity for the new show?

16. In a study of over 14,000 high school students, reported in September, 2000, it was found that teens, both boys and girls, who were members of sports teams were less apt to engage in a number of unhealthy activities. For example, the athletes were less likely to (a). use drugs, (b). smoke, (c). carry weapons, (d). have unhealthy eating habits, and (e). have sex. The authors implied that the stereotype of the clean-cut jock may be more than just a Hollywood legend. They concluded that because of the positive relationship between sports participation and positive health behaviors, physicians should actively encourage young people to take advantage of the opportunity to join sports teams (Pate et al., 2000). Comment on the implied causal relationship between athletics and healthy behaviors.

17. Studies have shown that whereas armed citizens shoot the wrong person 2% of the time, police officers do so 11% of the time (Poe, 2001). Does this mean that police officers are more careless and less vigilant than armed citizens?

18. Bullying is the repeated physical or psychological mistreatment by a peer who is physically or psychologically stronger than the victim. Dan Olweus cites evidence showing that the victims of bullying are more likely to attempt suicide than are children in general (and far more likely than are the bullies themselves). This appears to be the dramatic statistic that has encouraged schools to show more interest in providing programs aimed at reducing bullying. Does this mean that being bullied causes suicide attempts? What other explanations are possible?

19. Fatalities per million miles of driving are 5.1 for men and 5.9 for women. Be careful not to assume that this proves that men are better and safer drivers than are women. What else must be known to interpret this result?

Chapter

2

Graphs and Measures of Central Tendency

To be able to describe large amounts and, of course, small amounts of data, statisticians have created a series of tools or abbreviated symbols to give structure and meaning to the apparent chaos of the original measures. Thousands, even millions, of scores can be organized and neatly summarized by the appropriate use of descriptive techniques. These summaries allow for precise communication of whatever story the data have to tell.

During this initial foray into the realm of descriptive statistics, we shall concentrate on two major techniques: graphs and measures of central tendency.

GRAPHS

The process of creating a graph to embody a group of scores always begins with the creation of a distribution.

Distributions

To extract some meaning from original, unorganized data, the researcher begins by putting the measurement into an order. The first step is to form a distribution of scores. **Distribution** simply means the arrangement of any set of scores in order of magnitude. For example, the following IQ scores were recorded for a group of students: 75, 100, 105, 95, 120, 130, 95, 90, 115, 85, 115, 100, 110, 100, and 110. Arranging these scores to form a distribution means listing them sequentially from highest to lowest. Table 2.1 is a distribution of the list of IQ scores.

Frequency Distributions. A distribution allows the researcher to see general trends more readily than does an unordered set of raw scores. To simplify inspection of the data further, they can be presented as a **frequency distribution.** A frequency distribution is a listing, in order of magnitude, of each score achieved, together with the

TABLE 2.1 Distribution of IQ scores.

130
120
115
115
110
110
105
100
100
100
95
95
90
85
75

number of times that score occurred. Table 2.2 is a frequency distribution of the IQ scores. Note that the frequency distribution is more compact than the distribution of all the scores listed separately. The *X* at the top of the first column stands for the raw score (in this case, IQ), and the *f* over the second column stands for frequency of occurrence. As we can see, of the 15 people taking the test, 2 received scores of 115, 2 received 110, 3 scored 100, 2 scored 95, and 6 achieved unique scores.

In addition to presenting frequency distributions in tabular form, statisticians often present such data in graphic form. A graph has the advantage of being a kind of picture of the data. It is customary to indicate the actual values of the variable (raw scores) on the horizontal line, or *X* axis, which is called the **abscissa.** The frequency of

TABLE 2.2 IQ scores presented in the form of a frequency distribution.

X *(Raw Score)*	f *(Frequency of Occurrence)*
130	1
120	1
115	2
110	2
105	1
100	3
95	2
90	1
85	1
75	1

occurrence is presented on the vertical line, or *Y* axis, which is called the **ordinate.** It is traditional when graphing data in this way to draw the ordinate about 75% as long as the abscissa. Thus, if the abscissa is 5 inches long, the ordinate should measure about 3¾ inches.

Histograms or Bar Graphs

Figure 2.1 shows the data that were previously presented in tabular form arranged in a graphic form, called a **histogram.** To construct a histogram, a rectangular bar is drawn above each raw score. The height of the rectangle indicates the frequency of occurrence for that score.

Frequency Polygons

Figure 2.2 shows the same IQ data arranged in another commonly used graphic form called a **frequency polygon.** To construct a frequency polygon, the scores are again displayed on the *X* axis, and the frequency of occurrence is on the *Y* axis. However, instead of a rectangular bar, a single point is used to designate the frequency of each score. Adjacent points are then connected by a series of straight lines.

The frequency polygon is especially useful for portraying two or more distributions simultaneously. It allows visual comparisons to be made and gives a quick clue as to whether the distributions are representative of the same population. Figure 2.3 is a frequency polygon that displays the scores of two groups at once. When two distributions are superimposed in this manner, the researcher can glean information about possible differences between the groups. In this case, although the scores of the two groups do overlap (between the scores of 90 and 115), Group B outperformed Group A. Perhaps Group B represents a population different from that of Group A.

FIGURE 2.1 A histogram of IQ scores.

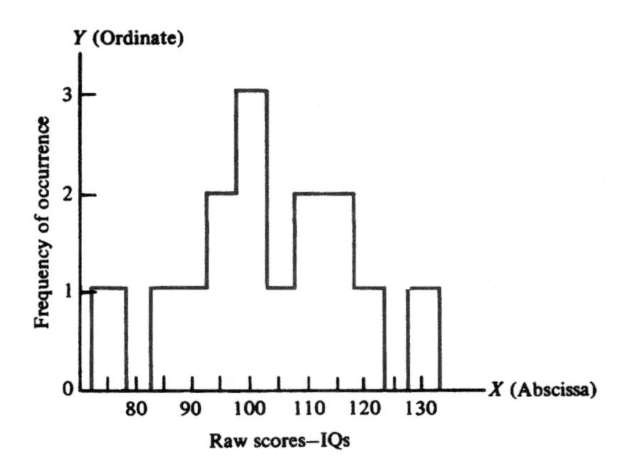

FIGURE 2.2 A frequency polygon.

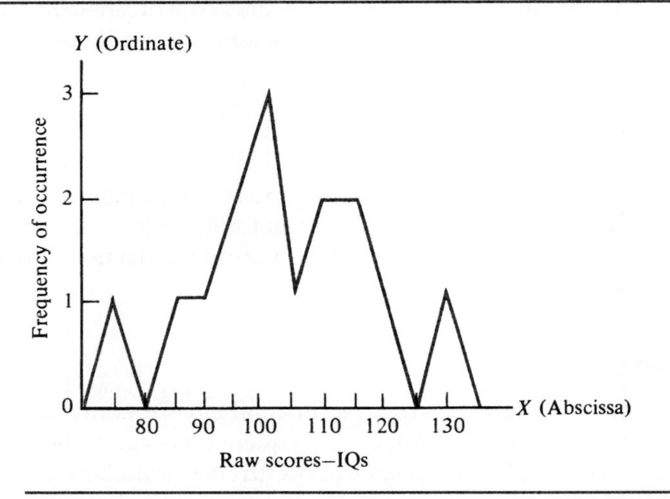

Perhaps the data show that for the trait being measured (in this case IQ), the difference between these groups is greater than would be predicted to occur by chance. Later in this book, statistical techniques for making that kind of conclusion in a more precise way will be presented.

Another method for displaying histograms has become increasingly popular, especially among statistical computer programs. It expresses the frequencies horizontally rather than vertically, with each asterisk representing an individual score. If we were graphing a distribution of arithmetic subtest scores from the WISC III, with the

FIGURE 2.3 A frequency polygon with two distributions.

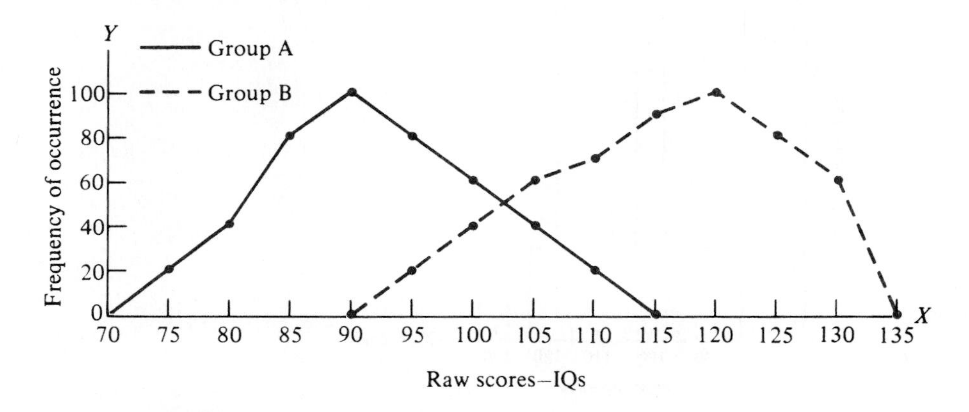

scaled scores on the left and the frequency of occurrence on the right, our histogram would appear as follows:

Score	*Frequency*
15	*
14	
13	****
12	***********
11	****************
10	********************
9	***************
8	************
7	*****
6	*
5	*

Finally, a very economical way of graphing data, called the **stem-and-leaf display,** presents only the first digit in the stem column, while in the leaf column we find the trailing digits. For example, if we were to graph the scores on a certain reading test, where the values ranged from a low score of 60 to a high score of 100, the stem column would present the digits 6 through 10, with the 6 representing scores in the 60s, the 7 for the 70s, and so on. The leaf column would then include the trailing digit, so that a 0 next to the 6 would indicate a score of 60, the 3 a score of 63, and the 5 a score of 65.

Stem	*Leaf*
6	035
7	13578
8	04455569
9	00145
10	0

Note that in this example, most of the scores were in the 80s—specifically, 80, 84, 84, 85, 85, 85, 86, and 89. Also, notice that there was only one person who scored 100.

Importance of Setting the Base of the Ordinate at Zero

In both the histogram and the frequency polygon, the base of the ordinate should ideally be set at zero (or at its minimum value); if it is not, the graph may tell a very misleading story. For example, suppose we are graphing data from a learning study that show that increasing the number of learning trials increases the amount learned. The number of trials is plotted on the abscissa and the number of correct responses on the ordinate. (See Figure 2.4.) The graph shows that by the fourth trial the subject made 8 correct responses, and that by the tenth trial the subject made 12 correct responses. These data are typical of the results obtained in many learning studies. A

FIGURE 2.4 A frequency polygon with the ordinate base at zero.

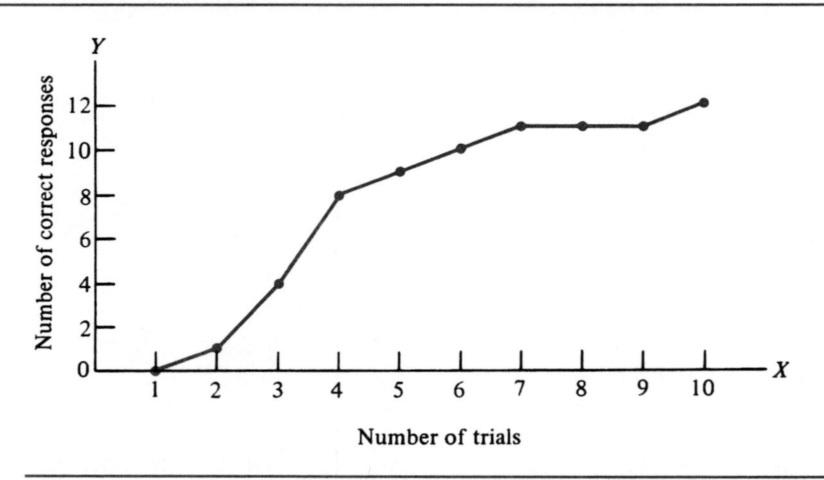

great deal of learning usually occurs during the first few trials, but as the number of trials increases, the rate of increase in learning levels off.

Suppose, however, that a statistician wants to rig the graph to create a false impression of what the data really reveal. The trick is to focus on one small area of the graph. (See Figure 2.5.) Now the data seem to tell a very different story about how the learning took place. It now appears that no learning took place before the fourth trial and that the great bulk of learning took place between the fourth and tenth trials. We know from the complete graph of the data that this is simply not true. In fact, the majority of the learning actually took place during the first four trials; after the fifth trial,

FIGURE 2.5 A frequency polygon with the ordinate base greater than zero.

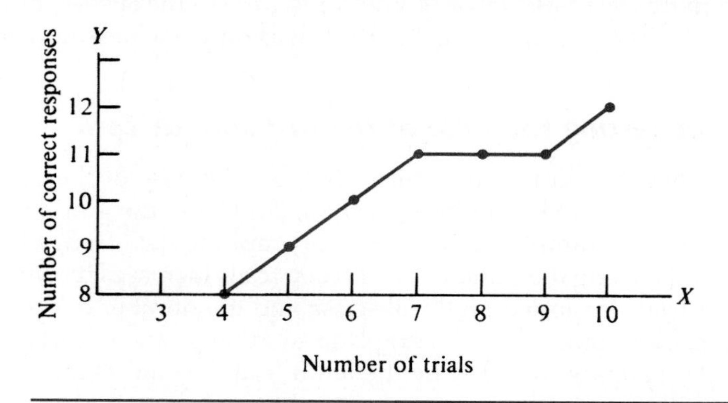

the rate of learning was beginning to top out and level off. This is a blatant instance of how statistics can be used to distort reality if the audience is naïve about statistical methods.

The type of deceit just described is variously known as a "Wow!" a "Gee whiz!" or an "Oh boy!" graph. Some corporations hire statistical camouflagers to devise this kind of tinsel for the sales and earnings graphs used in their annual reports. Insignificant sales gains can be portrayed as gigantic financial leaps by adept use of a "Wow!" graph. Another instance of this chicanery was perpetrated by a certain corporation, anxious to create the image of being an equal opportunity employer. It adroitly used a "Wow!" graph (see Figure 2.6) to prove to the world that it had a liberal and compassionate stance toward the hiring of minorities.

The graph shows time, as months of the year, on the abscissa and the number of minority persons employed on the ordinate. It seems to show only a modest increase in the number of minority persons employed throughout most of the year, and then ("Wow!") a dramatic increase during the final two months. This increase is, of course, more illusory than real; it reflects the hiring of only six new people, a real increase of less than 1%. If the ordinate is set at zero and the data are graphed again, you need an electron microscope to find the increase. (See Figure 2.7.)

Whenever a graph is presented in which the base of the ordinate is not set at zero, be on the alert. The stage has been set for a sleight-of-hand trick. Without question, the most serious error in graphing is the use of a value other than zero as a base. Where there is no visual relationship to zero, the values are without perspective and thus have no meaning. It's like someone telling you that the temperature outside is 40°, not indicating the scale—Fahrenheit or Celsius—or whether it's 40° above or below zero. You might dress to prevent frostbite and end up suffering heatstroke.

FIGURE 2.6 Example of a "Wow!" graph. (Note: the ordinate base is greater than zero.)

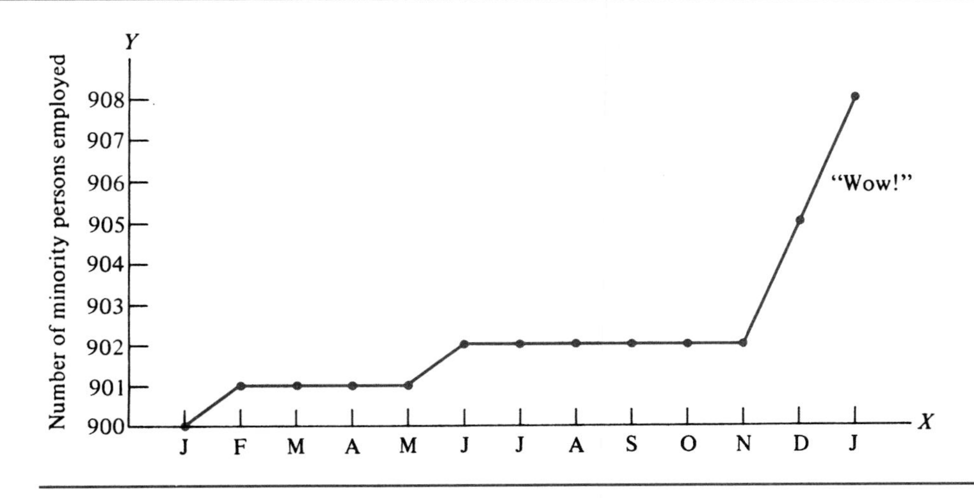

FIGURE 2.7 Graph of the data in Fig. 2.6 with the ordinate base equal to zero.

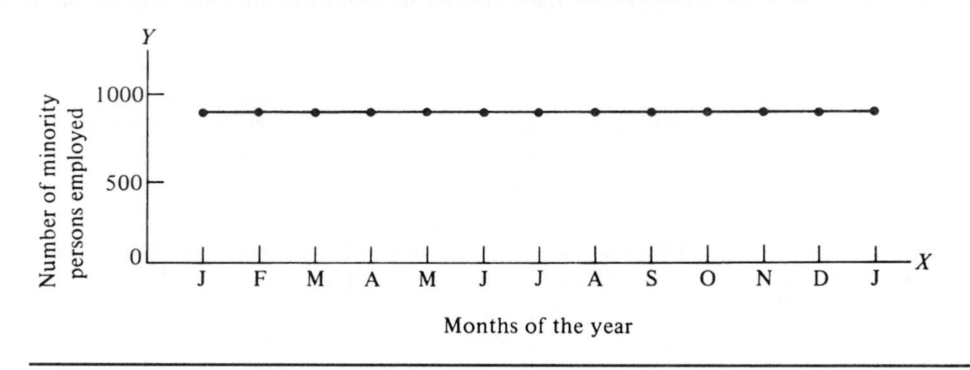

MEASURES OF CENTRAL TENDENCY

Tools called **measures of central tendency** are designed to give information concerning the average, or typical, score of a large number of scores. If, for example, we were presented with all the IQ scores of the students at a college, we could utilize the measures of central tendency to give us some description of the typical, or average, intellectual level at that school. There are three methods for obtaining a measure of the central tendency. When used appropriately, each is designed to give the most accurate

estimation possible of the typical score. Choosing the appropriate method can sometimes be tricky. The interpretation of the data can vary widely, depending on how the typical score has been found.

The Mean

The most widely, though not always correctly, used measure of central tendency is the **mean,** symbolized as *M*. The mean is the arithmetic average of all the scores. It is calculated by adding all the scores together and then dividing by the total number of scores involved. In some articles, especially those published a few years ago, the mean may be expressed as $\bar{X}$.

The formula for the mean is as follows:

$$M = \frac{\Sigma X}{N}$$

In the formula, *M*, of course, stands for the mean. The capital letter *X* stands for the raw score, or the measure of the trait or concept in question. The Greek capital letter Σ (sigma) is an operational term that indicates the addition of all measures of *X*. This is usually read as "summation of." Finally, the capital letter *N* stands for the entire number of observations being dealt with. (It is important that you follow the book's use of capital letters, since lowercase letters often mean something quite different.) Thus, the equation tells us that the mean (*M*) is equal to the summation (Σ) of all the raw scores (*X*) divided by the number of cases (*N*).

In Table 2.3, the mean of the distribution of IQ scores from Table 2.1 has been calculated. It happens that the mean is an appropriate measure of central tendency in this case because the distribution is fairly well balanced. Most of the scores occur in the middle range, and there are no extreme scores in one direction. Since the mean is calculated by adding together all of the scores in the distribution, it is not usually influenced by the presence of extreme scores, *unless* the extreme scores are all at one end of the range. The mean is typically a stable measure of central tendency, and, without question, it is the most widely used. Table 2.4 provides a calculation of the mean in which the result is not a whole number.

Interpreting the Mean. Interpreting the mean correctly can sometimes be a challenge, especially when either the group or the size of the group changes. For example, the mean IQ of the typical freshman college class is about 115, whereas the mean IQ of the typical senior class is about 5 points higher. Does this indicate that students increase their IQs as they progress through college? No, but since the size of the senior class is almost always smaller than the size of the freshman class, the two populations are not the same. Among those freshmen who never become seniors are a goodly number with low IQs, and their scores are not being averaged in when their class later becomes seniors.

The Mean of Skewed Distributions. In some situations, the use of the mean can lead to an extremely distorted picture of the average value of a distribution of scores.

TABLE 2.3 Calculation of the mean from a distribution of raw scores.

X
130
120
115
115
110
110
105
100
100
100
95
95
90
85
75

$\Sigma X = 1545$

$$M = \frac{\Sigma X}{N} = \frac{1545}{15} = 103$$

TABLE 2.4 Calculation of the mean from a distribution of raw scores in which the result is not a whole number.

X
72
71
70
68
68
68
65
63

$\Sigma X = 545$

$$M = \frac{\Sigma X}{N} = \frac{545}{8} = 68.125 = 68.13$$

TABLE 2.5 Calculation of the mean of a distribution of annual incomes.

X
$10,000,000.00
20,000.00
20,000.00
19,500.00
19,400.00
19,400.00
19,400.00
19,300.00
19,000.00
18,500.00
18,000.00
18,000.00
17,600.00

$\Sigma X = \$10,228,100.00$

$$M = \frac{\Sigma X}{N} = \frac{10,228,100.00}{13} = \$786,776.92$$

For example, look at the distribution of annual incomes in Table 2.5. Note that one income ($10,000,000.00) is so extremely far above the others that the use of the mean income as a reflection of averageness gives a highly misleading picture of great prosperity for this group of citizens. A distribution like this one, which is unbalanced by an extreme score at or near one end, is said to be a **skewed distribution.**

Figure 2.8 shows what skewed distributions look like in graphic form. In the distribution on the left, most of the scores fall to the right, or at the high end, and there are only a few extremely low scores. This is called a negatively skewed, or skewed to the left, distribution. (The skew is in the direction of the tail-off of scores, not of the majority of scores.) The distribution on the right represents the opposite situation. Here most of the scores fall to the left, or at the low end of the distribution, and only a very

FIGURE 2.8 A graphic illustration of skewed distributions.

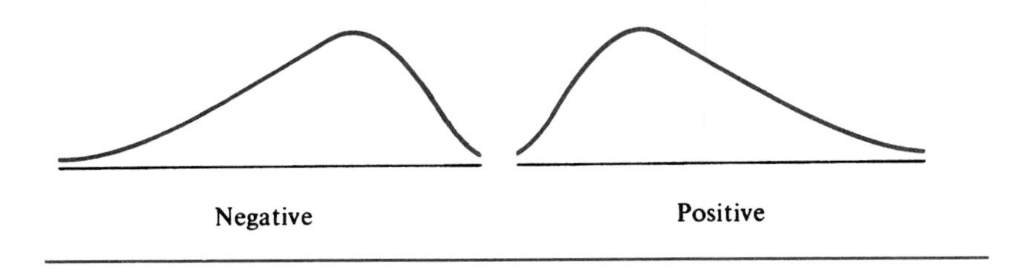

few scores are high. This is a positively skewed, or skewed to the right, distribution. Remember, label skewed distributions according to the direction of the tail. When the tail goes to the left, the curve is negatively skewed; when it goes to the right, it is positively skewed.

Thanks, Ralph. A rather dramatic example of how the use of the mean distorts "averageness" in a skewed distribution can be seen in a University of Virginia press release, undoubtedly meant as tongue in cheek. During an analysis of how well the members of the college's 1983 graduating class fared in the job market, it was discovered that the highest salaries were earned by graduates of the Department of Rhetoric and Communications Studies, where the beginning average pay was $55,000 a year. It should be pointed out that one student, the 7-foot 4-inch-tall basketball player Ralph Sampson, had a starting salary of well over $1 million. Perhaps the mean height for those same graduates was 6 feet 6 inches.

And if Ralph Sampson distorted the mean income at the University of Virginia, one can only imagine how the world's richest people might affect the averages of the colleges they attended, assuming they all went to college. Every year *Forbes* spotlights the richest persons in the world, and in the late 1990s, its list only included what *Forbes* calls the "working rich," which excludes such "idlers as royalty and dictators." The list is as follows (with wealth valued in BILLIONS):

Bill Gates (Microsoft)	$51.0
Sam Walton's family (Wal-Mart Stores)	48.0
Warren Buffett (Berkshire Hathaway)	33.0
Paul Allen (Microsoft)	21.0
Kenneth Thomson (Thomson Corp. of Canada)	14.4
Forest Mars (Mars candy bars)	13.5
J. & R. Pritzker (financiers)	13.5
Alwaleed Alsaud (Saudi Arabian construction and banking)	13.3
Lee Shau Kee (Hong Kong real estate)	12.7
The Albrecht family (German retail)	11.7
Steven Ballmer (Microsoft)	10.5
The Mulliez family (French retail)	10.3

Had royalty been allowed on the list, the Sultan of the oil-rich Emirate of Brunei would have been in third place with a fortune of $36 billion (*Forbes*, July 1998).

Anytime you want to remember which way income distributions are skewed, think of the people on that list and what would happen to the financial averages in your neighborhood if any of them moved *right* next door. The averages would skew to the right. Also, remember that these values are in *billions* of dollars. To put this in perspective, consider the following: If in the year A.D. 1 you had a billion dollars (and immortality) and you then spent a thousand dollars each and every *day* since then, by the year 2000 you'd still have over 200 years to go before running out of funds. As one U.S. senator once said during a budget debate, "A billion here, a billion there, and pretty soon you're talking about real money."

The Median

The **median** is the exact midpoint of any distribution, or the point that separates the upper half from the lower half. In fact, the median (symbolized as Mdn) is a much more accurate representation of central tendency for a skewed distribution than is the mean. Whereas the mean income of the distribution in Table 2.5 is $786,776.92, the median income is $19,400.00, a much more descriptive reflection of the typical income for this distribution. Since income distributions are almost always skewed toward the high side, you should be on the alert for an inflated figure whenever the mean income is reported. The median gives a better estimation of how the typical wage earner is actually faring.

As a memory trick for recalling the definition of the median, think of the median strip dividing a highway. The same width of road lies both to the left of the median strip and to the right.

Calculating the Median. To calculate the median, the scores must first be arranged in distribution form, that is, in order of magnitude. Then, count down (or up) through half of the scores. For example, in Table 2.5 there are 13 income scores in the distribution. Therefore, count down 6 scores from the top, and the seventh score is the median (there will be the same number of scores above the median point as there are below it). Whenever a distribution contains an odd number of scores, finding the median is very simple. Also, in such distributions the median will usually be a score that someone actually received. In the distribution in Table 2.5, three persons really did earn the median income of $19,400.

If a distribution is made up of an even number of scores, the procedure is slightly different. Table 2.6 presents a distribution of an even number of scores. The median

TABLE 2.6 Calculation of the median with an even number of scores.

	X	
	120	
	118	
	115	—114.5 Median
	114	
	114	
	112	
$\Sigma X =$	693	
$M =$	115.50 Mean	

is then found by determining the point that lies halfway between the two middlemost scores. In this case, the median is 114.5. Don't be disturbed by the fact that in some distributions of even numbers of scores nobody actually received the median score; after all, nobody ever had 2.8 children either.

The Median of Skewed Distributions. Unlike the mean, the median is not affected by skewed distributions (where there are a few extreme scores in one direction). In Table 2.7, for example, the median score is still found to be 114.5, even with a low score of 6 rather than one of 112 as reported in Table 2.6. The mean of this distribution, on the other hand, plummets to an unrepresentative value of 97.83. The mean is always pulled toward the extreme score in a skewed distribution. When the extreme score is at the high end, the mean is too high to reflect true centrality; when the extreme score is at the low end, the mean is similarly too low.

An Above-Average Scholar. In June of 1998, the senior class at Nauset High School on Cape Cod, Massachusetts, had among its members a woman who was 98 years old (*Springfield Union News,* June 16, 1998, pp. 1, 8). The inclusion of this woman's age increased the mean age of the graduating class from about 18 to well over 21, yet both the median and mode were totally unaffected. As another example, home sales nationally reached a median price in 1998 of $130,000 (as reported by the National Association of Realtors). Since the mean was $140,000 you should conclude that the distribution of home prices is skewed to the right, or has an Sk+.

The Mode

The third, and final, measure of central tendency is called the **mode** and is symbolized as Mo. The mode is the most popular, or most frequently occurring, score in a distribution. In a histogram, the mode is always located beneath the tallest bar; in a fre-

TABLE 2.7 Calculation of the median with an even number of scores and a skewed distribution.

	X	
	120	
	118	
	115	—114.5 Median
	114	
	114	
	6	
$\Sigma X =$	587	
$M =$	97.83 Mean	

FIGURE 2.9 The location of the mode in a histogram (left) and a frequency polygon (right).

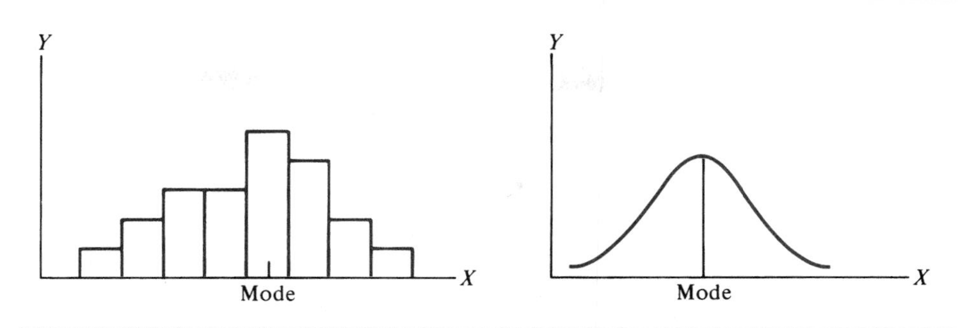

quency polygon, the mode is always found directly below the point where the curve is at its highest. This is because, as was pointed out previously, the Y axis, or the ordinate, represents the frequency of occurrence. (See Figure 2.9.)

Finding the Mode. When the data are not graphed, just determine which score occurs the most times, and you've got the mode. For example, in the distribution shown in Table 2.8, the score of 103 occurs more often than does any other score. That value is the mode. In Table 2.9, a frequency distribution of the same data is given. Here, to find the mode, all you have to do is to note which score (X) is beside the highest frequency value (f). The mode is a handy tool since it provides an extremely quick method for obtaining some idea of a distribution's centrality. Just eyeballing the data is usually enough to spot the mode.

TABLE 2.8 Finding the mode of a distribution of raw scores.

X	
110	
105	
105	
103	Mode
103	Mode
103	Mode
103	Mode
101	
101	
100	
100	
98	
95	

TABLE 2.9 Finding the mode when a frequency distribution is given.

X	f
110	1
105	2
103 ← Mode	4
101	2
100	2
98	1
95	1

Bimodal Distributions. A distribution having a single mode is referred to as a **unimodal distribution.** However, some distributions have more than one mode. When there are two modes, as in Figure 2.10, the distribution is called **bimodal.** (When there are more than two modes, it is called multimodal.) Distributions of its type occur when scores cluster together at several points, or if the group being measured really represents two or more subgroups.

Assume that the distribution in Figure 2.10 represents the running times (in seconds) in the 100-yard dash for a large group of high school seniors. There are two modes—one at 13 seconds and the other at 18 seconds. Since there are two scores that both occur with the same high frequency, it is probable that data about two separate subgroups are being displayed. For example, perhaps the running times for boys are clustering around one mode and the speeds for girls are clustering around the other.

Interpreting the Bimodal Distributions. Whenever a distribution does fall into two distinct clusters of scores, extreme care must be taken with their interpretation. Neither the mean nor the median can justifiably be used, since a bimodal distribution cannot be adequately described with a single value. A person whose head is packed in ice and whose feet are sitting in a tub of boiling water cannot be appropriately characterized as being in a tepid condition *on the average.*

FIGURE 2.10 Graph of bimodal distribution of running times.

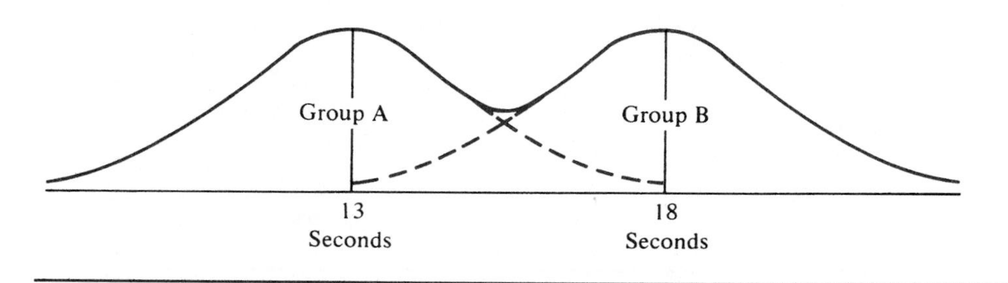

FIGURE 2.11 Graph of responses to an attitude questionnaire showing two modes.

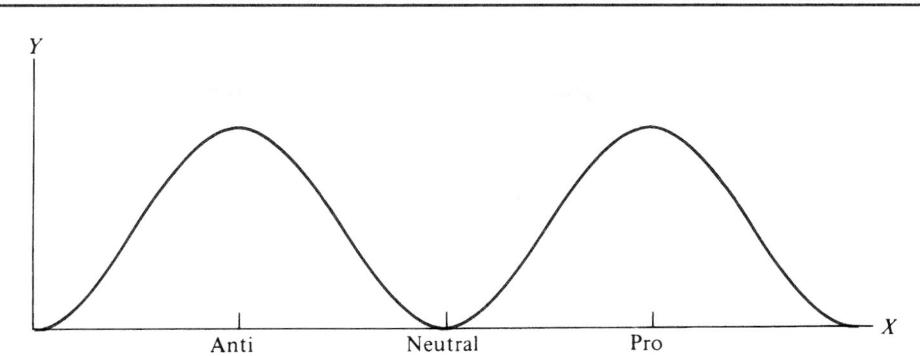

Bimodal distributions should not be represented by the use of a single average of the scores. Suppose that a group of individuals completed a certain attitude questionnaire. The resulting scores fell in two decidedly different clusters, half the group scoring around a "pro" attitude and the other half around an "anti" attitude. The use of either the mean or the median to report the results would provide a highly misleading interpretation of the group's performance, for in either case the group's attitude would be represented as being neutral. Figure 2.11 shows such a bimodal distribution. The two modes clearly indicate how divided the group really was. Note, too, that while *nobody* scored at the neutral point, using either the mean or the median as a description of centrality would imply that the typical individual in the group was indeed neutral. Thus, when a distribution has more than one mode, the modes themselves, not the mean or the median, should be used to provide an accurate account.

APPROPRIATE USE OF THE MEAN, THE MEDIAN, AND THE MODE

Working with Skewed Distributions

The best way to illustrate the comparative applicability of the three measures of central tendency is to look again at a skewed distribution. Figure 2.12 shows the income distribution per household in the United States in 1998 (*Barron's*, 1998). Like most income distributions, this one is skewed to the right. This is because the low end has a fixed limit of zero, while the sky is the limit at the high end. Note that the exact midpoint of the distribution, the median, falls at a value of $50,000 a year. This is the figure above which 50% of the incomes fall and below which 50% fall. Because there is a positive skew, the mean indicates a fairly high average income of $61,000. This value, however, gives a rather distorted picture of reality, since the mean is being unduly influenced by the few families at the high end of the curve whose incomes are in the

FIGURE 2.12 Distribution of income per household in the United States.

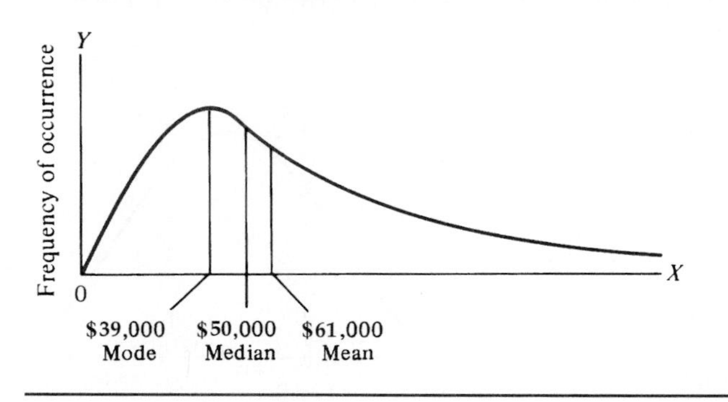

millions. Finally, the modal income, which is $39,000 per year, seems to distort reality toward the low side. Although the mode does represent the most frequently earned income, it is nevertheless lower than the point separating the income scores into two halves. In the case of a skewed distribution, then, both the mean and the mode give false, though different, portraits of typicality. The truth lies somewhere in between, and what's in between is the median. Thus, in a positively skewed distribution the order of the three measures of central tendency from left to right is first the mode, the lowest value; then the median, the midpoint; and finally the mean, the highest value. A negatively skewed distribution simply reverses this order. (See Figure 2.13.) The point to remember for skewed distributions is that the mean is always located toward the tail end of the curve.

FIGURE 2.13 A negatively skewed distribution.

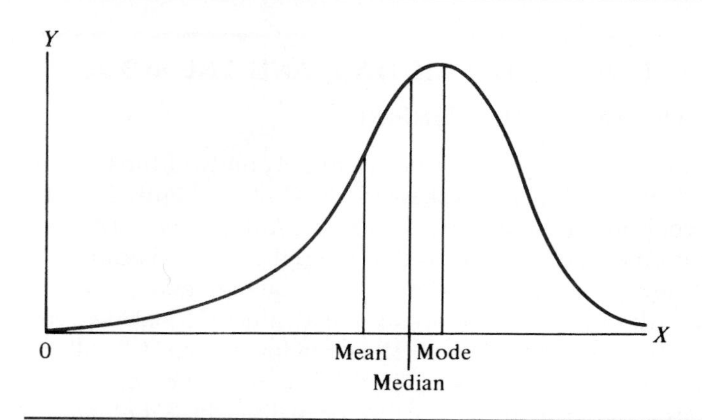

The figures shown above, as previously stated, apply to household incomes. Let's look now at individual incomes and how they compare with one's educational level. The mean income levels by educational attainment for 1998 were as follows:

Did not finish high school	$15,011
High school grad	$22,154
College BA	$38,112
Higher degrees	$61,317

Source: Springfield Union News, Monday, June 25, 1998.

This shouldn't be taken to prove that increasing education is the direct cause of the higher incomes, although in many cases it surely is. Other factors must be taken into account. Perhaps wealthier families are more apt to send their children to college and, because of their wealth and power, are more likely to have the connections needed to supply their offspring with high-paying positions. For example, if one of your relatives is the CEO of a major corporation, you might have the opportunity to join the company as a young executive, perhaps even as vice president of some department and then, of course, work your way up.

A mnemonic device often used for this purpose is to remember that as you go up the slope of a skewed curve, the measures of central tendency appear in alphabetical order: first the mean, then the median, and finally the mode. In using this memory prod, you must remember that you're going *up* the slope and that you're going up on the gentle slope, *not* the steep side. Picture yourself at a ski area, riding the chairlift up the novice slope.

Assessing Skewness

Although there are techniques for establishing a precise value for the amount of skewness (Sk) present in any unimodal distribution of scores (see the Appendix), a somewhat rough but easy method for spotting Sk is simply to compare the mean and median. Remembering that in skewed distributions the mean always "tips toward the tail," then the mean can be used as a quick indicator of skewness. If the mean lies to the left of the median, the distribution is skewed left. If it lies to the right of the median, the distribution is skewed to the right. Also, the greater the distance between the mean and median, the greater is the total amount of Sk. For example, if you are told that the mean price for a new home in the United States is $100,000 and that the median price is $80,000, then you may conclude that the distribution of prices is skewed to the right (which it really is), or it has an Sk+. Or, if the median golf score for the members of a certain country club happened to be 95, but the mean was only 85, then the distribution of golf scores would show a negative skew, or Sk–. In general, the greater the discrepancy between the mean and the median, the greater is the skewness.

By incorrectly using such descriptive statistics as the measures of central tendency, some extremely interesting bits of con artistry can be performed. As previously mentioned, if a distribution is skewed to the right, the mode reflects the lowest value of central tendency while the mean portrays the highest. With such a distribution, a researcher could report on a before-and-after study and make it appear that group performance had increased in the "after" condition, even though *every score in the entire group had remained the same.* How? The researcher simply reports the modal score for the results of the "before" condition and the mean score for the results of the "after" condition. Does it matter, then, which measure of central tendency is used? You bet it does!

When to Use the Median: An Example

The following is a listing of the over-the-counter drug sales for one year during the mid-1990s. These are the factory sales figures (in millions) for the fifteen best-sellers (Kline & Co., as reported in the *New York Times,* Friday, September 24, 1998, page d5):

> Advil 360, Alka-Seltzer 160, Bayer 170, Benadryl 130, Centrum 150, Excedrin 130, Halls 130, Metamucil 125, Mylanta 135, One Touch 220, Robitussin 205, Sudafed 115, Tums 135, Tylenol 855, Vicks 350.

Calculate the three measures of central tendency, mean, median, and mode, and indicate which measure of central tendency is the most appropriate for this set of scores. Note that since the question is asking for the median as well as the mean and mode, it would be best to first sort the data into an ordered sales distribution, such that

Tylenol	855
Advil	360
Vicks	350
One Touch	220

TABLE 3.2 Computational method.

X	X^2
10	100
8	64
6	36
4	16
2	4
$\Sigma X = 30$	$\Sigma X^2 = 220$

$N = 5$

$$M = \frac{\Sigma X}{N} = \frac{30}{5} = 6.00$$

$$SD = \sqrt{\frac{\Sigma X^2}{N} - M^2}$$

$$SD = \sqrt{\frac{220}{5} - 6.00^2}$$

$$SD = \sqrt{44 - 36}$$

$$SD = \sqrt{8} = 2.828 \text{ or } 2.83$$

You are urged to use the computational method whenever you work on standard deviations. It is easier, takes less time, and fits perfectly with the capabilities of your calculator. Later, when we cover some of the tests of significance, it will be assumed that this method is being used. Why, then, was the deviation method presented? Simply because it gives a clearer picture (since you must calculate each deviation score) of the concept of the standard deviation and its reliance on the variability of all the scores around the mean.

Using the Standard Deviation. The standard deviation is an extremely useful descriptive tool. It allows a variability analysis of the data that can often be the key to what the data are communicating. For example, suppose that you are a football coach and that you have two halfbacks with identical gain averages of 5 yards per carry. Back A has a standard deviation of 2 yards, and Back B has a standard deviation of 10 yards. Back A is obviously the more consistent runner—most of the time he gains between 3 and 7 yards, but he almost always gets a gain. When only 2 or 3 yards are needed, Back A is unquestionably the choice. While Back B, having a much larger standard deviation, is more likely to make a longer gain, he, alas, is also more likely to lose substantial yardage. If your team is behind, and it's fourth down, 10 yards to go, and time is running out, Back B is more likely to provide the necessary last-minute heroics.

Or suppose that you are a manufacturer of flashlight batteries. Testing of large numbers of two types of batteries shows that each has the same average life—25 hours. Battery A, however, has a standard deviation of 2 hours, while Battery B has a standard deviation of 10 hours. If you plan to guarantee your battery for 25 hours, you had better decide to manufacture Battery A—there won't be as many consumer complaints or returns. Battery B, however, has more chance of lasting at least 35 hours but is also more likely to fail after only 15 hours or less.

The standard deviation is indeed a valuable descriptive tool. Its use does, however, require that the researcher have at least interval data.

The Variance

The **variance** is the third major technique for assessing variability. Once you can calculate a standard deviation, you can easily calculate the variance because the variance, V, is equal to SD^2:

$$V = SD^2$$

$$V = \frac{\Sigma x^2}{N} = \frac{\Sigma X^2}{N} - M^2$$

Thus, calculating the variance is precisely the same as calculating the standard deviation, without taking the square root.

Since the variance has such a straightforward mathematical relationship with the standard deviation, it must also be a measure of variability that tells how much all the scores in a distribution vary from the mean. Conceptually, therefore, it is the same as the standard deviation. Heterogeneous distributions with a great deal of spread have relatively large standard deviations and variances; homogeneous distributions with little spread have small standard deviations and variances. It may seem to be redundant to have two such variability measures, one that is simply the square of the other. However, there are situations in which working directly with the variances allows for certain calculations not otherwise possible. We shall see later that one of the most popular statistical tests, the F ratio, takes advantage of this special property of the variance.

No Variability Less Than Zero

Although it may seem to be stating the obvious, all measures of variability must reflect either some variability or none at all. There can *never be less than zero variability.* The range, the standard deviation, and the variance should never be negative values.

Consider the distribution of values in Table 3.3. Note that in this rather unlikely situation of all scores being identical, the range, the standard deviation, and the variance must equal zero. When all scores are the same, there simply is no variability. The value of zero is the smallest value any variability measure can ever have. If you ever calculate a negative standard deviation or variance, check over your math and/or the batteries in your calculator.

TABLE 3.3 Measures of variability for a distribution of identical scores.

X	X^2
10	100
10	100
10	100
10	100
10	100
10	100
$\Sigma X = 60$	$\Sigma X^2 = 600$

$M = 10$

$R = 0$

$$SD = \sqrt{\frac{600}{6} - 10^2}$$

$$SD = \sqrt{100 - 100}$$

$$SD = \sqrt{0} = 0$$

$$V = SD^2 = 0$$

GROUPED DATA

Before the days of electronic calculators and computers, statisticians typically used what were called grouped-data techniques. These had the advantages of making large numbers of scores more manageable and making values easier to calculate, but they also had the disadvantages of losing track of a given individual's score and losing some degree of accuracy.

Here's how it worked. Assume that we have the following set of 20 scores: 10, 10, 9, 10, 12, 4, 10, 9, 11, 15, 6, 13, 8, 2, 19, 16, 14, 8, 5, 6. Arranging them in distribution form gives the following: *X*: 19, 16, 15, 14, 13, 12, 11, 10, 10, 10, 10, 9, 9, 8, 8, 6, 6, 5, 4, 2.

Now *group* the data into class intervals, in this case with a width of 3, and select the midpoint of the interval as the score, X. Then identify the number of scores, *f* (frequency), falling within each interval. Finally, multiply that frequency by the midpoint of the interval, of *fX*. It's best to use odd-valued intervals (3, 5, 7, etc.), and then center each interval so that the midpoint is a multiple of the width.

Interval	X	f	fX
17–19	18	1	18
14–16	15	3	45
11–13	12	3	36
8–10	9	8	72
5–7	6	3	18
2–4	3	2	6
		$N = \Sigma f = 20$	$\Sigma fX = 195$

Next, add the *fX* column; in this case, the sum equals 195. Then, calculate the mean:

$$M = \frac{\Sigma fX}{N} = \frac{195}{20} = 9.75$$

Now, if we had not grouped the data but had instead simply added the scores, 19, 16, 15, 14, and so on, we would have obtained a value of 197 for ΣX and a mean of 9.85. The reason for the different results is that the grouped-data technique assumes that the mean of the scores within a given interval is equal to the midpoint of that interval. Since this assumption is not always true, the grouped-data technique usually produces results that are slightly in error. In this example, the grouped-data technique underestimated the mean by .10

The standard deviation is also available from grouped data. Using the values from the preceding example, we calculate it as shown in the following.*

Interval	f	X	fX	XfX
17–19	1	18	18	324
14–16	3	15	45	675
11–13	3	12	36	432
8–10	8	9	72	648
5–7	3	6	18	108
2–4	2	3	6	18
			$\Sigma fX = 195$	$\Sigma fX^2 = 2205$

$$M = \frac{\Sigma fX}{N} = \frac{195}{20} = 9.75$$

$$SD = \sqrt{\frac{\Sigma fX^2}{N} - M^2} = \sqrt{\frac{2205}{20} - (9.75)^2} = \sqrt{110.25 - 95.06} = \sqrt{15.19} = 3.90$$

*To obtain the values in the last column, multiply the *X* value by the *fX* value (18 × 18 = 324, 15 × 45 = 675, etc.). *Do not* simply square the *fX* value.

If we had worked out the standard deviation from these data using the raw score formula

$$SD = \sqrt{\frac{\Sigma X^2}{N} - M^2}$$

we would have obtained a value of 4.11. The grouped-data technique, therefore, underestimated the value of the standard deviation by .21.

GRAPHS AND VARIABILITY

The concept of variability, or dispersion, can be further clarified through the use of graphs. Look at the two frequency distributions of IQ scores shown in Figure 3.2. Both distributions have precisely the same mean, 100, and the same range, 60; but they tell dramatically different stories. The distribution in Figure 3.2a shows a large number of people having both very low and very high IQ scores—about 50 with IQs of 70 and another 50 with IQs of 130. There is a slight bulge in the middle of the distribution. However, it shows that only about 100 persons scored at the mean of 100. (Remember that the height of the curve represents the number of occurrences.) In the distribution in Figure 3.2b, there are very few people who scored IQs of 70 and very few who scored 130. The great majority (about 200) scored right around the mean, between the IQs of 90 and 110. Therefore, even though both distributions do indeed have the same range and mean, the one in Figure 3.2b represents a far more homogeneous group than does the one in Figure 3.2a. The distributions differ in terms of their standard deviations. While Figure 3.2a has many scores deviating widely from the mean and thus has a large standard deviation, Figure 3.2b has most of the scores clustering tightly around the mean and thus has a small standard deviation.

FIGURE 3.2 Two frequency distributions of IQ scores.

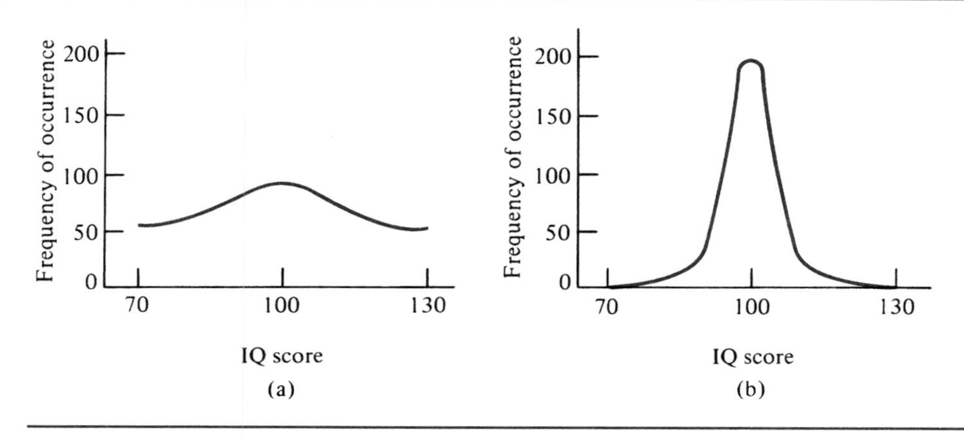

Kurtosis

Statisticians use the term **kurtosis** to describe how peaked or how flat a curve is. The distribution in Figure 3.2a is in the shape called platykurtic. A **platykurtic** curve represents a distribution with a relatively large standard deviation. Figure 3.2b shows a **leptokurtic** curve, which always indicates a relatively small standard deviation. This may sound like extremely pompous and stuffy jargon, but precise terms are necessary to convey precise meanings. Also, you can amaze your friends by working these terms into your conversation at lunch tomorrow.

A curve that is about halfway between platykurtic and leptokurtic extremes is called **mesokurtic.** The importance of the mesokurtic distribution will be discussed at length in the next chapter. It is called the normal curve and is the most important curve we will confront all semester. It is the statistician's "dream" curve, bell shaped and perfectly symmetrical.

Statisticians use the symbol Ku to describe kurtosis as follows: negative Ku values define platykurtic distributions, whereas positive Ku values indicate leptokurtic distributions. A Ku value of zero shows a mesokurtic distribution. For more information on how these values are computed, see the Appendix, Part A.

We have been reminded of two amusing mnemonics attributed to William Sealy Gossett. Platykurtic curves, like platypuses, are squat with short tails, whereas leptokurtic curves are high with long tails, like kangaroos—noted for their "lepping" (Balanda & MacGillivray, 1988).

Relationship Between the Standard Deviation and the Range

The relationship between the standard deviation and the range can best be illustrated in the context of kurtosis. First, however, it must be clearly understood that the range always results in a value larger than the value of the standard deviation. Since the range describes the entire width of a distribution, or the difference between the highest and the lowest scores, the value of the range must always be greater than that of any other measure of variability.*

The standard deviation, on the other hand, describes how much all the scores vary from the mean. Therefore, the standard deviation, which takes into account many more than just the top and bottom scores, must result in a value less than the value of the range. The greater the spread of scores around the mean, however, the more closely the value of the standard deviation will approach the value of the range. In a platykurtic distribution, the value of the standard deviation equals a fairly large proportion of the range. The more tightly the scores cluster around the mean—that is, the more leptokurtic a distribution is—the smaller will be the standard deviation relative to the range.

---

*Although the range equals about six times the standard deviation when the distribution is normal, the range must always be larger than the standard deviation. No matter what shape the distribution takes, the range must *always* be *at least* twice the size of the standard deviation, unless there is no variability, as when $R = 0$.

Is It Reasonable? A couple of very handy checks on the previous calculations—checks that will really prevent gross errors—are based on ensuring that the resulting values are at least reasonable. After all, even when using a calculator or computer, it's always possible to hit the wrong key. Just take a quick time-out and do an "is it reasonable?" check.

1. After calculating the mean, look back at the distribution and make sure that the mean's value is within the range of scores.
2. Also, examine the value of the standard deviation. This is a very important checkpoint, for the standard deviation has certain limits that it cannot exceed. The standard deviation can never be less than zero (it can't be negative) or greater than half the range.
3. When looking at a distribution of scores, it is always wise to spend a few moments screening the data to make sure that blatant entry errors are caught before the analysis proceeds. For example, with a distribution of SAT scores, if one were to find a score of 55 included, it can easily be spotted as an entry error, since a score of 55 is outside the range of possible scores. Had the value of 55 not been caught, it would have skewed the distribution, in this case negatively. Similarly, if a distribution of human adult height scores were found to include values such as 2 inches or 15 feet, then data entry is obviously to blame.

Assessing Kurtosis—The "1/6" Rule

In the Appendix you will find an equation for establishing the exact value for the kurtosis (Ku) of any unimodal frequency distribution. However, at this point you can simply use the following benchmark for coming up with a quick Ku evaluation. Since the standard deviation of a normal distribution is approximately one-sixth of its range, you may use this "1/6" rule as a kurtosis guidepost. If the standard deviation of a unimodal distribution is less than one-sixth of the range, the distribution tends to be leptokurtic. Similarly, if the standard deviation is more than one-sixth of the range, the distribution is platykurtic. For example, a unimodal distribution with a standard deviation of 50 and a range of 600 is clearly leptokurtic (since the ratio is only 1/12). Or a distribution with a standard deviation of 200 and a range of 600 is platykurtic (since the ratio is 1/3). Thus, the "1/6" rule aids us in clarifying the concept of "relatively large" or "relatively small" standard deviations. Remember, for a quick kurtosis evaluation, simply look at the values of the standard deviation and range. Divide the range by 6 and use this value as a kind of marker or *standard value,* SV. This standard value provides an approximation of what the standard deviation should be if the distribution were normal. Now compare the actual standard deviation with the SV, and if the *SD* is greater than the SV, then the distribution is platykurtic; if it's smaller, then it's leptokurtic. This does not mean that any distribution whose *SD* is equal to 1/6 of the range is absolutely normal, but it does mean that if the *SD* is not equal to 1/6 of the range, the distribution cannot be normal. Thus, the 1/6 rule is a necessary but not sufficient condition for normality.

KURTOSIS: SOME EXAMPLES

If you are dealing with a distribution where the highest score is 100 and the lowest score is 10, the range has to be 90. At this point you should know that the standard deviation could not possibly be greater than 45. If by accident you were to then calculate a standard deviation of 50 (probably because you forgot to take the square root), you should at least know that it is an impossible value and that you had better check and recheck your work.

On the basis of the 1/6 rule, evaluate the kurtosis of the following:

a. $M = 50$, Mdn = 50, Mo = 50, $R = 24$, $SD = 10$.

Ans: Ku–, or platykurtic because the *SD* of 10 is greater than the standard value of 4. ($SV = R/6$, or $24/6 = 4$).

b. $M = 100$, Mdn = 100, Mo = 100, $R = 36$, $SD = 3$

Ans: Ku+, or leptokurtic because the *SD* of 3 is less than the standard value of 6. ($SV = R/6$ or $36/6 = 6$).

c. $M = 80$, Mdn = 80, Mo = 80, $R = 18$, $SD = 12$

Ans: The kurtosis in this case cannot be evaluated because the *SD* has been miscalculated. "Gotcha," if you said platykurtic because even though the *SD* is greater than 1/6 of the range, it's too big. Remember, as previously shown, the *SD* can *never* be greater than half the range and 12 is certainly larger than 9.

Looking for Skewness

It was shown in Chapter 2 that skewness may be roughly approximated in any unimodal distribution by comparing the mean with the median, such that when the mean is higher than the median, we expect a positive skew, and when the mean is below the median, we expect a negative skew. However, sometimes the standard deviation can be an aid in spotting skewness, especially when the median is not reported. For example, the mean on a certain test turned out to be 15 and the standard deviation was shown to be 20. Furthermore, the range of scores was reported as 0 *to* 90. Since there can be no *negative scores* when the lowest possible score is zero, this distribution has to be skewed to the right. A normal distribution can only be approximated when the range of scores is equal to about six times the *SD*, and in this case the range would have to equal about 120. The scores would then have to fall between the mean minus 60 on the low end and the mean plus 60 on the high end, or a range from –45 to +75. Clearly, then, this distribution (whose range was 0 to 90) had to have an Sk+.

Questionnaire Percentages

Sometimes the results of a survey or questionnaire are given in percentage terms, especially when the question asks for a "yes," "no," or "agree–disagree" response. A quick method for obtaining the percentages of subjects answering in a certain way can be obtained without having to count out the percentages for each item. If the "yes" is scored as a 0, and the "no" as a 1, then

1. Calculate the arithmetic mean.
2. Multiply the mean by 100—which gives the percentage of "no" votes.
3. Subtract this percentage from 100% to get the percentage of "yes" votes. For example, 15 subjects answer a certain questionnaire item as follow (with a 0 scored for a "yes" and a 1 for a "no."

Item #5

0,1,1,1,1,1,1,0,1,1,1,1,0,1,1

1. $\Sigma X = 12$, and $M = 12/15 = .80$
2. (.80)(100) = 80% are "no" votes.
3. 100%–80% = 20% are "yes" votes.

SUMMARY

Whereas measures of central tendency define averageness or typicality in a distribution of scores, variability measures describe how scores differ. The three major measures of variability are the range, the standard deviation, and the variance.

1. The range (R) describes the entire width of the distribution and is found by subtracting the lowest score from the highest score. Two variations on the range theme are presented. The interquartile range is the difference between the first quartile (the 25th percentile) and the third quartile (or the 75th percentile). The interdecile range is the difference between the first decile (the 10th percentile) and the 9th decile (the 90th percentile).

2. The standard deviation (SD) describes variability in terms of how much all of the scores vary from the mean of the distribution. Therefore, when the standard deviation is known, precise variability statements can be made about the entire distribution.

3. The variance (V or SD^2), like the standard deviation, measures variability in terms of all of the scores in the distribution. The variance is simply the square of the standard deviation.

In a graph of a unimodal frequency distribution, variability is related to the kurtosis of the curve. Kurtosis is a description of the curvature (peakedness or flatness) of the graph. When a distribution has a small standard deviation (less than $R/6$), its curve has a leptokurtic shape—many scores clustering tightly around the mean. When a distribution has a large standard deviation (greater than $R/6$), its curve has a platykurtic shape—many of the scores spread out away from the mean.

Finally, when a distribution takes on the "ideal" bell shape of the normal curve, the distribution of scores is graphed as mesokurtic. Distributions that either are, or are close to being, mesokurtic in shape enjoy a special relationship between the range and the standard deviation. In this situation the standard deviation equals about one-sixth of the range.

Key Terms

deciles
deviation score
interdecile range
interquartile range
kurtosis
leptokurtic
measures of variability
mesokurtic
percentile ranks
percentiles
platykurtic
quartile deviation
quartiles
range
standard deviation
variance

PROBLEMS

1. At Company X, a group of eight blue-collar workers was selected and asked how much of their weekly pay (in dollars) they put into their savings bank account. The following is the list of amounts saved: $12, $11, $10, $20, $1, $9, $10, $10. Calculate the mean, the range, and the standard deviation.
2. At Company Y, another group of eight blue-collar workers was selected and asked how much of their weekly pay they saved. The following is the list of amounts saved: $20, $18, $3, $2, $1, $16, $15, $8. Calculate the mean, the range, and the standard deviation.
3. Despite the facts that the two groups of workers in problems 1 and 2 saved the same mean amount and that the ranges of amounts saved are identical, in what way do the two distributions still differ? Which group is more homogeneous?
4. A group of seven university students was randomly selected and asked to indicate the number of study hours each put in before taking a major exam. The data are as follows:

Student	Hours of Study
1	40
2	30
3	35
4	5
5	10
6	15
7	25

For the distribution of study hours, calculate
a. the mean.
b. the range.
c. the standard deviation.

5. The grade-point averages for the seven university students selected above were computed. The data are as follows:

Student	GPA
1	3.75
2	3.00
3	3.25
4	1.75
5	2.00
6	2.25
7	3.00

For the GPA distribution, calculate
a. the mean.
b. the range.
c. the standard deviation.

6. The IQs of all students at a certain prep school were obtained. The highest was 135, and the lowest was 105. The following IQs were identified as to their percentile:

X	Percentile
107	10th
114	25th
118	50th
122	75th
129	90th

a. What is the range for the distribution?
b. What is the median?
c. What is the interquartile range?
d. What is the interdecile range?

7. A sample of 10 patients from a county forensic unit were tested for psychopathy using the Holden Psychological Screening Inventory (high scores indicating higher levels of psychopathy). Their scores were as follows: 10, 15, 14, 16, 20, 15, 15, 15, 13, 17. Find the mean, median, mode, range, and standard deviation.

8. The following is a list of LSI scores taken from a sample of 10 inmates on their first day of incarceration at a county correctional facility. The LSI (or Levels of Service Inventory) is a measure of inmate security risk, the higher the scores the greater the risk: 2, 8, 5, 7, 5, 9, 4, 8, 4, 8. Find the mean, median, mode, range, and standard deviation.

9. A researcher investigating a new product for clearing up acne selects a random sample of 10 teenagers, gives them the facial product to use and asks that they report back how many days it took for the facial condition to clear up. The results (in days) were as follows: 20, 8, 10, 14, 15, 14, 12, 11, 14, 13. Find the mean, median, mode, range, and standard deviation.

10. A clinical psychologist is interested in assessing the prevalence of MDD (Major Depressive Disorders) among a group of fourth-grade students. A random sample of 13 students was selected and given the Children's Depression Inventory (CDI), a self-report instrument that measures levels of depression. Scores above 13 are said to indicate a major depressive disorder. The scores were as follows. 8, 10, 11, 7, 13, 4, 8, 7, 9, 3, 15, 10, 10. Find the mean, median, mode, range, and standard deviation.

11. Calculate the range and the standard deviation for the following set of scores: 10, 10, 10, 10, 10, 10, 10.

12. Under what conditions can a negative standard deviation be calculated?

13. What is the relationship between the relative size of the standard deviation and the kurtosis of a distribution?

14. For the following distributions, evaluate each for the type of kurtosis.
   a. $M = 50$, Mdn = 50, Mo = 50, $R = 100$, $SD = 2$.
   b. $M = 500$, Mdn = 500, Mo = 500, $R = 600$, $SD = 100$.
   c. $M = 100$, Mdn = 100, Mo = 100, $R = 60$, $SD = 25$.

15. If a distribution is mesokurtic and $R = 120$, find the approximate value of the standard deviation.

16. For a quick evaluation of kurtosis, what are the two key measures of variability that you must know?

17. If a given mesokurtic distribution has a mean of 100 and a standard deviation of 15, find the approximate value of the range.

18. Using the data in problem 17, calculate the variance.

True or False. Indicate either T or F for problems 19 through 26.

19. The standard deviation can only be computed with reference to the mean.

20. The distribution with the largest range necessarily has the largest standard deviation.

21. The 9th decile must always coincide with the 90th percentile.

22. The interquartile range includes only the middlemost 25% of the distribution.

23. In any distribution, the median must fall at the 50th percentile.

24. The interdecile range includes only the middlemost 10% of the distribution.

25. The standard deviation can never be greater than the range.

26. The mean is to central tendency as the standard deviation is to variability.

COMPUTER PROBLEMS

C1. For the following set of IQ scores, find the mean, median, range, and standard deviation. Evaluate the distribution for possible skewness.

100, 100, 86, 100, 101, 110, 100, 100, 145, 114, 101, 55, 99, 103, 97, 90, 100, 96, 104, 100, 101, 92, 108, 119, 79

C2. The following is a list of verbal PSAT scores; find the mean, median, range, and standard deviation.

30, 40, 45, 35, 42, 65, 55, 48, 56, 46, 42, 54, 56, 51, 47, 50, 51, 50, 50, 47, 52, 53, 47, 44, 69, 35, 40, 45, 35, 42, 65, 55, 48, 56, 46, 42, 54, 56, 51, 47, 50, 51, 50, 50, 47, 52, 53, 47, 78, 80

Chapter

4

The Normal Curve and z Scores

If the mean and the standard deviation are the heart and soul of descriptive statistics, then the normal curve is its lifeblood. Although there is some dispute among statisticians as to when and by whom the normal curve was first introduced, it is customary to credit its discovery to the great German mathematician **Carl Friedrich Gauss** *(1777–1855). Even to this day, many statisticians refer to the normal curve as the Gaussian curve.*

Gauss is undoubtedly one of the most important figures in the history of statistics. He developed not only the normal curve but also some other extremely important statistical concepts that will be covered later.

THE NORMAL CURVE

So many distributions of measures in the social sciences conform to the **normal curve** that it is of crucial significance for describing data. The normal curve, as Gauss described it, is actually a theoretical distribution. However, so many distributions of people-related measurements come so close to this ideal that the normal curve can be used in generating frequencies and probabilities in a wide variety of situations.

Key Features of the Normal Curve

The normal curve (Figure 4.1) is first and foremost a unimodal **frequency distribution curve** with scores plotted on the X axis and frequency of occurrence on the Y axis. However, it does have at least six key features that set it apart from other frequency distribution curves:

1. In a normal curve, most of the scores cluster around the middle of the distribution (where the curve is at its highest). As distance from the middle increases, in either direction, there are fewer and fewer scores (the curve drops down and levels out on both sides).

FIGURE 4.1 The normal curve.

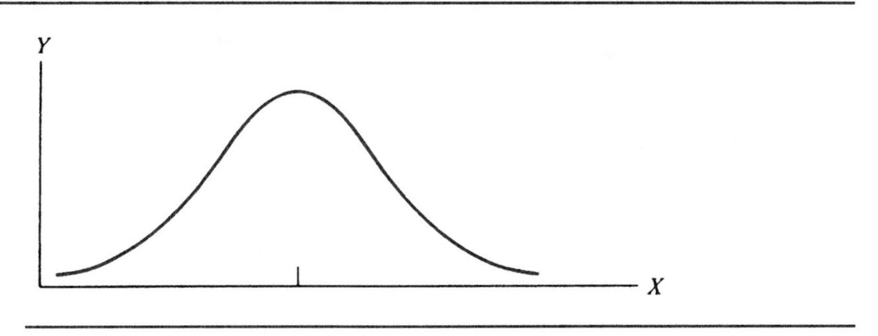

2. The normal curve is symmetrical. Its two halves are identical mirror images of each other; it is thus perfectly balanced.

3. In a normal curve, all three measures of central tendency—the mean, the median, and the mode—fall at precisely the same point, the exact center or midpoint of the distribution.

4. The normal curve has a constant relationship with the standard deviation. When the abscissa of the normal curve is marked off in units of standard deviation, a series of *constant* percentage areas under the normal curve are formed. The relationship holds true for all normal curves, meaning that if a certain percentage of scores is found between one and two standard deviation units above the mean, this same percentage is always found in that specific area of any normal curve. Also, because of the symmetry of the curve, that exact percentage is always found in the same part of the lower half of the curve, between one and two standard deviation units below the mean. This constancy of the percentage area under the curve is crucial to an understanding of the normal curve. Once the curve is plotted according to standard deviation units, it is called the *standard normal curve.*

5. With a normal curve, when you go out one full standard deviation unit from the mean, the curve above the abscissa reaches its point of inflection, that is, the point where the curve changes direction and begins going out more quickly than it goes down.

6. The normal curve is asymptotic to the abscissa. No matter how far out the tails are extended, they will never touch the X axis.

The standard normal curve (Figure 4.2) has a mean of zero and a standard deviation of 1.00. The standard deviation units have been marked off in unit lengths of 1.00 on the abscissa, and the area under the curve above these units always remains the same. Since, as has been stated, the mean and median of a normal curve (as well as the mode) always fall at exactly the same point on the abscissa under the normal curve, the

FIGURE 4.2 The standard normal curve.

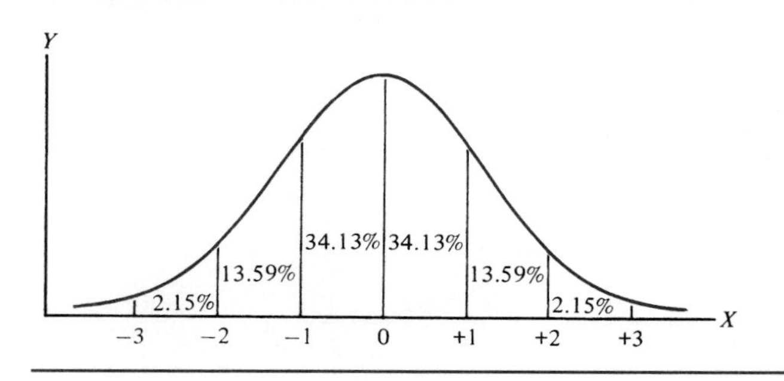

mean in this case is interchangeable with the median. The importance of this is due to the fact that since the median always divides any frequency distribution exactly in half, then when the mean and median coincide, the mean also divides this particular distribution in half, with 50% of the scores falling above the mean and 50% below the mean.

Further, the area under the curve between the mean and a point one standard deviation unit above the mean always includes 34.13% of the cases. Because of the symmetry of the curve, 34.13% of the cases also fall in the area under the curve between the mean and a point one standard deviation unit below the mean. Thus, under the standard normal curve, between two points, each located one standard deviation unit away from the mean, there are always 68.26% (twice 34.13%) of all the cases.

As we go farther away from the mean, between one and two standard deviation units, another 13.59% of the distribution falls on either side. Even though the distance away from the mean has been increased by another full standard deviation unit, only 13.59% of the cases are being added. This is because, as we can see from Figure 4.2, the curve is getting lower and lower as it goes away from the mean. Thus, although 34.13% fall between the mean and one standard deviation unit away from the mean, only 13.59% are being added as we go another length farther away from the mean, to the area included between one and two standard deviation units. Also, between the mean and a point two full standard deviation units away from the mean, a total of 47.72% (34.13% + 13.59%) of the cases can be found.

Including both halves of the curve, we find that between the two points located two standard deviation units above and below the mean, there will be 95.44% (twice 47.72%) of the distribution. Finally, the area under the curve between two and three standard deviation units away from the mean only holds 2.15% of the cases on each side. There are, then, 49.87% (47.72% + 2.15%) of the cases between the mean and a point three standard deviation units away from the mean. Almost the entire area of the curve, then—that is, 99.74% (twice 49.87%)—can be found between points bounded by three standard deviation units above and below the mean.

This may seem confusing. Before throwing in the towel, however, picture for a moment that we are installing a rug in the very small and odd-shaped room shown in Figure 4.3. We decide to begin measuring from the point marked with an arrow. To both the right and the left of the arrow, we measure 3 feet. The long side of this room

FIGURE 4.3 Carpeting an odd-shaped room.

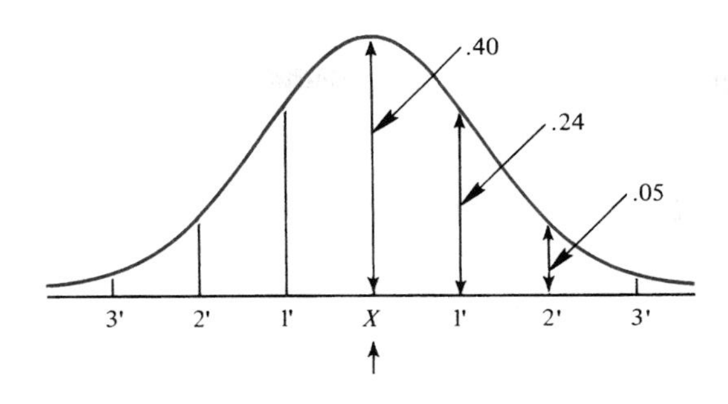

therefore measures 6 feet. We mark this distance off in 1-foot intervals, 3 to the right and 3 to the left of the point X. Taking great care, we now take measures of the other side of the room. We determine that between 1 foot to the right of X and 1 foot to the left of X, the area of the room will require roughly two-thirds, or 68%, of our rug. Now, this holds true whether the total area to be covered is 100 square feet, 10 square feet, or 1000 square feet—68% of the rug must fit in that precise area. Between 2 feet to the right and 2 feet to the left of X, 95% of the rug will be used. To cover the area between 3 feet to the right and 3 feet to the left of the arrow, virtually the entire rug is needed. Again, remember that these percentages hold true for this shape, whether the total area is the size of a gym or the size of a hall closet. (See the box on page 79 for an explanation of how the normal curve is drawn.)

z SCORES

In the preceding section we were, in fact, without naming them as such, discussing ***z* scores** (also called standard scores). The ***z* distribution** is a normally distributed set of specially scaled values whose mean is always equal to zero and whose standard deviation must equal 1.00. As seen, the vast majority of the normal curve's total area (99.74%) lies between the *z* scores of ±3.00. As we shall soon see, *z* scores are enormously helpful in the interpretation of raw score performance, since they take into account both the mean of the distribution and the amount of variability, the standard deviation. We can thus use these *z* scores to gain an understanding of an individual's relative performance compared to the performance of the entire group being measured. Also, we can compare an individual's relative performance on two separate normally distributed sets of scores.

The z Score Table

So far the discussion of *z* scores and the areas under the standard normal curve has assumed that the standard deviation units always come out as nice whole numbers. We learned, for example, that 34.13% of the cases fall between the mean and a point one

CARL FRIEDRICH GAUSS (1777–1855)

Gauss has been called Germany's greatest pure mathematician, and one of the three greatest who ever lived (the other two being Archimedes and Newton). His father was a bookkeeper who was in charge of the accounts for a local insurance company. Although his father was a popular man about town, at home he was crude, authoritarian, and uncouth. Carl and his father were never close, and Carl later said that he was constantly repelled by the man. An only child, Carl was totally devoted to his mother, an uneducated woman who is said to have been unable to read or write. However, her brother was educated and was especially talented in math. He soon noticed his nephew's extraordinary intellect and began working with the brilliant toddler in both math and art.

Carl's early development as a mathematician is probably still unequalled. For example, as a child he was one of those "perfect-pitch" mathematical prodigies who put his teachers and classmates into a state of total terror. It is said that he could add, subtract, multiply, and divide before he could talk. It has never been entirely clear how he communicated this astounding ability to his family and friends. Did he, like the "Wonder Horse," stamp out his answers with his foot? At the age of 3, when he presumably could talk, he was watching his father's payroll calculations and detected a math error. His loud announcement of his father's mistake did not go a long way toward smoothing relations between them. When he was 8, he startled his schoolmaster by adding all the numbers from 1 to 100 in his head, a feat which took him only a few minutes. The teacher had given the class this busywork assignment to do while he corrected papers. Obviously little Carl's lightning calculations changed the teacher's plans and created a minor rift between the two. His schoolmaster soon discovered that he couldn't keep up with Carl and so an assistant was hired to work with the young genius. At this point he became trained in algebra and calculus, sometimes working late into the evening to unravel difficult proofs. By age 11 he discovered a more sophisticated proof for the binomial theorem. At age 15, his talent became so widely known that the Duke of Brunswick took him under his wing, providing tuition money and enough other financial support for Carl to go to the University of Gottingen, where he graduated in 1798. The Duke continued with his financial aid and Carl was awarded his Ph.D. at the University of Helmstedt in 1799, after only one year of graduate study. During that year Gauss developed his famous normal curve of errors, now known as the normal probability curve or even as the Gaussian curve. Two years later he published his greatest masterpiece, *Disquisitiones Arithmeticai*. By 1806, however, Gauss lost his financial patron, the Duke of Brunswick, who was killed by Napoleon's army at the battle of Auerstadt. Gauss struggled to pay his bills for several years, but finally in 1835, despite repeated assertions of his aversion to teaching, he took a job as a professor at the University of Gottingen, where he remained until his death 20 years later. Called the "prince of mathematicians," he became a legend in his own time. During his last years he was showered with honors and prizes from associations and universities all over Europe. He was even honored in the United States, by both the Academy of Arts and Sciences in Boston and the American Philosophical Association in Philadelphia.

standard deviation unit away from the mean. But what happens in the more typical case when the z scores turn out not to be those beautiful whole numbers?

The z score table (Appendix Table A) should become well worn during this semester. It will be used to obtain the precise percentage of cases falling between any z score and the mean. Notice that the z score values in this table are arrayed in two directions, up and down in a column, and left to right in a row. The column at the far left gives the z scores to one decimal place. The second decimal place of the z score is given in the top row.

Example To look up a z score of, say, .68, run a finger down the far left column until it reaches .60. Then, using a ruler or the edge of a piece of paper to prevent your eyes from straying off the line, follow across that row until it intersects the column headed by .08. The value there is 25.17. This, like all the values in the table, represents a percentage—the percentage of cases falling between the z score of .68 and the mean. Next, look up the percentage for a z score of 1.65. Run down the far left column until 1.6, then across that row until it intersects the column headed by .05. The value there is 45.05, or 45.05%.

Because the normal curve is symmetrical, the z score table gives only the percentages for half the curve. This is all that is necessary, since a z score that is a given distance to the right of the mean yields the same percentage as a z score that is the same distance to the left of the mean. That is, positive and negative z scores of a given value give precisely the same percentages, since they deviate the same amount from the mean.

A Fundamental Fact About the z Score Table. It is extremely important that you learn the following: *The z score table gives the percentage of cases falling between a given z score and the mean.* Put the book down for a minute and write that sentence on a piece of paper. Write it several times; repeat it aloud in round, clear tones. Paste it on your bathroom mirror so that it greets you each morning. Memorizing that simple sentence virtually ensures success with z score problems by preventing most of the big, fat, egregious errors that so many students seem to commit automatically. Once again now, the z score table gives the percentage of cases falling between a given z score and the mean. It does *not* give a direct readout of the percentage above a z score, or below a z score, or between two z scores (although all of these can be calculated).

Drawing the Curve. When working with z scores, it is extremely helpful to draw the curve each time. It only takes a couple of seconds and is well worth the effort. Drawing the curve and locating the part of the curve the question concerns will give you a much clearer picture of what is being asked. It isn't necessary to waste time or graph paper being compulsive over how the curve looks. Just sketch a quick approximation of the curve, and locate the mean and z scores on the baseline. Remember to place all positive z scores to the right of the mean and negative z scores to the left. Also remember that the higher the z score, the farther to the right it is located, and the lower the z score, the farther to the left it is placed.

Spending a few seconds to draw the curve is crucial. This step is not a crutch for beginners; in fact, it is common practice among topflight statisticians. Although drawing the curve may appear initially to be a trivial waste of time, especially when working with only one z score at a time, it is a good habit to get into. A nice picture of the problem right in front of you makes it easier to solve.

The box on page 79–80 presents the equation for the normal curve, with a couple of examples of ordinates obtained from the z scores. Unless you have some math background, it isn't necessary to even look at this equation. Simply take the shape of the curve on faith.

From z Score to Percentage of Cases

Case A: To find the percentage of cases between a given z score and the mean.

—**Rule** Look up the z score in Appendix Table A and take the percentage as a direct readout.

—**Exercise 4a** Find the percentage of cases falling between each of the following z scores and the mean.

1. 1.25
2. .67
3. –2.58
4. .10
5. –1.62

Percentage of Cases Below a z Score. If someone has a z score of 1.35 and we want to know the percentage of cases below it, we first look up the z score in Appendix Table A. The table reveals that 41.15% of the cases occur between the z score of 1.35 and the mean. We also know that a z score of 1.35 is positive and therefore must fall to the right of the mean. (See Figure 4.4.) Further, we know that on a normal curve 50% of the cases fall below the mean, so we add 50% to 41.15% and get a total of 91.15%.

If the z score happens to be negative, for example, –.75, again we look up the z score in Appendix Table A, and read the percentage of cases falling between it and the mean, in this case 27.34%. This z score is negative and therefore must be below the mean. (See Figure 4.5.) Since 50% of the cases fall below the mean, and since we have just accounted for 27.34% of that, we subtract 27.34% from 50% and get 22.66%.

Case B: To find the percentage of cases falling below a given z score (percentile).

—**Rule B1** If the z score is positive, look up the percentage and add 50%.

FIGURE 4.4 Percentage of cases below a *z* score of 1.35.

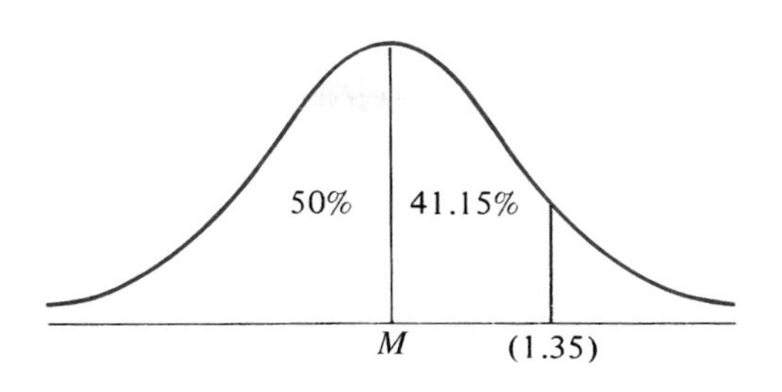

—**Rule B2** If the *z* score is negative, look up the percentage and subtract it from 50%.

—**Exercise 4b** Find the percentage of cases falling *below* the following *z* scores.

1. 2.57
2. –1.05
3. –.79
4. 1.61
5. .14

From z Score to Percentile

The percentile, as defined in Chapter 3, is that point in a distribution of scores below which the given percentage of scores fall. The median, by definition, always occurs precisely at the 50th percentile. Since on the normal curve the median and the mean coincide, then for normal distributions the mean must also fall at the 50th percentile. Since this is true, once the *z* score is known and the area below the *z* score found, the percentile can be easily found. Converting *z* scores into percentiles, then, is a quick and straightforward process.

When converting *z* scores into percentiles, remember the following points. First, all positive *z* scores *must* yield percentiles higher than 50. In the case of positive *z* scores, we always add the percentage from the *z* score table to 50%. Second, all negative *z* scores must yield percentiles lower than 50. In this case, we always subtract the percentage from the *z* score table from 50%. Third, a *z* score of zero, since it falls right

FIGURE 4.5 Percentage of cases below a *z* score of –.75.

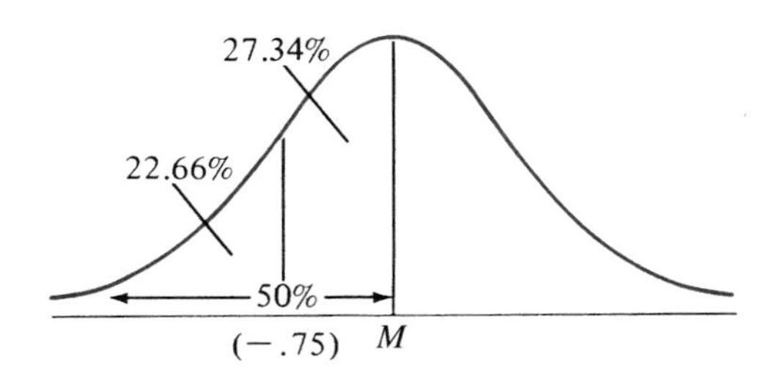

on the mean, must yield a percentile of 50. Finally, it is customary when working with percentiles (and this is the only exception to our round-to-two-places rule) to round to the whole number. Thus, for a z score of 1.35, which falls above 91.15% of all the cases, the percentile rank is stated as 91 or the 91st percentile.

—**Exercise 4c** Find the percentile for each of the following z scores.

1. .95
2. 1.96
3. –.67
4. –1.65
5. .50

In later chapters, we will be using a special, time-saving percentile table (Appendix Table B) for these kinds of problems. At this point, however, you really should work these out using Appendix Table A. The more you use the z score table now, the clearer the later sections on sampling distributions will appear.

Percentage of Cases Above a z Score. Many times, the area above the z score must be established. Suppose that we must find the percentage of cases falling *above* a z score of 1.05. First, we use Appendix Table A to get the percentage falling between the z of 1.05 and the mean. The table gives us a readout of 35.31%. Since this is a positive z score, it must be placed to the right of the mean. (See Figure 4.6.) We know that 50% of all cases fall above the mean, so we subtract 35.31% from 50% to get 14.69%.

If the z score is negative—for example, –.85—we look it up in the table and read 30.23% as the percentage of cases falling between the z score and the mean. Since all negative z scores are placed below the mean, 30.23% of cases fall above this z score and below the mean, that is, *between* the z score and the mean. (See Figure 4.7.) Also, 50% of the cases must fall above the mean. We add 30.23% to 50% and obtain 80.23% as the percentage of cases falling above a z score of –.85.

Case C: To find the percentage of cases above a given z score.

—**Rule C1** If the z score is positive, look up the percentage and subtract it from 50%.

—**Rule C2** If the z score is negative, look up the percentage and add it to 50%.

FIGURE 4.6 Percentage of cases above a z score of 1.05.

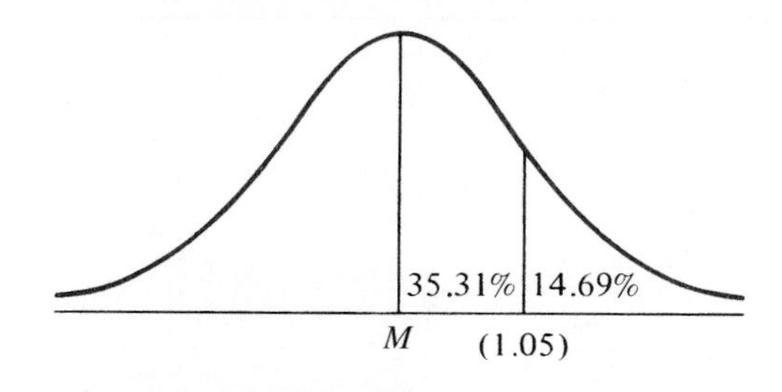

FIGURE 4.7 Percentage of cases above a z score of –.85.

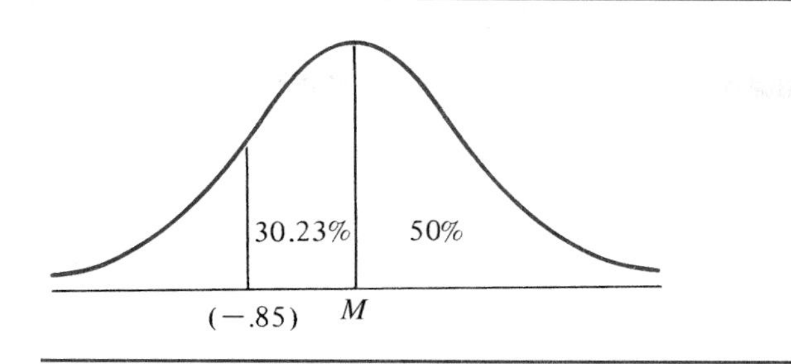

—**Exercise 4d** Find the percentage of cases falling *above* the following z scores.

1. –.15
2. 2.00
3. 1.03
4. –2.62
5. .09

Percentage of Cases Between z Scores. Finding the area under the normal curve between z scores is based essentially on the procedures already covered, except that we work with two z scores at a time. It is important to look up the percentages in the z score table *one at a time*. There are no creative shortcuts!

z Scores on Opposite Sides of the Mean

Example We want the percentage of cases falling between z scores of –1.00 and +.50.

First, as always, we draw the curve and locate the z scores on the X axis. Next, we look up each z score in Appendix Table A. For –1.00 we find 34.13%, and for +.50 we find 19.15%. Remember that each of these percentages lies between its own z score and the mean. Now add the two percentages and you discover that 53.28% (this is, 34.13% + 19.15%) of the cases fall between these z scores.

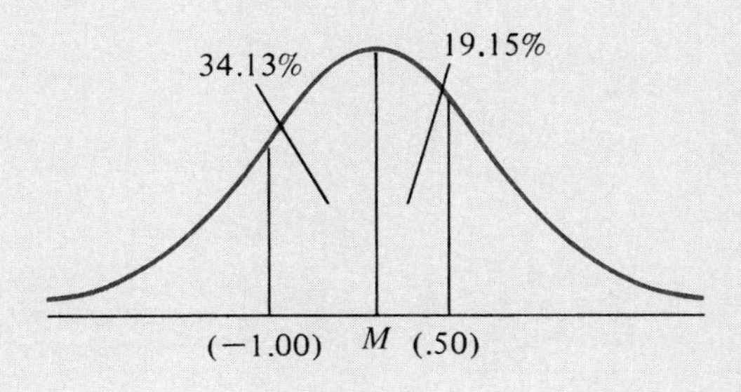

Example Find the percentage of cases falling between z scores of –2.00 and +1.65.

The percentages from Appendix Table A are 47.72% and 45.05%, respectively. Adding these percentages, we get a total of 92.77%.

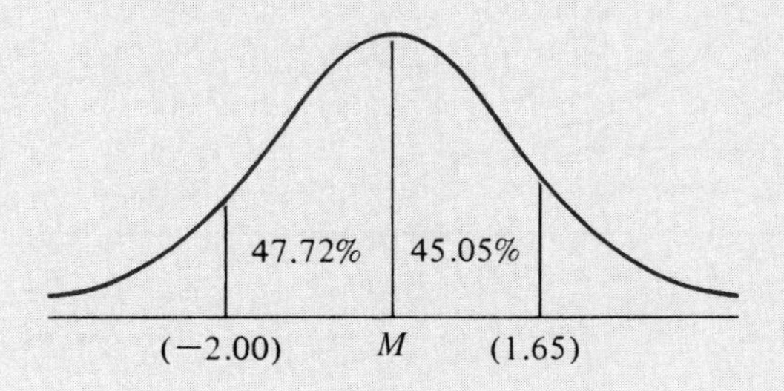

In each of these examples, the z scores were on opposite sides of the mean, one below the mean and the other above. The rule, then, to find the area between two z scores on opposite sides of the mean is to look up the two percentages and add them together. Add the percentages—*do not add* the z scores.

z Scores on the Same Side of the Mean. Many situations require establishing the area under the normal curve between z scores on the same side of the mean (z scores that have the same signs).

Example We wish to determine the percentage of cases between a z score of .50 and a z score of 1.00.

Again, we draw the curve and place the z scores on the X axis—.50 just to the right of the mean and 1.00 a little farther to the right. Look up the two percentages—19.15% for .50 and 34.13% for 1.00. Now, since 34.13% of the cases fall in the area from the mean to the z of 1.00, and since 19.15% of that total fall in the area from the mean to the z of .50, we must *subtract* to obtain 14.98% (34.13% – 19.15%), the percentage of cases between the z scores.

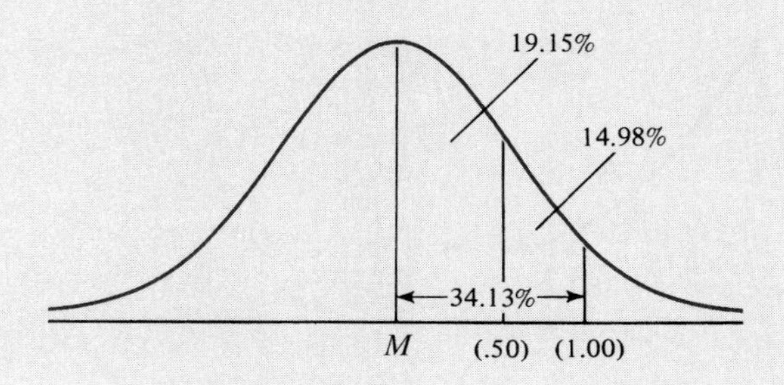

Example Find the area between the *z* scores of 1.50 and 1.00.

Appendix Table A gives 43.32% for 1.50 and 34.13% for 1.00. Subtracting those percentages gives us 9.19% of cases falling between a z of 1.50 and a z of 1.00.

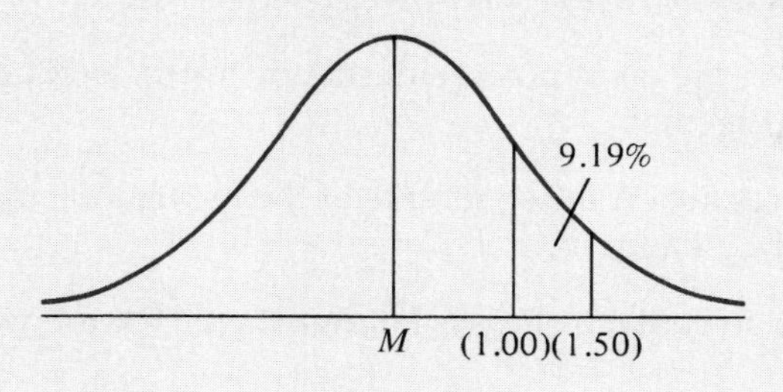

Again, a word of caution on this procedure. You must subtract the percentages, *not* the *z* scores. If, in the preceding example, we subtracted 1.00 from 1.50 and then looked up the resulting *z* score of .50 in Appendix Table A, our answer would have been 19.15%, instead of the correct answer of 9.19%. The reason for the error is that Appendix Table A gives the percentage of cases falling between the *z* score and the mean, not between two *z* scores, both of which are deviating from the mean. Remember, the area under the normal curve includes a much greater portion of the area around the mean where the curve is higher than when it drops down toward the tails of the distribution.

When working with negative *z* scores, the procedure is identical. For example, the percentage of cases falling between *z* scores of –1.50 and –.75 is 15.98% (43.32 – 27.34). (See Figure 4.8.)

FIGURE 4.8 Percentage of cases between *z* scores of –1.50 and –.75.

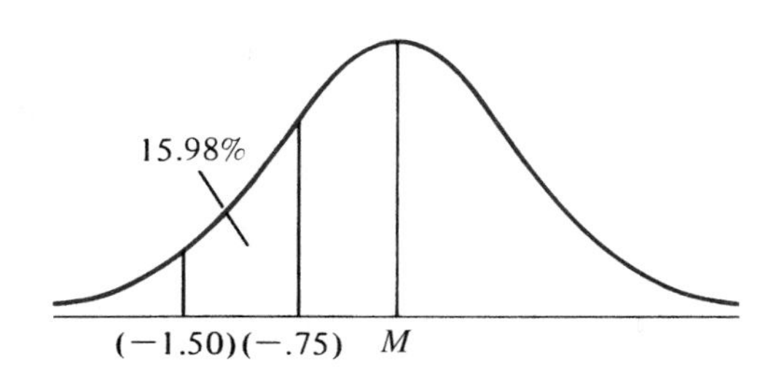

The rule, then, to find the percentage of cases falling between z scores both on the same side of the mean is to look up the two percentages in Appendix Table A and subtract the smaller from the larger.

Case D: To find the percentage of cases *between* two z scores.

—**Rule D1** If the z scores are on opposite sides of the mean, look up the two percentages and *add* them together.

—**Rule D2** If the z scores are on the same side of the mean, look up the two percentages and *subtract* the smaller from the larger.

—**Exercise 4e** Find the percentage of cases falling *between* the following pairs of z scores.

1. –1.45 and 1.06
2. –.62 and .85
3. .90 and 1.87
4. –1.62 and –.17
5. –1.65 and –1.43

TRANSLATING RAW SCORES INTO z SCORES

At this point you are fairly familiar with the z score concept. We have found the percentage areas above, below, and between various z scores. However, in all these situations, the z score has been given; it has been handed to you on a silver platter. Using exactly the same procedures, however, we can find the areas under the normal curve for any raw score values. This is actually the more common situation—to find the percentage of cases falling above, below, or between the raw scores themselves. To accomplish this, the values of the mean and the standard deviation for the raw score distribution must be known. Remember, without this knowledge, the raw scores are of little use. If the mean and the standard deviation are known, we can subtract the mean from the raw score, divide by the standard deviation, and obtain the z score. This gives us the opportunity to use the characteristics of the normal curve.

The z score equation

$$z = \frac{X - M}{SD}$$

defines the z score as a translation of the difference between the raw score and the mean (M) into units of standard deviation (SD). The z score, then, indicates how far the raw score is from the mean, either above it or below it, in these standard deviation units. A normal distribution of any set of raw scores, regardless of the value of the mean or the standard deviation, can be converted into the standard normal distribution, in which the mean is always equal to zero and the standard deviation is always equal to one.

THE EQUATION FOR THE NORMAL CURVE

The equation for the normal curve is as follows:

$$y = \frac{1}{\sqrt{2\pi}} e^{-(z^2/2)}$$

where y = the ordinate
$\pi = 3.14$
$e = 2.72$ (the base of the Napierian or natural log)
z = the distance from the mean in units of standard deviation

If $z = 1.00$, then

$$y = \frac{1}{\sqrt{(2)(3.14)}} 2.72^{-(1.00^2/2)}$$

$$= \frac{1}{\sqrt{6.28}} 2.72^{-(.50)}$$

$$= \left(\frac{1}{2.51}\right)(.61) = (.40)(.61)$$

$$= .24$$

Thus, the height of the ordinate when $z = 1.00$ is .24. Or, if $z = .50$, then

$$y = \frac{1}{\sqrt{(2)(3.14)}} 2.72^{-(.50^2/2)}$$

$$= \frac{1}{\sqrt{6.28}} 2.72^{-(.13)}$$

$$= \left(\frac{1}{2.51}\right)(.88) = (.40)(.88)$$

$$= .35$$

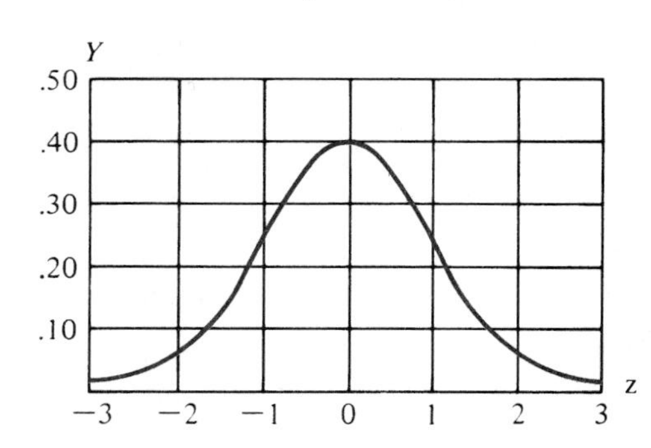

With a z score of .50, the height of the ordinate is .35.

The graph of the normal curve is said to be asymptotic to the abscissa. This means that the curve never touches the X axis, no matter how far it extends from the mean.

Although there are many different normal curves, each with its own mean and standard deviation, there is only one standard normal curve, where the mean is equal to zero and the standard deviation is equal to 1.00. All normal curves can be fitted exactly to the standard normal curve by translating the raw scores into units of standard deviation.

For finding the area between two *z* scores, calculus could be used to integrate the normal curve equation between the scores. Doing integrations for each normal curve problem is rather spine-chilling to contemplate, but fear not. All the integrations have been completed and the results are presented in convenient form in Appendix Table A.

z Scores and Rounding

Since the *z* score table, as we have seen, is a two-decimal-place table, it will be convenient to calculate the *z* scores by rounding to two places—that is, when dividing the numerator by the denominator, carry the calculations out to three places to the right of the decimal and then round back to two places to the right of the decimal. If the value in the third place is a 5 or higher, raise the value in the second place by 1.*

Using *z* Scores

The *z* scores are enormously helpful in the interpretation of raw score performances, since they take into account both the mean of the distribution and the amount of variability, its standard deviation. For example, if you get scores of 72 on a history test and 64 on an English test, you do not know, on the face of it, which score reflects the better performance. The mean of the scores on the history test might have been 85 and the mean on the English test only 50; in that case, you actually did worse on the history test, even though your raw score was higher.

The point is that information about the distribution of all the scores must be obtained before individual raw scores can be interpreted meaningfully. This is the reason for using *z* scores in this situation because they take the entire distribution into account. The *z* score allows us to understand individual performance scores relative to all of the scores in the distribution.

Areas Under the Normal Curve

As has been shown, if the abscissa of the normal curve is laid out in units of standard deviations (the standard normal curve), all the various percentage areas under the curve can be found. Suppose that a researcher selects a representative sample of 5000 persons from the adult population of the United States. Each person is given an IQ test, and each of the 5000 resulting IQ scores is duly recorded. The scores are added together and divided by the number of cases, or 5000, to find the mean of this distribution,

---

*Although some statistics texts maintain that a value of 5 to the right of the decimal should always be rounded to the nearest even number, this book uses the convention of rounding a 5 to the next highest number. This is consistent with the rounding program built into most modern calculators and computer programs.

which turns out to be 100. Then, all the IQ scores are squared, the squares added, and the sum divided by N. From this value, the square of the mean is subtracted, and finally the square root is taken. The resulting value of the standard deviation turns out to be 15. With the values for the mean and the standard deviation in hand, the normal curve is accessible.

As shown in Figure 4.9, the mean of 100 is placed at the center of the X axis, and the rest of the axis is marked off in IQ points. Since we found that the standard deviation of this IQ distribution is 15, *one standard deviation unit is worth 15 IQ points.* Note that this standard deviation value of 15 was not presented to the researcher as though sprung full grown out of the head of Zeus. It was calculated in the same way as the other standard deviations were in Chapter 3.

Since one standard deviation unit is equal to 15 IQ points, then two standard deviation units must equal 30 IQ points and three must equal 45 points. Each of the standard deviation units can now be converted into its equivalent IQ, ranging from 55 to 145. Also, since we know the percentage areas for standard deviation units in general, we know the percentage areas for the IQs.

Is finding the standard deviation important? It is absolutely crucial. This can best be demonstrated by a quick review of what was known and what is now revealed by having found the standard deviation and the mean of the IQ distribution.

We Knew	*Now Revealed*
1. 68.26% of the cases fall between z scores of ±1.00.	1. 68.26% of the cases fall between IQs of 85 and 115. (See Figure 4.10a.)
2. 95.44% of the cases fall between z scores of ±2.00.	2. 95.44% of the cases fall between IQs of 70 and 130. (See Figure 4.10b.)
3. 99.74% of the cases fall between z scores of ±3.00.	3. 99.74% of the cases fall between IQs of 55 and 145. (See Figure 4.10c.)

FIGURE 4.9 Normal distribution of IQ scores showing mean and standard deviation units.

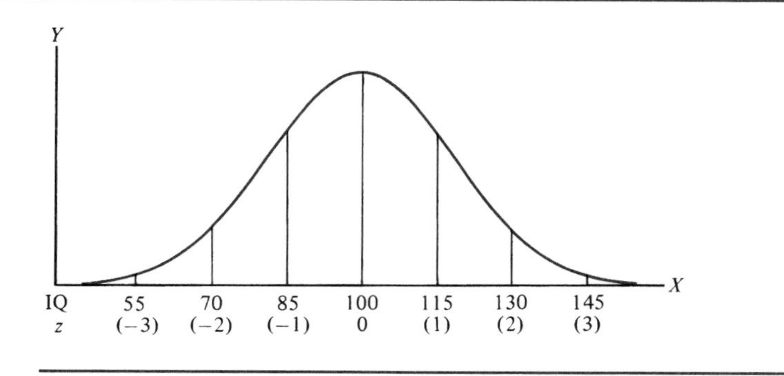

FIGURE 4.10 Percentage areas of the normal curve of IQ scores.

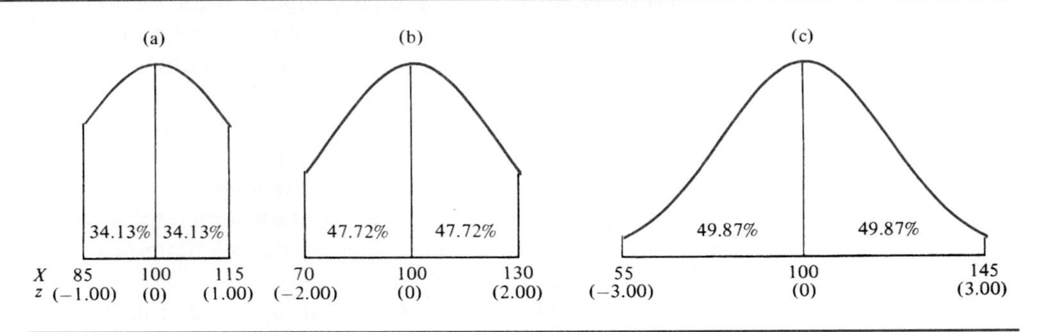

The conversions of IQs into z scores can be shown as follows:

$$z = \frac{X - M}{SD}$$

For an IQ of 115 then,

$$z = \frac{115 - 100}{15} = 1.00$$

And for 85,

$$z = \frac{85 - 100}{15} = -1.00$$

This procedure, then, yields z scores of +2.00 for an IQ of 130, +3.00 for an IQ of 145, –2.00 for an IQ of 70, and –3.00 for an IQ of 55.

z SCORE TRANSLATIONS IN PRACTICE

Let us now review each of the z score cases, only this time placing each case in the context of a normal distribution of *raw scores.*

Example *Case A: Percentage of Cases Between a Raw Score and the Mean.* A normal distribution of scores has a mean of 440 and a standard deviation of 85. What percentage of scores will fall between a score of 500 and the mean?

We convert the raw score of 500 into its equivalent z score.

$$z = \frac{X - M}{SD} = \frac{500 - 440}{85} = \frac{60}{85} = .71$$

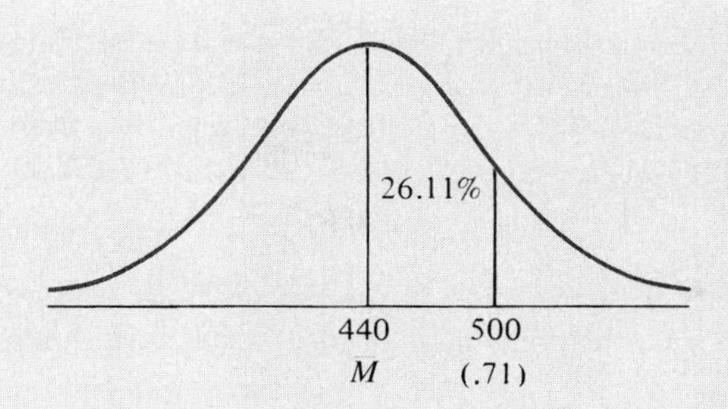

Then we draw the curve, placing all values on the abscissa and keeping the z score in parentheses to prevent confusing it with the raw score. We look up the z score in Appendix Table A, and take the percentage as a direct readout; that is, 26.11% of the scores fall between 500 and the mean.

—**Exercise 4f** On a normal distribution with a mean of 48 and a standard deviation of 6.23, find the percentage of cases between the following *raw scores* and the mean.

1. 52
2. 42
3. 55
4. 37
5. 40

Example *Case B: Percentage of Cases Below a Raw Score.* On a normal distribution of adult male weight scores, with a mean of 170 pounds and a standard deviation of 23.17 pounds, what percentage of weights fall below a weight score of 200 pounds?

We convert the raw score into its equivalent z score.

$$z = \frac{200 - 170}{23.17} = \frac{30}{23.17} = 1.29$$

Then we draw the curve, placing all values on the abscissa.

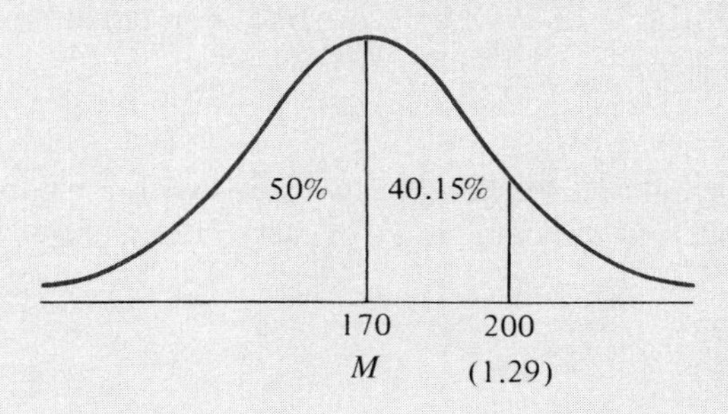

Since the z score is positive (above the mean), we look up the percentage in Appendix Table A and add it to 50%. That is, 40.15% + 50% = 90.15%, which is the percentage of cases below the score of 200. If the z score had been negative, then we would have looked up the percentage in Appendix Table A and subtracted it from 50%.

—**Exercise 4g** On a normal distribution of raw scores with a mean of 103 and a standard deviation of 17, find the percentage of cases *below* the following raw scores.

1. 115
2. 83
3. 127
4. 105
5. 76

Example *Case C: Percentage of Cases Above a Raw Score.* On a normal distribution of systolic blood pressure measures, the mean is 130 and the standard deviation is 9.15. What percentage of systolic measures is above 150?

We convert the raw score into its z score equivalent.

$$z = \frac{150 - 130}{9.15} = \frac{20}{9.15} = 2.19$$

We draw the curve, placing all values on the abscissa (see the accompanying figure).

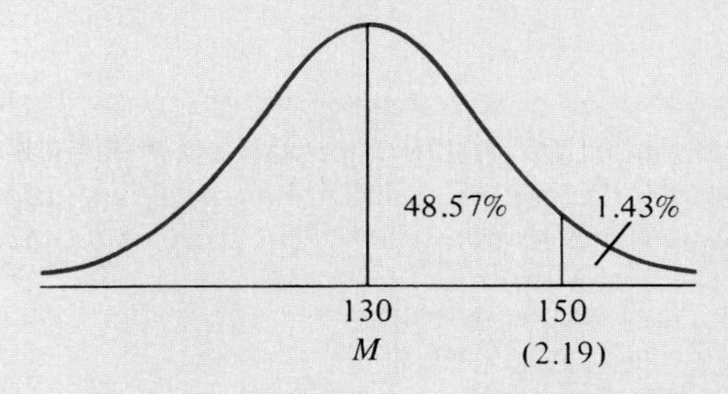

Since the z score is positive (above the mean), we look up the percentage in Appendix Table A and subtract it from 50%; that is, 50% – 48.57% equals 1.43%, the percentage of systolic measures above 150. If the z score had been negative, we would have used the Case C2 rule.

—**Exercise 4h** On a normal distribution of raw scores with a mean of 205 and a standard deviation of 60, find the percentage of cases falling above the following raw scores.

1. 290
2. 210
3. 167
4. 195
5. 250

Example *Case D: Percentage of Cases Between Raw Scores.* On a normal distribution of adult female height scores, the mean is 65 inches and the standard deviation is 3 inches. Find the percentage of cases falling between heights of 60 inches and 63 inches.

We convert both raw scores into their equivalent z scores.

$$z = \frac{60 - 65}{3} = \frac{-5}{3} = -1.67$$

$$z = \frac{63 - 65}{3} = \frac{-2}{3} = -.67$$

Then we draw the curve, placing all values on the abscissa.

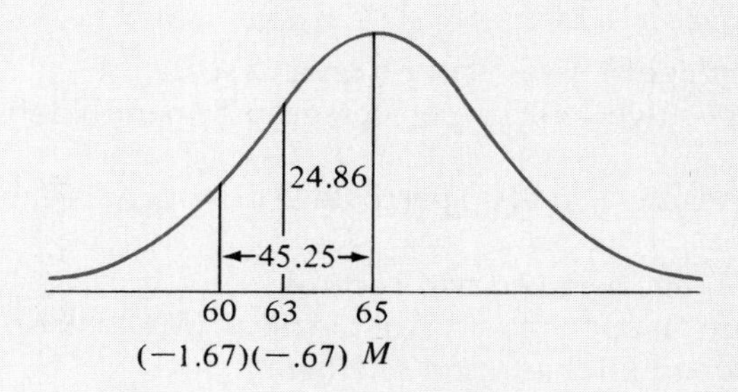

Since both z scores are on the same side of the mean, use Case D2 rule. Look up the percentage in Appendix Table A for each z score, and then subtract the smaller from the larger; that is, 45.25% − 24.86% equals 20.39%, the percentage of heights between 60 inches and 63 inches. If the two z scores had been on opposite sides of the mean, we would have looked up the two percentages and added them together.

—**Exercise 4i** A normal distribution of raw scores has a mean of 68 and a standard deviation of 3.06. Find the percentage of cases falling between each of the following pairs of raw scores.

1. 60 and 72
2. 65 and 74
3. 70 and 73
4. 62 and 66
5. 59 and 67

The Normal Curve in Retrospect

Whenever a distribution of values assumes the shape of the normal curve, the z score table, as we have clearly seen, allows us to do some truly remarkable computations. We can find the percentage of cases falling above, below, or between any z score values. This will become a central issue during later discussions of both probability and inferential statistics.

Remember, however, that as remarkable as these percentages are, they apply only when the distribution is normal—not skewed, or bimodal, or leptokurtic, or any other shape. If, for example, more than or less than 68% of the scores in a distribution fall in the area within a standard deviation unit of the mean, the distribution simply is not normal, and the *z* score techniques do not apply. In short, when the distribution is normal, use the *z* scores. When it isn't, don't!

z Score Percentage Rules

1. Case A Rule: Percentage of cases between a given *z* score and the mean (*M*). Look up directly in Appendix Table A.
2. Case B Rule: Percentage of cases below a given *z* score (percentile).
   - B1. If the *z* score is positive, look up percentage in Appendix Table A and add 50%.
   - B2. If the *z* score is negative, look up percentage in Appendix Table A and subtract from 50%.
3. Case C Rule: Percentage of cases above a given *z* score.
   - C1. If the *z* score is positive, look up percentage in Appendix Table A and subtract from 50%.
   - C2. If the *z* score is negative, look up percentage in Appendix Table A and add 50%.
4. Case D Rule: Percentage between two *z* scores.
   - D1. If the *z* scores are on opposite sides of the mean (*M*), look up the two percentages in Appendix Table A and *add* them.
   - D2. If the *z* scores are on the same side of the mean (*M*), look up the two percentages in Appendix Table A and *subtract* the smaller from the larger.

The following matrix is to be used in conjunction with the *z* Score Percentage Rules 2, 3, and 4 as described above. Each rule's corresponding case letter and number is indicated. The numbers in parentheses represent the portion of the curve in the drawing.

Finding Percentages of Cases Under the Normal Curve

***Score* (X) *in relation to the Mean* (M)**

Question		ABOVE	BELOW
	Percentage ABOVE X	(1) Subtract Table A value from 50% **Case C1**	(3) Add Table A value to 50% **Case C2**
	Percentage BELOW X	(2) Add Table A value to 50% **Case B1**	(4) Subtract Table A value from 50% **Case B2**

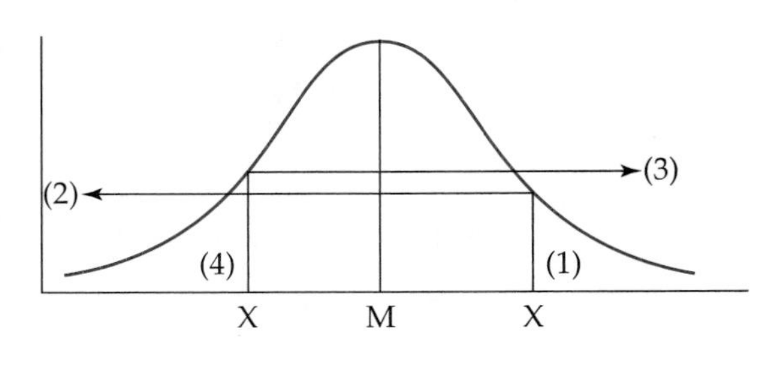

***Percentage Between Scores* (X)**

If the scores are on:	Then obtain Table A value and:
Opposite sides of mean (Case D1)	Add two percentages together
Same side of mean (Case D2)	Subtract smaller percentage from larger

Fun with your calculator.

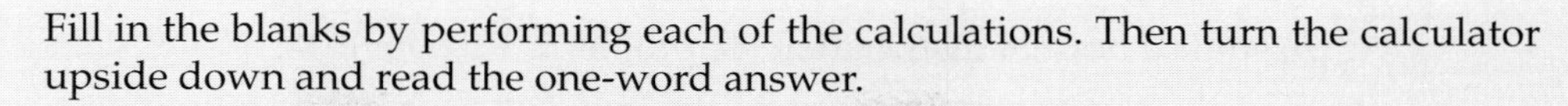

Fill in the blanks by performing each of the calculations. Then turn the calculator upside down and read the one-word answer.

1. The normal curve is shaped like a (19 × 19 × 19) + 879 _____
2. The range can never be 49833/9 than the standard deviation _____.
3. The z score + 1.96 should be imprinted on your brain's frontal (13 × 300) – 93. _____
4. Learning to calculate and understand the standard deviation should put you in a state of 193123/3.50 _____
5. When rounding your answers, don't (200 × 19) – 293 track of the decimal point. _____
6. When a distribution is negatively skewed, the value of the mean will be 94129/17 than the value of the median.
7. The use of the mean as a measure of central tendency in a severely skewed distribution is an example of how to tell (354 × 15) + 7 _____ with statistics.
8. The first time you do a statistical analysis using SPSS, you will (250 × 250) – (118.7 × 60) _____ the creator of the program.
9. Be systematic when working out an equation. Don't be a (3 × 2000) + 2075 _____.
10. And finally from the world of music a (770 × 2) + 1540 _____ can often be described as an "ill wind that nobody blows good."

SUMMARY

The normal curve is a unimodal frequency distribution, perfectly symmetrical (the mean, the median, and the mode coincide at a single point), and asymptotic to the abscissa (the curve never touches the *X* axis). The equation for the normal curve provides the curve with a constant relationship to the standard deviation. When the curve is plotted on the basis of standard deviation units, it is called the standard normal curve. Appendix Table A contains all the specific percentage areas for the various standard deviation units.

Standard scores, or *z* scores, result when a raw score is translated into units of standard deviation. The *z* score specifically defines how far the raw score is from the mean, either above it or below it, in these standard deviation units.

By looking up a *z* score in Appendix Table A, we can find the exact percentage of cases falling between that *z* score and the mean. A series of rules is outlined for quickly solving the various *z* score problems in conjunction with Appendix Table A. When problems are posed in terms of raw scores, convert the raw score to its equivalent *z* score and use the procedures outlined earlier.

Key Terms and Names

frequency distribution curve
Gauss, Carl Friedrich
normal curve
z distribution
z score

PROBLEMS

Assume a normal distribution for problems 1 through 14.

1. Find the percentage of cases falling between each of the following z scores and the mean.
   a. .70
   b. 1.28
   c. –.42
   d. –.09
2. Find the percentage of cases falling below each of the following z scores.
   a. 1.39
   b. .13
   c. –.86
   d. –1.96
3. Find the percentage of cases falling above each of the following z scores.
   a. 1.25
   b. –1.65
   c. 1.33
   d. –.19
4. Find the percentage of cases falling between the following pairs of z scores.
   a. –1.52 and .15
   b. –2.50 and 1.65
   c. –1.96 and –.25
   d. .59 and 1.59
5. The mean IQ among the members of a certain national fraternity is 118 with a standard deviation of 9.46. What percentage of the members had IQs of 115 or higher?
6. The mean resting pulse rate for a large group of varsity athletes was 72 with a standard deviation of 2.56. What percentage of the group had pulse rates of 70 or lower?
7. The mean grade-point average among freshmen at a certain university is 2.00 with a standard deviation of .60. What percentage of the freshman class had grade-point averages
   a. below 2.50?
   b. above 3.50?
   c. between 1.70 and 2.40?
   d. between 2.50 and 3.00?
8. Using the data from problem 7, find the percentile rank for the following grade-point averages.
   a. 3.80
   b. .50
   c. 2.10
   d. 3.50

9. The mean serum cholesterol level for a large group of police officers is 195 mg/dl (milligrams per deciliter) with an *SD* of 35.50. What percent of the officers had cholesterol levels of
   a. more than 220?
   b. less than 180?
10. The mean diastolic blood pressure reading for the entire student body at a large western university was 85.00 mm Hg (millimeters of mercury) with an *SD* of 7.65. What percent of the student body measured
   a. between 70 and 80?
   b. between 90 and 100?
11. Road tests of a certain compact car show a mean fuel rating of 30 miles per gallon, with a standard deviation of 2 miles per gallon. What percentage of these autos will achieve results of
   a. more than 35 miles per gallon?
   b. less than 27 miles per gallon?
   c. between 25 and 29 miles per gallon?
   d. between 32 and 34 miles per gallon?
12. On a normal distribution, at what percentile must
   a. the mean fall?
   b. the median fall?
   c. the mode fall?
13. With reference to which measure of central tendency must z scores always be computed?
14. Approximately what percentage of the scores on a normal distribution fall
   a. below the mean?
   b. within ±1 *SD* unit from the mean?
   c. within ±2 *SD* units from the mean?

True or False. Indicate either T or F for problems 15 through 24.

15. An individual score at the 10th percentile is among the top tenth of the distribution.
16. A negative z score always yields a negative percentile.
17. A positive z score always yields a percentile rank above 50.
18. A negative z score always indicates that the raw score is below the mean.
19. Between z scores of ±3.00 under the normal curve can be found almost the entire distribution.
20. A z score of zero always indicates that the raw score coincides with the mean.
21. In a normal distribution, it is sometimes possible for the mean to occur above the 50th percentile.
22. Normal distributions must always be unimodal.
23. All unimodal distributions are normal.
24. For a normal distribution, 50% of the z scores must be positive.

Chapter

5

z Scores Revisited: T Scores and Other Normal Curve Transformations

At this point we have learned to solve a great number of z score problems. However, in each case the answer has been in percentage terms. That is, the mean and the standard deviation have both been given, and the job was to calculate the frequency of occurrence (percentages) for various areas of the curve. We have consistently solved the z score equation, $z = (X - M)/SD$ for the unknown value of z. Many z score problems, however, take a different form. Though the z score equation, of course, remains the same, in this chapter it will be shifted around to solve for a variety of other unknowns.

OTHER APPLICATIONS OF THE *z* SCORE

From z Score to Raw Score

Just as knowledge of the mean, the standard deviation, and a raw score can be used to generate a ***z* score,** so too can knowledge of the *z* score produce the raw score. Thus, since $z = (X - M)/SD$,

$$X = zSD + M$$

This important variation on the *z* score theme provides specific information about an individual's performance, stated in terms of the trait being measured.

Example We know that on a weight distribution with a mean of 150 pounds and a standard deviation of 17 pounds, a certain individual had a *z* score of 1.62. We can utilize this information to specify the measurement of that individual in terms of the units for the trait being measured, in this case, pounds.

$$X = zSD + M$$
$$= (1.62)(17) + 150 = 27.54 + 150$$
$$= 177.54 \text{ pounds}$$

If the individual scores below the mean (negative z score), the same procedure is used.* Assume the same weight distribution ($M = 150$, $SD = 17$), and determine how much an individual weighs who has a z score of $-.65$.

$$X = zSD + M$$
$$= (-.65)(17) + 150 = -11.05 + 150$$
$$= 138.95 \text{ pounds}$$

—**Exercise 5a** On an IQ distribution (normal), with a mean of 100 and a standard deviation of 15, find the actual IQs (raw scores) for individuals with the following z scores.

1. .56
2. 2.73
3. −1.64
4. −.10
5. 2.58

From Percentile to z Score

If you know a percentile, you can easily determine the z score, and as we have seen, once you know the z score, you can always get the raw score. We learned that a **percentile** specifies the exact percentage of scores at or below a given score. We know, for example, that a percentile of 84 means that 84% of the cases fall at or below that point on the baseline of the curve. Then, by using Appendix Table B, we can quickly convert any percentile into its equivalent z score.

THE PERCENTILE TABLE

Appendix Table B is a separate percentile table that has been specially constructed for this book. Notice that next to each percentile can be found its equivalent z score. Also notice that all the percentiles in the first column, from the 1st to the 49th, have nega-

*If your calculator cannot handle a negative number placed first in a series of added terms, as in −11.05 + 150, then simply change the order of the values and enter the 150 first, 150 − 11.05. Since addition is what mathematicians call commutative, it makes no difference in which order you choose to enter the terms.

tive z scores, which again reinforces the fact that z scores that fall below the mean (those below the 50th percentile) must be given negative signs. In the second column, the z scores are all positive, since all percentiles from the 51st on up are, of course, above the mean percentile of 50.

—**Exercise 5b** Find the z score for each of the following percentiles.

1. 95th percentile
2. 10th percentile
3. 39th percentile
4. 55th percentile
5. 75th percentile

From Percentiles to Raw Scores

It is obvious from the preceding that if we have a percentile, we know the z score. All it takes is the use of Appendix Table B. Also, we have previously seen how the raw score can be determined from the z score ($X = zSD + M$). Thus, whenever a percentile is known, the raw score can be accurately determined.

Example A normal distribution has a mean of 250 and a standard deviation of 45. Calculate the raw score for someone who is at the 95th percentile.

First, the z score for a percentile of 95 is equal to 1.65. We find this by looking up the 95th percentile in Appendix Table B. Now, using the z score equation, we solve it for the raw score, X.

$$X = zSD + M$$

Then plug in all known values:

$$X = (1.65)(45) + 250 = 74.25 + 250$$
$$= 324.25$$

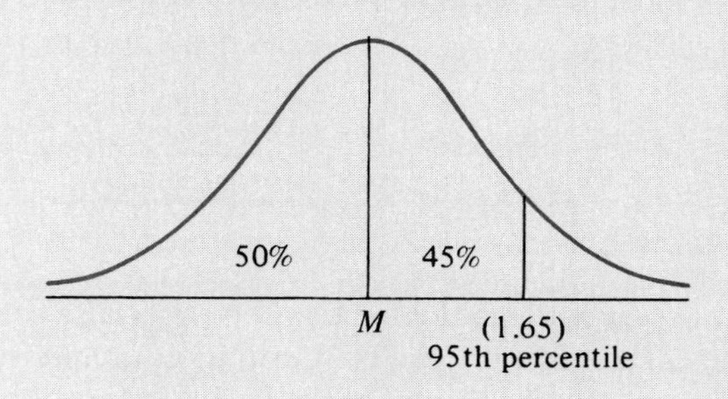

Example Using the data from the preceding example ($M = 250$, $SD = 45$), calculate the raw score for someone who is at the 10th percentile.

Look up the 10th percentile in Appendix Table B and get the *z* score of –1.28. Then plug all known values into the equation for X.

$$X = zSD + M$$
$$= (-1.28)(45) + 250 = -57.60 + 250$$
$$= 192.40$$

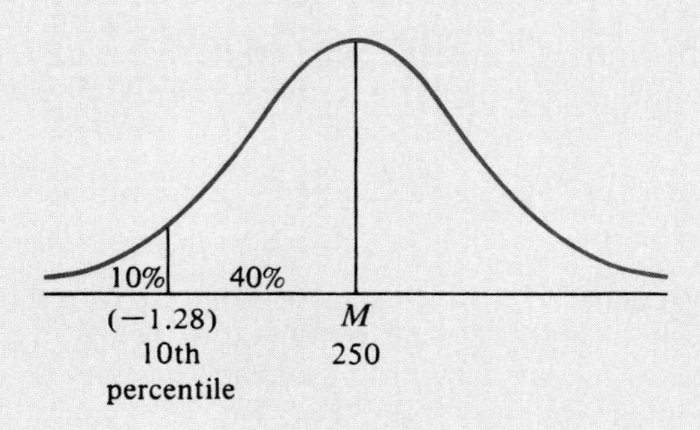

—**Exercise 5c** On a normal distribution with a mean of 1500 and a standard deviation of 90, find the raw score for each of the following percentiles.

1. 45th percentile
2. 55th percentile
3. 15th percentile
4. 79th percentile
5. 5th percentile

Although percentiles are intuitively easy to understand, they can lead to some serious distortions, especially for scores at the very high and low ends of the normal distribution. Percentiles tend to compress the tails of the distribution, making it appear that persons who have extreme scores are closer together than they really are. For example, let's take a male adult height distribution, whose mean is known to be 69 inches and whose *SD* is 3.85. We are going to compare the heights of two persons, one at the 50th percentile with another at the 54th percentile. From Appendix Table B we find the two *z* scores, which are 0 and .10, respectively. Thus, we solve for *X* to get the actual height values, which for the 50th percentile would be

$$X = (z)(SD) + M \quad \text{or} \quad (0.0)(3.85) + 69 = 0 + 69 = 69$$

And then, for the 54th percentile,

$$X = (z)(SD) + M \quad \text{or} \quad (.10)(3.85) + 69 = .39 + 69 = 69.39$$

for a difference of .39 inch between the two. Thus, between percentiles of 50 and 54 (scores around the middle of the distribution), the difference in height is only about one-third of an inch. Now let's compare the heights of two persons who in percentile units will also differ by 4 points, but who will both be out in the far right tail of the distribution with percentiles of 95 and 99, respectively. From Appendix Table B, the

z scores are 1.65 for the 95th percentile and 2.41 for the 99th percentile. Again, solving for X, the 95th percentile yields

$$X = (z)(SD) + M \quad \text{or} \quad (1.65)(3.85) + 69 = 6.35 + 69 = 75.35$$

And for the 99th percentile,

$$X = (z)(SD) + M \quad \text{or} \quad (2.41)(3.85) + 69 = 9.28 + 69 = 78.28$$

for a difference of almost 3 full inches. Thus, the height difference out in the tail of the distribution is roughly 10 times greater than the same percentile difference near the center.

Example On an IQ distribution with a mean of 100 and an *SD* of 15, compare the differences in IQ points between two persons at the 50th and 54th percentiles with those at the 95th and 99th percentiles.

For the 50th percentile, the IQ will be

$$X = (z)(SD) + M \quad \text{or} \quad (0.0)(15) + 100 = 0 + 100 = 100$$

And for the 54th percentile, the IQ will be

$$X = (z)(SD) + M \quad \text{or} \quad (.10)(15) + 100 = 1.50 + 100 = 101.50$$

for a difference of one and a half IQ points.

Comparing the 95th and 99th percentiles, we first get the IQ for the 95th percentile:

$$X = (z)(SD) + M \quad \text{or} \quad (1.65)(15) + 100 = 24.75 + 100 = 124.75$$

And then, for the 99th percentile, the IQ will be

$$X = (z)(SD) + M \quad \text{or} \quad (2.41)(15) + 100 = 36.15 + 100 = 136.15$$

for a difference of 12 IQ points, or a difference that is eight times greater than the difference shown above.

Examples On an SAT verbal distribution with a mean of 500 and an *SD* of 100, compare the difference of 6 percentage points for SAT scores between persons at the 47th and 53rd percentiles with those at the 1st and 7th percentiles.

Answer: SAT scores for the 1st and 7th percentiles are 259 and 352, for a large difference of almost 100 points. SAT scores for the 47th and 53rd percentiles are 492 and 508, for a much smaller difference of only 16 points.

On an adult female weight distribution with a mean of 153 pounds and an *SD* of 30, compare the difference of 3 percentage points for weights between persons at the 47th and 50th percentiles with those at the 96th and 99th percentiles.

Answer: Weights for the 47th and 50th percentiles are 150.60 and 153, for a small difference of less than 3 pounds. Weights for the 96th and 99th percentiles are 205.50 and 225.30, for a much larger difference of almost 20 pounds.

From z Score to Standard Deviation

Just as we learned to flip the z score equation around to solve for the raw score, so too we can solve it for the standard deviation. Since $z = (X - M)/SD$, then

$$SD = \frac{X - M}{z}$$

Therefore, if we are given the raw score, the mean, and the z score (or the percentile from which the z score can be found), we can calculate the standard deviation.

Example On a normal distribution, the mean is equal to 25. A certain individual had a raw score of 31 and a z score of 1.68. The standard deviation would thus be calculated as follows:

$$SD = \frac{X - M}{z} = \frac{31 - 25}{1.68} = \frac{6}{1.68}$$
$$= 3.57$$

Instead of being presented with the z score, we might instead be given the percentile.

Example On a normal distribution, the mean is 820. A certain individual has a raw score of 800 and is at the 43rd percentile. Find the standard deviation.

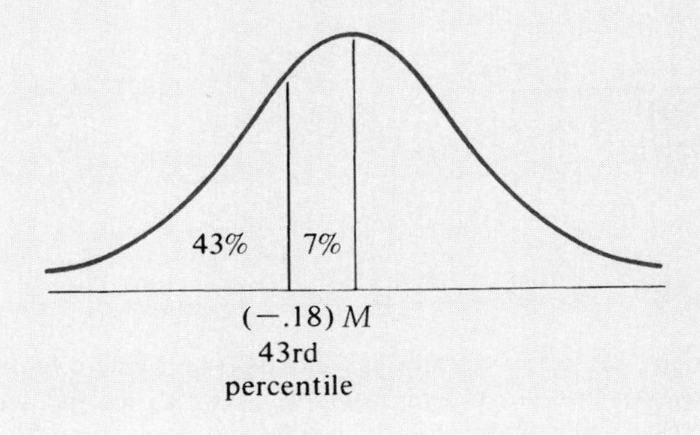

First, we find the z score for a percentile of 43. Looking in Appendix Table B, we find that z score to be –.18. Using the known values, we calculate the standard deviation.*

$$SD = \frac{X - M}{z} = \frac{800 - 820}{-.18} = \frac{-20}{-.18}$$

$$= 111.11$$

The Range

Now that we have found the value of the standard deviation, a close approximation of the range may easily be calculated. Since the range is roughly six times the *SD* (see page 54), the range for the distribution above should be (6)(111.11), or 666.66. Also, knowing the values for both the mean and the range allows for the determination of the highest and lowest scores in the distribution. Since under the normal curve, one-half of the range falls above the mean and one-half below it, the range is divided by 2, and the resulting value is then added to and subtracted from the mean. Thus, dividing 666.66 by 2 equals 333.33. By adding 333.33 to the mean of 820, the highest score in the distribution is estimated to be 1153.33, and then by subtraction, the lowest score should be approximately 486.67.

Example The mean on a normal distribution is 500. A certain individual has a raw score of 600 and is at the 85th percentile. Find the standard deviation, the range, and both the highest and lowest scores.

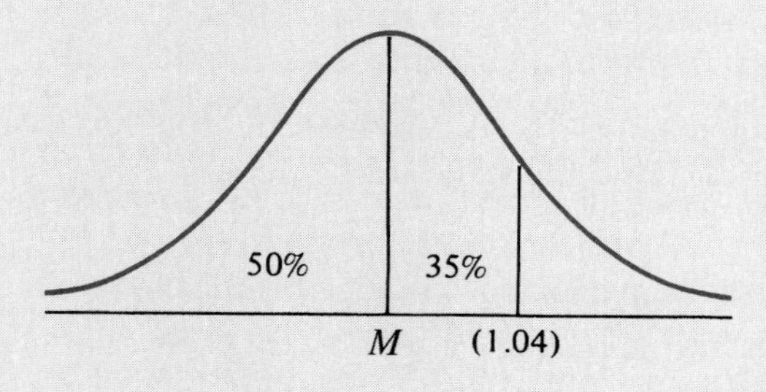

*Dividing a negative value by another negative value always yields a positive value. Also, never forget the fact that all standard deviations *must be positive.* You can have negative z scores and even negative raw scores, but you should never, never get a negative standard deviation.

From the percentile, the z score is found to be 1.04. Next, plug the known values into the equation for the *SD:*

$$SD = \frac{X - M}{z} = \frac{600 - 500}{1.04} = \frac{100}{1.04}$$

$SD = 96.15$

$R = (6)(96.15) = 576.90$

$R/2 = 288.45$

Highest score = 500 + 288.45 = 788.45

Lowest score = 500 − 288.45 = 211.55

—**Exercise 5d** On a normal distribution with a mean of 115, find the standard deviation for the following raw scores.

1. 120 (at the 60th percentile)
2. 100 (at the 30th percentile)
3. 110 (at the 47th percentile)
4. 85 (at the 7th percentile)
5. 140 (at the 83rd percentile)

The ability to determine the standard deviation from a z score can be especially important to the guidance counselor. Perhaps students' scores on a certain standardized test are returned in the form of raw scores and percentiles, but no information regarding the standard deviation is provided. All that is necessary are the values for one student to calculate the standard deviation for the entire distribution.

Example On a verbal SAT distribution with a mean of 500, a certain student received a raw score of 540 and is at the 65th percentile.

The z score for the 65th percentile is found to be .39. Then the standard deviation can be calculated.

$$SD = \frac{X - M}{z} = \frac{540 - 500}{.39} = \frac{40}{.39} = 102.56$$

From calculations like those in the preceding example, a guidance counselor can make up a chart showing precisely what raw scores (even those not actually received by the counselor's students) correspond to the various percentiles. Since both the standard deviation and the equation, $X = zSD + M$, are available, they can be used to generate the entire distribution.

Example For a certain test with a mean of 440 and a standard deviation of 91.35, a chart can be constructed by finding the z score for each percentile and then solving for X.

For the 90th percentile, the z score is 1.28. Solve for the value of X.

$$X = zSD + M = (1.28)(91.35) + 440$$
$$= 116.93 + 440 = 556.93$$

The following figure illustrates the chart.

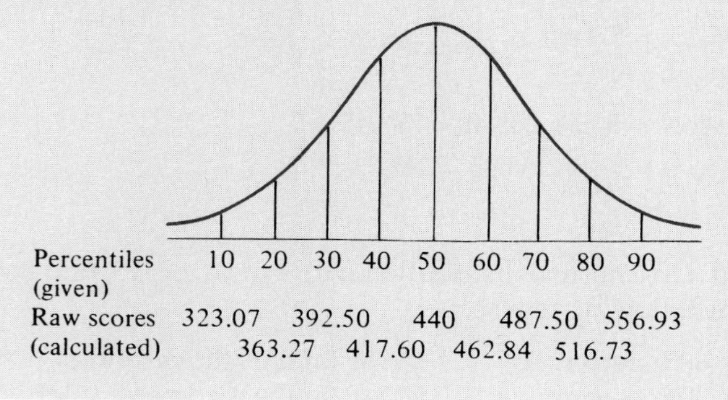

From z Score to the Mean

Since $z = (X - M)/SD$, it can be shown that

$$M = X - zSD$$

Any time a raw score, z score, and the standard deviation are all given, the mean is readily available. Also, since we have shown that a percentile can be quickly and easily converted to a z score, the mean can be found whenever the percentile or the z score is given (assuming the SD is known). Also, since the mean, median, and mode all fall at the same point under the normal curve, once the mean is known, so are the median and mode.

Example A normal distribution has a standard deviation of 15. A raw score of 122 falls at the 90th percentile. Find the mean of the distribution.

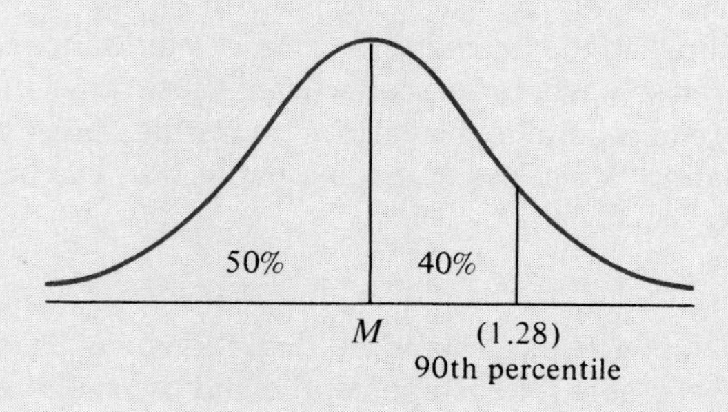

First, we find the z score for the 90th percentile by looking in Appendix Table B. The correct z score is 1.28. Using the equation $M = X - zSD$, we plug in all known values.

$$M = 122 - (1.28)(15) = 122 - 19.20$$
$$= 102.80$$ (which because the distribution is normal is also the value for the median and mode)

Next, we consider the same type of problem in a situation where the raw score falls to the left of the mean.

Example On a normal distribution the standard deviation is 90. A raw score of 400 falls at the 18th percentile. Find the mean.

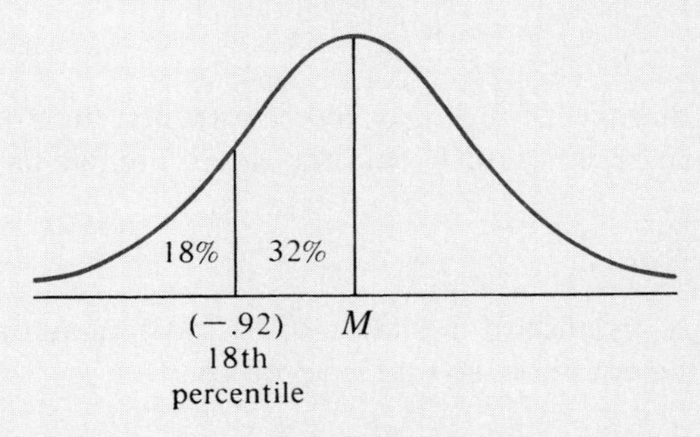

From Appendix Table B, we establish the z score for the 18th percentile as −.92. Plugging all known values into the equation yields the mean.*

$$M = X - zSD = 400 - (-.92)(90)$$
$$= 400 - (-82.80)$$
$$= 400 + 82.80$$
$$= 482.80$$

—**Exercise 5e** Find the mean for each of the following normal distributions.

1. $SD = 10$, $X = 50$, percentile rank = 35
2. $SD = 50$, $X = 300$, percentile rank = 76
3. $SD = 3$, $X = 15$, percentile rank = 95
4. $SD = 23$, $X = 74$, percentile rank = 34
5. $SD = 100$, $X = 1600$, percentile rank = 85

---

*Note that when subtracting a negative product, we simply change the sign to positive: −(−82.80) becomes +82.80

T SCORES

In a great number of testing situations, especially in education and psychology, raw scores are reported in terms of ***T* scores.** A *T* score is a converted *z* score, with the mean always set at 50 and the standard deviation at 10. This ensures that all *T* scores are positive values (unlike *z* scores, which are negative whenever the raw score is below the mean).

The *T* score, like the *z* score, is a measure of an individual's performance relative to the mean. The *T* score is the measure of how far a given raw score is from a mean of 50 in standard deviation units of 10. Therefore, a *T* score of 60 is one full standard deviation unit above the mean of 50. A *T* score of 30 is two full standard deviation units below the mean of 50. (See Figure 5.1.)

Thus, *T* score values range from 20 to 80, giving us a range value (*R*) of 60. (In those exceedingly rare instances when a *z* score falls more than 3 *SD* units away from the mean, the *T* score will obviously fall outside the 20 to 80 range.)

Calculating T *Scores*

Calculating a *T* score is just as easy as finding a raw score when the *z* score, the standard deviation, and the mean are known. To find the raw score, we used the following equation:

$$X = zSD + M$$

Since the *T* score assumes a standard deviation of 10 and a mean of 50, these values can be substituted into the equation, so that it becomes

$$T = z(10) + 50$$

With this equation, if the *z* score is known, finding the *T* score becomes automatic. It is not necessary to know the actual mean or standard deviation of the distribution of scores from which the *z* score was derived—just the *z* score itself is sufficient. Any other information about the distribution is superfluous. Information as to the mean

FIGURE 5.1 Normal curve showing *T* scores and standard deviation units.

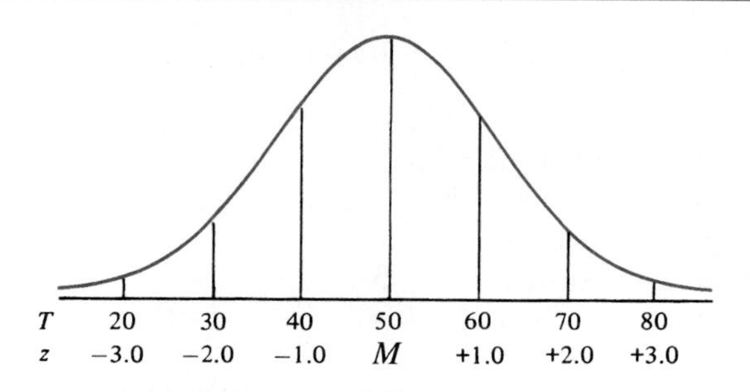

and the standard deviation is needed to create the z score in the first place; but once z is known, so is T.

Example On a normal distribution, an individual has a z score of 1.30. Find the T score.

$$T = z(10) + 50 = (1.30)(10) + 50 = 13 + 50$$
$$= 63$$

Or, on a normal distribution, an individual has a z score of –.50. Find the T score.

$$T = z(10) + 50 = (-.50)(10) + 50 = -5 + 50$$
$$= 45$$

Applications of T *Scores*

Comparing Two Raw Scores. The transformation of raw scores to T scores can be very helpful, since T (like the z score) provides a standard by which performances on different tests can be directly compared.

Example Suppose that an individual takes two different IQ tests. The mean of both tests is 100. The standard deviation for the first test is 15, and the standard deviation for the second test is 18. On the first test, the individual received an IQ score of 110. On the second test, the IQ score received was 112. Using T scores, find out whether the individual did better on the second test.

First, calculate the z score for the raw score on the first test.

$$z = \frac{X - M}{SD} = \frac{110 - 100}{15} = \frac{10}{15}$$
$$= .67$$

Then convert that to a T score.

$$T = z(10) + 50 = .67(10) + 50 = 6.70 + 50$$
$$= 56.70$$

Do the same for the score on the second test.

$$z = \frac{X - M}{SD} = \frac{112 - 100}{18} = \frac{12}{18}$$
$$= .67$$
$$T = z(10) + 50 = (.67)(10) + 50 = 6.70 + 50$$
$$= 56.70$$

In each case, the T score was 56.70. The individual's performance on the two tests was thus identical. (This, of course, had to be the case, since the z scores were identical.)

—**Exercise 5f** On a normal distribution with a mean of 500 and a standard deviation of 95, find the T score for each of the following raw scores.

1. 400
2. 610
3. 330
4. 775
5. 550

From T to z to Raw Scores

Sometimes it is necessary to convert T scores back into raw scores. A guidance counselor or teacher may be presented with a set of student scores in the form of T values. To make this particular conversion, the mean and the standard deviation of the raw score distribution must be known. The procedure involves first translating the T score back into its z score equivalent using the following equation.

$$z = \frac{X - M}{SD} = \frac{T - 50}{10}$$

Then, the raw score equation, $X = zSD + M$, is used to complete the conversion.

Example On a normal distribution with a mean of 70 and a standard deviation of 8.50, a certain student reports a T score of 65 and wishes to know the equivalent raw score.

Find the z score.

$$z = \frac{T - 50}{10} = \frac{65 - 50}{10} = \frac{15}{10}$$

$$= 1.50$$

Then, using the known mean and standard deviation values, convert the z score to the raw score.

$$X = zSD + M = (1.50)(8.50) + 70 = 12.75 + 70$$

$$= 82.75$$

For a T score of less than 50, a score below the mean, follow the same procedure, being very careful regarding the minus signs.

Example Using the data from the previous example ($M = 70$, $SD = 8.50$), find the raw score for a student whose T score was 37.

$$z = \frac{T - 50}{10} = \frac{37 - 50}{10} = \frac{-13}{10}$$

$$= -1.30$$

Then solve for X.

$$X = zSD + M = (-1.30)(8.50) + 70 = -11.05 + 70$$

$$= 58.95$$

—**Exercise 5g** The following are from a normal distribution with a mean of 35 and a standard deviation of 4.50. Find the raw score for each T score.

1. $T = 43$
2. $T = 67$
3. $T = 25$
4. $T = 31$
5. $T = 78$

A Thumbnail Sketch

If algebra gives you night sweats and hot flashes, try using your thumb on the following diagram:

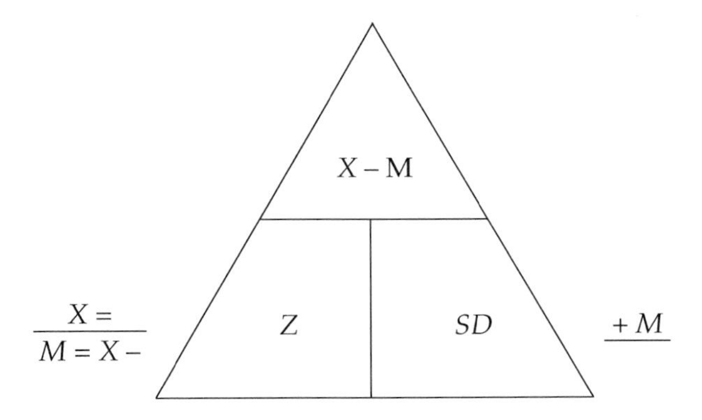

Use your thumb to cover the unknown and let the diagram SOLVE the equation. For example, if you want to find the Z score, cover the Z and read out what's left over, which will then show as $X - M$ over the SD. Now, write out the equation you just uncovered, that is

$$Z = \frac{X - M}{SD}$$

Similarly, if the SD is needed, use the same procedure. Cover the SD and read $X - M$ over Z.

$$SD = \frac{X - M}{Z}$$

If X, the raw score, is needed, cover the numerator (which contains the element to be found) and follow the equation shown below, Thus, for X the equation tells us that X is equal to $(Z)(SD) + M$.

And if its M that is needed, again cover the numerator (which also contains the element to be found and follow the equation below that shows M as equal to $X - (Z)(SD)$.

A Thumbnail for the T score

Using the same method as shown above, if you want the Z score cover the Z and read the equation as

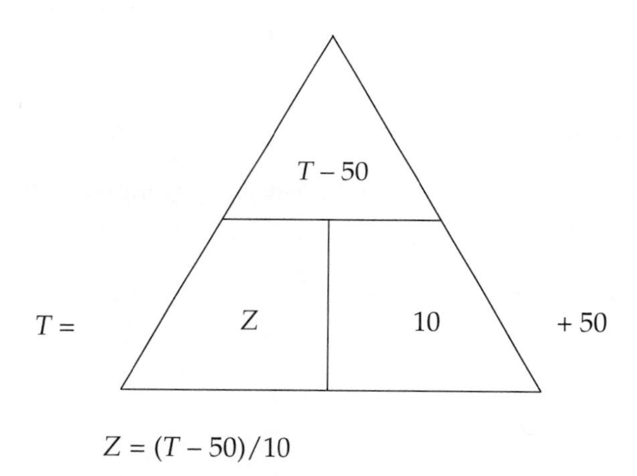

$Z = (T - 50)/10$

Or to get the T score, cover the $T - 50$, and read the equation as $T = (Z)(10) + 50$

The T score is thus a Z score that is converted to a different scale, a scale that typically goes from 20 to 80.

NORMAL CURVE EQUIVALENTS

Another increasingly popular standardized score is the **NCE (normal curve equivalent) score.** Like the T score, the NCE is based on the normal curve and is calculated by setting the mean at 50 and the standard deviation at approximately 21. Because of its larger standard deviation, the range of NCE scores is wider than for *T* scores, approximating 0 to 100.

STANINES

Stanines, like *z*, *T*, and NCE scores, are also based on the normal curve; but unlike the others, stanines divide the distribution into units of nine intervals (whereas *z*, for example, divides the distribution into six intervals). The first stanine is equal to a *z* score of –1.75, and the scale then increases by *z* units of roughly .50 until it reaches the 9th stanine, which is equal to a *z* of +1.75. The mean of the stanine distribution must equal 5 and the standard deviation must equal approximately 2. By examining Table 5.1, we can see that stanine 5 (the middlemost interval) contains 20% of the cases, stanines 4 and 6 contain 17%, stanines 3 and 7 contain 12%, stanines 2 and 8 contain 7%, and, finally, stanines 1 and 9 both contain 4%.

Table 5.1 shows the percentage of cases within each of the stanine *intervals;* but if you now look at Table 5.2, you can determine the percentile score for any specific stanine score. Notice that a child who scored at the 6th stanine, for example, would have performed better than 60% of those taking the same test; or a child who was at stanine 4.12 would have scored at the 33rd percentile. Many standardized tests in the field of educational psychology report scores on the basis of stanines.

To assist you in converting scores from one form to another, Table 5.2 allows for quick transition among several of the most popular scales. This table includes raw score equivalents for the SAT (Scholastic Assessment Test), the NCE (normal curve equivalent) scores, and the WIQ (Wechsler IQ Tests, which include the WAIS, WISC, and WPPSI). All conversions are based on the following formula: score = $(z)(SD)$ + mean. Since SAT scores only come in units of 10 points, a 608 rounds to 610 and a 482 to 480.

TABLE 5.1 Stanines divide the distribution into units of nine intervals.

Stanine	*Percentage in Each Interval (Rounded)*	*z Score*
1	4	–1.75 and below
2	7	–1.25
3	12	–0.75
4	17	–0.25
5	20	+0.00
6	17	+0.25
7	12	+0.75
8	7	+1.25
9	4	+1.75 and above

TABLE 5.2 Conversion table: the SAT, NCE, and WIQ.

Percentile	z	T	*Stanine*	*SAT*	*NCE*	*WIQ*	
1	–2.41	25.90	0.18	259	1.0	63.85	
2	–2.05	29.50	0.90	295	6.7	69.25	
3	–1.88	31.20	1.24	312	10.4	71.80	
4	–1.75	32.50	1.50	325	13.1	73.75	1st Stanine
5	–1.65	33.50	1.70	335	15.4	75.25	
6	–1.56	34.40	1.88	344	17.3	76.60	
7	–1.48	35.20	2.04	352	18.9	77.80	
8	–1.41	35.90	2.18	359	20.4	78.85	
9	–1.34	36.60	2.32	366	21.8	79.90	
10	–1.28	37.20	2.44	372	23.0	80.80	1st Decile
11	–1.23	37.70	2.54	377	24.2	81.55	2nd Stanine
12	–1.18	38.20	2.64	382	25.3	82.30	
13	–1.13	38.70	2.74	387	26.3	83.05	
14	–1.08	39.20	2.84	392	27.2	83.80	
15	–1.04	39.60	2.92	396	28.2	84.40	
16	–1.00	40.00	3.00	400	29.1	85.00	
17	–0.95	40.50	3.10	405	29.9	85.75	
18	–0.92	40.80	3.16	408	30.7	86.20	
19	–0.88	41.20	3.24	412	31.5	86.80	
20	–0.84	41.60	3.32	416	32.3	87.40	2nd Decile
21	–0.81	41.90	3.38	419	33.0	87.85	
22	–0.77	42.30	3.46	423	33.7	88.45	
23	–0.74	42.60	3.52	426	34.4	88.90	3rd Stanine
24	–0.71	42.90	3.58	429	35.1	89.35	
25	–0.67	43.30	3.66	433	35.8	89.95	1st Quartile
26	–0.64	43.60	3.72	436	36.5	90.40	
27	–0.61	43.90	3.78	439	37.1	90.85	
28	–0.58	44.20	3.84	442	37.7	91.30	
29	–0.55	44.50	3.90	445	38.3	91.75	
30	–0.52	44.80	3.96	448	39.0	92.20	3rd Decile
31	–0.50	45.00	4.00	450	39.6	92.50	
32	–0.47	45.30	4.06	453	40.1	92.95	
33	–0.44	45.60	4.12	456	40.7	93.40	
34	–0.41	45.90	4.18	459	41.3	93.85	
35	–0.39	46.10	4.22	461	41.9	94.15	
36	–0.36	46.40	4.28	464	42.5	94.60	
37	–0.33	46.70	4.34	467	43.0	95.05	
38	–0.31	46.90	4.38	469	43.6	95.35	
39	–0.28	47.20	4.44	472	44.1	95.80	
40	–0.25	47.50	4.50	475	44.7	96.25	4th Decile & Stanine
41	–0.23	47.70	4.54	477	45.2	96.55	
42	–0.20	48.00	4.60	480	45.8	97.00	
43	–0.18	48.20	4.64	482	46.3	97.30	
44	–0.15	48.50	4.70	485	46.8	97.75	
45	–0.13	48.70	4.74	487	47.4	98.05	

TABLE 5.2 *Continued*

Percentile	*z*	T	*Stanine*	*SAT*	*NCE*	*WIQ*	
46	–0.10	49.00	4.80	490	47.9	98.50	
47	–0.08	49.20	4.84	492	48.4	98.80	
48	–0.05	49.50	4.90	495	48.9	99.25	
49	–0.03	49.70	4.94	497	49.5	99.55	
50	0.00	50.00	5.00	500	50.0	100.00	Median (5th Decile)
51	0.03	50.30	5.06	503	50.5	100.45	
52	0.05	50.50	5.10	505	51.1	100.75	
53	0.08	50.80	5.16	508	51.6	101.20	
54	0.10	51.00	5.20	510	52.1	101.50	
55	0.13	51.30	5.26	513	52.6	101.95	
56	0.15	51.50	5.30	515	53.2	102.25	
57	0.18	51.80	5.36	518	53.7	102.70	
58	0.20	52.00	5.40	520	54.2	103.00	
59	0.23	52.30	5.46	523	54.8	103.45	
60	0.25	52.50	5.50	525	55.3	103.75	6th Decile & Stanine
61	0.28	52.80	5.56	528	55.9	104.20	
62	0.31	53.10	5.62	531	56.4	104.65	
63	0.33	53.30	5.66	533	57.0	104.95	
64	0.36	53.60	5.72	536	57.5	105.40	
65	0.39	53.90	5.78	539	58.1	105.85	
66	0.41	54.10	5.82	541	58.7	106.15	
67	0.44	54.40	5.88	544	59.3	106.60	
68	0.47	54.70	5.94	547	59.9	107.05	
69	0.50	55.00	6.00	550	60.4	107.50	
70	0.52	55.20	6.04	552	61.0	107.80	7th Decile
71	0.55	55.50	6.10	555	61.7	108.25	
72	0.58	55.80	6.16	558	62.3	108.70	
73	0.61	56.10	6.22	561	62.9	109.15	
74	0.64	56.40	6.28	564	63.5	109.60	
75	0.67	56.70	6.34	567	64.2	110.05	3rd Quartile
76	0.71	57.10	6.42	571	64.9	110.65	
77	0.74	57.40	6.48	574	65.6	111.10	7th Stanine
78	0.77	57.70	6.54	577	66.3	111.55	
79	0.81	58.10	6.62	581	67.0	112.15	
80	0.84	58.40	6.68	584	67.7	112.60	8th Decile
81	0.88	58.80	6.76	588	68.5	113.20	
82	0.92	59.20	6.84	592	69.3	113.80	
83	0.95	59.50	6.90	595	70.1	114.25	
84	1.00	60.00	7.00	600	70.9	115.00	
85	1.04	60.40	7.08	604	71.6	115.60	
86	1.08	60.80	7.16	608	72.8	116.20	
87	1.13	61.30	7.26	613	73.7	116.95	
88	1.18	61.80	7.36	618	74.7	117.70	
89	1.23	62.30	7.46	623	75.8	118.45	8th Stanine
90	1.28	62.80	7.56	628	77.0	119.20	9th Decile

(continues)

TABLE 5.2 *Continued*

Percentile	z	T	Stanine	SAT	NCE	WIQ	
91	1.34	63.40	7.68	634	78.2	120.10	
92	1.41	64.10	7.82	641	79.6	121.15	
93	1.48	64.80	7.96	648	81.1	122.20	
94	1.56	65.60	8.12	656	82.7	123.40	
95	1.65	66.50	8.30	665	84.6	124.75	
96	1.75	67.50	8.50	675	85.9	126.25	9th Stanine
97	1.88	68.80	8.76	688	89.6	128.20	
98	2.05	70.50	9.10	705	93.3	130.75	
99	2.41	74.10	9.82	741	99.0	136.15	
100	3.00	80.00	11.00	800	100	145.00	

GRADE-EQUIVALENT SCORES: A NOTE OF CAUTION

Grade-equivalent scores (GEs) are based on relating a given student's score on a test to the average scores found at the same time of year for other students of roughly the same age and in a particular grade. For example, assume that in September a large, representative sample of third-graders (called the norming group) produced an average score of 30 on a certain arithmetic test. If a given student is then tested and receives a score of 30, he or she would be assigned a grade-equivalent score of 3.0. If the child did somewhat better than that and had a score of, say, 3.4, it would indicate a performance equal to a third-grade student in the fourth month (December) of the school year.

Grade-equivalent scores are typically reported in tenths of a year, so that a score of 5.9 refers to the ninth month (June) of the fifth grade, and a score of 0.0 to the first day of kindergarten. Thus, the scores range from 0.0 (or sometimes K.0) through 12.9, representing the 13 years of school from kindergarten through grade 12. The first of September is given on the scale as 0, the end of September as .1, the end of October as .2, and so on until the end of June, as .9.

Grade-equivalent tables are provided for the major nationally standardized tests, and although they are quick and easy to convert (simply look up the raw score in the test publisher's table), they must be interpreted with some caution. First, children do not all grow and develop at the same yearly rate, never mind the same monthly rate. A seemingly bright child, for example, might suddenly underperform the norms by a few months and yet may quickly catch up and even outperform the norms several months later. Second, a precocious child's high score in some area shouldn't be reason to have that child skip a grade or two. A third-grader's 7.0 on a given test doesn't mean that child is now ready for a fast promotion to junior high. What it does mean is that the third-grader has certainly conquered third-grade material and, in fact, has done as well as a seventh-grader when measured on a *third-grade* test. However, there are undoubtedly many things the seventh-grader has learned and is expected to know that are simply not even part of a third-grader's consciousness and, of course, don't appear on a third-grade test (Sprinthall, Sprinthall, & Oja, 1998).

THE IMPORTANCE OF THE *z* SCORE

In this and Chapter 4, we have kept the focus on the *z* score. We have twisted it, flipped it forward and backward, and studied all its variations, perhaps, you may feel, ad nauseam. The reasons for this scrutiny will become clear as we move into other areas, such as probability and the sampling distributions. This has definitely been time well spent. A thorough understanding of the *z* score concept will make these other topics far easier to understand. Much of the material to come will rest, like an inverted pyramid, on the perhaps tiny, but stable, base provided by the *z* score. Over and over again, we will refer to the discussions and examples found in this chapter.

By now perhaps even the skeptics among you are ready to admit that the math involved in doing these problems is not too difficult. If you follow the text step-by-step, the math will *not* get harder. If you and your trusty calculator can handle these *z* score problems, you can take the rest of the book in stride.

SUMMARY

The *z* score may be used to obtain the raw score in any normal distribution of scores—by solving the *z* score equation for *X*, where the mean and standard deviation of the raw score distribution are known ($X = zSD + M$). Further, a percentile score may be converted into a *z* score and, thus, also used to generate the raw score. All negative *z* scores yield percentiles less than 50 and translate into raw scores below the mean. Positive *z* scores, on the other hand, yield percentiles above 50 and raw scores above the mean.

The *z* score equation may also be used to find the standard deviation, whenever the *z* score, the raw score, and the mean of the raw score distribution are known. That is, since $z = (X - M)/SD$, then $SD = (X - M)/z$. This is especially useful to the psychologist or guidance counselor who works with standardized tests. Simply knowing the raw score and percentile rank (from which the *z* score can be found) of a single student, along with the mean of the distribution, allows for the calculation of the standard deviation. Also, once the standard deviation is known, the range can be approximated by using the equation $R = 6(SD)$. Knowing the range, in turn, allows for the estimation of both the highest and lowest scores in the distribution. Similarly, if needed, the mean of the raw score distribution can be found whenever the raw score, the *z* score, and the standard deviation are known.

T scores are *z* scores that have been converted to a distribution whose mean is set at 50 and whose standard deviation is equal to 10. *T* scores range in value from 20 to 80, since 20 is three standard deviation units below the mean, and 80 is three standard deviation units above. If the *z* score is known, the *T* score can be obtained without having any further information. Finally, if the *T* score is given, it can be used to find the *z* score. Other measures that are based on the normal curve are normal curve equivalents (NCEs) and stanines. The NCE is calculated by setting the mean at 50 and the standard deviation at approximately 21. Because of its larger standard deviation, the range of NCE scores is wider than for *T* scores and approximates 0 to 100. Stanines divide the distribution into units of nine intervals (whereas z, for example, divides the distribution into six intervals). The mean of the stanine distribution must equal 5, and the standard deviation must equal approximately 2. Many standardized tests in the field of educational psychology report scores on the basis of stanines. Finally, grade-equivalent scores (GEs) are also popular in the field of education. GEs are based on the average score found for students in a particular grade at both the same age and time of year. GEs are reported in tenths of a year such that a GE of 6.9 refers to the ninth month (June) of the sixth grade. Thus, a third-grade child who achieved the same score as the average sixth-grader during June would be awarded a GE of 6.9. The GE should be viewed with some caution, however, since many children experience occasional temporary educational lags and since a GE that may appear to show that a certain child in the third grade is capable of doing, say, sixth-grade work really indicates that this child scored as well as the average sixth-grader who had taken a *third-grade* test.

Key Terms

Grade-equivalent scores (GEs)
NCE (normal curve equivalent) score
Percentile
Stanine
T score
z score

PROBLEMS

Assume a normal distribution for problems 1 through 13.

1. The employees at a certain fast-food restaurant have a mean hourly income of $10.00 with a standard deviation of $2.10. Find the actual hourly wages for the employees who have the following z scores:
   a. 2.50
   b. 1.56
   c. −.97
   d. −1.39

2. The mean verbal SAT score at a certain college is 450, with a standard deviation of 87. Find the SAT scores for the students who scored at the following percentiles.
   a. 10th percentile
   b. 95th percentile
   c. 5th percentile
   d. 45th percentile

3. The grades of the students in a large statistics class are normally distributed around a mean of 70, with an *SD* of 5. If the instructor decides to give an A grade only to the top 15% of the class, what grade must a student earn to be awarded an A?

4. Suppose the instructor in the class in problem 3 (where the mean was 70 and the *SD* was 5) decided to give an A grade only to the top 10% of the class. What grade would the student now have to earn to be awarded an A?

5. The mean height among adult males is 68 inches. If one individual is 72 inches tall and has a z score of 1.11,
   a. find the *SD*.
   b. find the range.
   c. find the highest score.
   d. find the lowest score.

6. If a weight distribution for female college students has a mean of 120 pounds, and one individual who weighed 113 pounds was found to be at the 27th percentile,
   a. find the *SD*.
   b. find the range.
   c. find the highest score.
   d. find the lowest score.

7. A student has a grade-point average of 3.75, which is at the 90th percentile. The standard deviation is .65. Give the values for
   a. the mean.
   b. the median.
   c. the mode.

8. A certain brand of battery is tested for longevity. One battery is found to last 30 hours and is at the 40th percentile. The standard deviation is 13.52 hours. What is the mean life of the tested batteries?

9. Find the equivalent T scores for the following z scores.
   a. .10
   b. –1.42
   c. 2.03
   d. –.50

10. On a normal distribution of English achievement test scores, with a mean of 510 and a standard deviation of 72, find the T score for each of the following raw scores.
    a. 550
    b. 400
    c. 610
    d. 500

11. On a normal distribution of SAT scores, with a mean of 490 and a standard deviation of 83, find the SAT score for each of the following T scores.
    a. $T = 55$
    b. $T = 32$
    c. $T = 65$
    d. $T = 47$

12. On any normal distribution, the middlemost 68.26% of the cases fall between
    a. which two z scores?
    b. which two T scores?

13. A score that falls at the first quartile (25th percentile) must receive
    a. what z score?
    b. what T score?

True or False. Indicate either T or F for problems 14 through 20.

14. If the z score is zero, the T score must be 50.
15. The standard deviation of the T score distribution must equal the standard deviation of the raw score distribution.
16. A T score of less than 50 must yield a negative z score.
17. All T scores greater than 50 must yield percentiles that are greater than 50.
18. A normal distribution with a mean of 50 and a standard deviation of 10 must have a range that approximates 60.
19. If the mean of a set of normally distributed values is 11.83, the median is also equal to 11.83.
20. If the mean of a normal distribution is equal to 100, with a range of 50, the highest possible score must be approximately 125.

For problems 21 through 25, use Table 5.2.

21. Find the T score for someone whose stanine was 4.18.
22. Find the Wechsler IQ for someone whose NCE was 42.5.
23. Find the z score for someone whose stanine was 6.04.
24. Find the SAT score for someone whose T score was 49.
25. Find the stanine for someone whose NCE was 35.80.

Using material from chapters 2 through 5, indicate what is WRONG with at least one value from the following:

26. A certain distribution of scores has a high value of 25, a low value of 5, a range of 20, a mean, median, and mode of 27, and an SD of 4.
27. A set of sample scores shows a range of 50, a mean of 25, a median of 25, a mode of 25, and an SD of 27.
28. A normal distribution produces a range of 600, a mean of 500, a median of 450, a mode of 400, and an SD of 100.
29. A set of sample scores is skewed to the right. The range is 600, the mean is 500, the median is 600, the mode is 700, and the SD is 50.
30. A set of scores is normally distributed with a mean, median, and mode all equal to 100. The range is 90 and the SD is 20.

Chapter

6

Probability

By the time you arrived at college, you had made literally thousands of probability statements, both consciously and unconsciously. Lurking behind virtually every decision you have ever faced has been an overt or covert probability estimate. This chapter, then, deals with a subject with which you are already thoroughly familiar; this generalization does not apply only to the card players or horse race bettors among you. Every time you cross the street in traffic (Can I make it?), decide whether or not to carry an umbrella (Will it rain?), choose a space to park your car (Is it safe?), or buy one brand of appliance rather than another (Will it last?), you are estimating probabilities. Even your selection of a date (Can I trust this person?) or a college to attend (Will I be accepted?) is affected by the ever present intrusion of probabilities.

Life may be like a box of chocolates, and we really don't know exactly what we're going to get. But even Forrest Gump can make probability predictions, with regard to the selection of chocolates as well as other types of life activities.

Perhaps you and a friend flipped a coin this morning to determine who would buy coffee. You certainly were aware at the time of the probability of your being the one to have to pay. Because of this familiarity with probability, it is therefore fondly hoped that you will know as much about the concept after you finish this chapter as you do now. If asked to specify the exact probability of your having to buy the coffee in the coin-flipping example, you might say "50%" or "fifty–fifty," or perhaps something like "even-up." All these statements are essentially correct, even though they are not stated in the precise terms used by statisticians. Our next step is to take your current knowledge of probability and translate it into statistical terminology.

THE DEFINITION OF PROBABILITY

Probability is defined as the number of times a specific event, s, can occur out of the total possible number of events, t. In general, classic probability theory assumes the following criteria:

1. Each specific event, s, must have an equally likely outcome. That is, in assessing the probability of tossing a head with a fair coin, the probability of a head turning up must be equal to the probability of tossing a tail, since tossing a tail is the only other possible specific event that could occur.

2. The p (probability) assigned to each specific event must be equal to or greater than zero *or* equal to or less than 1.00. Since a p value of zero states that the event cannot occur and a p of 1.00 states that it must occur, no probability value may fall outside these limits.

3. The sum of the probabilities assigned to all the specific events must equal 1.00. For example, if the p of flipping a head is .50 and the p of tossing a tail is .50 (the only two possible outcomes), the sum of these p values must and does equal 1.00.

Classical Model

When the selection is based on t (the total number of *equally likely* and mutually exclusive events) and the number of specific events has the value s, then the probability of s is the ratio of s to t. Using the letter p for probability, we can say that

$$p = \frac{s}{t}$$

Thus, if we take an unbiased coin, the probability of selecting a head from a fair toss is 1/2, since the coin contains a head on one side out of the total of two possible sides, with each side being equally likely (again assuming it is an unbiased coin). Statisticians typically do not express probability statements in fraction form, however. Instead, the fraction is divided out and the probability is written in decimal form:

$$p = \frac{1}{2} = .50$$

"LISTEN UP, GUYS, OF COURSE WE CAN'T WIN, BUT PLEASE COVER THE POINT SPREAD."

Probability theory does have a very practical side.

Thus, the probability of your being correct when calling heads is .50. This value is a constant for each and every coin flip ever made. Even if you have called it wrong for 20 straight flips, that does not increase, or in any way affect, the probability of your being a winner on the 21st flip. Each coin flip is totally independent of every other coin flip. You may remember your long string of losses, *but the coin doesn't remember.* When events are independent of one another, they are just that—independent. They cannot influence or be influenced by one another. The classical model is based on deductive logic. That is, you don't have to observe a long series of coin flips or dice tosses to establish probabilities of this type.

Long-Run Relative Frequency, or Empirical, Model

But what if the events are not equally likely? What if the coin is not fair? What if it has somehow been "loaded" or weighted in such a way that it no longer presents equally likely outcomes? Is probability theory now no longer of use? Quite the contrary! In this case, statisticians speak of the long-run frequency, or empirical, model of probability. This model is based on the rather simple notion that what you see (often enough) is what you get. Thus, the results of the "loaded" coin are examined over a large number of tosses until a fairly constant trend begins to emerge. The more outcomes that are observed, the higher is the likelihood that the results will start stabilizing at some constant value. (See Table 6.1.)

Let's look at the results of observing the outcomes of tossing a coin that has been weighted in such a way that tails are now more likely to occur than are heads. This is typically the way coins are loaded, since it has also been observed that when given the choice, people are more apt to call heads than tails. As can be seen, as the number of repetitions increases, the relative frequency begins to converge toward, in this case, the constant value of .33. Thus, in the long run, heads will lose two-thirds of the time.

The equation for establishing the probability of the relative frequency model is fs (the observed frequency of the specific event) divided by N (the number of observed repetitions).

$$p = \frac{fs}{N}$$

TABLE 6.1 Relative frequency of tossing a head with a single "loaded" coin.

Number of Heads	*Number of Repetitions*	*Relative Frequency*
6	10	.60
8	20	.40
28	100	.28
70	200	.35
331	1000	.33
659	2000	.33

BLAISE PASCAL (1623–1662)

Blaise Pascal is credited with the first clear formulations of probability theory, and his insights continue to be used by statisticians even today. Pascal also invented what many call the world's first computer, a calculating device that is still in existence and can be seen today in a French museum.

Pascal was born in France in 1623. His mother died when he was only three years old. His father, although highly educated, was both dictatorial and, to put it kindly, a little off-beat. He took full charge of young Blaise's education in what could now be called an early experiment in home schooling. He taught Blaise only those things he considered important, and science was definitely not on the list. He specialized in ancient languages, especially Latin. By age 12, however, Blaise found a math book and secretly taught himself geometry. By age 16, he had written a book titled *The Geometry of Conics,* and because of this publication, he was discovered by the great French mathematician, philosopher, and rationalist, Rene Descartes. Although the study of math was seen to be too scientific for his father's tastes, he finally, under great pressure, relented and let the boy study math with Descartes.

In 1642, at the tender age of 19, Blaise invented the first digital calculator, called the Pascaline, a device that directly foreshadowed the modern computer. He also invented the syringe, the hydraulic press, and the barometer (Adamson, 1995).

In the meantime, his sister had become strongly influenced by the Jansenists, an offshoot of Catholicism that had been branded heresy by the Jesuits. Through his sister, Blaise also became a Jansenist and began spending much time debating religious issues. In fact, in 1664, he joined a French monastery, and although most of his work from then on involved religious treatises, he did not totally neglect his math. He wrote a series of replies to the Jesuits and argued that the only perfect knowledge came through Christian revelation. Reason can just go so far, but faith has no limits (Mortimer, 1959).

Pascal also retained his faith in numbers, and he continued to analyze the problem earlier posed by his gambling friend, the Chevalier de Mère (see page 16). He began to wonder how the pot should be divided if a game had to end early. Pascal said that the expectation of each player should be based on the number of specifically favorable outcomes divided by the total possible number of outcomes. That is, s(specific)/t(total). To ease the computations, Pascal devised his famous triangle where each number in the triangle is based on the sum of the two numbers just above it.

Using Pascal's triangle, notice that the sum of the two numbers just above each number equals that number. For example, take the number 2 in row 3. The numbers above are 1 and 1. Or take the number 20 in row 7, and notice that the two numbers above it are 10 and 10. (To further amplify, please turn to the special section on probability at the back of the book and look over the

Row	*N* coins		Combinations
1		1	
2	1	1 1	2
3	2	1 2 1	4
4	3	1 3 3 1	8
5	4	1 4 6 4 1	16
6	5	1 5 10 10 5 1	32
7	6	1 6 15 20 15 6 1	64
8	7	1 7 21 35 35 21 7 1	128

Pascal's Triangle

binomial function.) For each row, starting with row two, the number of coins and the probabilities for each combination of heads or tails are shown in that row. For example, starting in row two for one coin, the probability is identical for either a head or a tail and the total number of combinations must be 2. In row three with two coins, the number of combinations adds up to 4, and so on. If one were a gambler, the probabilities for the various coin tosses become instantly apparent and the bet could be adjusted accordingly. For example, in row three, using two coins, there is a total of four combinations, one head, a head and a tail, a tail and a head, or one tail. That means that there are four ways—HH, HT, TH, and TT—that the two coins could be flipped. If, as was assumed, the probabilities of heads and tails are equally likely, the *probabilities* of these outcomes are then 1/4, 1/2, 1/2, and 1/4. It is rather strange that although gambling had been popular for thousands of years, nobody had formulated anything concerning the regularity of various events (even though they must have been known by inveterate gamblers, at least at the empirical level). For the ancients, events were mysterious and governed by the whim of the gods. Later, the Christians believed that events occurred as a result of God's will, and it would be considered heretical to theorize that events could occur according to the blind laws of mathematical probability. Thus, people like Pascal (and even Descartes) had trouble with church doctrine. In a famous line, Pascal stated that "the heart hath its reasons which reason knows not of." He meant by this that when theological issues are involved, the most that reason can do is allow us to make a wager, a bet that God exists. Pascal's bet was—"If God does not exist, one will lose nothing by believing in him, while if he does exist, one will lose everything by not believing"—an ironic twist, coming as it does from a probability theorist who had a strong belief in God and in mathematics (Bernstein, 1997). Pascal died at the early age of 39, and yet despite the brevity of his life, he has gone down in statistical history as one of its most brilliant thinkers, a mathematical genius whose efforts created the foundation on which modern probability theory rests.

This equation is based on the assumption that N is indeed a large number. Looking back at Table 6.1, you would have really been faked out if you had established the probability after observing only 10 trials. Despite the .60 value after only 10 trials, in the *long run* the probability stabilized at .33. Even against a tail-loaded coin, by calling heads you might win in the short run, but in the long run you'd be wiped out. As many a gambler has ruefully said, "You can beat a race, but not the races."

Thus, the relative frequency model does not assume equally likely outcomes, but it does demand mutually exclusive events; that is, each repetition still produces a distinctly separate outcome. The relative frequency model can aid the statistician in trying to assess future probabilities that are based on past histories. For example, a quarterback on a football team breaks his leg, and the coach wants a prediction as to when he can put him back in the lineup. Certainly, we can't simply count the number of times this player has broken his leg, but we can consult with others and establish a rate of recovery time based on a large number of other athletes with similar injuries. The greater the similarity among the cases, the higher will be the accuracy of the prediction for any specific case. This is essentially the logic employed by insurance companies. Obviously, they can't count the number of times you died at age 20, but they can make probability statements based on the similarities they find between you and 20-year-olds who have died.

Examples of the relative frequency model are common in sports. Thus, baseball managers plan pitching strategies around the past experience presented by each hitter, and football coaches determine defenses on the basis of what the opposing team usually does in a given situation. In the National Football League, empirical data gathered over the years have shown that on third down, the probability of making a first down is .80 when there are 3 yards or less to go but drops to a mere .20 when there are 8 or more yards to go (*New York Times,* September 6, 1994, page B12). Similarly, although it may not sound intuitively logical, in major league baseball if a pitcher *walks* the first batter in an inning, that batter has a better than even chance of scoring, typically running as high as .70. Thus, if you were to make an even bet that the base runner would score, and if you made enough bets, in the long run you'd have to be a winner.

Conditional Probability

The difference between independent and conditional probabilities can be dramatically illustrated by considering the grisly game of Russian roulette. If a six-chamber revolver is used and a bullet is placed in just one of those chambers, the probability of having the gun go off is 1/6 (.167). If the cylinder is fairly spun between trigger pulls, the probability remains at 1/6, since the act of spinning the cylinder should keep the events independent. However, if the cylinder is not spun, then the probabilities become conditional. The probability of the gun's going off on the first pull is 1/6 (.167), on the second pull 1/5 (.20), on the third pull 1/4 (.25), on the fourth pull 1/3 (.333), on the fifth pull 1/2 (.50), and finally on the last pull (and it would definitely be the last pull), the probability becomes a booming 1.00.

Gambler's Fallacy

Gamblers often convince themselves after a long series of losses that the law of averages makes them more certain to be winners. Sometimes after a losing streak, a gambler becomes certain that the probability must swing the other way, to the point, finally, of assuming the next bet is a "sure thing," $p = 1.00$. The gambler is assuming, incorrectly, that having lost the previous bet is somehow going to change the outcome of the next bet, or that the probabilities are conditional—when in fact they are independent. This is known as the **gambler's fallacy,** and we researchers and research consumers must be just as aware of it as the inveterate gambler should be. In fact, if a coin were to come up heads 100 straight times, it is now more, not less, likely to again come up heads on the next toss . . . because on a probability basis the coin is probably loaded.

And don't ever be fooled into thinking that after you're lucky enough to win some money in a casino that you are now playing with "their" money. Once you win, it's no longer "their" money. It's your money, and if you keep playing long enough, you will lose it and then it will be "their" money. Remember, the odds are set so that in the *long run* the house has to win and you have to lose.

While it is true that if an unbiased coin is flipped an infinite number of times, heads will turn up 50% of the time, that does not change the probability for any single coin flip. The so-called law of averages is true only in the long run. This is why casino operators in Atlantic City and Las Vegas have to be winners. The casino is making many more bets than is any individual player. The overall probability in favor of the casino is much more likely, then, to conform to mathematical expectation than is that for the individual player, who, though seemingly involved in an extended series of bets, is really only playing in a short run. As Greg Kimble puts it, "the laws of chance or principles of probability apply, not to single events but to large numbers of them" (Kimble, 1978). Also, as John Scarne, an adviser to many casinos, counsels the individual player, "You can beat a race, but not the races" (Scarne, 1961). As for the "law of averages," Scarne says that gamblers "don't understand that the important word in that phrase is not 'law' but 'averages.' The theory of probability is a mathematical prediction of what may happen on the average, or in the long run, not a law which states that certain things are inevitable." The best statistical advice is to stay away from the casinos. In the long run you have to lose, and in the short run, you may even risk becoming addicted. As Daniel Smith points out,"Like other addicts, the problem gamblers begin to self-medicate. Just as alcoholics reach for a drink, problem gamblers turn to gambling to ease the effects of a bad day or to lift an unsettled mood" (Smith, 2000, p. 11). For those gamblers who believe they have the perfect system, the casinos have one word—welcome.

People can certainly have hot streaks for a short run, but over the long run the more typical probabilities have a way of asserting themselves. For example, you often see a major league baseball player hitting for an average as high as .450 over the first few weeks of the season, but nobody in the history of the game has ever batted .450 for an entire season. In fact, in the spring of 1994 there were so many incredibly high batting averages that on April 25 Red Sox slugger Mo Vaughn was hitting .386 and yet was ranked only 7th in the American League.

To sum up, then, so-called laws of chance are really descriptions of what generally happens in the long run, not enforcers of what must happen in the short run. This is why the gambler who drives to Atlantic City in a $60,000 Mercedes often returns home as a passenger in a $400,000 bus.

Probability Versus Odds

It is important to distinguish probability from the related concept of **odds.** Probability states the number of times a specific event *will* occur out of the total possible number of events; odds, on the other hand, are usually based on how often an event will *not* occur. For example, if we toss a normal, six-sided die, the probability of any one of the faces turning up is one out of six.

$$p = \frac{s}{t} = \frac{1}{6} = .17$$

Odds, on the other hand, are typically stated in terms of the chances against a specific event occurring. (In this case, the odds are 5 to 1—there are five sides that will not come up for each single side that will.)

Example Suppose that we wish to determine the probability of selecting a diamond from a normal deck of playing cards, that is, 52 cards divided into 4 suits—diamonds, clubs, hearts, and spades. The probability of the specific event, a diamond, is, therefore, one out of four (the total number of suits is four).

$$p = \frac{1}{4} = .25$$

The odds against a diamond selection, on the other hand, are 3 to 1, since there are three suits you don't want to select for the single suit you do want to select.

To establish the relationship between probability and odds, we can run through two more conversions. If the probability of an event occurring is 1 out of 20, or .05, the odds against it are 19 to 1. If the probability of an event occurring is 1 out of 100, or .01, the odds against it are 99 to 1. To check, notice that when you sum the odds (99 + 1), the result must equal the denominator in the probability statement, in this case 100.

PROBABILITY AND PERCENTAGE AREAS OF THE NORMAL CURVE

The normal distribution is often referred to as the normal probability curve, since the characteristics of the normal curve, as outlined in Chapter 4, allow its use in making probability statements. The curve, as presented so far, has been set forth in terms of percentage areas. For example, we learned that roughly 68% of the cases fall within ±1 standard deviation unit of the mean, and so on. The *z* score table (Appendix Table A) is outlined in terms of percentage areas. For this reason, all the *z* score problems given so far have been based on percentages, or frequencies of occurrence. When we did all those problems concerning the area between, above, or below given *z* scores, our answers were always expressed as frequencies—32% fell here, 14% there, and so on.

There is, however, a direct and intimate relationship between percentage and probability. Specific percentage areas are based on the fact that the total percentage area equals 100%. When expressed in decimal form, probability is based on the fact that the total number of events equals 1.00. Thus, the total number of events in both cases is, in fact, 100% of all the events. This means that 25% of the possible events is equivalent to a probability of .25 or 25/100. Or, looking at it the other way, if an event has a probability of .25, this means it will occur 25% of the time. As percentages vary from 100 to zero, so do probabilities vary from 1.0 to zero. Thus, if an event is certain to occur—that is, will occur at a frequency of 100%—its probability of occurrence is 1.00. If an event cannot occur, its probability is zero.

Converting Percentages to Probability Statements

The conversion of percentages to probability statements is quick and easy. We simply divide the percentages of cases by 100 and *drop the percentage sign.*

$$p = \frac{\%\text{ of cases}}{100\%}$$

Thus, if the frequency of occurrence is 90%, then

$$p = \frac{90}{100} = .90$$

We divide through by 100 because we are indicating the specific event in percentage terms (the numerator) and thus must also express the total possible number of events (the denominator) in percentage terms, which is, of course, *100% of all the events.*

It's as simple as that. All you need is both ends of a pencil—the lead point to move the decimal two places to the left and the eraser to get rid of the percentage sign. You must erase the percentage sign. Do *not* write a probability value like .10%. This makes no sense. The problems will be solved *either* in percentage terms *or* in probabilities, so it is not necessary to create new combinations.

COMBINATIONS, PERMUTATIONS, AND STATISTICAL INFERENCE

As noted in the Preface, this book does not include problems based on combinations and permutations. Although a more thorough understanding of probability may involve these topics, the choice was made to jump you ahead directly into the area of inferential statistics and hypothesis testing, using traditional inferential techniques such as t, r, F, and chi square. You can read the literature in your field without an extensive mathematical grounding in combinations and permutations, but you cannot make sense of the literature without familiarity with the statistical tests of significance.

To give you at least a nodding acquaintance with the general notions involved, however, consider the following.

Combinations refer to the number of ways a given set of events can be selected or *combined,* without regard to their order or arrangement. Suppose, for example, that we have six groups of subjects, and we wish to compare each group with each of the other groups. How many two-group comparisons are required?

$$C_{nr} = \frac{n!}{r!\,(n-r)!}$$

where C_{nr} = the total number of combinations of n events, taken r at a time, and ! = factorial, or $n(n-1)\,(n-2)(n-3)\ldots$ Thus,

$$C_{nr} = \frac{6!}{2!\,(6-2)!} = \frac{6 \times 5 \times 4 \times 3 \times 2 \times 1}{(2 \times 1)(4 \times 3 \times 2 \times 1)}$$

$$= \frac{720}{(2)(24)} = \frac{720}{48}$$

$$= 15$$

There are a total of 15 ways to compare all the pairs taken from six groups.

A Winning Combination?

In those states with lotteries, the largest payoff (Big Bucks, Lotto Bucks, Megabucks, etc.) typically results when the drawing is based on a series of consecutive numbers, taken r ways. For example, a player might be given a card with the numbers 1 through 36 and be asked to select any six different numbers. The total possible combinations, therefore, are

$$C_{nr} = \frac{n!}{r!\,(n-r)!} = \frac{36!}{6!\,(36-6)!} = \frac{36!}{(6!)(30!)} = 1,947,792$$

At a dollar a card, then, for an investment of \$1,947,792, you could cover every possible combination.

Now (assuming that the logistical problem, no small problem, of buying and filling out almost 2 million cards could be solved), when the jackpot reaches \$3 million, \$4 million, or even \$10 million, is it necessarily a profitable venture to invest this much money to ensure a win? Not really, since it is possible that many players could have winning cards, in which case the jackpot would be split. Playing all the combinations does assure you of a win, but in no way does it guarantee the size of the win. Also, any change, even a seemingly slight change, in the total number of items to be selected from can dramatically alter the chances of winning. In one state, the rules for the lotto-type game were changed from the player selecting 6 numbers out of 44 possible numbers to the player having to select 6 numbers out of 48. This change almost doubled the chance of losing. Using the preceding formula, 6 out of 44 yields 7,074,468 possible combinations, where selecting 6 out of 48 jumps the possible combinations to 12,277,228. But then again, the probability of winning drops to a flat-out zero if the player fails to buy a ticket. You have to be in it to win it!

The lure of the lottery is that it provides instant and effortless wealth. But the probability of winning is so small that in reality it becomes a tax on the stupid. You may have to be in it to win it, but if you refuse, you just can't lose.

Permutations refer to the number of *ordered* arrangements that can be formed from the total number of events. For example, assume that an intramural softball league has eight teams. We wish to determine the total possible number of ways the teams could finish:

$$p_n = n!$$

where p_n = the total number of ordered sequences of n events and n = the total number of events. Thus,

$$p_n = 8! = 40{,}320$$

There are a total of 40,320 possible orders of finish for an eight-team softball league.

z Scores and Probability

All of those *z* score problems, where we calculated percentage areas, can now be rephrased as probability problems. The only thing that will change is the manner in which we express the answers.

Example Let's say that we are given a normal distribution with a mean of 500 and a standard deviation of 87. We are asked to find the percentage of cases falling between the raw scores of 470 and 550. The same problem can be translated into probability form: State the probability that any single score falls somewhere between the raw scores of 470 and 550.

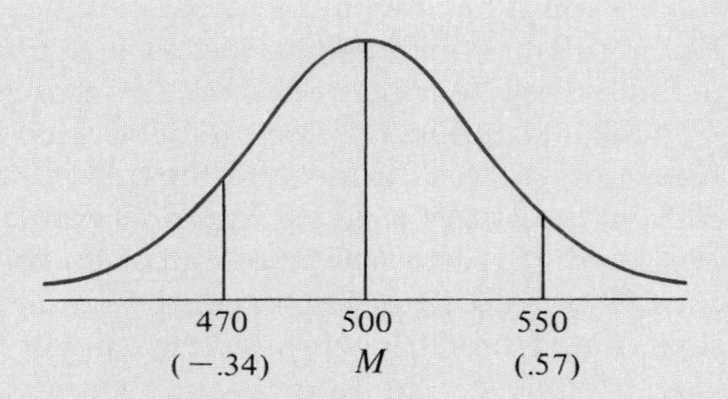

We draw the curve and calculate the two z scores:

$$z_1 = \frac{X - M}{SD} = \frac{550 - 500}{87} = .57$$

$$z_2 = \frac{X - M}{SD} = \frac{470 - 500}{87} = -.34$$

Then, we look up the percentages for each z score in Appendix Table A. Since the two scores fall on either side of the mean, we add the percentages (13.31% + 21.57%) for a total of 34.88% (Rule D1, page 86). This is the answer to the original frequency question. To answer the probability question, we simply divide the percentage by 100.

$$p = \frac{34.88}{100} = .348 = .35$$

Example Another frequency-type problem can be translated to probability. With a mean of 68 and a standard deviation of 3.50, what is the probability that any single score will fall between 70 and 72?

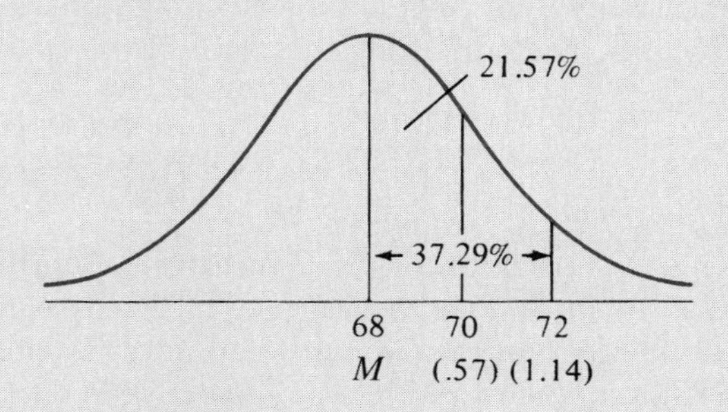

The two z scores are

$$z = \frac{72 - 68}{3.50} = 1.14$$

$$= \frac{70 - 68}{3.50} = .57$$

From Appendix Table A, we find 37.29% for 1.14 and 21.57% for .57. We subtract the percentages (Rule D2) to get 15.72%. Then, it can be found that the probability is as follows:

$$p = \frac{15.72}{100} = .157 = .16$$

—**Exercise 6a** On a normal distribution with a mean of 25 and a standard deviation of 2.42, find the following.

1. The probability that any single score will fall above a raw score of 27
2. The probability that any single score will fall below a raw score of 30
3. The probability that any single score will fall between raw scores of 20 and 28
4. The probability that any single score will fall between raw scores of 21 and 24
5. The probability that any single score will fall between raw scores of 29 and 32

Probability and the Single Score

Up to this point we have talked about the probability of finding someone scoring between this score and that score, or above a certain score, or below a given score, but what about the probability of a specific score being obtained? To avoid having to solve the normal curve equation for the ordinate, and yet still arrive at very close approximations, we can make use of the characteristics of a continuous distribution. In the z distribution, or for that matter any continuous distribution, we tend to round the values at some convenient level. Because the number of decimal places chosen for any continuous measure is in fact arbitrary, statisticians describe each continuous value as falling within an interval of values. Thus an IQ of, say, 115 is assumed to be somewhere in an interval ranging from 114.50 to 115.50. If one wants the probability of someone scoring exactly an IQ of 115, we can arrive at an extremely close approximation by getting the z scores for 114.50 (which with a mean of 100 and an *SD* of 15 translates to a z of .97) and 115.50 (which translates to a z of +1.03). We look up the frequencies for each of these z scores, 33.40% and 34.85%, and then obtain the frequency between them by subtraction, 1.45%, and convert to probability: $1.45/100 = .0145$, or .01. By now you realize that z scores falling closer to the mean have a higher probability of occurring than those falling farther out in the tails. Thus, the IQ score of 115 should have a higher probability than an IQ of, say, 135. To find the probability for an IQ of 135, find the z

scores for 134.50 (2.30) and 135.50 (2.37), look up the percentage areas for each (48.93% and 49.11%), subtract (.18%), and then convert to the tiny probability of only .18/100 = .0018, or .002.

—**Exercise 6b** On a normal distribution of adult male height scores, with a mean of 69 inches and an *SD* of 3 inches, find the approximate probability of selecting at random an adult male who is exactly

a. 72 inches tall
b. 65 inches tall
c. 62 inches tall
d. 70 inches tall

COMBINING PROBABILITIES FOR INDEPENDENT EVENTS

Two rules are of extreme importance when probabilities are combined. We'll call them the "ADD-OR" rule and the "MULT-AND" rule.

ADD-OR Rule

To obtain the probability of one event OR another event occurring, assuming mutually exclusive events, the separate probabilities should be added (hence the label "ADD-OR"). For instance, if you were to flip two coins and wanted the probability of obtaining a head or a tail (which, of course, is a certainty), you would simply add the two separate probabilities:

$$p = .50 + .50 = 1.00$$

If the problem involves playing cards instead of coins, we can just as easily find the "or" probabilities of selecting, say, a king of spades, or a ten of hearts. Since there is only one king of spades and only one ten of hearts in a deck of 52 cards, the probability of selecting one or the other would be

$$p = 1/52 + 1/52 = 2/52 = .0384, \quad \text{or rounded} \quad = .04$$

The same rule also applies to z scores. Suppose that we wished to find the probability of selecting at random someone whose IQ was below 85 or someone whose IQ was above 115. Since the IQ distribution has a mean of 100 and an *SD* of 15, the two z scores in question would be –1.00 and +1.00. Using Appendix Table A, we find 50.00% – 34.13%, or 15.87%, falling below a z score of –1.00, and 50.00% – 34.13%, or 15.87%, falling above the z of +1.00. Since this is an "either-or" situation, ADD the percentages (15.87% + 15.87% = 31.74%). To finish the problem, convert the percentage to a probability value (see page 121).

$$p = .3174 \text{ or } .32$$

Thus, there are 32 chances out of 100 that the person selected will have an IQ either below 85 or above 115.

MULT-AND Rule

When attempting to find the probability of one event AND another event occurring, we multiply the separate probabilities (thus the label MULT-AND). For example, if we were to flip a coin twice and wanted the probability of getting a head on one toss AND a tail on the second toss, the MULT-AND rule says

$$.50 \times .50 = .25$$

There are thus 25 chances out of 100 that on two coin flips you could obtain both a head and a tail.

Going back to our playing card example, the same rule applies. For example, assume that we wanted to find the probability of selecting an ace and a spade from the deck of cards. We know that there are 4 aces in the deck of 52 cards, so its probability is $4/52 = 1/13$. We also know that there are 13 spades in the deck, yielding a probability of $13/52 = 1/4$. By multiplying, we get

$$p = 1/13 \times 1/4 = 1/52$$

which is the probability of selecting an ace and a spade, or, in this case, the ace of spades.

In using cards or dice, of course, we are applying the classical model of probability. This rule, however, applies whether the classical model or the empirical model is used, as long as the events are independent of each other. For example, since it is known that every individual is genetically unique, scientists are now able to identify suspects in criminal investigations by the use of DNA "fingerprinting." The procedure involves the extraction of DNA, called a "probe," from small bits of blood, hair, tissue, or semen found at the scene of a crime and the matching of that specimen to the DNA taken from the cells of a suspect. The statistician then determines the probability of a false match. Since it has been determined that in the population at large, getting a match with one probe (Probe A) has a frequency of 1 out of every 1000, AND from a second probe (Probe B) a frequency of 5 out of 1000, AND from Probe C 100 in a 1000, AND from Probe D 200 in a 1000; then the separate probabilities .001, .005, .10, and .20 are multiplied, and it becomes obvious that the DNA of only one person in every 10 million persons would produce this particular four-probe print $(.001)(.005)(.10)(.20) = .0000001$. As the number of probes increases, the probability continues to decline, but the result is still a probability statement and not an affirmation of absolute truth.

We can, of course, use the same rule on the z distribution. For example, to find the probability of selecting someone whose IQ is below 85 and on the next draw someone whose IQ is above 115, you multiply the separate probabilities:

$$.1587 \times .1587 = .025, \quad \text{or} \quad \text{rounded} = .03$$

There are, therefore, only about 3 chances out of 100 that one could randomly select someone whose IQ is below 85 and on the next draw someone whose IQ is above 115.

Bill the Accountant. Bill was a business major in college, intelligent and rather conservative politically, but he did have something of a wild side. He liked to party and he liked to drink. During one of his sprees, the campus security caught him repeatedly driving his car across the football field and leaving huge ruts in his wake. For that infraction, Bill was suspended for one full semester. He did return to college and later graduated. Which of the following has a higher probability?

a. Bill is now working as an accountant.
b. Bill is now working as an accountant and his continued drinking problem has cost him several jobs over the years.

The answer is that answer a is more probable, since a single statement (or event) is always more probable than the joint combination of several events. Notice that answer b is saying that Bill is an accountant AND that Bill still drinks AND that his drinking has interfered with his career. It's like asking, Which is more probable—rolling a six with one die or getting three sixes on three consecutive rolls? If you can't answer that one, don't go near a casino.

Logic of Combinations

If you're having a problem understanding these combination concepts, relax. You're not alone. Just take a short time-out to think about the logic of these combinations. One reason for difficulty in this area seems to be that the phrase "one event AND another event" sounds intuitively very much (probably too much) like a problem in addition (like, How much is 4 AND 4?). The key to this puzzle is to try to remember that "either-or" problems increase probability (through addition, as in .50 + .50 = 1.00), whereas "one-event-and-another" problems decrease probability (through the multiplication process, as in .50 × .50 = .25). Since probability values cannot be greater than 1.00, in this latter situation you're in effect multiplying fractions, which will then reduce the size of the resulting value. The logic of this scenario is that one event OR another increases the likelihood, whereas one event AND another decreases the likelihood.

Although undoubtedly apocryphal, the following should help to dramatize the logic of these combinations. It seems that a young couple had just bought their first home and immediately insured it with a "fire-and-theft" policy. A week later the house was burglarized and all the rooms were trashed. The couple called the insurance company, only to find out that the fire-AND-theft policy didn't cover them. In fact, for them to collect, a burglar would have had to break into their home while it was ablaze—certainly an unlikely set of circumstances. Now, if they had only bought fire-OR-theft policy, a more likely combination, they would have been covered.

A REMINDER ABOUT LOGIC

If you have been able to do the problems presented throughout this chapter, you should have no difficulty with the problems at the end. You must think each one out carefully before plunging into the calculations. Remember, this course has more to do with logic than with math. When appropriate, always draw the curve and plot the values on the abscissa. This small step can prevent big logic errors. For example, if the problem states that someone has a score of 140 and is at the 25th percentile, it is illogical to come up with a mean of less than 140 for the distribution. Or, if a problem asks for a probability value for an event occurring between scores that are both to the right of the mean, it is illogical to come up with a value of .50 or higher. With the curve drawn out and staring back at you, you cannot make these kinds of logical errors.

If you can do the problems so far, you may now revel in the luxury of knowing that you have a working knowledge of probability—at least of probability as it relates to the normal curve. For the next few chapters, that is all you need to know about probability. Estimating population means, deciding whether or not to reject the chance hypothesis, establishing the alpha level—all these depend on a working knowledge of the probability concept. Once you have mastered this chapter, you are ready for inferential statistics.

SUMMARY

Probability, p, is equal to the number of times a specific event can occur out of the total possible number of events, s/t. Statisticians typically divide out this fraction and express probability in decimal form. Thus, a probability of 1 out of 2, 1/2, is written as .50.

Whenever events are independent of one another, as in a succession of dice throws or coin flips, the result of any one of these events is not conditional on the result of any of the preceding events. Assuming that independent events are conditional is known as the gambler's fallacy.

Whereas probability specifies the outcome in terms of the number of times an event will occur out of the total possible number of events, odds are stated in terms of how often an event will not occur as compared to how often it will. Thus, if the probability of an occurrence is one out of two or .50, the odds against it are 1 to 1.

The percentage areas of the normal curve can be used to obtain probability values. If an event occurs at a frequency of 50%, the probability that the event will occur is 50 out of 100, or .50. To obtain probability values from the percentages given in Appendix Table A, simply move the decimal two places to the left (divide by 100) and erase the percentage sign.

All the z score problems previously solved in frequency (percentage) terms can be quickly and easily translated into probability statements. Thus, if 68% of the cases fall between z scores of ± 1.00, then the probability that a single case will occur in that same area is .68.

When probabilities are combined, two rules are needed: (1) to get the probability of one event or another on successive occasions, add the separate probabilities, and (2) to get the probability of one event and another on successive occasions, multiply the separate probabilities.

Key Terms gambler's fallacy
odds
probability

PROBLEMS

1. A certain event occurs one time out of six due to chance.
   a. What is the probability of that event occurring?
   b. What are the odds against that event occurring?
2. A given event occurs at a frequency of 65%. What is the probability that the event will occur?
3. Assume that a deck of 52 cards is well shuffled.
   a. What is the probability of selecting a single card, for example, the ace of hearts?
   b. What are the odds against selecting that single card?
   c. If that card were selected and then replaced in the deck, and the deck shuffled, what is the probability of selecting it a second time?

   Assume a normal distribution for problems 4 through 14.
4. A certain guided missile travels an average distance of 1500 miles with a standard deviation of 10 miles. What is the probability of any single shot traveling
   a. more than 1525 miles?
   b. less than 1495 miles?
   c. between 1497 and 1508 miles?
5. The mean weight among women in a certain sorority is 115 pounds with a standard deviation of 13.20 pounds. What is the probability of selecting a single individual who weighs
   a. between 100 and 112 pounds?
   b. between 120 and 125 pounds?
6. On the WISC III, the mean on the arithmetic subtest is 10.00 with a standard deviation of 3.00. What is the probability of selecting at random one person who scored above 11.00 *or* another person who scored below 8.00?
7. Using the data from problem 6, what is the probability of selecting at random two persons, one of whom scored above 11.00 *and* another who scored below 8.00?
8. The mean systolic blood pressure taken from the population of student nurses in a large metropolitan hospital was found to be 120 mm Hg (millimeters of mercury) with an *SD* of 7.00. What is the probability of selecting at random someone whose systolic pressure was 130 or higher?

9. The mean on the Woodcock Johnson Test of Cognitive Ability (WJ-RCOG) is 100 with an *SD* of 15. What is the probability of selecting at random someone who scored between 92 and 110?

10. The mean of the Need Aggression scale of the Edwards Personal Preference Schedule (EPPS) is 11.55 with an *SD* of 4.50. What is the probability of selecting at random someone who scored 9 or less?

11. The mean number of filled cavities found among a large group of prison inmates at a county correctional center was 9.5 with an *SD* of 3.15. What is the probability of selecting at random someone who had more than 12 filled cavities?

12. A group of young women trying out for the college softball team is asked to throw the softball as far as possible. The mean distance for the group is 155 feet with a standard deviation of 14.82 feet. What is the probability of selecting at random one woman who can throw the ball
    a. 172 feet or more?
    b. 160 feet or less?
    c. between 145 and 165 feet?

13. The freshman basketball coach's office has a doorway that is 77 inches high. The coach will recruit only those student athletes who have to duck to enter the office. Of all the freshmen who wish to play basketball, the mean height is 75 inches with a standard deviation of 2.82 inches. What is the probability that any single individual in this group will be recruited; that is, what is the probability of any of the would-be recruits standing 77 inches or taller?

14. Among the shoppers at a certain women's store, dress sizes are normally distributed around a mean of 12.06. The standard deviation is 3.03. What is the probability of a customer entering the store who needs a size 18 or larger?

15. A certain school district has a population of 500 high school seniors, 275 girls and 225 boys. On a single draw, what is the probability of selecting
    a. a boy?
    b. a girl?
    c. a boy or a girl?

16. Using the data from problem 15, what is the probability, on successive draws, of selecting both a boy and girl?

17. In tossing a single coin and a single die, what is the probability of getting
    a. a head on the coin and a three on the die?
    b. a tail on the coin or a three on the die?
    c. heads or tails on the coin and a six on the die?

18. At a certain racetrack, there are a total of eight horses in a given race. The horses are all alleged to be evenly matched.
    a. What is the probability of winning your bet by selecting a single horse that will *win* the race?

b. By betting that single horse to win, what is the probability that you will *lose*, that is, that one of the other horses will win?
c. What is the probability of winning your bet if you buy a ticket on a given horse to place, that is, to come in either first or second?
d. What is the probability of winning your bet if you buy a ticket on a given horse to show, that is, to come in either first, or second, or third?

19. At a certain casino, there are a large number of one-armed bandits (slot machines). On every slot machine there are three wheels that operate independently, and on each wheel there is a series of seven pictures of various kinds of fruit—an orange, a grape, a banana, and so on.
a. What is the probability that on a single spin of the wheel three oranges will appear?
b. What is the probability that on a single spin of the wheel no oranges will appear?

Fill in the blanks in problems 20 through 24.

20. When an event cannot occur, its probability value is ____.

21. When an event must occur, its probability value is ____.

22. The middlemost 50% of the scores in a normal distribution fall in such a way that ____% of the scores lie to the left of the mean.

23. On a normal distribution, ____% of the scores must lie above the mean.

24. The ratio *s/t* (specific event over total number of events) defines what important statistical concept?

True or False. Indicate either T or F for problems 25 through 32.

25. If a given event occurs at a frequency of 50%, its probability of occurrence is .50.

26. Probability values can never exceed 1.00.

27. If a coin is flipped 10 times and, by chance, heads turn up 10 times, the probability of obtaining heads on the next flip is .90.

28. A negative z score must yield a negative probability value.

29. The more frequently an event occurs, the higher is its probability of occurrence.

30. To find the probability of selecting one event or another, the separate probabilities should be added.

31. To find the probability of selecting one event and another, the separate probabilities should be multiplied.

32. Probabilities may take on negative values only when the events are occurring less than 50% of the time.

UNIT

II

Inferential Statistics

Chapter

7

Statistics and Parameters

All of us have, at one time or another, been warned about jumping to conclusions, about forming hasty generalizations. "One swallow doesn't make a summer," as the old adage has it. The statistician puts this message differently: "Never generalize from an N *of one." Or, as someone once said, "There are no totally valid generalizations—including this one!" All of this is good advice. It is best to be alert to the dangers of the "glittering generality," which sometimes seems to have a validity of its own, even though it is based on scanty, or even fallacious, evidence. Hasty conclusions and unfounded generalizations are indeed dangerous—they often are used to produce "the big lie."*

However, some generalizations are safer than others. For example, you are playing cards and betting against the chance that your opponent will draw an ace from the deck. You will undoubtedly feel more comfortable about your bet if you are holding two aces in your hand than if you are holding none. Furthermore, you will positively be oozing confidence if you happen to be holding three aces. (Of course, if you have all four aces, nothing is left to chance and the probability of your winning the bet is a perfect 1.00.)

In the game of predictive, or inferential, statistics, the statistician rarely, if ever, holds all four aces. Therefore, every inference must be couched in terms of probability. Inferential statistics is not designed, and could not be, to yield eternal and unchanging truth. Inferential statistics, however, does offer a probability model. When making a prediction, the statistician knows beforehand what the probability of success is going to be. This does give the statistician a tremendous edge over the casual observer.

GENERALIZING FROM THE FEW TO THE MANY

The main thrust of inferential statistics is based on measuring the few and generalizing to the many. That is, observations are made of a small segment of a group, and then, from these observations, the characteristics of the entire group are inferred.

For many years, advertisers and entertainers asked, "Will it play in Peoria?" The question reflected their tacit belief that if an ad campaign or showbiz production were successful in Peoria, Illinois, it would be successful throughout the United States. Peoria was seen as a kind of "magic" town, whose inhabitants echoed precisely the attitudes and opinions of the larger body of Americans in general. The importance of the question for us lies not in its acceptance that the citizens of Peoria truly reflect the American ethos but in its implication of a long-held belief that by measuring the few, a picture of the many will clearly emerge.

KEY CONCEPTS OF INFERENTIAL STATISTICS

Population

A **population,** or universe, is defined as an entire group of persons, things, or events having at least one trait in common. The population concept is indeed arbitrary, since its extent is based on whatever the researcher chooses to call the common trait. We can talk of the population of blondes, or redheads, or registered voters, or undergraduates taking their first course in statistics, or whatever we choose as the common trait. Furthermore, since the definition specifies *at least* one trait in common, the number of shared traits can be multiplied at will. It is possible to speak of a population of blond, left-handed, 21-year-old, male college students majoring in anthropology and having exactly three younger siblings.

Adding too many common traits, however, may severely limit the eventual size of the population—even to the point of making it difficult to find anyone at all who possesses all of the required traits. As an example, consider the following list of traits specified to qualify for a certain scholarship: "Must be of Italian extraction; live within a radius of 50 miles from Boston; be a practicing Baptist; be attending a college in North Dakota; be a teetotaler; and be planning to major in paleontology." We read such a list in awe and wonder if anyone ever qualifies. Obviously, the more traits being added, the more you limit the designated population, but as your population becomes increasingly limited, so too does the group to which your findings can rightfully be extrapolated.

You can't study the population of 6-year-old females living in Greenwich Village, New York, and then generalize these findings to include 50-year-old male rug merchants working the streets of Teheran. Nor should you do a study on albino rats learning their way through a maze and then quickly extrapolate these findings to cover the population of U.S. college sophomores.

A population may be either *finite,* if there is an exact upper limit to the number of qualifying units it contains, or *infinite,* if there is no numerical limit as to its size. Also, when all members of the population have been included, it is then called a *census* and the values obtained are called census data.

Parameter

A **parameter** is any measure obtained by having measured an entire population. If we gave IQ tests to every single college student in the country, added the IQ scores, and divided by the total number of students, the resulting mean IQ score would be a parameter for that *specific* population, that is, the population of college students. Furthermore, we could then compute the median, mode, range, and standard deviation for that vast distribution of IQ scores. Each of the resulting values would be a parameter.

Obviously, when a population is of any considerable size, parameters are hard to get. It would be difficult, if not impossible, to measure the IQ of every American college student. Even if it were possible, the time and expense involved would be enormous. Remember, to obtain an official parameter, every student would have to be found and given the IQ test. Because they are often so difficult to obtain, most parameters are estimated, or inferred. There are very few situations where true parameters are actually obtainable. As was pointed out in Chapter 1, an exception is in the area of political polling, where the results do become vividly clear, but not until after the election. One final point is that if a parameter is known, there is no need at all for inferential statistics—simply because there's nothing left to predict.

Sample

A **sample** is a *smaller* number of observations taken from the *total* number making up the population. That is, if Omega College has a total student population of 6000, and we select 100, or 10, or even 2 students from that population, only the students so selected form our sample. Even if we select 5999 students, we still only have a sample. Not until we included the last student could we say we had measured the total population. Obviously, for purposes of predicting the population parameters at Omega College, a sample of 5999 is far more accurate than is a sample of 10. After all, with 5999 students already measured, how much could that last student change the values of the mean or the standard deviation?

Therefore, *other things being equal* (which they rarely are), the larger the sample, the more accurate is the parameter prediction. Be very careful of that statement, since "other things being equal" is a powerful disclaimer. Later, we will see that where other things weren't equal, some positively huge samples created unbelievably inaccurate predictions. So far, we have simply defined the sample; we have yet to address the important issue of what constitutes a good or a bad sample.

Statistic

A **statistic** is any measure obtained by having measured a sample. If we select 10 Omega University students, measure their heights, and calculate the mean height, the resulting value is a statistic. Since samples are usually so much more accessible than populations, statistics are far easier to gather than are parameters.

This, then, is the whole point: We select the sample, measure the sample, calculate the statistics, and, from these, infer the population parameters. In short, inferential statistics is the technique of predicting unknown parameters on the basis of known statistics—the act of generalizing to the many after having observed the few.

TECHNIQUES OF SAMPLING

To make accurate predictions, the sample used should be a **representative sample** of the population. That is, the sample should contain the same elements, should have the same overall coloration from the various subtle trait shadings, as the population does. As we mentioned previously, the nurse who takes a blood sample for analysis is almost positive that the sample is representative of the entire blood supply in the circulatory system. On the other hand, the researcher who takes a sample of registered voters from a given precinct cannot be absolutely certain that accurate representation is involved, or even that everyone in the sample will actually vote on election day. The goal for both the nurse and the political forecaster, however, is the same—to select a sample truly reflective of the characteristics of the population. The fundamental rule when applying a sample statistic is that it must be a consistent estimator of the population parameter. That is, as the sample size increases, the sample measure should converge to the true population measure, and this will only occur if the sample is truly representative of the population to which its value is being extrapolated (Foster & McLanahan, 1996).

A good, representative sample provides the researcher with a miniature mirror with which to view the entire population. There are two basic techniques for achieving representative samples: random sampling and stratified or quota sampling.

Random Sampling

Random sampling is probably one of the media's most misused and abused terms. Newspapers claim to have selected a **random sample** of their readers; TV stations claim to have interviewed a random sample of city residents; and so on. The misapplication results from assuming that random is synonymous with haphazard—which it is not.

Random sampling demands that each member of the entire population has an equal chance of being included. And the other side of that coin specifies that no members of the population may be systematically excluded. Thus, if you're trying to get a random sample of the population of students at your college, you can't simply select from those who are free enough in the afternoon to meet you at the psychology laboratory at 3:00. This would exclude all those students who work in the afternoon, or who have their own labs meet, or who are members of athletic teams. Nor can you create a random sample by selecting every *n*th person entering the cafeteria at lunchtime. Again, some students eat off campus, or have classes during the noon hour, or have cut classes that day in order to provide themselves with an instant minivacation—the reasons are endless. The point is that unless the entire population is available for selection, the sample cannot be random.

To obtain a random sample of your college's population, you'd have to go to the registrar's office and get a list of the names of the entire student body. Then you'd clip out each name individually and place all the separate names in a receptacle from which you can, preferably blindfolded, select your sample. Then, *you'd go out and find each person so selected.* This is extremely important, for a sample can never be random if the subjects are allowed to select themselves. For example, you might obtain the list of entering students, the population of incoming freshmen, and, correctly enough so far, select from the population of names, say, 50 freshmen. You would then place requests in each of their mailboxes to fill out an "attitude-toward-college" questionnaire, and, finally 30 of those questionnaires would be dutifully returned to you. Random sample? Absolutely not, since you would have allowed the subjects, in effect, to select themselves on the basis of which ones felt duty bound enough to return those questionnaires. Those subjects who did comply may have differed (and probably did differ) systematically on a whole host of other traits from the subjects who simply ignored your request. This technique is really no more random than simply placing the questionnaire in the college newspaper and requesting that it be clipped, filled out, and returned to you. The students who exert the effort to cooperate with you almost surely differ in important ways from those who do not. Nor are college newspapers the only offenders. Even prestigious *Time* magazine offers its readers the opportunity to participate in this type of nonscientific polling when it publishes clip-out, mail-in questionnaires for its *Time* Survey features (see, for example, *Time,* 1990).

Another blatant example of the self-selected sample is when a TV station asks its audience to respond to some seemingly momentous question (such as whether viewers believe that Elvis is still alive) by dialing, at a cost of 50 cents, a 900 number. Clearly, only the most zealous viewers will then bestir themselves to get up, go to the phone, and be willing to pay for the privilege of casting a vote. In fact, some organized groups are willing to amass sizable telephone bills in order to sway a vote that might later be used to influence public opinion. In one example, a 900 line was used presumably to tap public attitudes toward Donald Trump. A three-day "Trump Hot Line" was set up by a national newspaper, *USA TODAY,* and callers were asked to say which statement they agreed with: "Donald Trump symbolizes what makes the U.S.A. a great country," or "Donald Trump symbolizes the things that are wrong with this country." The results seemed clearly to show that "The Donald" was indeed an American folk hero, with 80% of the callers agreeing with the first statement. A later analysis of the calls, however, revealed a darker side: 72% of those calls came from just two phone numbers, with most of them made during the final hour of the poll. It was later discovered that the phone blitz was arranged by the president of a large insurance company who wanted the world to admire Trump's "entrepreneurial spirit."

It was a true random selection that the U.S. Selective Service System used to tap young men for the draft. All the days of the year were put in a huge fishbowl and a certain number of dates pulled out. If your birthday fell on one of the dates selected, you were called! Whether or not you went is another story, but since everyone's birth date was included in the fishbowl, the *selection* was random. (Also, note that the government did not wait for you to go to them; they came to get you.)

Along with the Selective Service System, another extremely efficient government operation is the Internal Revenue Service. When the IRS asks its big computer in West Virginia to spit out 2% of the income tax returns for further investigation, the selection is again strictly random, since a copy of every taxpayer's return is coded into the computer. Note that the IRS also does not wait for the sample to select itself. It chooses you!

Compare these examples with a TV station's attempt to select a random sample by interviewing every tenth person who happens to be walking down the main street on a Tuesday afternoon. Is every resident downtown that day, that hour, strolling along the main thoroughfare? Or, a comparable example is if the college newspaper interviews every tenth student entering the library on a Thursday afternoon. This cannot possibly be a random sample. Some students might be in class, or at work, or starting early on their round of weekend parties, or back in their rooms doing statistics problems. As long as any students are excluded from the selection process, the sample cannot, by definition, be random.

Stratified, or Quota, Sampling

Another major technique for selecting a representative sample is known as **stratified,** or **quota, sampling.** This is sometimes combined with random sampling in such a way that once the strata have been identified, random samples from each subgroup are selected. You've probably guessed that this is then called *stratified* random sampling. To obtain this type of sample, the researcher must know beforehand what some of the major population characteristics are and, then, deliberately select a sample that shares these characteristics in the same proportions. For instance, if 35% of a student population are sophomores and, of those, 60% are majoring in business, then a quota sample of the population must reflect those same percentages.

Some political forecasters utilize this sampling technique. They first decide what characteristics, such as religion, socioeconomic status, or age, are important in determining voting behavior. Then they consult the federal government's census reports to discover the percentage of individuals falling within each category. The sample is selected in such a way that the same percentages are present in it. When all this is done carefully and with real precision, a very few thousand respondents can be used to predict the balloting of over 100 million voters.

Sampling Error

Whenever a sample is selected, it must be assumed that the sample measures will *not* precisely match those that would be obtained if the entire population were measured. Any other assumption would be foolhardy indeed. To distinguish it from the sample mean, M, the mean of the population is designated by the Greek letter mu, μ. **Sampling error** is, then, the difference between the sample value and the population value.

$$\text{sampling error} = M - \mu$$

Note that this is a normal, expected deviation. We expect that the sample mean might deviate from the population mean. Sampling error is *not a mistake!* Sampling error is conceptually different from the error committed by a shortstop booting a grounder. It is an expected amount of deviation. Also, sampling error should be random; it can go either way. The sample mean is just as often below the population mean as it is above it. If we took the means of 100 random samples from a given population, then 50% of the resulting sampling errors should be positive (the sample mean overestimating the population mean) and the other 50% should be negative. Applying the relationship between percentage and probability, as defined in Chapter 6, we can say here that the probability that a given sampling error will have a plus sign is the same as the probability that it will have a minus sign, that is, $p = .50$.

Outliers

When one or two scores in a large random sample fall so far from the mean—say, three or four standard deviation units away—they are called **outliers.** Outliers indicate either that the distribution is not normal or that some measurement error has crept in. Even the most scrupulously careful researcher is bound, sooner or later, to encounter a few of these outliers. When they occur, they may increase the *SD* to more than one-sixth of the range, so that the distribution suddenly looks more platykurtic (flat) than it should. Also, unless they're balanced off by occurring at both ends of the distribution, a couple of outliers can definitely cause skew. When it is clear that outliers were not produced by bias, most researchers feel legitimately comfortable in discarding them. More will be said on this topic in Chapters 10 and 11.

Bias

Whenever the sample differs systematically from the population at large, we say that **bias** has occurred. Since researchers typically deal in averages, bias is technically defined as a constant difference, in one direction, between the mean of the sample and the mean of the population. For example, suppose that the mean verbal SAT score at your college is 500, and yet you select samples for your particular study only from those students who, because of poor performance on the English placement test, have been assigned to remedial English courses. It is likely that the average verbal SAT scores among your samples are *consistently lower* than the mean of the population as a whole. This biased selection would be especially devastating if your research involved anything in the way of reading comprehension, or IQ, or vocabulary measures. Your results would almost certainly underestimate the potential performance of the population at your college.

Bias occurs when most of the sampling error loads up on one side, so that the sample means are consistently either over- or underestimating the population mean. Bias is a *constant sampling error in one direction.* Where there is bias, no longer does the probability that M is higher than μ equal .50. The probability now might be as great as .90 or even 1.00 (or, on the other side, might be as low as .10 or even zero).

Assume that John Doe, a major in pseudoscience, is engaged in a research project in which he has to estimate the height of the average American college student. He wants a sample with an N of 15. Late in the afternoon, he runs over to the gym, having

heard there were some players out there on the gym floor throwing a ball into a basket. They obligingly stop for a few minutes, long enough for John to measure each of their heights. The sample mean turns out to be 81 inches. Undaunted, our hero then repeats this procedure at a number of colleges, always going to the gym in the afternoon and always coming up with mean heights of 77 inches or more. Since the parameter mean height for male college students is actually about 69 inches, the sample means John obtained were constantly above the population mean. This is bias "writ large." This story may seem a little contrived. For a true history of the effects of bias, we can turn to the saga of political forecasting of presidential elections.

Political Polling

The first known political poll took place in 1824. The candidates for president that year were John Quincy Adams and Andrew Jackson. A Pennsylvania newspaper attempted to predict the election's outcome by sending out reporters to interview the "man on the street." The data were tallied and the forecast made—Jackson to win. Jackson, however, lost. The science of political forecasting did not get off to an auspicious start.

***Literary Digest* Poll.** Throughout the rest of the 1800s other newspapers in other cities began to get involved in political polling, and the results were decidedly mixed. In 1916, an important magazine, *Literary Digest,* tried its hand at political forecasting. Over the next several presidential elections, the magazine enjoyed a fair degree of success. The technique was simple: send out sample ballots to as large a group of voters as possible, and then sit back and hope that the people would be interested enough to return the ballots. The measured sample had to separate itself from the selected sampie by choosing whether or not to mail in the ballots. Of course, a mail-in poll can be accurate, by luck, but the *Digest*'s luck ran out.

The candidates during the Depression year 1936 were Alf Landon, Republican, and Franklin D. Roosevelt, Democrat. Beginning in 1895, the *Literary Digest* had compiled a list of prospective subscribers, and by 1936, there were over 20 million names on the list. The names, however, were of people who offered the best potential market for the magazine and its advertisers, or middle- and upper-income people. To "correct" for any possible subscriber bias, the magazine also collected a list of names selected from various telephone books throughout the country and also from names provided by such sources as the state registries of motor vehicles. Among the persons polled were every third registered voter in Chicago and, believe it or not, every registered voter in Allentown, PA. This list became known as the famous "tel-auto public." In all, the *Digest* sent out more than 10 million ballots and got back 2,376,523 (almost 25%). Said the *Literary Digest:*

> Like the outriders of a great army, the first ballots in the great 1936 Presidential Campaign march into the open this week to be marshalled, checked and counted. Next week more states will join the parade and there will be thousands of additional votes. Ballots are pouring out by tens of thousands and will continue to flood the country for several weeks. They are pouring back too, in an ever increasing stream. The homing ballots coming from every state, city, and hamlet should write the answer to the question, "Roosevelt or Landon?" long before Election Day (Gallup & Rae, 1968, p. 42).

On October 31, 1936, the *Literary Digest* wrote its answer—Landon in a landslide. The final returns in the *Digest*'s poll of 10 million voters were Landon 1,293,669 and Roosevelt 972,897. They went on to predict that Landon would carry 32 states to Roosevelt's 16 and receive 370 electoral votes to only 162 for Roosevelt. The actual vote on election day went the other way around. Roosevelt received 27 million votes (over 60%) to Landon's 16 million votes. Following the election, the magazine's next issue had a blushing, pink cover with the cutesy caption "Is our face red!" The humiliation, however, must have been too great to bear, since the *Digest* soon went out of business.

What went wrong?—Bias, which is a constant sampling error in one direction. It produced a sample in which Republicans were markedly overrepresented.* Also, as is the case with all mail-in polls, most members of the sample were allowed to select themselves. Not everyone who received one returned a ballot, but those who did probably had more fervor about their choice than did the population at large.

Perhaps the major finding from the whole fiasco was that in 1936 Democrats did not have phones, did not drive cars, and certainly did not read or respond to the *Literary Digest.*

---

*It is now assumed that telephone owners are representative of the voting population. One poll in 1996, conducted by the CBS network, was based on a random sample of fewer than 1500 telephone owners throughout the entire country. The margin of error was considerably less than 3%.

Gallup Poll. In 1936, a new statistical star arose. His name was George Gallup, and he headed the American Institute of Public Opinion, headquartered in Princeton, New Jersey. Gallup correctly predicted Roosevelt's win over Landon, even though the margin of victory was off by a full 7 percentage points. In the next two elections Gallup also predicted the winner, Roosevelt over Willkie in 1940 and Roosevelt over Dewey in 1944. In both instances, Gallup reduced his margin of error, to 3% in 1940 and to a phenomenally low 0.5% in 1944.

Gallup, unlike the *Literary Digest*, did not allow the sample to select itself; he went to the sample. He stressed representative sampling, not necessarily large sampling. Using the quota technique described earlier in this chapter, he chose fewer than 3000 voters for inclusion in his sample; compare this with the enormous sample of over 2 million voters used by the *Literary Digest*. Some of the population parameters used by Gallup were geographic location, community size, socioeconomic level, sex, and age.

In 1948 the Republicans, as they had in 1944, again chose Thomas E. Dewey, and the Democrats chose Harry S. Truman. Because of Gallup's three previous successes, his poll was now highly regarded, and his predictions were awaited with great interest. During the last week of September, a full six weeks before the election, Gallup went to press and predicted a win for Thomas Dewey. His poll was considered infallible, and one Chicago newspaper even printed a headline proclaiming Dewey's victory before the final vote was completely counted.

Harry Truman, of course, won, and statistical analysis learned its second hard lesson in a dozen years. We now know that Gallup made two crucial mistakes. First, he stopped polling too soon. Six weeks is a long time, and voters do change their minds during a political campaign. If the election had been held when Gallup took his final count, Dewey might indeed have won. Second, Gallup assumed that the "undecided" were really undecided. All those individuals who claimed not yet to have made a decision were placed in one pile and then allocated in equal shares to the two candidates. We now know that many persons who report indecisiveness are in reality, perhaps only at an unconscious level, already leaning toward one of the candidates. Questions such as "Who are your parents voting for?" or "Who are your friends voting for?" or "How did you vote last time?" may often peel away a voter's veneer of indecision.

In 1952 Gallup picked Eisenhower, although he badly underestimated the margin of victory. Perhaps by then, having been burned twice when predicting Republican wins, the pollsters no longer trusted the GOP and therefore underplayed the data.

In 1960, however, Gallup became a superhero by predicting Kennedy's razor-thin victory over Nixon. Gallup's forecast even included the precise margin of Kennedy's win, which was 0.5%. Polling had come of age. Since 1960 most of the major political forecasters have done a creditable job, and none use a sample much larger than 3000 voters. This is truly remarkable, considering that the voting population is over 100 million.

Sources of Polling Bias

Over the years it has become obvious that the pollsters' mistakes, when they made them, lay in getting nonrepresentative samples. Their samples, in fact, were biased toward the

Republicans. This is understandable, considering that Republicans are typically more affluent and, therefore, easier for the pollsters to get at. Republicans are more likely to live in the suburbs, or in apartment houses with elevators, or, generally, in places where the pollsters themselves feel safe. High-crime areas are more apt to trigger that ultimate pollster cop-out of going to the nearest saloon and personally filling out the ballots. This practice, known in the trade as "curbstoning," is definitely frowned on.

The following data indicate the bias toward Republicans found in some early Gallup Poll results:

Year	Predicted Republican Vote (%)	Actual Republican Vote (%)	Error in Favor of Republicans (%)
1936	44	38	+6
1940	48	45	+3
1944	48	46	+2
1948	50	45	+5

When polling household members via the telephone, it is important that the pollster keep calling back until the person who was chosen answers the phone. It has been found that when a telephone rings in a U.S. household, it will be answered by women more than 70% of the time. Even the U.S. Census Bureau has admitted that its counting procedures may leave something to be desired. In 1980 and 1990, the census again admitted a serious undercount of African Americans, 4.5% in 1980 and 5.7% in 1990 (Passell, 1991). Undercounting is a matter of grave concern to local government officials, especially in several northeastern states, for federal funding to cities and the reapportioning of the U.S. House of Representatives and state legislatures depend on these census figures. Counting everyone is admittedly not an easy job. For example, getting an accurate count of the homeless is very difficult—they're never home. Despite efforts to ease the problem by making statistical adjustments to the census data, the uneasy feeling remains that some minorities are still undercounted in our population (Fienberg, 1993).

Exit Polling and Bias

Exit polling, a method of predicting the vote after a certain number of persons have already voted, has also had its share of problems. Although intuitively logical (after all, predicting the vote after people vote doesn't seem like a mind-bending leap of faith), there is always the suspicion that those who vote before 4 PM, when the exit pollsters are waiting to pounce, may systematically differ from those who vote later that evening. This, of course, may introduce an element of bias.

An example of blatant selection bias comes from the famous Kinsey report on male sexual behavior (Kinsey et al., 1948). In it, Kinsey reported that a whopping 37% of his sample had at least some overt homosexual experience—to the point of orgasm. What only the careful reader would discover, however, is that an enormous 18% of his 5,300 male sample had spent time in jail, hardly a representative sample of the general adult male population, especially in the 1940s. The interested reader should see the James H. Jones biography of Kinsey (Jones, 1997).

In 1987, Shere Hite published the results of her survey of U.S. women (Hite, 1987). Among the more provocative findings was Hite's assertion that 70% of America's women, married at least five years, were involved in extramarital affairs. Although that result may have been true of the sample, there is no way of knowing whether this sample accurately reflected the population of American women, since the sample was totally self-selected. Hite mailed 100,000 questionnaires to women's groups, subscribers to certain women's magazines, and other groups and received only 4500 returns, or less than 5%. Could it not be that the 5% who did take the time and trouble to respond differed systematically from the 95% who did not? Or, worse, it is a virtual certainty that the 100,000 women who Hite originally selected were not representative of the millions of women who were not subscribers to those particular magazines or members of the targeted women's groups.

As powerful as the inferential statistical tests are and they are, nothing can compensate for faulty data. There's an old computer adage, which says "GIGO," or garbage in, garbage out. Shaky data can only result in shaky conclusions.

SAMPLING DISTRIBUTIONS

Each distribution presented so far has been a redistribution of individual scores. Every point on the abscissa has always represented a measure of *individual* performance. When we turn to sampling distributions, this is no longer the case. This, then, requires a dramatic shift in thinking. In **sampling distributions,** each point on the abscissa represents a measure of a group's performance, typically the arithmetic average, or mean performance, for the group. Everything else remains the same. The ordinate, or Y axis, still represents frequency of occurrence, and the normal curve remains precisely as it was for all of those z score problems. The spotlight, however, now turns to the mean performance of a sample.

Mean of the Distribution of Means

Assume that we have a large fishbowl containing slips of paper on which are written the names of all the students at Omega College—in all, 6000 students identified on 6000 slips of paper. Thus, $N_p = 6000$, or the number of individuals making up the entire population. We shake the bowl and, blindfolded, draw out 30 names. The group of 30, so selected, forms a sample (any number less than the population taken from the population). Even more important (since all 6000 names had an equal chance of selection), this group forms a random sample. Next, we search the campus, find the 30 students selected, and give each an IQ test. When we add the scores and divide by 30 ($N_s = 30$, the size of the sample group), the resulting value equals the mean performance for that sample group. It turns out to be a mean of 118 ($M_1 = 118$). Now we can't expect a sample mean of 118 to be identical to the population mean. The concept of sampling error tells us that we must expect the sample mean to deviate from the population mean.

Next we go back to the fishbowl, draw out 30 more names, find these students, give the IQ tests, and again calculate the mean. This time it turns out to equal 121 (M_2 = 121). We continue this process, selecting and measuring random sample after random sample until there are no more names left in the bowl. Since we started with 6000 names and drew them out 30 at a time, we select our last sample on our 200th draw.*

We thereby create a long list of sample means, 200 in all. Since each value is based on having measured a sample, it is, in fact, a list of statistics.

$$
\begin{aligned}
M_1 &= 118 \\
M_2 &= 121 \\
&\vdots \\
M_{200} &= 112
\end{aligned}
$$

We now add these mean values, divide by the number of samples, and calculate a kind of supermean, the overall mean of the distribution of sample means, or M_M.

---

*Technically, to create a sampling distribution, we should replace each sample taken out and make an infinite number of selections. But since the concept of infinite selection is hard to grasp (and impossible to accomplish), the example is designed to give a more vivid picture of the reality of a sampling distribution. (See the box on page 148.)

Is this an end in itself, or just a means to an end?

$$M_M = \frac{\Sigma M}{N}$$

This value is, and must be, a parameter. Although it was created by measuring successive *samples,* it required measuring *all* the samples, and therefore the entire population, to get it. That is, the mean of the distribution of means, M_M, is the same value that would have been obtained by measuring each of the 6000 subjects separately, adding their IQs, and dividing by the total N. The mean of the distribution of sample means is equal to the parameter mean μ.

$$M_M = \mu$$

Therefore, if we actually followed the procedure of *selecting successive samples* from a given population until the population was exhausted, we would, in fact, have the parameter μ, and there would be nothing left to predict. Why, then, do we go through such a seemingly empty academic exercise? Because later we will be predicting parameters, and it is crucial that you have a clear idea of exactly what values are then being inferred.

We now have a measure of central tendency in the mean of the sampling distribution of means, and we must next get a measure of variability.

INFINITE VERSUS FINITE SAMPLING

The two sampling distributions, the distribution of means and the distribution of differences, are based on the concept that an infinite number of sample measures are selected, with replacement, from an infinite population. Because the selection process is never ending, it is obviously assumed that all possible samples will be selected. Therefore, any measures based on these selections will be equivalent to the population parameters. Because the theoretical conceptualization of infinite sampling, replacement, and populations can be rather overwhelming, this text for purposes of illustration discusses populations of fixed size (finite) and sampling accomplished without replacement. The important message that this should convey is that combining sample measures does not equal population parameters until every last measure, from every last sample, has been counted. For sample measures to be the equivalent of population measures, all possible samples must be drawn until the population is exhausted.

Although sampling distributions are, in fact, theoretical distributions based on infinite random selection with replacement from an infinite population, for practical purposes of illustration, we treat the population as though it were finite.

The noted psychologist and statistician Janet T. Spence acknowledges that although the theoretical distribution of random sample means is indeed based upon the assumption of drawing an infinite number of samples from an infinite population, "it is quite satisfactory for illustrative purposes to go through the statistical logic with a finite population and a finite number of samples" (Spence et al., 1976). Also, if the sample is large enough, or, indeed, if the population is infinite, the difference between sampling with or without replacement becomes negligible.

Standard Deviation of the Distribution of Means

When distinguishing between samples and populations, different symbols are used for their means, *M* for the sample and μ for the population. We make a similar distinction regarding variability. The true standard deviation of all the sample scores is designated, as we have seen, as *SD,* whereas the actual standard deviation of the entire population is always assigned the lowercase Greek letter sigma, σ.

The calculations involved in obtaining either standard deviation are identical, since each is producing a true (not an estimated) value. The difference between them is simply a reflection of what each value is describing. The *SD,* when used in this fashion, becomes a statistic (a sample measure), and σ, because it is a population measure, is a parameter. We can, therefore, calculate the standard deviation of the distribution

"I'M BEGINNING TO UNDERSTAND ETERNITY, BUT INFINITY IS STILL BEYOND ME."

of means just as we did when working with distributions of individual scores. That is, we can treat the means as though they are raw scores and then use the same standard deviation equation.

$$\sigma_M = \sqrt{\frac{\Sigma M^2}{N} - M_M^2}$$

The standard deviation of the distribution of means is equal to the square root of the difference between the sum of the squared means divided by the number of means and the mean of the means squared. (Know what we mean?) Just think about it. You may never have to do it, but you should at least think through the concept so you will know what could be done. Again, to obtain this standard deviation, all the sample means in the entire population had to be included. This value, like the mean of the means, is a parameter (since the whole population, though taken in small groups, had to be measured to get it). If we obtained the actual standard deviation of the distribution of means with the method just described, we would again have achieved a parameter and would have nothing left to predict.

Comparing the Two Standard Deviations. In comparing the standard deviation of the distribution of sample means with the standard deviation of the distribution of raw scores, we find a dramatic difference. The standard deviation of means is much smaller. This is because the standard deviation of means, like all standard deviations, is measuring variability, and the distribution of means has far less variability than does the distribution of individual scores. When shifting from the distribution of individual scores to the distribution of sample means, *extreme scores are lost through the process of averaging.* When we selected our first sample of 30 students from the population of Omega College students, 1 or 2 of those subjects might have had IQs as high as 140. But when these were averaged in with the other IQs in the sample (perhaps 1 student was measured at only 100), the *mean* for the sample turned out to be 118. The distribution of sample means will be narrower than the distribution of individual scores from which the samples were selected. All the measures of variability will be similarly more restricted, meaning a smaller range as well as a smaller standard deviation (Asok, 1980).

Standard Error of the Mean

The standard deviation of the distribution of means provides a measure of sampling error variability. We expect sample means to deviate from the population mean, but the precise amount of this deviation is unknown until the standard deviation is calculated. This particular standard deviation is thus called the **standard error of the mean,** since it expresses the amount of variability being displayed between the various sample means and the true population mean. It is a measure of sample variability, and although it certainly could be calculated by using the method just shown (that is, by treating each sample mean as though it were a raw score and then using the standard deviation equation), it may also be found in a less cumbersome fashion.

Had we not wanted to continue selecting random sample after random sample, we could also have calculated the standard error of the mean on the basis of the entire population's distribution of raw scores. With the population of M's, and a sample of sufficient size, the equation becomes

$$\sigma_M = \frac{\sigma_x}{\sqrt{N}}$$

That is, to get the parameter standard error of the mean, we could divide the standard deviation of the population's distribution of raw scores by the square root of the number of scores in the sample.

Central Limit Theorem

That the sampling distribution of means may approach normality is a crucial consideration. If it were not the case, we could no longer utilize Appendix Table A. Perhaps you wondered why that particular curve is normal, or perhaps you just accepted it, took it on faith. It is true, and the **central limit theorem** describes this very important fact. The theorem states that when successive random samples are taken from a single population, the means of these samples assume the shape of the normal curve, regardless of whether or not the distribution of individual scores is normal. Even if the distribution of individual scores is skewed to the left, or to the right, the sampling distribution of means approaches normality, and thus the z score table can be used.

The central limit theorem applies only when all possible samples have been randomly selected from a single population. Also, the distribution of sample means more closely approaches normality as the sample sizes become larger. To understand this point, assume for a moment that your samples are made up of only one person each. In this farfetched case, the distribution of sample means is identical to the distribution of individual scores. Then, if the distribution of individual scores is skewed, so too is the sampling distribution. If each sample is sufficiently large, however (typically around 30 cases), then we can rest assured that our sampling distribution will be approximately normal.

GALTON AND THE CONCEPT OF ERROR

When reading about or using the term *standard error,* it must be kept firmly in mind that, to a statistician, the word *error* equals deviation and does not imply a mistake. Hence, a standard error always refers to a standard deviation, or to a value designating the amount of variation of the measures around the mean. It was originally called "error" by the German mathematician Carl Friedrich Gauss, the father of the normal curve. In the early nineteenth century, Gauss had used the term error to describe the variations in measurement of "true" physical quantities, such as the measurement errors made by astronomers in determining the true position of a star at a given moment in time. The idea behind the current use of the concept of the "standard error" was largely formulated during the latter half of the nineteenth century by Sir Francis Galton (1822–1911).

Galton noted that a wide variety of human measures, both physical and psychological, conformed graphically to the Gaussian, bell-shaped curve. Galton used this curve not for differentiating true values from false ones but as a method for evaluating population data on the basis of their members' variation from the population mean. Galton saw with steady clarity the importance of the normal curve in evaluating measures of human traits. Thus, to Galton, the term error came to mean deviation from the average, not falsity or inaccuracy. The greater the error among any set of measures, then, the greater is their deviation from the mean, *and the lower their frequency of occurrence.* Said Galton, "there is scarcely anything so apt to impress the imagination as the wonderful form of cosmic order expressed by the Law of Frequency of Error. The law would have been personified by the Greeks and deified, if they had known of it" (Kevles, 1984).

Importance of the Sampling Distribution Parameters

Why are the parameters μ and σ_M important? Because the sampling distribution of means can approach a *normal distribution,* which allows us to use the percentages on the normal curve from the z score table. This puts us in the happy position of being able to make probability statements concerning the distribution of means, just as we previously did with the distribution of individual scores. We have already done problems like the following.

Example On a normal distribution with a mean of 118 and a standard deviation of 10, what is the probability of selecting someone, at random, who scored 125 or above?

$$z = \frac{X - M}{SD} = \frac{125 - 118}{10} = \frac{7}{10} = .70$$

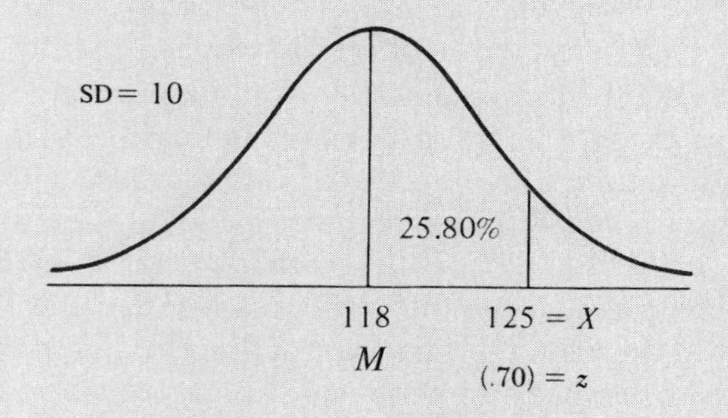

From Appendix Table A, z of .70 yields 25.80% falling between the z score and M. Since 50% fall above the mean, the percentage above the z score of .70 is 50% – 25.80% = 24.20%.

$$p = \frac{\text{percentage}}{100} = \frac{24.20}{100} = .24$$

Example On a sampling distribution of means with a known population mean of 118(μ) and a known standard error of the mean of 10, what is the probability of selecting a single sample whose mean (M) would be 125 or above?

Since in the two examples, the values of X, M, and SD are the same as M, μ, and σ_M, then the same probability value will result:

$$z = \frac{M - \mu}{\sigma_M} = \frac{125 - 118}{10} = .70$$

$$p = .24$$

Example The sampling distribution of means yields an overall mean weight for professional football players of 230 pounds, with a known standard error of the mean of 10 pounds. What is the probability of selecting a sample of football players whose mean weight would be 200 pounds or less?

$$z = \frac{M - \mu}{\sigma_M} = \frac{200 - 230}{10} = -3.00$$

The z score, –3.00, shows a percentage of 49.87 falling between it and the mean, leaving only 0.13% (far less than 1%) falling below that point.

$$p = .0013$$

Thus, there is almost no chance of finding a random sample of players whose mean weight is 200 pounds or less. The probability of selecting an individual player weighing less than 200 pounds is undoubtedly higher, but selecting a whole group of players, a random sample, that averages out in that weight category is extremely remote. This again illustrates the important point that variation among sample means is considerably less than it is among individual measures. It also illustrates the importance of random sampling when utilizing the z score table. Certainly a group of football players could be culled out, perhaps a group of wide receivers, whose mean weight would be less than 200, but that would be a biased, nonrepresentative sample of the population at large.

These examples demonstrate two important points to remember. First and foremost, you can do them. If you can work out a probability statement for a distribution of individual scores, you can just as easily do the same thing for a sampling distribution. Second, the reason for needing those two key parameters, the mean and the standard deviation of the distribution of means, becomes clear. With them, the probability values built into the normal curve are again available to us.

BACK TO *z*

The theory underlying parameter estimates rests solidly on the firm basis of z score probability. For example, on one of the Woodcock Reading Mastery Tests (for word-passage comprehension), the national population mean is 100 with a population standard deviation of 15. Suppose we now select from a certain school district a random sample of 36 youngsters, all of whom are of the same age and grade level, and discover that the mean for this sample is 95. The school department indicates that this small difference of only 5 points is trivial, probably due to sampling error, and that the children in this district are still doing at least as well as the national norms. The parents, however, are extremely upset and demand answers as to why their children are not reading as well as the norms indicate for the rest of the nation. To shed some light on this controversy, a consulting statistician is hired to do an analysis of these data and, hopefully, to come to some conclusions.

Thus, the statistician is being presented with four values: the mean of the population, the standard deviation of the population, the mean of the sample, and the size of the sample.

1. Mean of the population = $\mu = 100$.
2. Standard deviation of the population = $\sigma = 15$.
3. Mean of the sample = $M = 95$.
4. Size of the sample = $N = 36$.

First, the statistician calculates the standard error of the mean:

$$\sigma_M = \frac{\sigma_x}{\sqrt{N}} = \frac{15}{\sqrt{36}} = \frac{15}{6} = 2.50$$

Because the population parameters are known and because the distribution of reading scores is known to approach normality, the statistician decides to use the z equation for the distribution of means and to assess the probability of obtaining a difference of 5 points or more on a distribution whose population mean is known to be 100:

$$z = \frac{M - \mu}{\sigma_M} = \frac{95 - 100}{2.50} = -2.00$$

Looking up the z score of –2.00, we find that 47.72% of the cases fall between that point and the mean. Thus, we see that the total area included by z scores of ±2.00 is

47.72% + 47.72%, or 95.44%. Therefore, the area *excluded* by these z scores contains only 4.56% of all the sample means, thus producing a probability value of less than .05 of finding a difference in either direction that is this large or larger. We say "in either direction" because if the distribution of means is indeed normal, a sample mean of 105, or a value that would be 5 points greater than μ, could just as easily have been selected. If a sample is chosen randomly, its mean has the same probability of being less than μ as it has of being greater than μ. The probability is therefore extremely slight that this sample of children is really part of the same population whose national mean is known to be 100.

The statistical consultant now says that more precise terminology should be used. To say that the probability is "extremely slight" would be just too vague and subjective. What is needed is a clear line of demarcation, where on this side you have a sample that represents the population and on that side you don't. The statistician tells us that that line has been drawn, and, by convention, it is set at a probability value of .05 or less. Thus, since our sample mean of 95 is in an area of the curve whose frequency is 5% or less, then it can be concluded that it probably could not have come from a population whose mean was known to be 100, or that the two means, sample and population, differ by too much to expect them to be part of the same normal distribution of sample means. The extreme 5% of the curve's area, remember, is that area excluded by z scores of ± 1.96.

Any z value, then, that falls beyond the ± 1.96 values leads to a conclusion that the mean is not part of the known population. The statistician thus has concluded that this sample of youngsters is underperforming with regard to the national reading norms. Now, as the statistician further advises us, this decision may not always be correct. There are, after all, 5% of the sample means that, on the basis of chance alone, really do fall way out there under the tails of the curve.

This was not greeted as good news by the school administrators, however, since the odds still favor the statistician's decision. Although it's possible that a sample mean could deviate from the true mean by z's as large as ± 1.96, the chances are extremely remote that our particular sample mean could have been so far away from the true mean. In fact, these chances are so rare that the statistician advises us to reject that possibility. Using this type of logic, we are going to be right far more often than wrong. In fact, this limits our chance of being wrong to an exact probability value of only .05 or less. Later in the chapter, and in the next, we shall see that on some occasions, a researcher may wish to limit the probability of being wrong to an even more stringent .01 or less.

The z Test

When both the parameter mean and the *parameter standard deviation* are known, as in the preceding example, the ***z* test** can be used to determine the probability that a given sample is representative of the known population. In this way an inference is being drawn about the population being represented by that sample. It was known that the overall population mean was 100, or that $\mu = 100$. It was also known that the sample mean was 95; on that basis, one could guess that this sample might represent a population whose mean might equal 95. As the next chapter will show, this predicted value of μ_1 is called a *point estimate,* since that was the population that had been sampled. We can, therefore, test the chance hypothesis (called the *null hypothesis*) that $\mu_1 = \mu$. The null hypothesis, symbolized as H_0, states that the mean of the population of students we selected from is equal to the mean of the national population in general. Thus, the null hypothesis is suggesting that despite the blatant fact that there is a difference of 5 points between the population mean being estimated by our sample and the known population in general, this difference is only due to chance, or random sampling error—that in reality there is no difference. The alternative hypothesis is that $\mu_1 \neq \mu$, or that the population mean of our youngsters is not equal to the population mean as a whole.

$$z = \frac{M - \mu}{\sigma_M} = \frac{95 - 100}{2.50} = -2.00$$

Since the z test is yielding a value that, as we have seen, could have occurred less than 5 times in a 100 on the basis of chance alone, we decided to reject the null hypothesis and suggest instead that the alternative hypothesis is true—that the sample mean is different from the mean of the population of students at large. (In Chapter 8, this will be referred to as a *significant difference.*)

In summary, then, we have

Null hypothesis:	H_0: $\mu_1 = \mu$	where μ_1 = hypothesized population mean and μ = known population mean
Alternative hypothesis:	H_a: $\mu_1 \neq \mu$	
Critical value of z at .05:	$z_{.05} = \pm 1.96$	
Calculated value of z test:	$z = -2.00$ reject H_0	

With the z test, the null hypothesis is rejected whenever the calculated value of z is equal to or greater than ±1.96, since its probability of occurring by chance is .05 or less. At the more stringent .01 level of probability, the critical value of z needed to reject the null hypothesis is 2.58 (the z scores that exclude the extreme 1% of the distribution).

Sample Size and Confidence

Every sample, as we have seen, no matter how rigorously selected, will contain some sampling error. Although random selection allows for the assumption that sampling errors will be random, they still occur. One issue facing the researcher, especially the survey researcher, is how large the group must be to provide a certain level of confidence that the selected sample comes as close as possible to truly mirroring the population. If the measures in question can be assumed to be normally distributed in the population, one solution is to use z scores to establish a confidence interval. Thus, to be .95 certain, we use the z scores of ±1.96 (since between these z scores the normal curve contains 95% of all the cases). The equation thus becomes

$$\text{sample size} = \left[\frac{(z)(SD)}{\text{precision unit}}\right]^2$$

If we wished to estimate the height of adult U.S. males, we could plug in the values as follows, where $z = \pm 1.96$: Let us assume an *SD* of 3 inches (which could be known from previous research or could even be estimated from the sample standard deviation) and that the precision unit for this problem will be set at .50 inch or less.

$$\text{sample size} = \left[\frac{(\pm 1.96)(3.00)}{.50}\right]^2 = \left[\frac{\pm 5.880}{.50}\right]^2 = \pm 11.760^2$$

$$= 138.298, \quad \text{or} \quad 138 \text{ men}$$

Or, if we wished to be .99 confident, we could substitute z scores of ± 2.58 and get

$$\text{sample size} = \left[\frac{(\pm 2.58)(3.00)}{.50}\right]^2 = \left[\frac{\pm 7.74}{.50}\right]^2 = \pm 15.480^2$$

$$= 239.630, \quad \text{or} \quad 240 \text{ men}$$

From z to t: A Look Ahead

The concepts introduced in the previous section, such as the null hypothesis, the alternative hypothesis, critical values, and so on, will be more fully covered in the next chapter. The z test, although a perfectly valid statistical test, is not as widely used in the behavioral sciences as are some of the tests that follow. After all, how often are we really going to know the parameter values for the population's mean and standard deviation? However, the z test is still a conceptually important test for you to understand. It helps to reinforce your existing knowledge of z scores and also to lay the groundwork for that supercrucial test to be covered in the next two chapters, the t test.

SOME WORDS OF ENCOURAGEMENT

This chapter has taken you a long way. Some sections, as in the beginning of the chapter, read as quickly and easily as a novel. Some sections you may have to read and reread. If that happens, you should not be discouraged. Rome wasn't built in a day. We will be reading about and using these terms throughout the rest of the book. If you feel that you have yet to conquer this material completely, there is still plenty of time left. But do give it your best shot now. Memorize some of the definitions: sample, population, random sample, statistic, parameter, and standard error of the mean. The definitions are exact; they must be because communication among researchers must be extremely precise. Do the problems now. Putting them off may make future problems look like a fun-house maze—easy to enter but almost impossible to get through.

SUMMARY

Statistical inference is essentially the act of generalizing from the few to the many. A sample is selected, presumed to be representative of the population, and then measured. These measures are called statistics. From these resulting statistics, inferences are made regarding the characteristics of the population, whose measures are called parameters. The inferences are basically educated guesses, based on a probability model, as to what the parameter values should be or should not be.

Two sampling techniques used by statisticians to ensure that samples are representative of the population are (1) random—where every observation in the population has an equal chance of selection—and (2) stratified—where samples are selected so that their trait compositions reflect the same percentages as exist in the population.

Regardless of how the sample is selected, the sample mean, M, must be assumed to differ from the population mean, μ. This difference, $M - \mu$, is called sampling error. Under conditions of representative sampling, sampling error is assumed to be random. When it is not random, when a constant error occurs in one direction, the resulting deviation is called bias. Biased sample values are poor predictors of population parameters.

When successive random samples are selected from a single population and the means of each sample are obtained, the resulting distribution is called the sampling distribution of means.

The standard deviation of the sampling distribution of means is called, technically, the standard error of the mean, a value that can be found by measuring the entire population, either individually or as members of sample groups. Since the shape of the sampling distribution of means is assumed to approximate normality (central limit theorem), z scores may be used to make probability statements regarding where specific means might fall. Finally, when the population parameters, mean and standard error of the mean, are known, the z test may be used to assess whether a particular hypothesized population mean could be part of the same sampling distribution of means whose parameter mean is known. The null, or chance, hypothesis states that the means represent the same population, while the alternative hypothesis states that they represent different populations.

Inferential statistical techniques do not provide a magic formula for unveiling ultimate truth. They do provide a probability model for making better than chance predictions, in fact, in most cases, far better.

Key Terms

- bias
- central limit theorem
- outlier
- parameter
- population
- random sample
- representative sample
- sample
- sampling distributions
- sampling error
- standard error of the mean
- statistic
- stratified, or quota, sampling
- *z* test

PROBLEMS

1. On a certain standardized math test, the distribution of sample means has a known parameter mean of 50 and a known standard error of the mean of 4. What is the probability of selecting a single random sample whose mean could be 52 or greater?
2. The distribution of sample means on a standardized reading test yields a known population mean of 75 and a standard error of the mean of 5. What is the probability of selecting at random a single sample whose mean could be 70 or lower?
3. The standard deviation of the population of individual IQ scores is 15. A given random sample of 40 subjects produces a mean of 105.
   a. Calculate the standard error of the mean.
   b. Test the hypothesis, via the z test, that the sample mean of 105 is still representative of the population whose mean is known to be equal to 100.
4. The standard deviation of the population of individual pulse rates is 5.00 beats per minute. A given random sample of 100 subjects yields a mean pulse rate of 73 beats per minute.
   a. Calculate the standard error of the mean.
   b. Test the hypothesis, via the z test, that the sample mean of 73 could be representative of a population whose mean is known to be equal to 76.
5. A researcher is planning to use a special phonics reading technique (combining letter sounds with songs) to increase reading speed and comprehension among 10-year-old dyslexic children. The researcher selects what appears to be a random sample of 50 dyslexic children, but she wants to be assured that this sample is indeed representative of the general population of dyslexics for that age group. A nationally standardized test for dyslexia is available that shows a parameter mean of 50 and a population standard deviation of 7.50. When measured on this test, the sample produces a mean of 54.
   a. Calculate the standard error of the mean.
   b. Does the selected sample represent the population of dyslexics as measured on this test?
6. A researcher is interested in discovering whether hyperactive adolescents are more impelled to seek new sensations than are children not so afflicted. A sample of 50 hyperactive 17-year-olds is selected and given the Zuckerman Sensation Seeking Scale. The mean score calculated for this sample is 22. The scale has a nationally standardized mean of 20 and a standard deviation of 4.
   a. Calculate the standard error of the mean.
   b. Does this sample represent the national norms regarding sensation-seeking needs?
7. A researcher wishes to estimate the weight of U.S. adult females. The standard deviation is known to be 30 pounds. How large a sample should the researcher attempt to select in order to be .95 certain that the error will be 5 pounds or less?
8. For problem 7, how large should the sample be in order to be .95 certain that the error will be 3 pounds or less?

9. A researcher wishes to estimate the weight of U.S. adult males. The standard deviation is known to be 34 pounds. How large a sample should the researcher attempt to select in order to be .95 certain that the error will be 5 pounds or less?
10. For problem 9, how large should the sample be in order to be .95 certain that the error will be 3 pounds or less?
11. Suppose that the researcher wanted to be .99 certain that the weights of the sample of men would stay within the error of 5 pounds or less. How large would this sample have to be?
12. Suppose that the researcher wanted to be .99 certain that the weights of the sample of men in problem 11 would stay within the error of 3 pounds or less. How large should this sample be?

Fill in the blanks in problems 13 through 25.

13. Statistic is to sample as _________ is to population.
14. The total number of observations sharing at least one trait in common is called the _________.
15. That branch of statistics wherein estimates of the characteristics of the entire group are made on the basis of having measured a smaller group is called _________.
16. Any measurement made on the entire population is called a(n) _________.
17. Any measurement made on a sample is called a(n) _________.
18. The difference between the sample mean and the population mean ($M - \mu$) is called _________.
19. A method of sampling in which every observation in the entire population has an equal chance of being selected is called _________.
20. Every point on the abscissa of a sampling distribution of means represents a(n) _________.
21. A nonrepresentative sample is often due to _________, which is constant sampling error in one direction.
22. If successive random samples are taken from the population, the standard deviation of the resulting distribution of means is called _________.
23. The central limit theorem states that the shape of the entire distribution of sample means will tend to approximate _________.
24. The three major assumptions of the central limit theorem are: _________, _________, and _________.
25. When samples are selected randomly from a single population,
    a. the probability of selecting a sample whose mean is higher than the population mean is _________.
    b. the probability of selecting a sample whose mean is lower than the population mean is _________.

True or False. Indicate either T or F for problems 26 through 30.

26. Every sample measure is assumed to contain bias.
27. Every sample measure is assumed to contain sampling error.
28. If sample sizes are large, bias is eliminated.
29. The null hypothesis for the z test states that the hypothesized population mean is different from the known population mean.
30. To calculate the z test, the population's parameter values must be known.

For each of the terms in problems 31 through 36, indicate whether it is a statistic or a parameter.

31. Sample mean (M).
32. The range of population values.
33. Population mean (μ).
34. Standard deviation of the entire sampling distribution of means.
35. Standard deviation of the sample.
36. The range of sample scores.

Chapter

8

Parameter Estimates and Hypothesis Testing

As we have seen, knowing the mean and the standard error of the sampling distribution of means is of critical importance. These values allow us to use the z score table, which in turn permits us to make probability statements regarding where specific samples might fall. We have also seen, however, that to obtain the mean and the standard deviation of this important distribution, we must select every last sample and measure each individual in all the samples—in short, we must measure the entire population. Whenever we do so, the resulting values are the parameters. When the parameters are known, there is nothing left to predict—there is no need for inferential statistics.

The job of the statistician, then, is to estimate these important parameters, to predict their values without measuring the entire population. This is done by measuring a random sample, calculating the resulting values, called statistics, and using only these statistics for inferring the parameters.

ESTIMATING THE POPULATION STANDARD DEVIATION

Up to now, whenever we had to calculate a standard deviation, from either a set of sample scores, *SD*, or an entire population of scores, σ, there was really no problem. We simply squared the scores, added the squares, divided the sum by *N*, subtracted the square of the mean, and extracted the square root.

$$SD = \sqrt{\frac{\Sigma X^2}{N} - M^2} \quad \text{or} \quad \sigma = \sqrt{\frac{\Sigma X^2}{N} - u^2}$$

The result, in either case, was the true value of the standard deviation, for that particular distribution of scores. However, when attempting to estimate the standard deviation of a population, we are simply not in possession of all the scores (if we were, there would be nothing to estimate).

In fact, in obtaining an estimated standard deviation, not only do we not have all the scores, we don't even have the exact value of μ, the population mean. This does present a problem, especially with very small samples. Why? Because had we used any value other than the mean when calculating the standard deviation, our result would have taken on a somewhat higher value. For example, in the following distribution, the *deviation* method will produce a standard deviation of 2.828.

X	x	x^2
10	4	16
8	2	4
6	0	0
4	−2	4
2	−4	16
$\Sigma X = 30$		$\Sigma x^2 = 40$

$$SD = \sqrt{\frac{\Sigma x^2}{N}} = \sqrt{\frac{40}{5}} = \sqrt{8} = 2.828$$

Suppose, however, that instead of correctly subtracting the mean of 6 each time, we had incorrectly chosen a different value—say, 5—or any other number we may choose:

X	x	x^2
10	5	25
8	3	9
6	1	1
4	−1	1
2	−3	9
		$\Sigma x^2 = 45$

$$SD = \sqrt{\frac{45}{5}} = \sqrt{9} = 3.000$$

The result shows a higher value, one of more variability than this sample of five scores actually possesses. The problem is therefore that when estimating a population standard deviation on the basis of a sample mean, the resulting predicted value is in all likelihood too low, since the population mean is almost certainly a value different from the mean of the sample. Because of sampling error, we expect the sample mean to deviate somewhat from the mean of the population. The conclusion, then, is that the standard deviation of the set of sample scores systematically underestimates the standard deviation of the population and is therefore called a biased estimator. To correct for this possible flaw, statisticians have worked out a special equation for estimating the population standard deviation, usually called s, the *unbiased estimator:*

$$s = \sqrt{\frac{\Sigma x^2}{N - 1}}$$

By reducing the denominator by a value of 1, the resulting value of the standard deviation is slightly increased. Using the previous distribution, we now get

X	x	x^2
10	4	16
8	2	4
6	0	0
4	–2	4
2	–4	16
$\Sigma X = 30$		$\Sigma x^2 = 40$

$$s = \sqrt{\frac{\Sigma x^2}{N - 1}} = \sqrt{\frac{40}{4}} = \sqrt{10} = 3.162$$

Since, as was pointed out in Chapter 3, the deviation method can become unwieldy when there is a large number of scores and/or the mean is something other than a whole number, the following computational formula has been derived:

$$s = \sqrt{\frac{\Sigma X^2 - (\Sigma X)^2/N}{N - 1}}$$

X	X^2
10	100
8	64
6	36
4	16
2	4
30	220

$$s = \sqrt{\frac{\Sigma X^2 - (\Sigma X)^2/N}{N - 1}} = \sqrt{\frac{220 - 30^2/5}{5 - 1}} = \sqrt{\frac{220 - 900 / 5}{4}} = \sqrt{\frac{220 - 180}{4}}$$

$$= \sqrt{\frac{40}{4}} = \sqrt{10} = 3.162$$

Thus, whereas the sample standard deviation for these same data had yielded a value of 2.828, the unbiased estimated standard deviation produces the slightly higher value of 3.162. As the sample size increases, however, the effect of the $(N - 1)$ factor diminishes markedly. With 50 or 60 scores, the two techniques produce virtually the same result. This is, after all, as it should be, since the greater the number of random sample scores, the greater is the likelihood of the sample mean coming ever closer to the population mean. This would obviously lead us to expect less of a discrepancy between the sample and population standard deviations.

Anytime you want to convert from one form to another, simply use the following:

$$s = SD\sqrt{\frac{N}{N-1}} \qquad \text{and} \qquad SD = s\sqrt{\frac{N-1}{N}}$$

A final note of caution is now in order. When checking the calculated value of the estimated population standard deviation, we must reexamine our old rule (Chapter 3) that it never exceed a value of half the range. An estimated standard deviation may, when taken from very small samples, be slightly higher than half the range of the sample scores. After all, the fact that the variability in the population is likely to be larger than the variability among the scores found in tiny samples is the very reason for using this estimating technique in the first place. Thus, although the estimated standard deviation of the population scores may, at times, be somewhat greater than one-half the range of sample scores (among extremely small samples, as much as 72% of the sample range), it is still never greater than one-half the true range existing in the population.

An Important Note on Rounding. Although some statisticians maintain that a value of 5 to the right of the decimal point should always be rounded to the next even number, others suggest that only every other 5 should be rounded up. This book uses the convention of rounding a 5 to the next highest number, consistent with the rounding features built into most modern calculators and computer programs. Because the z score table is rounded to two places, in our discussion of z scores (Chapters 4 through 7) we rounded answers to two decimal places. However, from this point on, we will round our calculations to three places; most of the tables from this point on are three-place tables. Our answers will then become more consistent with computer output. Students have been known to lament that when doing a problem by hand, always rounding to two places, they find slightly different answers when using the computer. Despite the fact that both answers are identical within rounding error, the small difference does cause some confusion. For those instructors who wish to maintain a consistent, two-place rounding system throughout, the answers for Chapters 8 through 19 are given in the back of the book in both forms. However, whatever your choice, you must be consistent. If you choose two-place rounding, for example, every calculation within the formula must be so rounded.

Examples of two-place rounding:

1.253 = 1.25
17.569 = 17.57
6.999 = 7.00
6.949 = 6.95

Examples of three-place rounding:

17.6590	=	17.659
15.1325	=	15.133
0.0136	=	0.014
7.0009	=	7.001

ESTIMATING THE STANDARD ERROR OF THE MEAN

The **estimated standard error of the mean** is a statistic that allows us to predict what the standard deviation of the entire distribution of means would be if we had measured the whole population. Since it is a statistic, the estimated standard error of the mean can be calculated on the basis of the information contained in a single sample, namely, its variability and its size.

Using the Estimated Standard Deviation to Find SE_M

The estimated standard error of the mean is equal to the estimated population standard deviation divided by the square root of the sample size. (The symbol SE_M is used here to avoid confusion with σ_M, which represents the parameter.)

$$SE_M = \frac{s}{\sqrt{N}}$$

Example A random sample of elementary school children was selected and measured on the information subtest of the WISC III intelligence test. Their scores were as follows:

X	X^2
10	100
9	81
8	64
7	49
7	49
7	49
7	49
6	36
5	25
4	16
$\Sigma X = 70$	$\Sigma X^2 = 518$

$$M = \frac{\Sigma X}{N} = \frac{70}{10} = 7.00$$

$$s = \sqrt{\frac{\Sigma X^2 - (\Sigma X)^2/N}{N-1}} = \sqrt{\frac{518 - 70^2/10}{10-1}}$$

$$= \sqrt{\frac{518 - 4900/10}{9}} = \sqrt{\frac{518 - 490}{9}} = \sqrt{\frac{28}{9}}$$

$$= \sqrt{3.11} = 1.76$$

$$SE_M = \frac{s}{\sqrt{N}} = \frac{1.76}{\sqrt{10}} = \frac{1.76}{3.16} = .56$$

Using the Sample Standard Deviation: An Alternate Route

If your instructor prefers the use of *SD*, the actual standard deviation of the sample, then the equation for the estimated standard error of the mean becomes the following:

$$SE_M = \frac{SD}{\sqrt{N-1}}$$

$$M = \frac{\Sigma X}{N} = \frac{70}{10} = 7.00$$

$$SD = \sqrt{\frac{\Sigma X^2}{N} - M^2} = \sqrt{\frac{518}{10} - 7.00^2}$$

$$= \sqrt{51.80 - 49.00} = \sqrt{2.80} = 1.67$$

$$SE_M = \frac{SD}{\sqrt{N-1}} = \frac{1.67}{\sqrt{10-1}} = \frac{1.67}{\sqrt{9}} = \frac{1.67}{3} = .556$$

$$= .56$$

Even though the actual standard deviation of the sample is a biased estimator, in this equation the correction is made by subtracting the value of 1 in the denominator of SE_M allowing the estimated standard error of the mean to remain unbiased.

Either way, regardless of the alternative chosen, we obtain the same value for the estimated standard error of the mean.

The following is an exercise that should help reinforce the concept that the underlying distribution of individual scores has more variability than does the distribution of sample means:

It is an easy matter to find either the s or the SD, once the standard error of the mean is known. Keep firmly in mind that the s or SD must be larger than the standard error of the mean. For example, suppose a researcher announces that with a sample size of 5, the standard error of the mean is equal to .510. Find the s (estimated population standard deviation) or (if your instructor prefers) the SD (true standard deviation of the sample scores).

$$s = SE_M \sqrt{N} \qquad SD = SE_M \sqrt{N - 1}$$

$$s = .510\sqrt{5} = (.510)\,(2.236) = 1.140 \qquad SD = .510\sqrt{5 - 1} = (.510)\,\sqrt{4} = (.510)\,(2) = 1.020$$

For the following, find the estimated population standard deviation.

a. A researcher measures the heights of a sample of 50 college women.
The $SE_m = .440$
ANS: $s = 3.111$ or ($SD = 3.080$)

b. A researcher collects 100 IQ scores from a sample of high school seniors.
The $SE_m = 1.508$
ANS: $s = 15.080$ or ($SD = 15.004$)

c. A researcher collects Verbal SAT scores from a sample of 1000 high school seniors.
The $SE_m = 3.190$
ANS: $s = 100.877$ or ($SD = 100.826$)

Notice that as the sample sizes increase, the values for s and SD become increasingly close.

ESTIMATING THE POPULATION MEAN: INTERVAL ESTIMATES AND HYPOTHESIS TESTING

With the estimated standard error of the mean safely in hand, we can now turn to the problem of inferring the population mean, or the true mean of the entire distribution of means. Two ways in which this can be done are interval estimation and using a point estimate to test a hypothesis.

To predict where the true mean of the population, μ, might be, a point estimate may be made, whereby a single sample value is used to estimate the parameter. A random sample is selected and measured, and the sample value is used to infer a population value. Thus, the single inferred, or predicted, population value is called the **point estimate.** Due to sampling error, however, even this "best estimate" can never assure us of the true parameter value. Using the point estimate as a focal point, however, a range of possible mean values can be established within which it is assumed that μ is contained a given percentage of the time. The goal is to bracket μ within a specific interval of high and low sample means. A probability value can then be calculated that indicates the degree of confidence we might have that μ has really been contained within this interval. This estimated range of mean values is called the **confidence interval,** and its high and low values are called the *confidence limits.*

When the researcher wishes to estimate the actual point value of the population mean, μ, he or she selects a random sample and assumes the sample mean, M, represents the population mean, μ. Thus, the researcher is guessing a population parameter on the basis of a measured sample. Because of sampling error, this guess may, of course, not be valid. After all, sample means do deviate from parameter means. But because it is known (central limit theorem) that the distribution of sample means approaches normality, then most sample means will be fairly close to the population mean. As has been shown over and over again, under the normal curve, far more cases are going to lie near the center of the distribution than out under the tails.

Although the point estimate is a guess, it is nevertheless not a wild guess, but an educated guess. Inferring a parameter value from a sample measure is not just a shot in the dark. It may be a shot that doesn't always hit the bull's eye, but at least we can be sure that it is landing somewhere on the target. Why? Because the sample mean that has been selected *must* be one of the possible mean values lying along the entire sampling distribution of means. Hence, a point estimate of μ is a *hypothesized parameter.* This, in fact, was precisely the same logic used in the last chapter when a parameter mean for the population of children taking the reading test was inferred on the basis of a measured sample of those children. In that situation we used the z test to determine whether the population mean being predicted by the sample was the same as the known population mean in general. However, in that problem we were able to use z scores, since the standard deviation of the population was known and the shape of the distribution was known to be normal.

What about the real world, where all the population parameters are almost never known, and why does this matter? Because when selecting a sample, there is no absolute assurance of obtaining that beautiful, bell-shaped Gaussian curve, even though the population of sample means may indeed be normal. It must by now be fairly obvious that the smaller the sample, the less the chance of obtaining normality. It was exactly for that real world of population inferences that William Sealy Gossett, whom we first met in Chapter 1 and whom we will meet again in Chapter 10, constructed the t ratio. We now turn to the t distribution, a family of distributions each of which deviates from normality as a function of its sample size.

THE *t* RATIO

Degrees of Freedom and Sample Size

For the mean, **degrees of freedom (*df*)** are based on how many values are free to vary once the mean and the sample size are set. Let's say that we have five values ($N = 5$) and a mean of 3:

X
5
4
3
2
?
$\Sigma X = 15 = (N)(M) = (5)(3)$

Once the number of values and the mean are known, then ΣX is fixed, since $(N)(M) = \Sigma X$.

Thus, with an N set at 5 and ΣX set at 15, that last value is not free to vary. That last value must equal 1 (? = 1). So with five values, four of the five values are free to vary, but the last one is not, indicating that this distribution has $N - 1 = 4$ degrees of freedom. The larger the sample, naturally, the more the degrees of freedom and, assuming random selection, the closer the distribution gets to normality. The *t* ratio, unlike the *z* score, allows us to use smaller samples and compensate for their possible lack of normality via these degrees of freedom.

Two-Tail t Table

The two-tail *t* table, Appendix Table C, presents the critical values of *t* for probability values of both .05 and .01, each set for the various degrees of freedom. At the very bottom of the df (degrees of freedom) column, you'll see the sign for infinity, ∞. An infinite number of df identifies a situation in which the sample size is also infinite and a distribution that assumes normality. Under such conditions, the *t* distribution approaches the standard normal *z* distribution. Therefore, the values in the .05 and .01 columns are, respectively, 1.96 and 2.58. These, you may recall, are the very same values that exclude the extreme 5% and 1% of the normal curve, as illustrated in Table 8.1.

Thus, by chance, 95% of all *t* ratios will fall somewhere between ±1.96 and only 5% of all *t* ratios will be more extreme than ±1.96. Note, also, that as the df are reduced and as sample sizes get smaller, the *t* ratios needed to exclude those same extreme percentages of the curve get increasingly larger. For example, at 10 df we need a *t* ratio of at least ±2.228 to exclude the extreme 5% and ±3.169 to exclude the extreme 1%.

Single-Sample t Ratio (Parameter Standard Deviation Unknown)

Although the procedures that follow will seem very much like those used in the *z* test, there is one overwhelming difference. When the *z* was used, in Chapter 7, to compare a sample of children's reading scores with the population of reading scores in general, we knew for certain the population mean and also the real standard deviation of that distribution. In the next example, we will not know the population standard deviation.

TABLE 8.1 Two-tail *t* (selected values).

df	*.05*	*.01*
1	12.706	63.657
10	2.228	3.169
30	2.042	2.750
∞	1.960	2.576 = *z*

Example A researcher knows that the population mean among college students taking the new Social Conformity Test is a "neutral" 100. (Scores higher than 100 are assumed to represent more conformity than average, and those lower than 100 indicate less conformity.) A random sample of 30 students was selected, and the following values were recorded:

Sample mean:	$M = 103$
Estimated standard deviation of the population:	$s = 11$, or $SD = 10.830$
Sample size:	$N = 30$

From these values, we calculate the estimated standard error of the mean, either with s or SD.

$$SE_M = \frac{s}{\sqrt{N}} = \frac{11}{\sqrt{30}} = \frac{11}{5.48} = 2.01$$

$$\text{or with } SD,\ SE_M = \frac{10.830}{\sqrt{N-1}} = \frac{10.830}{\sqrt{29}} = \frac{10.830}{5.385} = 2.011$$

Either way, the answer is, of course, the same. This value is now used in the denominator of that special kind of z score that should be used whenever the standard error of the mean is being estimated from the sample scores, rather than known for certain.

$$t = \frac{M - \mu}{SE_M} = \frac{103 - 100}{2.011} = 1.491$$

The t ratio tells us specifically how far the sample mean deviates from the population mean in units of standard errors of the mean. With the t ratio, we will always be comparing *two* means—in this case, a known sample mean with an assumed population mean. Notice how closely the t formula resembles the formula used previously for the z test. The only difference is in the denominator. For the z test we were dividing the numerator by the parameter standard error of the mean, whereas with t we divide by an *estimate* of that important value.

Since the sample mean value of 103 is one of many possible means in an entire distribution of means, we may now ask what the probability is that a sample so chosen could deviate from the known μ of 100, in either direction, by an amount equal to 3 or more. We say "in either direction" because if the distribution of means is indeed normal, we could just as easily have selected a random sample mean of 97, or a value of 3 less than μ. When samples are chosen randomly, then sampling error itself is assumed to be random; that is, our sample mean has the same probability of being less than μ as it has of being greater than μ.

Parameter Estimates as Hypotheses

Whenever we assume what a parameter value might be, what we are doing, in effect, is guessing or *hypothesizing* its value. For our conformity problem we knew the parameter mean of 100. The null hypothesis would state that on the basis of chance, our sample mean of 103, now called our point estimate (μ_1), can be readily expected as one of the possible sample means in a distribution of means whose μ is equal to 100.

H_0, the **null,** or **"chance," hypothesis,** is written symbolically as

$$H_0: \mu_1 = \mu$$

where μ_1 represents the population mean being estimated by the sample and μ is the known mean of the population at large. Also, there is the alternative hypothesis, H_a, which states that the sample estimate for μ_1 is some value other than the true population mean. The alternative hypothesis is stating in this case that we cannot expect a sample mean of 103 if the true parameter mean is really equal to 100.

$$H_a: \mu_1 \neq \mu$$

Now, it is obvious that these two statements, H_0 and H_a, cannot both be simultaneously true. Either μ is equal to 100 or it isn't. We thus have to make a decision, called the statistical decision, to either reject H_0, in which case H_a is assumed to be true, or accept H_0, in which case H_a is assumed to be false. In either case, however, the statistical decision to reject or accept H_0 is a probability statement, and its accuracy can never be totally ensured. But making the decision using probability values of .05 or .01 certainly tips the scales in our favor; we're going to make the correct decision far more often than the incorrect one. Our statistical decision, however, never has absolute certainty. Although not infallible, scientists believe that proof without certainty is far better than certainty without proof.

Recall that our t had a value of 1.491.

$$t = \frac{M - \mu}{SE_M} = \frac{103 - 100}{2.011} = 1.491$$

Since we had 29 df, 30 – 1, Appendix Table C tells us that the critical value of t under the .05 column is equal to 2.045. We then compare our calculated value of t with the table value for the appropriate degrees of freedom. If the calculated t is equal to or greater than the table value of t, the null hypothesis is rejected. In this case, since 1.491 is neither greater than nor equal to the table value, our statistical decision is to accept the null hypothesis or, in other words, to state that μ_1 does equal μ. We conclude that there is no real difference between the population mean being estimated by our sample and the population mean that was known by the researcher.

Sign of the t Ratio

When making the t comparisons on Appendix Table C, the sign of the t ratio is disregarded. The importance of the t values in this table lies in their absolute values, since at this point we are testing whether a sample mean could have deviated by a certain amount from the population mean, not whether the sample mean was greater or less than μ, just whether it's different. A discussion of the other case, the directional or one-tail t, will be presented in Chapter 10.

Writing the t *Comparison*

As we just saw, when the table value of t was taken from the .05 column in Table C, it was found to be ± 2.045 and our calculated t value was 1.491. We would therefore conclude that a single sample t was computed comparing the sample mean with a known population mean of 100. No significant difference was found ($t(29) = 1.491$, $p > .05$). The null hypothesis was accepted, meaning that this sample could be representative of the known population. The statement that $p > .05$ means that with an alpha level of .05, the probability of committing a Type 1 error is GREATER than 5 chances in 100.

If on the other hand, the calculated t value had been higher—say 2.500—(which of course is greater than 2.045) we could then have said a significant difference was found ($t\ (29) = 2.500$, $p < .05$). In this case stating that $p < .05$ means that the alpha level probability of committing the Type 1 error was less than 5 chances out of 100. The null hypothesis then would have been rejected and we could conclude that the difference between the sample mean and the true population mean is so great that we must reject chance and conclude that this sample mean is probably not part of a distribution of means whose overall μ was indeed 100.

Significance

An extremely important statistical concept is that of **significance.** The research literature has many statements like "the means were found to be significantly different," or "the correlation was determined to be significant," or "the frequencies observed differed significantly from those expected." *Do not read too much* into the term "significance." Because of the connotations of the English language, the reader sometimes assumes that anything that has been found to be significant must necessarily be profound, heavy, or fraught with deep meaning. The concept of significance is really based on whether or not an event could reasonably be expected to occur strictly as a result of chance. If we decide that we can exclude chance as the explanation, we say that the event is significant. If we decide that the event is the result only of chance, it is considered not significant. There are therefore many significant conclusions in the research literature that are not especially profound. Some may even be trivial. It is necessary to understand the word *significant* to mean "probably not just a coincidence." Sir Ronald Fisher (see his biography in Chapter 12) told us many times that the basic intent of testing the null hypothesis is to answer the question, "Could this have been a coincidence?" When the null hypothesis is rejected, we assume that the results shown are probably not due to the vagaries of chance.

THE TYPE 1 ERROR

Researchers must be ever aware that the decision to reject the null hypothesis, or chance, involves a degree of risk. After all, in the area of inferential statistics, chance never precludes anything; it positively insists that somewhere, sometime, someplace some very freaky and far-out things will indeed occur.

Currently, some writers refer to certain rare events as having "defied chance." This is really an erroneous way to phrase the "slight probability" concept, for in inferential statistics nothing can be said to defy chance. One researcher in the area of extrasensory perception wrote that a certain subject guessed 10 cards correctly out of a deck of 25 cards made up of five different suits and that chance could not possibly explain this phenomenal string of successes. Chance, in this instance, predicts that only 5 cards should be correctly identified. While it is true that this particular subject did beat the chance expectation, this is certainly no reason to suggest that chance has been defied. The probability of winning the Irish Sweepstakes is far lower than the probability of guessing 10 out of 25 cards, but someone does win it, probably without the aid of any occult forces. The odds against any four people being dealt perfect bridge hands (13 spades, 13 hearts, 13 diamonds, and 13 clubs) are, indeed, astronomical, being something on the order of 2,235,197,406,895,366,368,301,599,999 to 1, and yet such bridge hands do occur.*

In inferential statistics, chance never totally precludes anything. Aristotle once said that it was probable that the improbable must sometimes occur. Or as the famous Chinese detective, philosopher, and critic of the human condition Charlie Chan put it, "Strange events permit themselves the luxury of occurring." In short, anytime we, as researchers, reject the possibility of chance (the null hypothesis), we must face the fact that we may be wrong—that the difference really was due to chance after all. For this reason, the researcher *always* prefaces the decision to reject chance with an important disclaimer. This is called the **Type 1 error,** and it occurs when the researcher erroneously rejects the null hypothesis. Please note that it can only occur when the null hypothesis is REJECTED. Thus, whenever the researcher makes the wrong decision by rejecting null (when it should have been accepted), the Type 1 error has been committed.

THE ALPHA LEVELS

Since we can never be absolutely certain that the rejection of null as the controlling factor is a correct decision, we must add a probability statement specifying the degree of risk involved. The probability of committing the Type I error is called **alpha** and it is typically set at either .05 or .01. Alpha should typically not be set higher than .05—a line has to be drawn somewhere and .05 is usually that line. Without it, someone could do a study and reject the null hypothesis with an alpha level of .50, meaning that the decision to reject chance in that case could have been wrong a rousing 50 chances out of 100. In this extreme example, it would have been easier to flip a coin than do the experiment.

Setting alpha at a maximum of .05 ultimately seems to be based on tradition rather than any hard-and-fast quantifiable rule. Back in 1925, Sir Ronald Fisher, the eminent

*These odds assume a random shuffle and dealing of the cards, a condition that is not usually met by most card players. Statisticians tell us that it takes at least seven of the standard riffle shuffles to get a good mix, whereas the average card player shuffles only three or four times.

British statistician and creator of ANOVA, said "It is convenient to draw the line of significance at about the level at which one can say: Either there is something in the treatment, or a coincidence has occurred such as can not occur more than once in twenty trials" (Fisher, 1925, p. 509).

Remember the alpha error by definition applies only when the null hypothesis has been rejected. If the null hypothesis is accepted, alpha is irrelevant. In a later chapter, the concept of the Type 2 error and its probability level, called *beta,* will be explained. This will become an especially important topic when we later discuss the "power" of a statistical analysis.

This discussion is intended to stress that chance can never be totally ruled out of any statistical decision. After all, when we reject the null hypothesis, it is entirely possible that we are wrong, that there really is no difference out there in the population. Similarly, when we accept the null hypothesis, we might also be wrong. Maybe there really is a significant difference in the population, even though we failed to detect it in the sample. Some researchers are so concerned about this issue that they dislike ever stating that the null hypothesis has been accepted. Instead, they prefer to state that they have *failed to reject* the null hypothesis. As long as you keep clearly in mind that accepting null is just as much a probability estimate as rejecting null, the phrase "accept null" will carry its intended meaning. Thus, rather than saying "failed to accept," we will consistently phrase this as *rejecting the null hypothesis,* even though there is still some small likelihood of being wrong. This is just one more attempt to loudly proclaim that statistical proof is probability proof, not a statement of eternal etched-in-stone truth.

The **single-sample *t*** has 3 main research applications:

1. The single-sample *t* can be used when a sample value is compared to a known population value in order to determine if the selected sample could or could not be representative of the known population. In this case the known population value has already been established on the basis of previous research and, with standardized tests, can be found in the test manual. The *t* ratio then compares the sample mean to the known parameter mean in an attempt to establish whether the two values may differ. If the *t* ratio does show a significant difference, the researcher may conclude that the sample does not represent the known population.

An educational researcher theorizes that a new training technique may be used to increase reading comprehension among dyslexics. The researcher selects a sample of 10 children that the school psychologist has identified as dyslexic, but the researcher wants to be sure that they represent the population of known dyslexics as shown on the nationally standardized test. The standardized test's manual indicates that the population mean for diagnosed dyslexics is 80. Before proceeding with the study, the researcher gives the test to the sample, and their scores are given in the following table.

X	X^2
98	9,604
85	7,225
82	6,724
81	6,561
81	6,561

Continued

Note: scores are continued from previous page.

X	X^2	
80	6,400	
79	6,241	
79	6,241	
78	6,084	
64	4,096	
$\Sigma X = 807$	$\Sigma X^2 = 65,737$	$M = \frac{\Sigma X}{N} = \frac{807}{10} = 80.700$

Using s,

$$s = \sqrt{\frac{\Sigma X^2 - [(\Sigma X)^2/N]}{N-1}} = \sqrt{\frac{65,737 - 807^2/10}{9}}$$

$$s = \sqrt{68.011} = 8.247$$

$$SE_M = \frac{s}{\sqrt{N}} = \frac{8.247}{\sqrt{10}} = 2.608$$

OR Using SD,

$$SD = \sqrt{\frac{\Sigma X^2}{N} - M^2} = \sqrt{\frac{65,737}{10} - 80.70^2}$$

$$SD = \sqrt{61.210} = 7.824$$

$$SE_M = \frac{SD}{\sqrt{N-1}} = \frac{7.824}{\sqrt{9}} = 2.608$$

$$t = \frac{M - \mu}{SE_M} = \frac{80.70 - 80}{2.608} = .268$$

$t_{.05(9)} = \pm 2.262$ Accept H_0; difference is not significant

Since the Table C value of t at the .05 level is found to be $\pm$ 2.262, this would be written as follows:

A single sample t was calculated comparing the sample mean of 80.70 with the known population mean of 80. No significant difference was found ($t(9) = .268$, $p >$ .05). The null hypothesis was therefore accepted, indicating that this sample probably could be representative of the known population.

Even though we are accepting null in this example, theoretically we are not proving it to be true. As R. A. Fisher stated:

> It should be noted that in theory the null hypothesis is never proven or established, but is possibly disproved in the course of experimentation. Every experiment may be said to exist only in order to give the facts a chance of disproving the null hypothesis (Fisher, 1935, p. 19).

Although, as Fisher said, accepting the null hypothesis does not prove it to be true, researchers usually act as if it were true. Robert Malgady argues that if an experiment is conducted using a new drug for treating depression, and if the results of the study fail to reject the null hypothesis, then the drug will certainly not be prescribed for persons suffering from this disorder. Here the null hypothesis would have stated that the two groups, experimental (drug) and control (placebo), do not differ with regard to alleviating depressive symptoms (Malgady, 1998), and the medical community would then agree.

2. In the second case, the sample values are compared with the chance level of performance on some set of values. Suppose, for example, that a researcher in ESP (extrasensory perception) wishes to find out if a sample of college sophomores can exhibit ESP by outperforming the chance level of expectation. A deck of ESP cards (known as Zener cards) is used. This deck includes 25 cards arranged in five suits, in this case symbols: a star, circle, square, plus sign and parallel wavy lines. Thus, on the basis of chance alone, the subjects should guess an average of 5 correct hits (1/5 times 25). The mean hits for the sample would then be compared to this chance prediction. A random sample of 10 subjects was selected and asked to "guess" which card would appear next as the experimenter went through the Zener deck one card at a time. The number of correct "hits" were as follows: 6, 4, 8, 7, 6, 3, 9, 4, 5, 4. The researcher now runs a single sample t test to see if the actual scores differed significantly from the chance expectation of 5 "hits."

$$M = 5.600$$

$$s = 1.955, \text{ or } SD = 1.855$$

$$SE_M = .618$$

$$t = \frac{M - \mu}{SE_M} = \frac{5.600 - 5.000}{.618} = \frac{.600}{.618}$$

$$t = .971$$

Thus, with 9 degrees of freedom and an alpha level of .05, the researcher would need a t ratio of 2.26 to reject the null hypothesis. The decision in this case, then, is to accept null. The results thus failed to show that this sample exhibited ESP.

3. In the third case the sample scores are compared to a standard value that is known to be true. In testing for the Müller-Lyer optical illusion, for example, subjects are asked to adjust the length of one line until it appears to be equal in length to a standard line. These estimated lengths, sample values, produced by the subjects may then be used to estimate a parameter mean that can be compared, via the t test, with the physically true length of the standard line. If the t ratio shows a significant difference, the researcher may conclude that a perceptual illusion has indeed occurred. A random sample of 10 subjects is selected and shown the following figure.

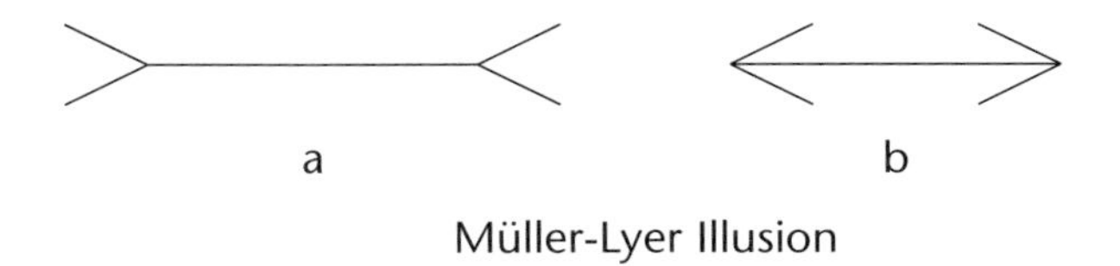

Müller-Lyer Illusion

They are then presented with a kind of slide rule that portrays the arrow shown in position "b" and asked to adjust its length to correspond with the length of a standard line "a." (In fact, both lines are identical in length.) The theory would suggest that if the illusion is valid, the subjects will overestimate the length of line "a" by setting "b" too high. The scores of the subjects, in inches, were as follows: 12, 14, 15, 11, 12, 13, 14, 16, 14, and 12. The actual physical length of line "a" was 10 inches.

$$M = 13.300$$
$$s = 1.567, \text{ or } SD = 1.487$$
$$SE_M = .496$$
$$t = \frac{M - \mu}{SE_M} = \frac{13.300 - 10.000}{.496} = \frac{3.300}{.496}$$
$$t = 6.653$$

Setting alpha at .01, a t ratio of 3.250 is needed to reject the null hypothesis with 9 degrees of freedom. Since the calculated value of t is clearly greater than the table value, the null hypothesis is rejected and the theory that a perceptual illusion has occurred is supported.

EFFECT SIZE

Some significant differences may be so slight as to serve no real useful purpose. The difference in visual acuity between persons who measure 20/20 versus persons who measure 20/30 or 20/15 may be statistically significant and yet make no difference at all regarding the visual skill that produces driving records or numbers of accidents. Although this will be a more important issue in the next chapter when discussing the two-sample t test, it should at least be mentioned here. For example, in the problem shown above where the researcher was attempting to determine if there really was such a phenomenon as the Müller-Lyer illusion, a t ratio of 6.653 was obtained. Since it led to a rejection of the null hypothesis, we know that it was significant, and, therefore, an effect did occur. But the question now is, how strong was the effect? To answer that, we must calculate what is called an **effect size,** or ES, and for the single-sample t test, the effect size is symbolized as "d." Regardless of sign, an effect size when reported as d is considered small at .20, medium at .50, and strong at .80 or greater.

For the single-sample t ratio, the effect size (ES) is calculated as

$$d = \frac{t}{\sqrt{N}}$$

Thus, for the Müller-Lyer problem shown above, the t ratio was 6.653 and the N (sample size) was 10.

$$d = \frac{t}{\sqrt{N}} = \frac{6.653}{\sqrt{10}} = \frac{6.653}{3.162} = 2.104 \text{ (a strong effect)}$$

It must be noted that some have argued against the use of effect-size estimates. For a full discussion of this issue, see Jacard (1998).

PROBLEMS

For the following single-sample t ratios, find d, the effect size.

a. With a t ratio of 2.65, and an N of 25, find the effect size.
ANS: $d = .53$ (a medium effect)

b. With a t ratio of 2.03, and an N of 50, find the effect size.
ANS: $d = .288$ (a small effect)

INTERVAL ESTIMATES: NO HYPOTHESIS TEST NEEDED

As stated earlier, population means may also be inferred by setting up a range of values within which there can be some degree of confidence that the true parameter mean is likely to fall; that is, once the sample mean has been used to predict that μ is equal to, say, 100 (the point estimate), the interval estimate then suggests how often μ will fall somewhere between two values, say, 95 and 105. In this case, the two values 95 and 105 would define the limits of the interval and are therefore called confidence limits. Predictions of confidence intervals are probability estimates. A confidence interval having a probability of .95, for example, brackets an area of the sampling distribution of means that can reasonably be assumed to contain the population mean 95 times out of every 100 predictions. Similarly, a confidence interval set at a probability level of .99 (a somewhat wider interval) can be assumed to contain the parameter mean 99 times out of every 100 predictions.*

Since the parameter can only be inferred in this case, not known for certain, we can never determine whether any particular interval contains it, as tempting as that prospect might be. With interval estimates, we neither accept nor reject the null hypothesis, since with this procedure we are not testing a hypothesis. There are some statisticians who would prefer to use this technique for that very reason, since they feel that hypothesis testing has not been the best approach to theory building and theory testing (Haig, 2000). The interested student will find that this issue has been fully addressed by the Task Force on Statistical Inference (Wilkinson, 1999).

As we have seen, it was Fisher who created NHST (null hypothesis significance testing), and those quantitative studies that were done prior to Fisher usually used point estimates and confidence intervals (Oakes, 1986). Today, despite some objections regarding the possible over-reliance on Fisher's techniques, most statisticians believe that null hypothesis testing and confidence intervals are not mutually exclusive and should be a part (along with meta-analysis) of every researcher's tool box (Howard, Maxwell, and Fleming, 2000).

---

*Although technically the probability of a parameter falling within a given interval is .50 (it either does or it doesn't), the concept of "credible intervals" allows for the definition of a probability value for a given parameter being contained within a calculated interval. For a further discussion of this technical issue, see Hays and Winkler (1975).

Confidence Intervals and Precision

Of course, the larger the predicted range of mean values, the more certain we can be that the true mean will be included. For example, if we are trying to predict the average temperature for the next Fourth of July, we can be very sure of being correct by stating limits of 120°F on the high side and 0°F on the low side. Though we may feel very confident about this prediction, we are not being at all precise. This is not the kind of forecast to plan a picnic around.

The same holds true for predicting the parameter mean—the greater the certainty of being right, the less precise the prediction. Typically, statisticians like to be right at least 95% of the time, which gives a prediction a 95% confidence interval. Some statisticians demand a confidence interval of 99%, but as we shall see, this level creates a broader, less precise range of predicted values. Confidence and precision must be constantly traded off; the more we get of one, the less we get of the other.

Also, the larger the sample size, the narrower becomes the confidence interval for a given set of values and thus the more precise the eventual prediction. This is due to the fact that as the sampling distribution approaches normality, the area out under the tails of the curve becomes smaller. Later, in chapter 10, there will be a discussion of an important statistical concept called "power". It will then become obvious that as the confidence interval narrows, there is an increase in power (see McClelland, 2000).

Calculating the Confidence Interval

Example With a sample of 121 cases, a sample mean of 7.00, and an estimated standard error of the mean of .56, between which two mean values can the true mean be expected to fall 95% of the time?

Since these are sample data, we must first find the t value that corresponds to the z of ±1.96, that is, the t values that include 95% of all the cases. Appendix Table C shows that for 120 df ($N - 1$), the t ratio at the .05 level is equal to ±1.980—the t value that excludes the extreme 5% and therefore *includes* the middlemost 95%. Because the sample is so large, this particular t value is extremely close to the 1.96 value found under the normal curve.

The predicted population mean, μ, falls within an interval bounded by the product of the table value of t, times the estimated standard error of the mean, plus the mean of the sample. We use the mean of the sample because that is our point estimate of μ.

$$\text{confidence interval} = (\pm t)(SE_M) + M$$

Since we don't have the population parameters in this equation, we can't use z or the population standard error of the mean, σ_M. But we do have the next best thing to these parameters: their estimates, t and SE_M. Using the previous data, we get at .95 and 120 df, C.I.

$= (\pm 1.980)(.56) + 7.00$

upper limit of C.I.

$= (+1.980)(.56) + 7.00 = 1.109 + 7.00 = 8.109$

lower limit of C.I.

$= (-1.980)(.56) + 7.00 = -1.109 + 7.00 = 5.891$

Thus, we are estimating that 95% of the time, the true mean value will fall within an interval bounded by 8.109 on the high side and 5.891 on the low side.

A frequency of 95% generates a probability value of .95. Therefore, at a probability level of .95, the confidence interval is between 5.891 and 8.109. If we wanted to predict at the .99 confidence interval, we would simply extend the range by going farther out into the tails of the distribution. For the examples that follow, remember that *two* mean values must be determined in order to establish the limits for each interval.

Example Find the .95 confidence interval for the following sample measures. A random sample of 25 persons yields a mean of 200 and an SE_M of 2.86.

at .95 and 24 df, C.I.

$= (\pm 2.064)(2.86) + 200$

upper limit of C.I.

$= (+2.064)(2.86) + 200 = 5.903 + 200 = 205.903$

lower limit of C.I.

$= (-2.064)(2.86) + 200 = -5.903 + 200 = 194.097$

The limits for the .95 confidence interval are therefore 205.903 and 194.097.

Example Find the .99 confidence interval for the following sample measures. A random sample of 41 persons yields a mean of 72 and an SE_M of 1.39.

at .99 and 40 df, C.I.

$= (\pm 2.704)(1.39) + 72$

upper limit of C.I.

$= (+2.704)(1.39) + 72 = 3.759 + 72 = 75.759$

lower limit of C.I.

$= (-2.704)(1.39) + 72 = -3.759 + 72 = 68.241$

The limits for the .99 confidence interval are therefore 75.759 and 68.241.

We are able to use our sample mean in the confidence interval equation because we know that it is one of the possible values in the entire distribution of sample means. Further, since the central limit theorem states that the distribution of sample means approaches normality, we know that most of the values in the distribution fall close to the true mean. Of course, it is possible that our particular sample mean is one of those rarities that falls way out in one extreme tail of the distribution—it is possible but not very probable. That, after all, is the main point of inferential statistics: Estimates are made on the basis of probability, not on the basis of proclaiming eternal truth. Thus, although we are at the .95 or .99 level of confidence for our prediction of the true mean, the possibility of being wrong always exists. When we are at the .95 confidence level, we will be wrong at a probability level of .05 or less, and when we are at the .99 confidence level, we will only be wrong at a probability level of .01 or less. There's no sure thing in statistics, but with chances like these, this technique sure beats flipping a coin.

Confidence Interval from Data

Using the data found back on page 176, we will now calculate a confidence interval. At the .95 level of confidence, we would look up the t value at the .05 alpha level (and, of course, at the .99 confidence level we would use the alpha of .01). This is because the probability of a given portion of the distribution falling within a certain confidence interval is equal to 1 – alpha. That is, confidence probability + the alpha level must always equal 1.00.

1. The mean was found to be 80.700
2. The standard error of the mean is 2.608

Placing these values in the equation, and recalling that with 9 degrees of freedom and alpha set at .05, Table C tells us that we needed a t ratio of 2.262.

$$\begin{aligned} \text{CI } .95 &= (\pm t)(SE_m) + M \\ &= (\pm 2.262)(2.608) + 80.700 \\ &= +5.899 + 80.700 = 86.599 \\ &= -5.899 + 80.700 = 74.801 \end{aligned}$$

Confidence and "the Long Run"

As we talk about being right or wrong regarding the various confidence intervals, we must remember that sample means and estimated standard errors are statistics—sample measures—and as such are subject to random variation. If we were to draw a second or third or fiftieth sample, we would expect, due to sampling error, to obtain slightly different values. Each of these different sample values could then produce a somewhat different confidence interval. However, if we were to use a confidence level of .99, then 99% of these predicted intervals should contain μ, while 1% will not. Thus, in the words of the noted American statistician Allen Edwards, "in the long run we may expect .99 of our estimates about μ, based upon a 99% confidence interval, to be correct" (Edwards, 1967, p. 208).

The Single Sample t *and the Confidence Interval*

Choosing to spotlight the single sample *t* and the confidence interval in the same chapter was, as you may have suspected, no accident. The two procedures bear a marked similarity and are, in fact, two sides of the same coin. With the *t* ratio, the attempt was made to determine whether a given sample mean could be representative of a population whose mean was known. This procedure demands that we test the null hypothesis. With the confidence interval, however, no hypothesis test was involved. A given sample mean was used as a point estimate, around which we attempted to bracket the true population mean.

Let's look back at the example of the single-sample *t* on page 178. In that problem we rejected the null hypothesis and stated that the sample mean of 13.300 could probably not come from the general population, whose mean was known to be 10. Using the data from that problem, a .99 confidence interval could be set up.

at .99 and 9 df, C.I.

$= (\pm 3.250)(.496) + 13.300$

upper limit of C.I.

$= +1.612 + 13.300 = 14.912$

lower limit of C.I.

$= -1.612 + 13.300 = 11.688$

As you now can see, the known mean of 10 does not fall within the .99 limits produced by the sample mean of 13.300. This was the decision the *t* test made previously and thus echoes the notion that the two means, 13.300 and 10, probably represent different populations. Notice also that had the known mean been within the confidence interval that is, anywhere between 14.912 and 11.688, the single-sample *t* would have led us to *accept the null hypothesis,* and the two means would have been evaluated as probably representing the same population.

SUMMARY

Since the central limit theorem states that under certain conditions the distribution of sample means approaches normality, then the estimated standard error of the mean may be used to make probability statements regarding where specific sample means might fall as well as where the true parameter mean might be found. Methods for predicting where the true mean of the population might fall are based on the *t* distributions, a family of distributions that are all expected to deviate from normality as a function of the sample size.

The *t* ratio may be used for assessing the likelihood of a given sample being representative of a particular population. The *t* ratio is therefore a test of the null hypothesis. The null, or "chance," hypothesis states that the sample is representative of the population, and the alternative hypothesis states that it is not. Since, in inferential statistics, chance never precludes anything, the decision to reject the null hypothesis could be erroneous—in which case a Type 1 error has been committed. The alpha level then sets the probability value for the Type 1 error, typically .05 or less.

The single sample *t* has 3 main research applications:

1. It can be used when a sample value is compared to a known population value in order to determine if the selected sample could be representative of that population.
2. Sample values can be compared with the chance level of performance on some set of values.
3. The sample scores are compared to a standard value that is known to be true.

When a *t* ratio is found to be significant, its effect size (ES) can be calculated using the equation for *d*. This tells us how strong the effect is, ranging, regardless of sign, from small effects of .20, through medium effects of .50, and on to strong effects of .80 or greater.

Point estimates are hypothesized parameters based on sample measures. The point estimate is a sample value that is supposed to provide for the "best" estimate of a single parameter value. Due to sampling error, however, even this "best" estimate can never fully assure us of the true parameter value. Researchers use interval estimation to predict the parameter mean by bracketing, between high and low values, a range of mean values within which the population mean is estimated to fall.

Key Terms

alpha level
confidence interval
degrees of freedom (df)
effect size
estimated standard deviation
estimated standard error of the mean
null, or "chance," hypothesis
point estimate
significance
single-sample *t* ratio
Type 1 error
unbiased estimate

PROBLEMS

1. A random sample of 25 elementary school children was tested on the Stanford–Binet, yielding a sample mean mental age of 6.25 years with an SE_m of .180. On the basis of this information, estimate whether this sample could represent a population whose mean is 6.00 years. Test H_0 at the .05 level.

2. A researcher in sports medicine hypothesizes a mean resting pulse rate of 60 beats per minute for the population of long-distance college runners. A random sample of 20 long-distance runners was selected, and it was found that the mean resting pulse rate for the sample was 65 with an SE_m of .690. On the basis of this sample information, estimate whether this sample could represent a population whose mean is 60. Test H_0 at the .01 level.

3. On a nationally standardized reading comprehension test, the norm for the population mean was designed to yield a scaled score value of 50 for the population of fifth-grade students. A random sample of 10 elementary school children was tested, yielding scores of 48, 42, 55, 35, 50, 47, 45, 45, 39, and 42. Estimate whether this sample might be representative of the population for whom the test was designed. Test the hypothesis at the .05 level.

4. A researcher knows that the national mean on the verbal SAT is 500. A sample of 121 seniors from a high school in an affluent suburb of Cleveland yields a mean of 519 and an estimated standard error of the mean of 8.86. Test the hypothesis at the .05 level that this sample is representative of the national population.

5. Based on the data in problem 4, use the given sample measures and
   a. produce a point estimate of the population mean.
   b. construct a .95 confidence interval for your estimate.
   c. Would the national mean of 500 be included within the interval?

6. Given a sample mean of 35, an N of 41, and an estimated standard error of the mean of .80,
   a. calculate the .95 confidence interval.
   b. calculate the .99 confidence interval.

7. A researcher selects a random sample of 10 persons from a population of truck drivers and gives them a driver's aptitude test. Their scores are 22, 5, 14, 8, 11, 5, 18, 13, 12, and 12.
   a. Calculate the estimated standard error of the mean.
   b. At the .95 probability level, calculate the confidence interval.

8. A set of 10 random sample scores is collected: 10, 16, 5, 9, 9, 11, 8, 12, 10, and 9.
   a. With alpha set at .05, determine whether this sample could be representative of a population whose parameter mean is known to be 8.13.
   b. Using the t value obtained in part (a) and the same alpha level of .05, determine whether a sample of infinite size could be representative of a known population whose mean is 8.13.

9. A random sample of nine recreational vehicles is selected from a manufacturer's production line. Each is given a road test, and the fuel efficiency is computed for each vehicle. The following miles-per-gallon rates were achieved by the vehicles: 15, 14, 7, 5, 5, 10, 12, 9, and 9.
   a. Calculate the estimated standard error of the mean.
   b. Calculate the .95 confidence interval.
   c. Calculate the .99 confidence interval.

10. A random sample of prison inmates was selected and checked for the number of filled cavities found for each man. The data follow, each score indicating the number of filled cavities: 7, 6, 5, 10, 9, 8, 15, 10, 10, 10. Could this sample represent a prison population whose parameter mean was 9.500?

11. A random sample of male college students who had tried out for the varsity basketball team yielded the following height scores, inches: 68, 72, 76, 70, 69, 73, 74, 70, 78,

70. If the population for adult males yields a parameter mean of 69, could this sample be representative of the known population?

12. In an attempt to test for the horizontal-vertical illusion (where the vertical line should be perceived as being longer than a horizontal line of equal length), a random sample of college freshmen was selected and asked to judge the length of a vertical line (line b) after having been told that the width of the horizontal line (line a) was 5 inches. The estimates (in inches) were as follows: 7, 7, 9, 7, 6, 7, 7, 8, 6, 6. Test the hypothesis at the .05 level the illusion occurred.

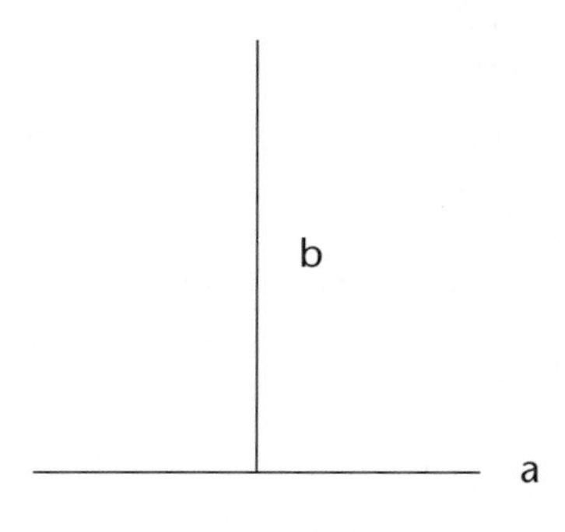

Horizontal-Vertical Illusion

13. If the t ratio in the previous problem was significant, calculate the ES (effect size) and indicate if it is strong, moderate, or weak.

14. The Epworth Sleepiness Scale (ESS) was designed to measure whether a person is getting enough sleep. The test asks such questions as how likely you are to doze off in a variety of situations, such as watching TV, or at a meeting, or even as a passenger in a car on a long ride. The test is normed to produce a mean score of 7.50 (Johns, 1992). A random sample of college students was selected and the following scores were obtained. 6, 10, 4, 3, 8, 12, 5, 7, 9, 4, 10, 2, 7, 5, 4, 7, 8, 11, 3, 7. Could this sample be representative of a population whose mean is 7.50?

15. A researcher selects a random sample of 10 college students, all of whom are majoring in philosophy. Each student is tested on the CTA, Critical Thinking Aptitude test, and their scores were as follows: 12, 5, 3, 7, 6, 6, 5, 2, 5, 6. Could this sample be representative of a population whose known parameter mean was 7.00?

Fill in the blanks in problems 16 through 28.

16. For the single-sample t ratio, degrees of freedom are equal to __________.

17. As the degrees of freedom increase (as sample sizes get larger), the t distributions approach what other important distribution? __________

18. When the calculated value of t is equal to or greater than the table value of t for a given number of degrees of freedom, the researcher should __________ the null hypothesis.

19. The statement that the sample is representative of the same parameter as that assumed for the population at large is called the __________ hypothesis.

20. To establish a "significant difference," what decision must first be made regarding the null hypothesis? __________

21. When the null hypothesis is rejected, even though it should have been accepted, the __________ error has been committed.

22. With a confidence interval of .95, the probability of not including the parameter mean is equal to __________ or less.

23. Assuming a sample of 12 subjects, what t value would be needed to reject H_0 at an alpha level of .01? __________

24. With infinite degrees of freedom, what t ratio is needed to reject H_0 at the .05 alpha level? __________

25. The range of mean values within which the true parameter mean is predicted to fall is called the __________.

26. For a given sample size, setting the alpha error at .01 rather than .05 makes it __________ (more or less) likely to reject the null hypothesis.

27. A point estimate is a hypothesized parameter value that is based on a(n) __________ value.

28. Changing the confidence interval from .95 to .99 __________ (increases or decreases) the range of mean values within which μ is assumed to fall.

True or False. Indicate either T or F for problems 29 through 33.

29. The larger the standard deviation of the sample, for a given N, the larger is the estimated standard error of the mean.

30. For a given value for the sample's standard deviation, increasing the sample size decreases the estimated standard error of the mean.

31. Because of sampling error, a point estimate can never be guaranteed to produce the true population parameter.

32. A point estimate is a sample value that is said to provide the "best" estimate of a single parameter value.

33. In the long run we may expect .99 of our estimates, based on a 99% confidence interval, to be correct.

COMPUTER PROBLEMS

C1. A science teacher constructs a 100-item, multiple-choice test, each item having four alternatives. Concerned that the sample of students she has this year may not be performing up to their abilities, she decides to check whether her students' scores on the test are any different from what could be expected on the basis of chance. If chance were operating, she assumes that the population mean on the test should equal 25. The scores earned by her students are as follows:

25	32	21	26	24	31	18
23	27	20	30	24	26	
42	26	12	38	24	34	
25	40	34	35	32	25	

a. With alpha set at .05, test the hypothesis that this sample could be representative of a population whose parameter mean is 25.

b. Using the t value obtained in part (a) and the same alpha level of .05, determine whether a sample of infinite size could be representative of a population whose mean is 25.

C2. A random sample of 25 male college students is selected and given the "Attitude Toward Women" scale (higher scores indicating a more enlightened attitude). The scores are

15	12	11	3	11	10	12
9	8	9	13	10	11	
12	8	9	9	12	13	
11	13	18	14	15	12	

a. Calculate the estimated standard error of the mean.

b. Test the hypothesis that this sample could be representative of a population whose mean is 10.

C3. The following lists the serum cholesterol levels for a group of 25 adult males. Test the hypothesis that this group could be representative of a population whose mean is 200. These measures are in units called milligrams per deciliter (mg/dl).

195	165	150	245	200	194	285
200	180	210	193	188	205	185
197	193	196	194	189	196	201
194	195	195	175			

C4. The following is a list of diastolic blood pressure readings for a group of 25 adult females. Test the hypothesis that this group could be representative of a population whose mean is known to be 80. The measures are in units called millimeters of mercury (mm Hg).

85	84	83	87	88	90	60
112	110	100	75	84	85	
85	86	90	80	87	83	
86	84	85	90	85	82	

C5. A random sample of eight-year-old students, all of whom tested as LD and in fact were all receiving special education services, was selected and given the DIAL (Developmental Indicator for the Assessment of Learning) test. The following are the raw scores on the Concepts section of the test. The population mean for this subtest is known to be 20 for the non-LD population. Test the hypothesis that this sample could be representative of the non-LD population.

38	25	23	22	22	20	20
18	18	19	19	19	19	20
2	15	16	17	17	18	

Chapter

9

The Fundamentals of Research Methodology

Before we introduce the major tests of statistical significance, we should take a short time-out from the numbers and take a look at what these numbers are going to be used for. Without some knowledge of research methodology, the bare calculation of statistical tests can be an empty exercise in mathematical procedures. Without some background as to how and when they are used and how their results are interpreted, statistical tests may just be a hodgepodge of sterile and meaningless calculations. This chapter, therefore, provides an overview of the fundamental concepts of research methodology. The importance of these fundamentals cannot be overemphasized. Basic knowledge of the logic of research procedures puts you in a position to understand a goodly portion of the statistical research in your field. Without this knowledge, research literature may be somewhat like a credit card debt—easy to get into, but almost impossible to get out of. Besides, the fundamental research methods are not difficult to comprehend.

It's the ability to analyze and understand the data that breathes warm life into cold numbers. Merely reading research conclusions, without attempting to find out if they are justified, puts you at the mercy of the researcher's often subtle and perhaps even unconscious biases. This chapter, then, is a mixture of one part technical procedures and one part honest skepticism.

We often hear the derogatory comment, "Statistics can be used to prove anything." Sometimes we are addressed with the equally cynical, and somewhat plaintive, reproach, "But that's only a statistical proof." If it were true that one can prove anything with statistics, then, of course, there would be a real question as to the value of statistical analysis. The fact is, however, that the only time one can prove anything *with statistics is when the audience is totally ignorant of statistical procedures. To the uninitiated, liars can indeed figure, and their figures may be seductively plausible. By the time you finish this chapter, they will not easily be able to lie to you.*

RESEARCH STRATEGIES

Although there are a variety of general research methods, such as naturalistic observation, descriptive, historical, case studies, and surveys, the emphasis in this chapter will be on the two major research strategies that demand quantitative inferential techniques of hypothesis testing.

The two major types of research strategies are called *experimental* and *post facto.* When performed correctly, both are perfectly legitimate, but they do lead to different types of conclusions. Experimental research offers the opportunity for drawing direct cause and effect conclusions, whereas post-facto research typically does not. In fact, post-facto research does not fully address the issue of cause and effect; it can't prove causation, nor can it disprove it. As we shall see later in the chapter, experimental methodology comes in two forms, the true experiment and the quasi experiment. Although the quasi experiment does not offer the same unambiguous analysis of possible causation as the true experiment provides, it does allow the researcher to make a stronger deductive argument regarding the possibility of causation than does the straight post-facto analysis. Thus, the problem boils down to this: If the researchers have used experimental methodology, then they are allowed to discuss the possibility of having isolated a direct causal factor, but if the research is post facto, they may not. And as a person new to the field, you must keep in mind when reading research that it's not enough just to read the report's conclusions. You must examine the entire methodology to be sure that the author's conclusions are indeed justified. Unfortunately you just can't take the author's word for it when it comes to the "conclusions" section of the report. You must bore in yourself and carefully read the "methods and procedures" sections with a critical eye. Above all, don't assume that something was done simply because it seems obvious to you that it should have been done. You'll soon learn that what ought to be isn't always what is!

Variables and Constants

A variable is anything that can be measured, and although it might seem to go without saying, that *varies.* A person's height, weight, shoe size, intelligence, and attitudes toward war may all be measured, and, in comparison with measures on other people, all have the potential to vary. Indeed, the fact that people differ on a whole host of personal measures gave rise in the late 1800s to the study of what Sir Francis Galton called "individual differences." No two people are exactly alike on all human measures (even identical twins reared together are not exact personality clones), and thus the study of individual differences became the major theme song in psychology, education, and the social sciences. But, just because people differ on a wide range of personal measures, it does not mean that each and every study of potential differences has allowed all these measures to vary. Or, to state it differently, potential variables are not always treated by the researcher as variables. This is, after all, the way it should be. Each single study should not be obliged to allow all possible measures to vary simultaneously. Some measures should be left as *constants* or measures that are not allowed to vary. Here's an example.

It is known to all researchers in the area of growth and development that among elementary school boys there is a dependable relationship between height and strength—that is, taller boys tend to be stronger than shorter boys. Now, although this is a perfectly valid piece of research evidence, and it certainly allows the researcher to make better-than-chance strength predictions using height as the predictor, it can lead to some fuzzy interpretations if other variables are not controlled. For example, on the basis of the height-strength correlation alone, one might speculate that longer muscles produce more strength than do shorter muscles. The problem with this interpretation so far is that all other measures have been free to vary, and it may be that some of these other measures correlate even higher with strength than does the height measure. One that comes quickly to mind is age. Certainly 12-year-old boys tend to be both taller and stronger than 6-year-old boys. What would happen, then, if we were to hold age as a constant?

Suppose that we were to select only boys of age 10 years 6 months and then attempt to assess the height-strength relationship. When this is done (and it has been), the original correlation between height and strength becomes minimal. Speculation now might be concerned with the maturity of the muscle, not its length, as the important component in strength. The point of all this is that when, as in this case, only subjects of the same age are used, age, although certainly measurable, is *no longer a variable.* In reading research articles, then, be especially careful to review how the subjects have been selected.

If the selection process has been designed to ensure that the sample is identical on some measured trait, then for that particular study the common trait is not a variable but a constant. This is also true of studies in which environmental conditions are under examination. For example, a researcher may suspect that warm-color room decors are more conducive to relaxation than are cool-color decors. Two groups of college students are selected, and one group is taken to a room with a red-yellow decor and the other to a blue-green room. The subjects are then connected to some biofeedback equipment, which is supposed to assess their levels of relaxation. Here again, certain potential environmental variables should be held constant if indeed the researcher hopes to get at the effects of room color. The groups should be tested at the same time of day and under the same conditions of illumination, temperature, noise level, and so on. That is, the researcher should make every effort to control as many potential environmental variables as is possible—in short, to convert these other variables into constants to minimize their possible influence. Of course, a study of this type should also be designed to control individual difference variables as well. For example, one group should not be composed only of overweight, phlegmatic persons and the other, underweight, tense persons. Much more will be said on this topic later under the heading "Confounding Variables."

INDEPENDENT AND DEPENDENT VARIABLES

Scientists identify two major classes of research variables (variables, not constants) as independent and dependent. It is typically assumed that the **independent variable**

(IV) precedes the **dependent variable** (DV). Thus, in any antecedent-consequent relationship, the independent variable is the antecedent and the dependent variable the consequent. Your ability to identify these variables is absolutely critical to your understanding of research methodology. In some studies, which as we will see later are called "experimental," the IV is truly antecedent in every sense of the word. In these studies, the IV is assumed to be the causal half of the cause-and-effect relationship, with the DV being the effect, or consequent, half. For example, in the study mentioned earlier on room color and relaxation, the room color (the presumed cause) would be the IV and the amount of resulting relaxation the DV. In another type of study, which will be called *post facto*, the IV is antecedent only to the extent that it has been chosen (sometimes arbitrarily) as the predictor variable, or the variable used from which to make the prediction. The DV in this type of research is the consequent variable to the extent that it is the variable being predicted. For example, in most of the TV-violence studies the researchers were trying to use type of TV viewing to predict the extent of viewer aggression. The IV, then, was the type of TV that was viewed and the DV would be the amount of overt aggression being displayed. In general, then, the IV-DV relationship is similar to an input–output relationship, with the IV as the input and the DV as the output. Also, the DV is usually some measure of the subject's behavior, and in this case, behavior is being defined in its broadest possible context. The DV might be a behavioral measure as obvious as the subject's performance on an IQ or attitude test, or as subtle as the subject's incidence of dental cavities, or even the measured amount of acetylcholine at the neural synapse.

Researchers distinguish between two categories of independent variables, manipulated and subject. Moreover, this distinction is absolutely crucial to an understanding of research methodology. A decision regarding which type of IV is involved in turn determines which type of research is being conducted—and, hence, the kinds of conclusions that are permissible. A manipulated IV is one where the researcher has actively changed the environmental conditions to which the sample groups are being subjected. That is, with a manipulated IV, the researcher determines which groups of subjects are to be treated in which particular ways. Perhaps one group of subjects is being tested under low-illumination conditions and the other group under high illumination. Illumination would thus be the independent variable; notice that it does and must vary, in this case from low to high. (If all subjects had been tested under identical conditions of illumination, then illumination could not be an independent or any other kind of variable.) Or, if a researcher in the field of learning were attempting to show that a fixed-ratio reinforcement schedule of 3 to 1 (3 correct responses for each reinforcement) leads to more resistance to extinction than does a continuous reinforcement schedule (reinforcing each response), then the type of reinforcement schedule that has been set up by the researcher would be the manipulated IV, while resistance to extinction would be the DV. Thus, whenever the experimenter is fully in charge of the environmental conditions (the stimulus situation in which the subject works) and these conditions are *varied*, then the IV is considered to have been manipulated.

On the other hand, if the subjects are assigned to different groups, or are categorized in any way, on the basis of trait differences they already possess, the IV is considered to be a *subject variable*, or as it is sometimes called, an assigned variable. Thus, a subject IV occurs when the researcher assigns subjects to different categories on the basis of a measured characteristic and then attempts to discover whether these assigned variables either correlate with or differ from some measure of the subject's response, this response measure again being the DV. For example, it might be hypothesized that women have higher IQs than do men. The subjects are selected and then categorized on the basis of gender, a subject IV. The two groups then both take IQ tests, and the IQ scores (the DV) are compared. Or a researcher wishes to establish whether college graduates earn more money than do noncollege graduates. The researcher selects a large sample of 35-year-old men, assigns them to categories on the basis of whether or not they graduated from college, and then obtains a measure of their incomes. Level of education would then be a subject IV and income level the DV. Notice that in this example, both age and sex were not allowed to vary and thus are constants. Subject characteristics—such as age, sex, race, socioeconomic status, height, and amount of education—may only be subject IVs. Such variables are simply not open to active manipulation. Some independent variables, however, can go either way, depending on how the researcher operates. For example, it might be hypothesized that fluoridated toothpaste reduces dental caries. The researcher might randomly select a large group of subjects and then *provide* half of them with fluoridated toothpaste, while also providing the other half with identical-appearing, nonfluoridated toothpaste. Then, perhaps a year later, the groups are compared regarding production of caries. In this case, then, the IV, whether or not the toothpaste contained fluoride, would be manipulated. If, however, the researcher had simply asked the subjects what kind of toothpaste they usually used and categorized them according to whether the toothpaste did or did not contain fluoride, then the IV would have been a subject variable. The DV, of course, would still be based on a comparison of the incidence of caries between the two groups.

THE CAUSE-AND-EFFECT TRAP

One of the most serious dangers lurking out there in the world of research is the so-called "cause-and-effect" trap. Too often the unwary reader of research is seduced into assuming that a cause-and-effect relationship has been demonstrated, when, in fact, the methodology simply doesn't support such a conclusion. For example, a study (which shall remain nameless, to protect both the innocent and the guilty) was conducted several years ago that purported to show that sleeping too long at night promoted heart attacks. Headlines throughout the country trumpeted this early-to-rise news, and America reset its alarm clocks. The message was stark—don't sleep too long or you'll die! The study that led to all this sound and fury, however, was done in the following way. A group of recent heart attack victims, all male, were questioned as to, among other things, their recent sleep habits. It was found that, as a group, they had been sleeping longer than had a comparison group of men of roughly the same age, weight, exercise patterns, and so on. That is, the independent variable, amount of sleep, was a subject variable and was *assigned after the fact*. For this reason, the study does not prove a causal link between sleep time and heart attacks. In fact, it's just as likely that the reverse is true—that is, that the same physiological conditions that led to the heart attack might also have led to feelings of fatigue and therefore to a desire to stay in bed longer in the morning. The point is that although both these explanations are possible, neither is proven by the study. When a cause-and-effect relationship is actually discovered, it must be unidirectional; that is, there must be a one-way relationship between the IV and the DV. When a light switch is flipped on, the bulb lights up, and this is unidirectional, since unscrewing the bulb doesn't move the light switch.

In a longitudinal study on over 1000 children, researchers discovered a link between breast-feeding and later cognitive growth (Horwood & Fergusson, 1998). Those who had been breast-fed, when tested later (up to age 18), had higher IQs, higher test scores on both verbal and math ability tests, and higher school grades. All the group differences were significant, but the largest differences occurred when the breast-feeding lasted 8 months or longer. The results were used to support the hypothesis that a certain fatty acid, DHA (docosahexaenoic acid), found in breast milk but not in bottle milk, increases the development of cognitive skills. The researchers did acknowledge, however, that the women who breast-fed were better educated, wealthier, in two-parent families, and less likely to smoke. Could other variables have intruded in this study? It's very possible. Perhaps these women also differed systematically in other important ways. Perhaps the women who breast-fed even had higher IQs, in which case the results might be better explained on the basis of an interaction between genetics and environment.

THEORY OF MEASUREMENT

According to an apostle, "the very hairs of your head are all numbered" (Matt. 10:30). Although that may be true, it's only part of the story. The history of civilization reveals that human beings began counting long before they began writing. Today, we are beset by such a dizzying array of measures, counts, estimates, and averages that modern life is beginning to resemble an ongoing problem in applied mathematics. Our society is so enamored of measures that current descriptions of life on this planet are as often conveyed by numbers as words.

Economists keep creating new financial ratios; psychologists, new personality assessments; sociologists, new sociometrics; and political scientists, new demographics—all at a mind-boggling pace. We have, as H. G. Wells put it, "fallen under the spell of numbers." How can we cope? By learning a few fundamental concepts of statistical measurements, we can soften and make more intelligible the noise from our numerical Tower of Babel.

Measurement is essentially the assigning of numbers to observations according to certain rules. When we take our temperature, or step on the bathroom scale, or place a yardstick next to an object, we are measuring, or assigning numbers in a prescribed way. Perhaps every student entering college has had the uncomfortable experience of being measured dozens of times, in everything from height and weight to need for achievement. Obviously, then, scoring an IQ test and reading a tape measure placed around your waist are examples of measurement. But so, too, are counting the number of times a politician's speech mentions the word inflation, or noting the order of finish of the horses at the Kentucky Derby. In each case, a number is being assigned to an observation. People are even measured as to such complex qualities as scientific aptitude and need for affiliation. The basic measurement theory involved, however, is the same as it is for something as precisely physical as being fitted for a new pair of ski boots.

Assigning the Numbers

The way in which the numbers are assigned to observations determines the scale of measurement being used. Earlier we noted that measurement is based on assigning numbers according to rules. The rule chosen for the assignment process, then, is the key to which measurement scale is being used.

The classification system that follows has become a tradition among statisticians, especially those involved in the social sciences, and is even followed in such major statistical computer programs as SPSS, as well as such standardized tests as the Graduate Record Examination. It was first introduced in 1946 by S. S. Stevens (Stevens, 1946). The Stevens system remains as an extremely handy set of rules for determining which statistical test should apply in specific research situations, and it will be used throughout the text of this book. (The student should be aware, however, that it has had its share of critics [Borgatta & Bohrnstedt, 1980].)

Nominal Scale: Categorical Data. Nominal scaling, or simply using numbers to label categories, is the lowest order of measurement. Of all the scales, it contains the least information, since no assumptions need be made concerning the relationships among measures. A nominal scale is created by assigning observations into various independent categories and then counting the frequency of occurrence within those categories. In effect, it is "nose-counting" data, such as observing how many persons in a given voting district are registered as Republicans, Independents, or Democrats. Or, it might be categorizing a group of children on the basis of whether or not they exhibit overt aggressive responses during recess, and then noting the frequencies or numbers of children falling within each category. The only mathematical rule involved in nominal scaling is the rule of equality versus nonequality; that is, the same number must be assigned to things that are identical, and different numbers must be assigned to things that differ from one another. The categories are thus independent of one another, or mutually exclusive, which means that if a given observation is placed in category number 1, it cannot also be placed in category number 2. In nominal scaling, then, we discover how many persons, things, or events have X or Y or Z in common. With nominal scaling, the concept of quantity cannot be expressed, only identity versus nonidentity. If we were to measure people according to gender, for example, by assigning a "1" to females and "0" to males, we aren't, of course, saying that females have more gender than males or that a classroom of students could possibly have an average gender of .75. Nominal scaling is simply a rule that arbitrarily substitutes, in this case, the number "1" for females and "0" for males.

Nominal scales may also be used in the design of experiments, where the number "1" might be assigned to one group and "0" to the other. The numerical value is, thus, again being used as a substitute for a verbal label.

Ordinal Scale: Ranked Data. Often it is not sufficient to know merely that *X* or *Y* is present. As inquiring social scientists, we wish to find out how much *X* or how much *Y*. The ordinal scale answers this need by providing for the rank ordering of the observations in a given category. Suppose that at a given schoolyard, 60% of the children were nominally categorized as aggressive. We may then examine that category alone and rank the children in it from most to least aggressive.

Mathematically, an ordinal scale must satisfy two rules: the equality/nonequality rule and also the greater-than-or-less-than rule. This means that if two individuals have the same amount of a given trait, they must be assigned the same number. Further, if one individual has more or less of a given trait than another individual, then they must be assigned different numbers. The main thing to remember about ordinal scaling is that it provides information regarding greater than or less than, but it does not tell how much greater or how much less. A good illustration is knowing the order of finish of a horse race, but not knowing whether the first-place horse won by a nose or by 8 furlongs. The distance between the points on an ordinal scale is, thus, unknown.

An example of ordinal scaling from sociology or political science is the ranking of a population on the basis of socioeconomic status, from the upper-upper class, through the middle class, and down to the lower-lower class. In the field of psychology or education, a researcher creates an ordinal scale by ranking a group of individuals on the basis of how much leadership ability each has exhibited in a given situation. Ordinal data are, therefore, rank-ordered data. Ordinal scaling defines only the order of the numbers, not the degrees of difference between them. It tells us that *A* is greater than *B* ($A > B$) or that *A* is less than *B* ($A < B$). It is a mistake to read any more into it than that, although sometimes it is the kind of mistake that is easy to make. It can be psychologically seductive to assume that a person who ranks, say, fifth, is about the same distance ahead of the one who ranks sixth as the person who ranks first is ahead of the one who ranks second. This psychological tendency must be restrained, however, when evaluating ordinal data. If someone were to tell you that a certain item costs "more than a dollar," you are only being told that the item costs anywhere from $1.01 to infinity.

In short, be careful of ordinal positions because at times they can be very misleading. You might be told that the *Daily Bugle* is the second-largest newspaper in the city, even though in reality it may, in effect, be a one-newspaper city (with second place going to a seventh grader who happens to own a copy machine). Or, it could be like the old joke where the freshman is told that he is the second-best-looking guy on the campus. All the others are tied for first.

Interval Scale: Measurement Data. A still further refinement of scaling occurs when data are in the form of an interval scale. In an interval scale, the assigning of numbers is done in such a way that the intervals between the points on the scale become meaningful. From this kind of scale, we get information not only as to greater-than-or-less-than status but also as to how much greater than or how much less than. Theoretically, the distances between successive points on an interval scale are equal. As a result, inferences made from interval data can be broader and more meaningful than can those made from either nominal or ordinal data. In general, the more information contained in a given score, the more meaningful are any conclusions that are based on that score. The Fahrenheit (or Celsius) temperature scale provides interval data. The difference between 80°F and 79°F is exactly the same as the difference between 35°F and 34°F. The thermometer, therefore, measures temperature in degrees that are of the same size at any point along the scale.

Psychologists have attempted to standardize IQ tests as interval scales. An IQ score of 105 is considered to be higher than a score of 100 by the same amount that a score of 100 is higher than a score of 95. Although there is some disagreement within the field of psychology on this point (some purists insist that IQ scores can form only an ordinal scale), the vast majority of researchers treat IQ data as interval data. The general opinion seems to be that if it quacks like a duck and walks like a duck, then it is a duck. There is no question, however, about such measures as height, weight, and income, which all form scales of equal intervals. As a matter of fact, weight and height have the added advantage of having an absolute zero value and thus form an even more sophisticated scale called a ratio scale.

Ratio Scale. When a scale has an absolute zero (as opposed to an arbitrary zero such as the 0 Fahrenheit or Celsius temperature measure), then valid ratio comparisons among the data can legitimately be made. That is, we can say that someone who is six feet tall is twice as tall as someone who is three feet tall. In any case, like interval data, data from a ratio scale do have equal interval distances between successive scale points with the added feature of an absolute, nonarbitrary zero point. The concepts in the social sciences rarely use ratio measures. No researcher can define absolute zero IQ or zero prejudice, or zero interest in politics. It is, therefore, incorrect to say that someone with an IQ of 100 is twice as intelligent as someone with an IQ of 50. However, the statistical tests of significance (which are presented in the chapters to follow) that have been designed to be used with interval data may also be used with ratio data.

In distinguishing among the various scales of measurement, remember that if what you are dealing with shows *only* that

1. One observation is different from another, it's nominal data.
2. One observation is greater (bigger, better, more) than another, it's ordinal data.
3. One observation is so many units (IQ points, degrees, SAT points) greater than another, it's interval data.
4. One observation is so many times larger, or heavier, or colder than another, it's ratio data.

Implications of Scaling. The social scientist's interest in measurement scaling is both acute and profound. The choice as to which statistical test can legitimately be used for data analysis rests largely on which scale of measurement has been employed. Further, the inferences that can be drawn from a study cannot, or at least should not, outrun the data being used. It is not correct to employ nominal data and then draw greater-than or less-than conclusions. Neither is it correct to employ ordinal data and then summarize in terms of how much greater or how much less.

Measurement scales are extremely important. Although most of the material in chapters you have covered so far was devoted to interval scaling, you should keep the other types of scales in mind. In Chapter 11 an ordinal test of correlation will be presented and Chapter 16 will be devoted entirely to the ordinal case. Nominal data will be the focus of interest in Chapter 13. It is very likely that your appreciation of these measurement distinctions will not be fully developed until you get a little further along in the book and begin testing ordinal and nominal research hypotheses. So, hang in there. There may be some patches of fog on the way up, but the view from the top is worth the climb.

Comparing Scores on the DV

In order to assess whether or not the IV had any impact, it is necessary to compare the dependent variable scores between the sets of scores. The analysis is thus focused on whether or not a *difference* can be detected between the DV measures found in the various groups. Two methods are available for analyzing these differences.

1. Between-Groups Comparisons. A between-groups comparison means exactly that, a comparison *between* separate and distinct groups of subjects. The DV scores of one (or more) experimental group(s) are compared with the DV measures obtained on the control group. The groups being compared must be completely *independent* of each other. That is, the selection of a subject to be a member of one group may in no way influence who is to be selected in a different group. Subjects are randomly selected or randomly assigned to membership in the experimental and control groups. Thus, there will be a separate DV score for each subject in the study, and in the statistical evaluation of the data the total number of scores being analyzed must equal the number of subjects in the study.

2. Within-Subjects Comparisons. A within-subjects comparison looks for differences, not between separate groups but *within* each of the separate subjects taken individually. For example, a group of subjects may be randomly selected and each person then individually measured on some variable, say weight. Then all the people are placed in a steam room for 30 minutes and immediately weighed again. In this study, the IV (treatment) would be whether the steam was present (and it is a variable, since initially the subjects don't get the steam and then at a later time they do get the steam), and the DV (outcome measure) is the measured weight. Notice that in this example, each subject's post-treatment weight is being compared with that same subject's pre-treatment weight. In short, the focus of analysis is on the change taking place within the subjects. In this case, the number of scores no longer equals the number of subjects, since each subject is being measured twice. This is called the before-after design. In some studies, however, the subjects are measured many times (a repeated measure design), and the number of scores far exceeds the number of subjects.

RESEARCH: EXPERIMENTAL VERSUS POST FACTO

Of the two basic research strategies, it's the experimental method that allows for the possibility of isolating a direct causal factor, and in the true experiment this causal factor should be unambiguously identified. How do you tell the difference between experimental and post facto? Closely examine the independent variable. If the IV has been actively manipulated, it is **experimental research.** If not, it's post facto. In the experimental method, then, the researcher always actively manipulates the independent (causal) variable to see if by so doing it produces a resulting change in the dependent (effect) variable. In post-facto research, on the other hand, the researcher does not manipulate the independent variable. Rather, the independent variable is *assigned* after the fact. That is, the subjects are measured on some trait *they already possess* and then *assigned to categories on the basis of that trait.* These subject-variable differences (independent variable) are then compared with measures that the researcher takes on some other dimension (i.e., the dependent variable).

Post-facto research precludes a direct cause-and-effect inference because by its very nature it cannot identify the direction of a given relationship. For example, suppose that a researcher discovers that among students there is a significant relationship between whether algebra is taken in the ninth grade and whether the student later attends college. Since taking algebra in the ninth grade was the student's own decision, or perhaps that of the parents, this IV was a subject variable and therefore had to have been assigned. The research strategy in this case then was post facto. Perhaps parents, or guidance counselors who view a student as college bound, then encourage that student to elect algebra and also discourage the student whose professed goal is to become a garage mechanic. Or perhaps a highly intelligent student, knowing that he/she eventually wishes to attend college, is self-motivated enough to elect ninth-grade algebra. Since the direction of the relationship is so ambiguous in post-facto studies, isolating a causal factor becomes very difficult. Although post-facto research does not allow for direct cause-and-effect inferences, it does provide the basis for better-than-chance predictions. Correlational research is one form of post-facto research and will be thoroughly covered in Chapter 11.

In the behavioral sciences, experimental research is sometimes called S/R research, since an (S) stimulus (independent variable) is manipulated and a corresponding change in a response (R, dependent variable) is sought. Similarly, post-facto research is sometimes called R/R research, since the (R) responses of a group of subjects are measured on one variable and then compared with their measured responses (R) on another variable.

THE EXPERIMENTAL METHOD: THE CASE OF CAUSE AND EFFECT

In the **experimental method** the relationship between the independent and dependent variables is unidirectional, since a change in the independent variable is assumed to produce a change in the dependent variable. The key to establishing whether research is experimental lies in discovering whether the IV is a treatment variable and has been actively manipulated. If it has, the method is indeed experimental; if not, the method is post facto.

In its simplest form, the experimental method requires at least two groups, an experimental group that is exposed to one level of the independent variable and a **control** or **comparison group** that is exposed to a different, or zero, level of the independent variable. The two groups, experimental and control, must be as much alike as it's humanly possible to make them. The two groups are then compared with regard to the outcome, or dependent variable, and if a significant difference exists between the two groups, the independent variable can be said to have caused the difference. This is because all the other potential variables existing among the subjects in the two groups are presumed to have been held constant, or *controlled.* In the next chapter we will learn one method, the independent t test, for determining whether a significant difference between the groups exists. For example, suppose you have been perusing the physiological literature and notice that a certain drug, magnesium pemoline, causes an increase in the production of one type of RNA in the cerebral cortex. Other studies then come to mind that seem to suggest that cortical RNA may be linked to human memory (through its role in protein synthesis). From reading all these studies and meditating on possible relationships, you begin to induce the hypothesis that perhaps the drug magnesium pemoline might lead to an increase in human memory. You decide to test the hypothesis by designing an experiment. First, you select a large group of students, and then by random assignment you deliberately attempt to create two groups that are as much alike as possible. Through this process of random assignment, you hope to control all those variables that might possibly relate to memory, such as IQ, age, grade-point average, and so on. One of the groups, the experimental group, is then given the drug, whereas the other (control) group is not. It would also be important that both groups be situated in identical environmental conditions—same type of room, same illumination, temperature, and so on. That is, the two groups should be identical in every respect, *except* that one receives the drug and the other does not. Ideally, the subjects should not be aware of which group they are in, for it is possible that if subjects knew they were in the experimental group, that in itself might affect them, perhaps make them more motivated. For this reason, when the members of the experimental group are given a capsule containing the drug, the subjects in the control group are given a nonactive capsule, called a placebo. Actually, the person conducting the experiment shouldn't even know which group is which. This prevents any possible experimenter bias, such as unconsciously encouraging one group more than the other.

Blind and Double-Blind

When the subjects are not aware of which group they represent, the experiment is said to be "blind." And when neither subjects nor experimenter are aware of which group is which, the experiment is said to be a **double-blind.** (Obviously someone has to know which group received the drug. Otherwise the results would be impossible to analyze.)

"IT WAS MORE OF A '<u>TRIPLE</u>-BLIND' TEST. THE PATIENTS DIDN'T KNOW WHICH ONES WERE GETTING THE REAL DRUG, THE DOCTORS DIDN'T KNOW, AND, I'M AFRAID, NOBODY KNEW."

Finally, both groups would then be given a memory test of some sort, and if the scores of the subjects in the experimental group average out significantly higher than those in the control group, a cause-and-effect relationship may legitimately be claimed.

In this example, whether or not the subjects received the drug would be the independent variable. This would be a *manipulated* independent variable, since it was the experimenter who determined which subjects were to receive how much of the drug. The subjects were not already taking the drug, nor were they given the opportunity to volunteer to take the drug. They were, in effect, being *treated* differently by the experimenter. The dependent variable in this study would be the subjects' measured memory scores.

Active Independent Variable

In experimental research, the independent variable

1. Is *actively* manipulated by the experimenter
2. Is the potential causal half of the cause-and-effect relationship.
3. Is always a stimulus—that is, some environmental change (treatment) that impinges on the subjects—in the fields of education, sociology, and psychology.

Dependent Variable

The dependent variable in experimental research (1) is always the potential effect half of the cause-and-effect relationship and (2) in the behavioral sciences is a measure of the subject's *response.*

CREATING EQUIVALENT GROUPS: THE TRUE EXPERIMENT

The recurring theme in this chapter is that in experimental research it is up to the researcher to keep the subjects in the experimental and control groups as nearly alike as possible. The reason is that if the groups of subjects were systematically different to begin with, then significant differences on the dependent variable would be difficult, if not impossible, to interpret. One might not be able to tell if these differences on the dependent variable resulted from the manipulation of the independent variable or were merely due to the fact that the groups differed at the outset on some important dimension. If one wished to study the effects of word training on reading speed, it would be obviously the height of folly to place all high IQ subjects in one group and low IQs in the other.

Internal Validity

An experiment that is tightly controlled, that has no systematic differences between the groups of subjects at the outset, is said to have **internal validity.** An experiment that is internally valid, then, allows the researcher to examine the pure effects of the independent variable, uncontaminated by any extraneous variables.

External Validity

We also make every effort to design an experiment that is externally valid. **External validity** asks the question, Can the results of this study be applied to organisms other than those participating in the study? One should not do a study on albino rats learning their way through a maze and then quickly extrapolate these findings to cover the population of U.S. college sophomores struggling through a first course in statistics. Such an extrapolation would violate the study's *external validity.*

Independent Selection

One way to create equivalent groups of subjects is to randomly select the subjects for the various sample groups *from the same population.* Groups formed in this way are said to be independent of each other, since the selection of a subject for one group in no way influences which subject is to be assigned to other groups. The random sample procedure gives us a high degree of confidence that the samples will be generally equivalent to each other (helping to promote internal validity), and that the results of the study may be generalized to the population from which the groups were selected (helping to promote external validity).

Randomized Assignments

Another technique, closely allied to this, is called *randomized* assignment of subjects. In this technique the sample is not originally selected randomly from a single population (since this is often a lot easier to talk about than do) but is, instead, presented to you as a fact. For example, your professor says that you can use her 9 o'clock section of intro psych students or your fraternity volunteers as a group to participate in your study. Now, although this is not the ideal way to obtain a research sample, it is, alas, often the best that you're going to do. The solution is to, strictly by random assignment, divide the large group into two smaller samples. This is called randomized assignment, since, although the original sample was not randomly selected, the random assignment process is still used to divide the whole group into two, it is hoped, equivalent groups. The theory here is that even though your original group may not truly represent the population, at least the two smaller randomly assigned groups do mirror each other—and are therefore equivalent. In passing, note that in either the random samples or randomized assignment techniques, the sample groups are independent of each other. That is, the selection of one subject for one group in no way influences which subject is to be assigned to the other groups.

Dependent (or Correlated) Selection

Another method of obtaining equivalent groups is simply to select one group and then use that group under different treatment conditions. The theory here is that no one on the planet is more like you than you, yourself, and so you are used in both the experimental and control conditions. This method, although seemingly both simple and pure, actually opens up a veritable Pandora's box of pitfalls and, as we shall see later, should be used sparingly, if at all. When studies are conducted in this manner, the subjects are obviously not independent of each other, since the same persons are used in each condition. The scores are now dependent and the analysis focuses on the within-subjects differences. Finally, the last major method of creating equivalent groups, also based on dependent selection, is to select one group (ideally by random selection), and then create a second group by matching the subjects, person for person, on some characteristics that might relate to the dependent variable. For example, if one were to test the hypothesis that the phonics reading system produces higher reading-comprehension scores than does the look-say technique, you would first attempt to list other variables that might influence the dependent variable, or in this case, reading comprehension. Such variables as age, IQ, grade, previous reading habits, and gender are strong possibilities. If one of your subjects is six years six months of age, has a WISC-III verbal IQ of 120, is in the first grade, and reads an average of two books a month, you would attempt to find another child who closely resembles this subject on these specific variables. The members of a given matched pair, of course, should always be assigned to separate groups. Notice that these groups, then, will not be independent of each other, since the selection of one subject for one group totally determines who will be selected for the other group.

DESIGNING THE TRUE EXPERIMENT

There are two major types of true experimental designs, and each has as its primary goal *the creation of equivalent groups of subjects.* This does not mean that all the subjects in both groups will be absolute clones of each other. There will always be individual differences *within* the groups, some subjects taller or smarter or stronger or whatever than other subjects within the same group. It does mean, however, that on balance the two groups average in the aggregate about equal on any characteristics that might influence the dependent variable.

1. The Between-Groups Experimental Design: After Only. In this design the subjects are either randomly selected from a single population or else placed in the separate sample groups by random assignment, or both. In any case, the presumption is that chance will ensure the equivalence of the groups. Further, the subjects in this design are usually measured on the dependent variable *only* after the independent variable has been manipulated. This is deliberately done in order to avoid the possibility of having the DV measures possibly affect each other. For example, suppose we wish to find out whether the showing of a certain motion picture might influence a person's racial attitudes. In this design, we only test the racial attitudes of the subjects *after* the IV has been manipulated, in this case, after the period of time when one group has seen the movie and the other group has not. The reasoning behind this is that perhaps if the racial attitudes had been tested before the movie, the pre-test itself may have influenced the way the film is perceived, perhaps heightening viewer awareness of the racial content, or sensitizing the viewers to the racial theme of the movie. When subjects are sensitized by the pre-test procedure, the problem of external validity becomes an important issue, since few experimenters would be willing to generalize their results only to pre-tested populations. Also, since the groups in the after-only design are created on the basis of random selection and/or random assignments, the groups are known at the outset to be independent of each other. Finally, although we have been dealing with the after-only design as though it were always a two-group design, one experimental and one control group, this design can be utilized just as readily on multigroup designs. The number of groups involved is a direct function of the number of levels of the independent variable being manipulated. If the IV is manipulated at only two levels, one group receiving zero magnitude of the IV (control group) and the other some specified magnitude of the IV (experimental group), then a two-group design is adequate, and if the data are at least interval, an independent *t* test (Chapter 10) could be used. But, as you will soon see when reviewing the research literature, many experiments are produced in which the IV is manipulated at three, four, or even five levels. Under these conditions, and again if the data are at least interval, some form of the analysis of variance (Chapter 12) could be used.

For example, a researcher may wish to know not just whether or not seeing a film about racial prejudice will then reduce anti-minority attitudes, but whether increasing the number of racially intense scenes within the movie has even more impact on viewer prejudice. For each different version of the movie, then, another experimental group must be added. To be truly an after-only design, however, none of the groups, experimental or control, should be tested on the prejudice scale before seeing the movie.

2. The Within-Subjects Experimental Design: Repeated Measures and Matched Subjects. The repeated-measures experimental design was originally borrowed from the physical sciences where it had a long and noble history. Physicists, for example, would measure something like the temperature of a piece of metal, then apply an electric current to the metal, and finally measure the metal's temperature a second time. If a significant difference occurred between the pre- and post-temperature readings, and all other variables had been held constant, the conclusion could be quickly drawn that the electric current caused an increase in the metal's temperature. Why not use the same approach in the behavioral sciences? After all, what could be more basic? On the surface, this design seems to be the most obvious from a commonsense point of view, and yet, as we shall soon see, it is also the one most fraught with the potential for dangerous ambiguities. Let's see how it works in the social sciences. In one form of this design, equivalent groups of subjects are formed by the simple expedient of using the same people twice. Could anything be more straightforward? It may be straightforward, but it sure can make the data interpretation a risky venture. Although, unfortunately, examples of this design can still be found in the literature, it should now be more kindly thought of as a historical curio. Why should we spend time discussing this dinosaur design? Because no other design offers the student the advantage of discovering so many research errors in one setting. It's a question of learning by bad example.

Advertisers constantly bombard us with examples of this basic design. You've seen the pictures in a before-after ad for a new diet: the woman looking fat, puffy, stupid, and sad in the "before" picture and then, miraculously, looking slim, proud, and exhilarated in the "after" pose. Now, look again at the pictures and count the number of variables that have changed other than body shape. The poses are entirely different, first slouched, then erect. The facial expressions vary, from pathetically depressed to joyously self-confident. The clothes differ, the lighting differs, the backgrounds differ, and on and on. In this example, trying to isolate the pure effects of the independent variable is like trying to find a needle in the proverbial haystack.

Let's look at how this design is presumed to work. In this design, a single group of subjects is selected, usually randomly, and then measured on the dependent variable. At this stage, the subjects are considered to be members of the control group. Then, the independent variable is manipulated, and these same subjects are again measured on the DV. At this point, the subjects are considered to be members of the experimental group. *In this design the subjects are obviously not independent of each other,* since "both groups" are composed of the same people. If the DV scores in the "after" condition differ significantly from those in the "before" condition, the independent variable is then assumed to have caused this difference. And this assumption, let us quickly add, is based on the allegation that nothing else (other than the IV) has changed in the lives of these people. It's this premise that so often contains this design's fatal flaw. Let's take an example. A researcher is interested in testing the hypothesis that teaching the "new math" increases math ability among sixth-grade elementary school children. A random sample of sixth graders is selected from the population of a large, metropolitan school district, and the chosen children are all given a standardized math ability test (measured on the DV). Then the IV is inserted, in the form of a 12-week "new math" program, replete with Venn diagrams and set theory. Then the math ability test is given again (DV), and the two sets of scores are compared. Let's assume that the second math measure was significantly higher than the first, thus seeming to substantiate the efficacy of the teaching program. But was that the only variable in the child's life that changed during that 12-week time span? Of course not! First, the child had 12 weeks in which to grow, mature, and also practice his/her math skills, perhaps using the new math during schooltime and the old math after school (when making subtractions from his allowance to see whether he can afford to buy a new baseball, or adding the total minutes of her piano-practice time in order to prove to mother that she can go outside to play). Also, just being selected for this new program may have caused the children to feel somehow special, therefore increasing their feelings of motivation.

As we've pointed out several times, another concern with the before-after design is the "pre-test" problem, that is, that the very act of measuring the subjects on the DV the first time might influence how they respond to the DV the second time. For example, the subjects might become *sensitized* to the test, and thus make the entire study overly artificial. That is, the act of taking the standardized math test the first time may have made some of the children more aware of what they didn't know or had forgotten and therefore prodded them into reviewing the rules on adding fractions, or whatever. When subjects become sensitized to the first test, they become more responsive to the second test, even without the manipulation of the independent variable. Sometimes, on the other hand, subjects become fatigued by the first test. To take a blatant example, assume that a researcher wishes to study the effects of drinking coffee on the behavior of hyperkinetic college students. A random sample of students is selected and told to report to the college's football field. They are then told to run eight laps and their running times are clocked. They are then given a cup of coffee to drink and told to run another eight laps. Their running times are again clocked and found to be significantly lower. Here we have the classic before-after design, running times being measured both before and after the introduction of the independent variable (coffee). However, the "before" measure so tired the students that the coffee probably had little to do with the lowered speed recorded in the "after" measure. This is the kind of study where the results could have been written before the data were even collected. Another hazard when using the before-after design results from the passage of time. Since some amount of time must elapse between the pre- and post-measures, perhaps the mere passage of time has changed the subjects in some important way. In short, literally dozens of other variables may have been impacting the lives of subjects in these before-after studies, and their changed scores may have really resulted from these uncontrolled variables *and not* from the independent variable. Suffice it to say here that whenever you come across a research study using a one-group, before-after design, be alert to the possibility of uncontrolled variables intruding on the results. When the subjects are measured on more than just two occasions, the design is typically called a repeated-measures or within-subjects design. Repeating the DV measures several times does to some extent alleviate a few of the problems inherent in the two-measure, pre-post design, and more will be said on this issue in Chapter 15 when we cover the paired *t* test and the within-subjects analysis of variance.

THE HAWTHORNE EFFECT

Perhaps the best and certainly most well-known example of the confounding effects that may occur in the one-group, before-after design can be found in the notorious research study conducted many years ago at the Hawthorne plant of the Western Electric Company (Roethlisberger & Dickson, 1939). The object of the study was simply to determine if increased illumination would increase worker productivity. The researchers went into one of the assembly rooms and measured the rate of worker productivity, increased the level of illumination, and then measured productivity rates a second time. Just as had been suspected, under conditions of increased illumination productivity did indeed go up. However, when the researchers later added a control group, that is, a room containing another group of workers whose illumination they only pretended to increase, they found to their dismay that productivity in this room also went up. What they really had discovered was that subjects will often improve their behavior merely because someone of seeming importance is paying attention to them. If there is no separate control group, the researcher can never know whether the subject's response has improved because of the manipulated IV or because the subject was flattered by the researcher's attention. When the increase is due to the heightened motivation of subjects generated by experimental conditions of attention and flattery, the change in behavior is said to be the result of the **Hawthorne effect.** Ambiguous results due to the Hawthorne effect are most common in studies where the researcher has used the before-after experimental design without an adequate control group. One researcher in the field of learning disabilities once complained that "any idea or finding which is unacceptable to anyone today can be explained away on the basis of the Hawthorne Effect" (Kephart, 1971). In point of fact, the only time results can be explained away on the basis of the Hawthorne effect is when the researcher carelessly fails to use an adequate control group. The Hawthorne effect should be viewed as an important warning to the researcher, since it encourages extreme caution when assigning specific causes to observed changes in behavior, especially in training studies using the before-after design.

The Matched-Subjects Experimental Design as an After-Only Technique

The matched-subjects, after-only design makes use of the aforementioned subject—matching technique for creating equivalent groups. The subjects in the control and experimental conditions are equated person for person on whatever variables the researcher assumes might possibly be related to the dependent variable. That is, each member of the experimental group has his/her counterpart in the control group. In other respects, however, this design is similar to the after-only technique, in that the DV is only measured after the introduction of the IV. (Again, this is done in order to prevent the first DV measure from, in any way, influencing performance on the second DV measure.)

As an example of a matched-subjects design, a researcher may wish to assess the effects of a behavior modification program on the responses of a group of incarcerated juveniles. A random sample of juveniles is selected from the population of names provided by the sheriff's department of boys who have been identified as having been especially disruptive. For each subject so selected, a matched subject is sought, also from the same population of names, who resembles the originally selected subject in a number of important respects, such as age, ethnicity, severity of problem, educational level, socioeconomic level, and previous history in the criminal justice system. Then, perhaps by a flip of the coin, the matched pairs of yoked subjects are divided into two groups. One group is chosen to receive a 6-week behavior modification program, while the other group is left to function on its own. At the end of the 6-week period, a panel of experts visits the detention center and judges all the subjects *in both groups* with respect to the extent of disruptive behavior. If the experimental group displays significantly fewer disruptive responses than the control group, then it may be concluded that the behavior modification program was indeed effective.

An experiment using the matched-subjects technique is not always easy to conduct, especially when large numbers of relevant variables are used. It may even become virtually impossible to find a suitable matched pair, who resemble each other closely on all the important variables. As a general rule, the more variables used in the matching process, the more difficult becomes the task of selecting the subject's control-counterpart. Another problem inherent in this design is caused by the fact that it's not always easy to know what the relevant matching variables should be.

REPEATED-MEASURES DESIGNS WITH SEPARATE CONTROL GROUPS

In order to correct for the obvious pitfalls inherent in the one-group, repeated-measures design, researchers now use an especially powerful variation of this design in which completely separate control groups are used. In fact, the use of a separate control group greatly minimizes the dangers inherent in the traditional before-after design. For example, many of the problems we encountered with the Hawthorne study could have been diminished or even eliminated had a separate control group been added. That is, instead of using a single group of workers, the researcher could have (and should have) selected one large group and then created two equivalent groups through random assignment. Both groups would then be measured on productivity levels, and then both groups would also enjoy the "heady" pride of being placed in an apparently special program. At this point, the experimental group gets the increased lighting, and the other group, the control group, continues with the plant's normal lighting. This should help in controlling the previously described uncontrolled variables, such as motivation, practice, and fatigue. When a separate control group is used, the analysis can be based on discovering how much each subject *changed* from the pre- to the post-measures. If the members of the experimental group change significantly more than do the members of the control group, then the independent variable is assumed to have caused this difference. One of the simplest and most effective ways to analyze the data from this design is to focus on the "change" scores, or the differences between the pre- and post-measures, since in many of these studies the control group can also be expected to show a change, usually an increase. The question, then, can be, *Which group changed the most*?

Although some researchers have challenged the use of "change" scores because of the possibility of lowered levels of reliability, statistical techniques have been devised to counter this possible threat (see Kerlinger, 1986, p. 311). One must keep in mind, however, that if one of the groups scores at the extreme end of the scale (either very high or very low) during the pre-condition, there just may not be room for much more of a change for this particular group.

Ceiling Effect

Another problem facing the researcher, known as the **ceiling effect,** occurs when scores on the DV are so high to begin with that there is virtually no room left for them to go up any further. On the other side is what might be called the *cellar effect,* where the DV scores are so low to begin with that there is little likelihood of any further decline. As a rather blatant example, suppose a researcher were trying to determine whether a certain preparation course is effective in increasing scores on the verbal Graduate Record Exam (GRE). Assume further that the only students who signed up for the course already had scores in the high 700s. Since the highest possible score is 800, there is very little room left for the course to effect a GRE gain. In one published study, researchers were interested in whether persons were able to judge their own physical attractiveness against a standard set by a panel of judges. The results showed that overestimation of one's physical attractiveness was greatest among low attractive persons and least among high attractive persons (Gurman & Balban, 1990). Notice that in this case, there was hardly any room left for those high in attractiveness to estimate at even higher levels, whereas those low in attractiveness had a great deal of room left in which to show an increase.

Floor Effect

The opposite of the ceiling effect is called the *floor effect.* It occurs when the lack of variability occurs as a result of a large number of scores clustering around the bottom of the scale. For example, in a study on attitudes toward dating aggression among high school students, it was found that the range of scores could be severely restricted because the majority of respondents reported that physical aggression was never justified (Smith-Slep et al., 2001)

Placebo Effect

A placebo is any treatment that is believed to be nonactive and therefore in theory should NOT impact the DV. Although, especially in medical studies, the placebo is used in an attempt to increase a study's interval validity, the placebo sometimes has an effect of its own. Even though the medical researcher is assured that the placebo is composed of inert substances, the individual taking the pill may sometimes change, often for the better, because of personal expectations. Because the pill is administered as though it is medicinal, the subjects assume they are being chemically treated and very often report symptom relief. In one study, in which several metanalyses covering antidepressant medications were reviewed, it was found that the effect sizes for the placebo groups were large enough to be comparable to those of the medicated groups (Kirsch & Lynn, 1999). In short, when people psychologically EXPECT to improve, they often do improve. To really nail down the impact of the medication, the experimental group would have to be shown to have gained significantly more than the placebo-control group. The placebo effect can occur in nonmedical studies as well, and can often lead to ambiguous results. A study comparing two teaching methods may fall prey to the placebo effect, especially if the subjects in the control group firmly believe they are in the treatment group and therefore become especially motivated to change.

The Separate Control Group

When setting up the separate control group for the repeated-measures design, the researcher can choose from the following:

1. Random Assignment. The control group may be independent of the experimental group, as when two separate random samples are selected from the targeted population, or when a nonrandom sample is divided into two groups by random assignment. The DV comparison in this situation is *between* the groups.

2. Matched Subjects. The separate control group may depend on the experimental group, as when the subjects in the two groups are matched on the basis of some variable(s) deemed relevant to the dependent variable. In this case each of the subjects in the experimental group is equated person for person with another subject in the control group. The DV comparison in this design is *within* subjects, since ideally the matched subjects should be so closely correlated as to be virtually a single person. In fact, Kerlinger suggests that the before-after design is really an extreme case of the matched-subjects design (Kerlinger, 1986). The matched-subjects technique is best used when the matching procedure takes place before the experiment begins, and the yoked subjects are then randomly assigned to the experimental and control groups.

Matched Groups

In an effort to simplify the matching process, some researchers have turned to the matched-groups design, where instead of matching subjects on a one-to-one basis, the entire group, *as a group,* is matched with another, similar group. Typically, this takes the form of matching group averages, so that one group whose average IQ is, say, 103 is matched with another group whose average IQ is also 103 (or close to it). Since groups with similar averages might still show systematic differences in variability, it is important that the groups also have approximately equal standard deviations. Statisticians treat matched groups as though they were independent.

REQUIREMENTS FOR THE TRUE EXPERIMENT

The following, as set down by Kerlinger (Kerlinger, 1986), are considered to be the basic requirements for the true experiment.

1. **The design must be able to answer the research question.** When setting out to design a true experiment, the researcher must keep constantly in mind what it is that he/she is able to prove. This means basically that the researcher must be clear regarding what the eventual statistical analysis will look like, and that it be *appropriate* to the overall design of the study. This must be done early, before the data are collected. Don't wait until the study is completed and then march up to the computer lab with a big pile of collected data and plaintively ask, "What do I do with it now?" By that time it is almost surely too late.

2. **The design must be internally valid.** The true experiment demands that extraneous variables have been controlled and that the experimenter has actively manipulated at least one IV. Further, the assignment of subjects to the various groups must be under the full control of the experimenter. In short, the true experiment must be designed in such a way as to ensure internal validity and the possibility of an unambiguous interpretation of the possible effects of the IV.

3. **The design should be as externally valid as possible.** The subjects in the study should be representative of the population to which the results are to be extrapolated, and the study should not be so artificially contrived as not to allow for any real-life translations. Of the three criteria, this is the most difficult to satisfy, since by its very nature some artificiality must be introduced whenever the researcher is involved in randomly assigning subjects to the various groupings.

The Quasi-Experiment

Before leaving our discussion of experimental research, there is one variation on this theme that has been gaining in popularity. It is called the *quasi-experimental* method. With this technique, the independent variable is still under the full, active control of the experimenter, but the control and experimental groups are no longer clearly representative of a single population. That is, the researcher may use "intact" groups, or groups that have already been formed on the basis of a natural setting and then, in some manner, subject them to different treatment conditions. Thus the quasi-experiment is an attempt to simulate the true experiment and is called by many researchers a compromise design (Kerlinger, 1986). For example, suppose that a researcher wants to assess the possibility of establishing a difference in the amount a student retains and the mode of presentation of the material to be learned. By a flip of a coin, it is decided that the students in Mrs. Smith's first-grade class will receive a visual mode of presentation, whereas the students in Mr. Shea's first-grade class (across the hall) will receive an auditory presentation of the same material. Measurement of the DV, scores on the retention test, is taken only after the groups have been differentially treated. This is done to prevent any pre-test sensitization of the material being learned and, in this case, therefore, would be an after-only design. Notice that the IV is being actively manipulated by the researcher; that is, the researcher has chosen which first-grade class is going to receive which treatment. However, the researcher did not create the groups on the basis of random assignment but instead used groups that had already been formed. It is true that these two groups may have differed on a number of characteristics, some that even may have been related to the DV, but for many real-world situations this is the best that one can do. School systems, clinics, even college deans may not always allow groups to be formed artificially on the basis of random assignment.

The quasi-experimental method may be set up according to after-only, before-after, or matched-subjects designs. The major problem with this type of research, however, focuses on the issue of how much equivalence the groups really have at the outset. The advantage of the quasi-experiment relates to its external validity. Since the groups have been naturally formed, extrapolation to real-world, natural settings becomes more convincing.

POST-FACTO RESEARCH

So far the discussion of the various experimental designs has focused on just that—experimental methodology (both true and quasi), where the independent variable has been manipulated by the experimenter. In **post-facto research,** on the other hand, the independent variable is not manipulated but *assigned* on the basis of a trait measure *the subject already possesses.* That is, rather than attempting to place subjects in equivalent or (in the case of the quasi-experiment) nearly equivalent groups and then doing something to one of the groups in hopes of causing a change in that group, the post-facto method deliberately places subjects in non-equivalent groups—groups that are known to differ on some behavioral or trait measure. For example, the subjects may be assigned to different groups on the basis of their socioeconomic class, or their sex, or race, or IQ scores, or whatever, and then the subjects are measured on some other variable. The researcher then attempts to ferret out either a correlation or a difference between these two variables. Thus, in the post-facto method, the researcher does not treat groups of subjects differently but instead *begins with a measure of the DV* and then retroactively looks at preexisting subject IVs and their possible influence on the DV. There's nothing wrong with this research. It's especially common in the social sciences. The problem lies not in the use, but in the misuse of the inferred conclusions. You simply shouldn't do a post-facto study and then leap to a cause-and-effect conclusion! That type of conclusion commits what the scientists call the *post-hoc fallacy,* which is written in Latin as "post hoc, ergo propter hoc" (translated as "because it came after this, therefore it was caused by this"). Many examples of this fallacy come readily to mind. We have all heard someone complain that it always rains right after the car is washed, as though washing the car caused it to rain. Or the traffic is always especially heavy when you have an important meeting to go to, as though all the other drivers know when the meeting has been planned and gang up in a contrived conspiracy to force you to be late. Or perhaps while watching a sporting event on TV, you have to leave your TV set for a few minutes. When you return, you discover that your favorite team has finally scored. You angrily hypothesize that the team deliberately held back until you weren't there to enjoy the moment. Some post-facto research conclusions have been very similar to these rather blatant examples.

Jerome Bruner once did a study in which children were selected from differing socioeconomic backgrounds and then compared with respect to their ability to estimate the sizes of certain coins (Bruner & Goodman, 1947). It was found that poor children were more apt to overestimate the size of coins than were wealthier children. In this example, socioeconomic class was the independent variable, clearly a subject variable, and was *assigned,* not manipulated. Bruner assigned the children to the economic categories on the basis of subject traits they already possessed. The dependent variable was the child's estimate of the coin size. What can be concluded from this study? Various explanations have been put forth. Bruner suggested that possibly the poorer children valued the coins more highly and thus overestimated their size. Another suggestion was that wealthier children had more experience handling coins and were thus more accurate in estimating their size because of familiarity. The point is that no causal variable can be directly inferred from this study.

Of what use, then, is post-facto research? Prediction! Even though causal factors may be difficult to isolate, post-facto studies do allow the researcher to make better-than-chance predictions. That is, being provided with information about the independent variable puts the researcher in the position of making above-chance predictions as to performance on the dependent variable. If, in a two-newspaper city, there is a dependable relationship between which newspaper a person buys (liberal or conservative) and the political voting habits of that person, then one might predict the outcome of certain elections on the basis of the newspapers' circulation figures. We can make the prediction without ever getting into the issue of what causes what. That is, did the liberal stance of the newspaper cause the reader to cast a liberal vote, or did the liberal attitudes of the reader cause the selection of the newspaper? We don't know, nor do we even have to speculate on causation in order to make the prediction. In short, accurate predictions of behavior do not depend on the isolation of a causal factor. One need not settle the chicken-egg riddle in order to predict that a certain individual might choose an omelet over a dish of cacciatore.

Researching via Post Facto

In the early stages of researching some new area, it is often necessary to use post-facto techniques and then, as information accumulates, follow up with the experimental method. Post-facto research is often quick and easy to do because the data may be already in hand, and it may also lead to educated speculation concerning independent variables, which might then be manipulated in experimental fashion.

Post-Facto Research and the Question of Ethics. It is also important to realize that for many kinds of research studies, post-facto techniques are the only ones that don't violate ethical principles. Post-facto techniques do allow a researcher to gather predictive evidence in areas that might be too sensitive, or possibly harmful to the subjects, to be handled experimentally. Suppose, for example, a researcher is interested in discovering whether the heavy use of alcohol increases a person's chances of committing a felony. To test this experimentally, one would have to select two groups randomly and then force one group to drink heavily on a daily basis, while preventing the other group from touching a drop. Then, if after a year or two, arrest differences were found between the two groups, it could be legitimately claimed that the use of alcohol influenced felony arrests. But in order to isolate a causal factor, the experimental subjects may have suffered in more ways than just the arrests. Suppose they developed liver problems, or delirium tremens, or whatever. Should a researcher be allowed to expose subjects to possible long-term damage merely for the sake of nailing down the causal factor? Of course not; the experimental method should be used only when the risks are minimal compared to the potential benefit to mankind.

A study of this type could have been more ethically handled by the post-facto method. The researcher would simply identify persons who are already known to be heavy drinkers and then compare their arrest records with a group of non-drinkers. In this case, the subjects themselves have chosen whether or not to drink, and the researcher simply finds out whether the two groups also differ on other measures.

Of course, no direct cause-and-effect statement is possible, for even if a significant difference is found, one cannot determine for certain the direction of the relationship. Perhaps A (drinking) caused B (felony arrests). Perhaps B (felony arrests) caused A (drinking). Or, perhaps X (unknown variable) caused both A and B. The X variable might be a depressed state of mind, or financial problems, which caused the person to both drink too much and have confrontations with the legal system. (For more on the problems of the "third variable," see Sprinthall, Schmutte, and Sirois, 1990.)

Experimental and Post Facto: A Final Comparison

The fundamental difference between the two methods, experimental and post facto, is in whether a cause-and-effect relationship may be unequivocally claimed. This difference is so enormous, however, that great care should be taken in identifying which method has been used before evaluating research findings. The bottom-line difference is this: When the experimental method has been used, the independent variable is a treatment variable and has been actively manipulated by the researcher. That is, when the independent variable has been actively manipulated, the researcher must have somehow *treated* the groups differently, subjected them somehow to different environmental conditions. If this has not been done, then the research is post facto. In the experimental method, the researcher attempts to make the subjects as much alike as possible (equivalent groups) and then treats them differently. In the post-facto method, the researcher takes groups of individuals who are already different on some measured variable and then treats them all the same (measures them all on some other variable).

Informed Consent

The researcher must keep in mind that subjects must give *informed* consent, not just consent. For example, an electrician on a ladder calls to his apprentice below and tells him that there are two wires on the ground and asks if he would mind picking one of them up. The apprentice says OK and picks up one of the wires. The electrician asks if he's OK. "I'm fine," says the apprentice. "That's good," says the electrician, "it must be the other wire that's carrying the 220 volts." Now that may be consent, but it's hardly informed. For the student researcher, the college usually provides a review committee whose approval is needed before any empirical research studies can be started.

COMBINATION RESEARCH

Sometimes both experimental and post-facto methods are *combined* in one study. This typically occurs when the research involves more than just a single independent variable. When at least one IV is manipulated (active) and at least one other IV is assigned (subject), the study is described as **combination research.** A researcher may wish to discover whether there is any difference in hyperactivity between older and younger children who are or are not taking a certain medication. In this situation, whether or not the child is taking the medication—say, Ritalin—is a manipulated IV (since the researcher will determine who gets the Ritalin and who gets the placebo), whereas age is clearly a nonmanipulated subject IV. Random samples of both younger and older hyperactive children are selected and then randomly divided into two groups: one to be given 10 mg of Ritalin and the other a placebo. The design would appear as follows:

	Ritalin	*Placebo*
Under age 15		
Over age 15		

In interpreting the results of such a study, the conclusions must be carefully thought out. Only the medication has the potential for being construed as a direct causal factor, whereas the age variable, although perhaps an important predictor, should not be interpreted as having an unambiguous causal impact. (More will be said on this in Chapter 12, especially regarding the possibility of an interaction effect among IVs.)

Qualitative Research

Although the focus of this chapter has been on quantitative research, which certainly accounts for the majority of studies in the social sciences, there is another approach that deserves mention. Quantitative research, as we have seen, is based on numerical data, whereas **qualitative research** is purely descriptive and therefore not really measurement based. The data in qualitative research are written descriptions of people, events, opinions, and/or attitudes (Sprinthall, Schmutte, & Sirois, 1990). Suppose, for example, that we wished to discover how psychology's laboratory techniques during the 1970s differ from the methods used today. The qualitative researcher might select a large group of experimental psychologists who have been in the business for at least 20 years and then record their answers to a series of questions about lab techniques. The result would be a nonstatistical, written description of the psychologists' perceptions of the changes. Or perhaps the researcher is interested in gathering information about the things people say, write, and do in their own natural settings. For example, in examining the school counseling process, the researcher would attempt to observe and then record descriptions of the various interactions between pupils and counselor, counselor and teachers, and so on. In some cases, both qualitative and quantitative research techniques are combined in a single study. Many quantitative surveys include global, nonstructured questions that are qualitative in nature, as well as the structured, more traditional items that are to be numerically scored.

There is some disagreement in the literature as to which is the better approach, quantitative or qualitative research, especially in the field of education. At times there may be heated debates among researchers regarding this question, complete with loud voices and even table pounding. However, the most important message is based on deciding which questions can be best answered by which method (Fraenkel, J. R. & Wallen N. E., 2000). For example, researchers interested in the quality of an instructional method rather than how often it occurs or how it can be measured may be better served by the use of qualitative methods. Therefore, the old debates about qualitative versus quantitative research tended to shed more heat than light on the issue. Today's researchers realize that fieldwork, narrative, and qualitative methods in general are all useful in trying to understand the world around us (Crotty, M., 1999).

RESEARCH ERRORS

Although there is probably an endless list of ways in which the researcher can create uninterpretable results when conducting a study, some of the most common and most serious errors will be presented in this section. These will all be errors in research strategy, not statistical analysis. In fact, these are errors that lead to uninterpretable findings regardless of the power or elegance of the statistical analysis.

Again, the presentation to follow is not expressly designed just to create cynics among you. It is to aid you in becoming more sophisticated research consumers whose practiced eye will be able to see through the camouflage and identify the flaws. It is also important to learn to spot research errors in the headlines and stories appearing in popular magazines, Sunday supplements, and daily newspapers as well as in the scientific journals. Face it. For many of you, the reading of scientific journals may not be top priority after you graduate from college. The approach used in this section will be essentially that of the case study, an approach that will, it is hoped, teach by example. The major message, which you will see repeated, is aimed at alerting you to pay at least as much attention to the researcher's methodology as to the researcher's conclusions.

Confounding Variables: Secondary Variance

As you have learned, in experimental research cause-and-effect conclusions are justified only when, all other things being equal, the manipulation of an independent variable leads to a concomitant change in the dependent variable. The fact that other things must be kept equal means that the experimental and control groups must be as nearly identical as possible. Other possibly influencing variables must be held constant, so that the only thing that really varies among the groups is the independent variable. When other factors are inadvertently allowed to vary, these factors are called confounding variables, and their presence stamps the research as "flawed".

Some of this may seem to be just plain common sense, but as we shall see, even among some of the most famous of the social science researchers, this "sense" hasn't always been so common. *Confounding variables are any extraneous influences that may cause response differences above and beyond those possibly caused by the independent variable.*

Failure to Use an Adequate Control Group

Clearly, then, confounding variables occur when the researcher makes a major mistake—the mistake of not having an adequate control group. Without fully obtaining the key experimental method ingredient—that is, equivalent groups of subjects—free-floating confounding variables are bound to be present. Without an equivalent control group, confounding variables will surely be lurking on the sidelines, just waiting to pounce on the conclusions and stamp them "flawed."

When a design has been flawed, the experiment automatically loses its internal validity. That is, no matter how beautifully representative the sample is, the results of the experiment become impossible to evaluate when secondary variance is allowed to seep into the study.

Case 1: Frustration Regression

In a study in the area of frustration, the famous Gestalt psychologist, Kurt Lewin, hypothesized that frustration, again in the form of goal blockage, would cause psychological regression in young children (Barker, Dembo, & Lewin, 1941). In this study the independent variable was manipulated by allowing the children to view, but not reach, a glittering array of brand-new toys. Here's how it went. The children were all placed in a single room and were allowed to play with serviceable, but obviously well-used, toys. Observers followed the children around and rated each child as to the level of maturity displayed by the child's play behavior. Next, a curtain was drawn, revealing the beautiful, shiny new toys; but, alas, a wire screen prevented the children from getting at these toys. Then the children were again assessed as to maturity level of their play with the used toys, a level, incidentally, that dropped considerably from the pre-measure.

Here it is again, a before-after experimental design, with the same children serving as their own controls. Remember, by the time the children were measured for the second time, they had been cooped up in that room for some time. Could boredom have set in? Could attention spans have begun to diminish? We simply can't tell from this study. Only with a separate control group (another group of children playing with similar toys for the same length of time, but with no view of the unobtainable toys) can one hope to eliminate some of the most blatant of the confounding variables found in this study.

EXPERIMENTAL ERROR: FAILURE TO USE AN ADEQUATE CONTROL GROUP

Case 2: Play Therapy

An example of a study where a control group was used, but the control group was not adequate, comes from the clinical psychology literature. In this study, an attempt was made to test whether nondirective play therapy might improve personality adjustment scores among institutionalized children (Fleming & Snyder, 1947).

All 46 residents of a children's home were tested on the Rogers Test of Personality Adjustment, and the 7 children with the lowest scores were selected for the special treatment. The other 39 children thus served as the separate control in this before-after design. However, the play therapy group was treated twice a week at a clinic located 10 miles from the institutional home. At the end of six weeks, both groups were again tested on the adjustment inventory. Two important confounding variables were allowed free rein in this study. First, the experimental group not only received the play therapy, but it also was treated to a 10-mile bus ride twice a week and hence an opportunity to leave the possible boredom of the institution's confines. Second, the two groups were not equivalent to begin with. Now, although it may seem especially fair to select only the most maladjusted children for the special treatment condition, it might very well be that these were the very children who would change the most, regardless of the independent variable, simply because on the measurement scale they had the most upward room in which to change. In Chapter 14 we will discuss something called "regression toward the mean," but suffice it to say here that persons who score low on any test are more apt, if they change at all, to change upward than are persons who score high. Finally, this study is triply flawed in that 16 of the control group children left the institution during the six weeks the study was in progress and could therefore not be retested. This fact produces the nagging concern that the departure of these children might have resulted from their having been perceived as becoming adjusted enough to go home.

Case 3: Group Decision

During World War II, Kurt Lewin did a series of studies on the topic of "group decision." The theoretical basis was that since a person's social attitudes are learned in a group situation (family, friends, school), then changing a person's attitude should also best take place in a group setting. In one study, a group of women was brought to a lecture and, in the impersonal role of being part of an audience, listened to an impassioned speaker (Lewin, 1952). The speaker exhorted the women to use less expensive and more plentiful cuts of meat. They were told that it was their patriotic duty and also that it was more healthful for their families. Another group of women was also brought together, although in a very different group atmosphere. This group informally sat around a table and heard a group leader raise the same points as the "lecture" group had heard, but in this "discussion group" setting, members were encouraged to participate, offer suggestions, and become generally involved. Several weeks later both groups were checked at home to determine whether they were indeed using the meat cuts that had been urged. Only 3% of the "lecture" group members had used the meat cuts, whereas 32% of the discussion group members had obliged. The difference was clearly significant, and it appeared as though the independent variable, lecture versus discussion group settings, did have an effect. However, Lewin used different people to lead the two groups. Perhaps the "discussion" leader had a more forceful personality, or was more believable, or whatever. The point is the independent variable was very much confounded.

In another variation of this study, Lewin again set up the same types of groups, one in the lecture and one in the discussion setting, but this time the same person, Dana Klisurich, conducted both groups (Sherif & Sherif, 1956). The women were urged to give their children orange juice, and again, several weeks later, investigators called on the homes of all the women and checked their refrigerators for the juice. The women who had participated in the discussion group were far more apt to have the juice in their homes. However, this time Lewin told the discussion group members that they would be checked on *but forgot to tell the lecture group!* That is confounding writ large. There is no way to tell whether the women had the juice because they had been in the discussion group or because they were told Lewin was going to check up on them. In fact, the children of the women in the experimental group may even have been deprived of getting any juice, since these mothers may have been saving it to show Kurt Lewin.

POST-FACTO ERRORS

Case 4: The Halo Effect

In many studies, especially correlational, the research becomes flawed due to the various evaluations of subjects being conducted by observers who know how the subjects scored on previous evaluations. Sometimes this knowledge is extremely intimate, since the same observer rated the subjects on several trait measures. In research, this error is called the **halo effect,** and it results from the very obvious fact that if an observer assigns a positive evaluation to a subject in one area, then, consciously or unconsciously, the observer tends to assign another positive evaluation when the subject is being measured in another area. In short, the subjects are having their trait measures generalized into a whole host of seemingly related areas. This is the reason why advertisers spend big money hiring well-known celebrities to do commercials. The viewer, it is hoped, will assume that because a certain individual is proficient with a tennis racket, he will also be an expert in determining which brand of razor to use.

The halo effect can be a hazard to both the researcher and the research consumer. The issue is especially acute in post-facto research testing the hypothesis of association. In one study, the investigators wished to study the impact of the halo effect on student grades (Russell & Thalman, 1955). The research was post facto; that is, the students' measurements on one trait were compared with the measures the same observers assigned to the students on other traits. The researchers asked the classroom teacher to assign personality ratings to each student and then compared these ratings with the grades the students received from the same teacher. The correlation was high and positive; that is, the more favorable the personality rating, the higher was the grade received. If this study had been designed to test the possibility of an independent relationship between personality and academic achievement, then the personality ratings should have been assigned by someone other than the person doing the grading. Also, this independent observer should not have even been made aware of what those students' grades had actually been.

In another example, a study was conducted at the Institute for Child Behavior in San Diego in which subjects were asked to rate the 10 persons "whom you know best" on two variables—happiness and selfishness. The results showed an inverse relationship; the more a person was judged to be happy, the less that person was seen as being selfish. The conclusion implied a link between the two variables, or that being unselfish (helping others) tended to create a state of personal happiness in the helper (Rimland, 1982). However, since the evaluation of a person's selfishness and happiness was made by the same observer, these results might be more readily explained on the basis of the halo effect. When you like someone, you may easily become convinced that that person abounds in a whole series of positive virtues, even in the face of contrary evidence.

Case 5: Smiling and Causation

In a study from the educational literature, the hypothesis was tested that students would achieve more academically when the classroom teacher spent more time smiling (Harrington, 1955). Observers visited a number of different classrooms and monitored the amount of time each teacher spent smiling. These results were then compared with the grades being received by the students in each of the classrooms. The results were significant—the more the teacher smiled, the higher were the student grades. This is, of course, post-facto research, since the teacher, not the experimenter, determined the amount of smile time. In fact, this would be a difficult study to conduct experimentally.

To manipulate the IV actively, the researcher would have to perhaps sit in the back of the room and, at random time intervals, signal the teacher that it was time to smile. This might obviously lead to a rather bizarre scene, where the teacher, in the middle of a vigorous reprimand, would suddenly have to break out in a broad grin. In any case, though, the study as conducted was post facto. Thus the results *cannot* tell us the direction of the relationship. It may well be, as the authors hypothesized, that smiling teachers produce achieving students. Or it may just as likely be that high-achieving students produce smiling teachers—teachers luxuriating in the fact that the student success rate is obvious proof of the teacher's own competence. Or it could also be that a third variable—say, the personality of the teacher—may have caused both the smiling and the high grades. Perhaps a smiling teacher is a happy optimist who always sees the best in everyone and is therefore more lenient when assigning grades.

META-ANALYSIS

Sometimes the behavioral researcher does not literally conduct his or her own study but instead does a thorough review of a number of studies, all generally focused on the same hypothesis. Despite the fact that these studies have often been conducted in different places—using different subjects, different statistical tests, and even, in some cases, different measures of the dependent variable—the researcher using **meta-analysis** combines the findings from all the studies by statistically integrating the various sets of results. Researchers using meta-analysis use statistics for estimating the effect size in order to predict what the actual population effects are. An effect size of zero indicates that the IV had no effect on the DV, whereas a .8 is considered to show a very strong effect. When the estimated effect size approaches .8, the hypothesis takes on added strength, since it shows that the effect has been shown to consistently cut across a variety of settings in the hands of a number of different researchers (Sprinthall, Schmutte, & Sirois, 1990). Meta-analysis has even been used to ferret out possible causal hypotheses, even among post-facto studies (Shadish, 1991).

Meta-analysis was originated by Gene Glass and Mary Lee Smith, and what they did may seem simple in retrospect, but at the time it was seen as a scientific breakthrough (Glass, 1976). They decided to amass all the studies they could find in a given area, and then determine the effect sizes for each of the studies taken separately. Next, they combined them in order to find the overall effect produced by all the studies taken together. Let's take an example. Suppose we wanted to examine the influence of Head Start on the intellectual development of a group of five-year-old, disadvantaged children who had just completed their second year in the program. The children are tested with the WISC-III and found to have a mean IQ of 95. It is also discovered that 34.13% of them scored between 95 and 110. Assuming a normal distribution, we can find the z score for the 110 by using Appendix Table A and looking up the 34.13%. We thus find that the z score equals exactly 1.00. Now, from Chapter 5 we can establish the SD by using the equation:

$$SD = \frac{X - M}{z}$$

Substituting the above values, we get,

$$SD = \frac{110 - 95}{1.00} = 15.00$$

We could then take a control group of children from the same neighborhood but who had never attended Head Start and find that they had a mean IQ of 90, on a distribution that also had an *SD* of 15. We next compare the two means and convert them to the standardized value, *d*, by dividing through by the *SD*. Recall from Chapter 8 that *d* is one of the symbols used for the effect size.

$$d = \frac{M_1 - M_2}{SD}$$

$$d = \frac{95 - 90}{15} = \frac{5}{15} = .33$$

The Head Start children, thus, had IQs that were one-third of a standard deviation higher than those in the control group. Now, let's assume that another Head Start study conducted in a different part of the country had a *d* value of .50, and still a third study yielded a *d* value of .10. For the meta-analysis, Glass and Smith tell us to average these effect sizes, or

$$\frac{.33 + .50 + .10}{3}$$

for an overall effect size of .31 (Glass, McGaw, & Smith, 1981).

Some researchers have become so enamored of meta-analysis that they now argue that it has become the only solid research tool. According to one researcher, meta-analysis has revealed how "little information there typically is in any single study. It has shown that contrary to widespread belief, a single primary study can rarely resolve an issue or answer a question" (Schmidt, 1996).

In their monumental study of the results of virtually all of the psychological and educational meta-analyses conducted up through 1993, Lipsey and Wilson have come up with the startling conclusion that just about all behavioral treatments seem to have an effect. "There is little in conventional reviews and past discussion of these treatment areas, either individually or collectively, that prepares a reviewer for the rather stunning discovery that meta-analysis shows nearly every treatment examined to have a positive effect" (Lipsey & Wilson, 1993, p. 1192). They further point out that the effect size is so overwhelmingly positive that it hardly seems plausible that it could really be presenting an accurate picture of the efficacy of all these various treatments. They examined a total of 302 meta-analysis studies, encompassing thousands of individual studies and well over a million subjects. How can such a finding be explained? Could it be that there is editorial bias in the selection of studies for publication, that studies which show statistical significance are more apt to be published? Or perhaps there is bias on the part of researchers who only send in for publication those results that have led to a rejection of the null hypothesis, while those studies that fail to show significance are left to languish in dust-covered file drawers. If that's the case, then studies chosen for a meta-analysis are not really representative of all the studies actually conducted. As Louis Hsu has said, "there is usually a bias against the inclusion of studies that do not yield statistically significant results as well as a bias in favor of the inclusion of studies that do yield statistically significant results" (Hsu, 2002). Or maybe, since so many treatment studies utilize quasi-experimental designs, the control and experimental groups are not really equivalent to begin with. This is especially true if the experimental group's pre-treatment status is *better* than that of the control group with which it is being compared. In such cases (which, alas, are all too common), their post-treatment scores might be higher regardless of whether the subjects really received effective treatment. Another possible explanation is that the studies are not always double-blind, and the experimental group knows that it's the experimental group, leading to such contaminants as the Hawthorne or placebo effect. Or, finally, perhaps every treatment that psychologists and educators dream up really does have a positive effect. Yeah, right!

Not all meta-analyses find significant effect sizes, however. In one important study on the effectiveness of the DARE (Drug Abuse Resistance Education) project for reducing drug use behavior, it was found that, for the eight studies rigorous enough to meet the criteria for evaluation, the effect sizes ranged from .00 to .11 when the DV was based on a reduction of drug *use* (Ennett et al., 1994). In fact when the DARE outcomes were compared with those for other programs (in which more emphasis was placed on social and general competencies and which relied on the use of interactive teaching strategies), the DARE outcome levels were significantly smaller. The DARE studies did show that the students knew more about drugs and the harmful effects of drugs and also had better attitudes toward the police, but as for later drug use there seemed to be virtually no effect. The authors of the study concluded by suggesting that DARE might even be taking the place of other more beneficial drug-use curricula that teenagers should in fact be receiving.

METHODOLOGY AS A BASIS FOR MORE SOPHISTICATED TECHNIQUES

This quick foray into the world of research methodology has, at one level, jumped ahead of our statistical discussion. At a more important level, however, it has prepared you for what is to come. Statistical techniques and research methodology must necessarily be unfolded in this leapfrog fashion. In presenting the inferential techniques in the following chapters, we will often refer to the methodological issues outlined in this chapter.

At this precise moment in time, you are not yet ready to conquer the world of research, but you can do something. You are now in a position to do more than sit back and admire. You are now in a position to get your hands dirty with the data.

SUMMARY

The two major types of research strategies are addressed in this chapter—experimental and post facto. The difference between the two can best be established by, first, identifying the IVs and noting the amount of control the researcher has over them. This issue is crucial, since when the researcher has full control over the IVs, then the potential exists for drawing cause-and-effect conclusions. In the true experiment the researcher has absolute control over the IVs, whereas in post-facto research this control is severely limited (since the IVs are assigned after the fact). The cause-and-effect trap (implying causation from post-facto studies) spotlights the extent of and reasons for certain types of fallacious analysis. The distinction between active IVs and subject IVs is stressed, the former identifying experimental research and the latter indicating post-facto research.

Measurement is the assigning of numbers to observations according to certain rules. The rules determine the type of scale of measurement being constructed. Nominal scale: Using numbers to label categories, sorting observations into these categories, and then noting their frequencies of occurrence. Ordinal scale: Rank- ordering observations to produce an ordered series. Information regarding greater-than or less-than status is contained in ordinal data, but information as to how much greater or less than is not. Interval scale: Scale in which the distances between successive scale points are assumed to be equal. Interval data do contain information as to how much greater than or how much less than. Ratio scale: Special interval scale for which an absolute zero can be determined. Ratio data allow for such ratio comparisons as one measure is twice as great as another. All statistical tests in this book that can be used for the analysis of interval data can also be used on ratio data.

The researcher focuses the analysis of experimental data on the possible differences that may be found among the DVs, or outcome measures. These DV comparisons may be *between groups,* where the sample groups are independently selected, or *within groups,* where the same group is measured under all experimental conditions. In post-facto research, the analysis may be focused on the differences among DV measures, or on the possible association (correlation) between the IV and DV.

In the true experiment, the researcher creates equivalent groups of subjects through the process of random assignment to the treatment conditions. Other methods used for providing sample-group equivalence are based either on using the same group under all conditions or matching the subjects on some variable known to be related to the DV. When groups are created on the basis of random assignment and are, thus, independent, the researcher may use the after-only experimental design, where the DV measures are taken only after the IV has been manipulated. This technique helps to prevent subjects from being sensitized to the IV. When the same group is used under all conditions, the design is called repeated-measures. When only one group is used, a number of possible research errors may intrude, including the famous Hawthorne effect. Whenever repeated-measures designs are used, the researcher must be alert to the possibility of threats to internal validity that result from uncontrolled contaminating variables. Problems inherent in matching designs are also pointed out, especially the problem of determining which variables should be used to create the matching condition.

Above all, a good experimental design should be internally valid, in order to ensure that the pure effects of the IV may be unambiguously examined. Experimenters should also make every effort to produce a high degree of external validity, so that the results of the study may be extrapolated to real-world populations.

Post-facto research, where subjects are assigned to groups after the fact, does not offer the potential for direct cause-and-effect conclusions. Despite this, however, better-than-chance predictions can still be made from carefully contrived post-facto studies. Post-facto studies are always open to the possibility of the post-hoc fallacy, "because it came after this, therefore it was caused by this."

Combination research is defined when the researcher uses a mixture of both experimental and post-facto techniques. That is, combination research is used when there are at least two IVs, one a manipulated IV and the second a subject IV. Interpretations of causality should be confined to the differences produced by the manipulated IV, whereas conclusions based on the subject IV should be restricted to prediction.

Experimental research always tests the hypothesis of difference, that is, that groups of subjects assumed to be originally equivalent now differ significantly on some measured trait. Since the subjects were the same before the independent variable was manipulated, if they now differ, the *cause* of this difference may be interpreted as resulting from the action of the independent variable.

Post-facto research sometimes tests the hypothesis of difference; that is, subjects are assigned to different groups on the basis of some original difference and then assessed for possible differences in some other area. Post-facto research may also test the hypothesis of association, that is, that a correlation exists among separate measures. In neither of these post-facto areas, however, should the results be interpreted as having isolated a causal factor.

Research errors may occur in both experimental and post-facto methodologies. In experimental research, a study will lack internal validity when the IV(s) cannot be unambiguously interpreted. Manipulated IVs become confounded, producing secondary variance, when their pure effects cannot be evaluated on a cause-and-effect basis. Failure to use an adequate control group is a major reason for the loss of internal validity. External validity results when a study's findings are generalizable to a real-life population in real-life conditions. When a laboratory study becomes so artificially contrived as to no longer reflect the real world, or when the subjects used in the study fail to represent the population, then a loss of external validity results.

Other major research errors are stressed. Among these are the halo effect and the Hawthorne effect. The halo effect occurs when a researcher measures a subject on one trait and is then influenced by that measure (either positively or negatively) when evaluating that same subject in a different area. This problem can be eliminated by using independent observers when measuring subjects on more than one trait. The Hawthorne effect occurs when a researcher, using a simple before-after design, assumes that a given difference is the result of the manipulated independent variable, when in fact the result may be due to the attention paid to the subjects by the experimenter. The judicious use of a separate control group can minimize this problem.

Key Terms

- ceiling effect
- combination research
- control group
- dependent variable (DV)
- double-blind study
- experimental method
- experimental research
- external validity
- halo effect
- Hawthorne effect
- independent variable (IV)
- internal validity
- meta-analysis
- post-facto research
- primary variance
- qualitative research
- secondary variance

PROBLEMS

Indicate both the type of research (experimental or post facto) and the hypothesis being tested (difference or association) for problems 1 through 6.

1. A researcher wishes to find out whether there is a difference in intelligence between men and women. A random sample of 10,000 adults is selected and divided into two groups on the basis of gender. All subjects are given the Wechsler IQ test. The mean IQs for the two groups are compared, and the difference is found to be not significant.

2. A researcher wishes to test the hypothesis that fluoride reduces dental caries. A large random sample of college students is selected and divided into two groups. One group is given a year's supply of toothpaste containing fluoride; the other group is given a year's supply of seemingly identical toothpaste without fluoride. A year later all subjects are checked by dentists to establish the incidence of caries. The difference is found to be significant.

3. A researcher wishes to test the hypothesis that persons with high incomes are more apt to vote than are persons with low incomes. A random sample of 1500 registered voters is selected and divided into two groups on the basis of income level. After the election, voting lists are checked; it is found that significantly more persons in the high-income group went to the polls.

4. A researcher wishes to establish that special training in physical coordination increases reading ability. A group of forty 10-year-old children is randomly selected. Each child is given a standardized "Reading Achievement Test" and then placed in a specially designed program of physical coordination training. After six months of training, the "Reading Achievement Test" is again administered. The scores show significant improvement.

5. A researcher wishes to test the anxiety-prejudice hypothesis. A random sample of 200 white college students is selected and given the Taylor Manifest Anxiety Test. They are then given a test measuring prejudice toward minorities. The results show a significant positive correlation—the more anxiety, the more prejudice.

6. A researcher wishes to test the subliminal perception hypothesis. A movie theater is chosen, and the sales of popcorn and Coca-Cola are checked for a period of two weeks. For the next two weeks, during the showing of the feature film, two messages are flashed on the screen every 5 seconds, each lasting only 1/3000 of a second (a point far below the human visual threshold). The alternating messages are "Hungry? Eat popcorn" and "Thirsty? Have a Coke." Sales are checked again, and it is found that popcorn sales increased by 60% and Coke sales by 55%.
   a. What kind of research is this, post facto or experimental?
   b. If experimental, what design has been used?
   c. What possible confounding variables might there be?
7. For problem 1, identify the independent variable, and state whether it was a manipulated or an assigned-subject variable.
8. For problem 2, identify the independent variable, and state whether it was a manipulated or an assigned-subject variable.
9. For problem 3, identify the independent variable, and state whether it was a manipulated or an assigned-subject variable.
10. For problem 4, identify the independent variable, and state whether it was a manipulated or an assigned-subject variable.
11. For problem 5, identify the independent variable, and state whether it was a manipulated or an assigned-subject variable.
12. For problem 6, identify the independent variable, and state whether it was a manipulated or an assigned-subject variable.
13. For problem 4, suggest possible confounding variables.
14. Of the six research examples, which seems most prone to the occurrence of the Hawthorne effect?
15. What is the major difficulty inherent in the use of the before-after experimental design?
16. What is the major difficulty inherent in the use of the matched-subjects design?
17. A researcher wishes to assess the possibility that a certain type of psychotherapy actually reduces neurotic symptoms. A random sample of subjects is selected from the population of names on a waiting list at a mental health clinic. For each subject selected, another subject is sought, also from the waiting list, who resembles the originally selected subject in a number of important respects, such as age, length of time on the waiting list, IQ, severity of symptoms, length of symptom duration, category of neurosis, and previous treatment history. By a flip of a coin, one group is chosen to receive six months of intensive psychotherapy, while the other group is told to continue waiting and not to seek treatment elsewhere. At the end of a six-month period, a panel of experts judges all the subjects with respect to the extent of symptomatology.
   a. What type of research is this, experimental or post facto?
   b. What is the IV and what is the DV?

c. What would you advise the researcher to do with respect to the judge's knowledge regarding which subject represented which group?
d. What other variables might you wish to match on?

True or False. Indicate either T or F for problems 18 through 30.

18. When an independent variable is actively manipulated, the research method must be experimental.

19. If a researcher uses the post-facto method and establishes a significant relationship, the possibility of a cause-and-effect relationship must be ruled out.

20. If a researcher establishes a unidirectional relationship, the research method used must have been experimental.

21. If a researcher establishes a significant correlation in order to predict college grades on the basis of high school grades, the high school grades are the independent variable.

22. The main purpose of the various experimental designs is to establish equivalent groups of subjects.

23. In experimental research, the independent variable always defines the differences in the conditions to which the subjects are exposed.

24. A study designed to test the effect of economic inflation on personal income should establish personal income as the dependent variable.

25. In correlational research, the independent variable is always manipulated.

26. Correlational research must always be post-facto research.

27. Meta-analysis attempts to combine many research studies and establish an overall effect size for the entire group of studies.

28. A one-group, repeated-measures design often runs the risk of falling prey to the Hawthorne effect.

29. Determining whether the IV is manipulated or assigned indicates whether the research is experimental or post facto.

30. In the quasi-experiment, the IV is always an actively manipulated variable.

For problems 31 through 35, indicate which scale of measurement—nominal, ordinal, or interval—is being used.

31. The phone company announces that area code 617 serves 2 million customers.

32. Insurance company statistics indicate that the average weight for adult males in the United States is 168 pounds.

33. Post office records show that 2201 persons have the zip code 01118.

34. The Boston Marathon Committee announces individual names with their order of finish for the first 300 runners to cross the finish line.

35. Central High School publishes the names and SAT scores for the students selected as National Merit Scholars.

Chapter

10

The Hypothesis of Difference

Up to this point, we have dealt exclusively with samples selected one at a time. Just as we previously learned to make probability statements about where individual raw scores might fall, we next learned to make similar probability statements regarding where the mean of a specific sample might fall. Note the emphasis of part of the definition of the estimated standard error of the mean: "an estimate made on the basis of information contained in a single sample." In many research situations, however, we must select more than one sample. Anytime the researcher wants to compare *sample groups to find out, for example, if one sample is quicker, or taller, or wealthier than another sample, it is obvious that at least two sample groups must be selected.*

Compared to What?

When the social philosopher and one-line comic Henny Youngman was asked how his wife was, he always replied, "Compared to what?" (Youngman, 1991). The researcher must constantly ask and then answer the same question. In Chapter 1, we pointed to the fallacious example used to argue that capital punishment does not deter crime. The example cited was that when pickpockets were publicly hanged, other pickpockets were on hand to steal from the watching crowd. Since pockets were picked at the hanging, so goes the argument, obviously hanging does not deter pickpocketing. Perhaps this is true, but we can still ask, "Compared to what?" Compared to the number of pockets picked at less grisly public gatherings? For such an observation of behavior to have any meaning, then, a control or comparison group is needed. We cannot say that Sample A is different from Sample B if, in fact, there is no Sample B. In this chapter, then, the focus will be on making comparisons between pairs *of sample means. The underlying aim of the entire chapter is to discover the logical concepts involved in selecting a pair of samples and then to determine whether or not these samples can be said to represent a single population. In short, we will select two independent samples, measure them on a dependent variable, and then ask if any difference found between the two sets of scores could be merely coincidental.*

SAMPLING DISTRIBUTION OF DIFFERENCES

Again, we need that large fishbowl containing the names of all the students at Omega University, a total population of 6000 students. We reach in, only instead of selecting a single random sample, this time we select a pair of random samples. Say we select 30 names with the left hand and another 30 names with the right. Again, we give each student selected an IQ test and then calculate the mean IQ for each sample group. These values, as we have seen, are statistics, that is, measures of samples. In Group 1 (the names selected with the left hand), the mean turns out to be 118, whereas in Group 2 (selected with the right hand), the mean is 115. We then calculate the *difference* between these two sample means.

$$M_1 - M_2 = \text{difference}$$
$$= 118 - 115 = +3$$

Back to the fishbowl, and another pair of samples is selected. This time the mean of Group 1 is 114, and the mean of Group 2 is 120. The difference between these sample means is calculated, 114 – 120 = –6. We continue this process, selecting pair after pair of sample means, until the population is exhausted, that is, until there are no more names left in the fishbowl. Since we started with 6000 names and drew out pairs of samples of 30 names, we end up in this instance with a long list of 100 mean difference values (30 × 2 × 100 = 6000).

Selection	M_1	–	M_2	=	*Difference*
1	118	–	115	=	+3
2	114	–	120	=	–6
.	.				
.	.				
.	.				
100	119	–	114	=	+5

Each of the values in the "Difference" column represents the difference between a pair of sample means. Since all pairs of samples came from a single population (the names in the fishbowl were of *all* the students at Omega University, and *only* those at Omega University), we expect that the plus and minus differences will cancel each other out. That is, the chance of Group 1 having a higher mean than Group 2 is exactly the same as that of Group 1 having a lower mean than Group 2. For any specific mean difference, then, the probability of that difference having a plus sign is identical to the probability of its sign being minus, that is, $p = .50$. After all, when selecting pairs of random samples, pure chance determines which names are included in Sample 1 and which in Sample 2.

The Mean of the Distribution of Differences

If we add the column of differences and divide by the number of differences, we can, theoretically, calculate the mean of the distribution. Furthermore, since all of the pairs of samples came from a single population, the value of this mean should approximate zero. (Since there will be about the same number of plus differences as there are minus differences, the plus and minus differences will cancel each other out.) To get this mean value of zero, however, we had to add the differences between *all* of the pairs of samples in the entire population. This mean value of zero, then, is a parameter, since to get it we had to measure the whole population. *Whenever the distribution of differences is constructed by measuring all pairs of samples in a single population, the mean of this distribution should approximate zero.*

Also, although there will be a few large negative differences and a few large positive differences, most of the differences will be either small or nonexistent; thus, *the shape of this distribution will tend toward normality.* (See Figure 10.1.)

Note that each point on the abscissa of this distribution of differences represents a value based on measuring a pair of samples. This, like the distribution of means, is a sampling distribution, and each point on the abscissa is a statistic. The only exception to the latter fact is that the very center of the distribution, the mean, is a parameter, since *all* pairs of differences in the entire population had to be used to obtain it. It must constantly be kept in mind that to create this normal distribution of differences with a mean value of zero, all pairs of samples had to come from the same population.

Random Selection with or without Replacement

As was stated before, when selecting a random sample from a population, classical inferential statistical theory assumes that the sampling has *occurred with replacement,* thus ensuring that the probability remains constant from selection to selection. To take an exaggerated example, if the total population were to number, say, only 10, then randomly selecting a specific individual from that population would yield a probability value of 1/10, or .10. By then putting that individual back into the population, the probability of selecting any specific individual on the next draw would still equal .10. If, however, the first person selected is not replaced, the population would now contain a total N of 9, and an individual's probability of selection would then become 1/9,

FIGURE 10.1 A distribution of differences, where $\mu_{M_1 - M_2}$ equals the population mean of the entire distribution of the differences between pairs of randomly selected sample means.

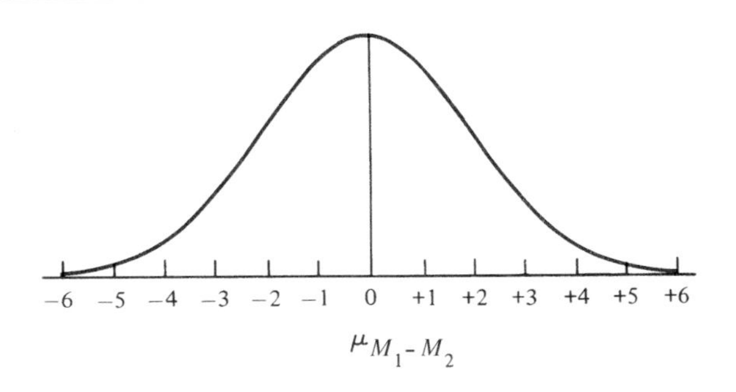

or .11. To continue this example, after 4 persons are taken out of the population, the probability on the next draw would be .60, and so on. To make the case as blatant as possible, if 9 persons are taken out of the population, the probability of that last survivor being selected would of course be a whopping 1.00. What then is the poor researcher to do?

Obviously, when selecting pairs of random samples for experimental purposes, if a given individual has been randomly selected to appear in the experimental group, that person cannot be placed back into the population for possible inclusion in the control group. If that were to happen, although the selection process would have been completely independent, the control and experimental groups would not be. The solution to this dilemma is to select samples from *large* populations where the ratio of sample size to population size is small. Under these conditions random sampling will ensure both constancy and independence with or without replacement.

Distribution of Differences with Two Populations

Assume that we now select our samples from two different populations; that is, we have two fishbowls, one containing the names of all the students at the Albert Einstein School of Medicine and the other the names of all the students enrolled at the Acme School for Elevator Operators. For each sample pair, Group 1 is selected from the population of medical students and Group 2 from the elevator operator trainees. It is a virtual certainty that in this case the mean IQs in Group 1 will be consistently higher than the mean IQs in Group 2.

Selection	$M_1 - M_2$ = *Difference*
1	125 − 100 = +25
2	120 − 105 = +15
3	122 − 103 = +19
⋮	⋮
100	130 − 100 = +30

Obviously, the mean difference for such a sampling distribution of differences does not approximate zero but is instead a fairly high positive value (or negative value if we had arbitrarily designated the medical students as Group 2). Thus, *whenever the mean of the sampling distribution of differences does not equal zero, we can be fairly certain that the sample pairs reflect different populations.*

Standard Deviation of the Distribution of Differences

As we have seen over and over again, whenever we have a normal distribution, the only values needed to make a probability statement are the mean and the standard deviation. To get the standard deviation of the distribution of differences, we square all the differences, add them, divide by N (the number of differences), subtract the mean difference squared, and take the square root ($M_1 - M_2$ designates the sampling distribution of differences).

$$\sigma_{M_1-M_2} = \sqrt{\frac{\Sigma(M_1 - M_2)^2}{N} - \mu^2_{M_1-M_2}}$$

That is, we can theoretically get the actual standard deviation of the distribution of differences in the same way that we calculated the standard deviation of a distribution of raw scores. We do this by treating all the differences as though they are raw scores and cranking them through the standard deviation equation. However, since the differences between all pairs of sample means in the entire population must be used to calculate this standard deviation, the resulting value is a parameter. Once it is known, there is nothing left to predict. This parameter is called the **standard error of difference.**

ESTIMATED STANDARD ERROR OF DIFFERENCE

Since it is exceedingly rare for anyone to measure all sample pairs in a given population, a technique has been devised for inferring the population parameters with the use of statistics. This, as has been pointed out, is really what inferential statistics is all about—measuring samples and inferring population parameters. The standard deviation of the distribution of differences can be predicted on the basis of the information contained in only two samples. The resulting value, called the **estimated standard error of difference,** can then be used to predict the true parameter standard error of the entire distribution. This estimated standard error of difference is based on the information contained in just two samples, and it is used to predict the value of the standard deviation of the entire distribution of differences between the means of successively drawn pairs of random samples. Since we measure only one pair of samples to generate this value, the estimated standard error of difference is a *statistic,* not a parameter. We will use SE_D to symbolize the estimated standard error of difference and to estimate the true parameter, $\sigma_{M_1 - M_2}$ (standard deviation of the distribution of mean differences). The equation for the estimated standard error of difference, based on independently selected pairs of samples, is

$$SE_D = \sqrt{SE^2_{M_1} + SE^2_{M_2}}$$

The estimated standard error of difference, therefore, is based on the estimated standard errors that have been obtained from each of the two samples. As stated, then, this standard error of difference estimates the variability of the entire distribution of differences. To get it, we combine, or pool together, the two variability estimates of the sampling distribution of means.

If this sounds like a lot of estimating, it's because it is. We rarely, if ever, know the true population parameters, and the main job of the statistician is to provide educated guesses as to what those parameters *probably* are. It has even been said, tongue in cheek, that if you ask a statistician what time it is, the reply would be an estimate.

Example Calculate the estimated standard error of difference for the following two independently selected samples.

Group 1: 17, 16, 16, 15, 14

Group 2: 13, 12, 11, 10, 9

Group 1		Group 2	
X_1	X_1^2	X_2	X_2^2
17	289	13	169
16	256	12	144
16	256	11	121
15	225	10	100
14	196	9	81
$\Sigma X_1 = 78$	$\Sigma X_1^2 = 1222$	$\Sigma X_2 = 55$	$\Sigma X_2^2 = 615$

1. Find the mean for each group.

$M_1 = 15.600 \qquad M_2 = 11.000$

2. Find the standard deviations for each group (by either method—see pages 163 and 164).

$s_1 = 1.400 \qquad s_2 = 1.581$

or

$SD_1 = 1.020 \qquad SD_2 = 1.414$

3. Estimate the standard errors of the means for each group—see pages 166 and 167.

$SE_{M_1} = .510 \qquad SE_{M_2} = .707$

4. Find the estimated standard error of difference for both groups combined.

$$SE_D = \sqrt{.510^2 + .707^2} = \sqrt{.260 + .500} = \sqrt{.760} = .872$$

Relationship Between the t Ratio and the z Score

We are familiar with the *z* score equation, which states that

$$z = \frac{X - M}{SD}$$

The z value tells us how far the raw score is from the mean in units of a measure of variability, or the standard deviation. Our total focus with z scores, then, is on the *raw score* and its relationship to the other values.

The equation for the t ratio is as follows:

$$t = \frac{(M_1 - M_2) - \mu_{M_1 - M_2}}{SE_D} = \frac{(M_1 - M_2) - 0}{SE_D}$$

Here our total focus is on the difference between the sample means and how this difference relates to the mean of the distribution of differences, also in units of variability.

What is the mean of the distribution of differences? We have already learned that it is equal to zero if both samples come from a single population. That, then, is the reason for including the zero in the numerator of the t ratio; zero is the expected mean of the distribution. In the z score equation, we translate our numerator difference into units of standard deviation. For the t ratio, since it is based on measuring samples, we do not have the true parameter standard deviation. However, we do have the next best thing—the estimated standard error of difference, which is estimating the true standard deviation of the distribution of differences. Therefore, the t ratio is in fact a kind of z score used for inferring parameters.

$$z = \frac{X - M}{SD} \qquad t = \frac{(M_1 - M_2) - 0}{SE_D}$$

The t ratio, then, tells how far, in units of the estimated standard error of the difference, the specific difference between our two sample means deviates from the mean of the distribution of differences, or zero. We can think of it as just a plain old z score wrapped up in a new package.

TWO-SAMPLE T TEST FOR INDEPENDENT SAMPLES

Everything in this chapter has been building toward the calculation of the t ratio for independent samples. This two-sample, independent t test allows us to make a probability statement regarding whether two independently selected samples represent a single population. By independently selected samples, we mean that the choice of one sample does not depend in any way on how the other sample is chosen. For instance, if the subjects in the two samples are matched on some relevant variable, or if the same sample is measured twice, we cannot use the independent t test for establishing any possible differences. (There are statistical tests available for such situations, as we shall see in Chapter 15.)

Calculating the t Ratio

The *t* **ratio** is calculated with the following equation:

$$t = \frac{(M_1 - M_2) - \mu_{M_1 - M_2}}{SE_D}$$

That is, *t* is a ratio of the difference between the two sample means and the population mean of the entire sampling distribution of differences to the estimated standard error of that distribution.

Notice that the actual numerator of the *t* ratio is $(M_1 - M_2) - \mu_{M_1 - M_2}$. Since we expect that the value of $\mu_{M_1 - M_2}$ (the mean of the entire distribution of differences) is really equal to zero when both samples have been selected from a single population, then our assumption is that $\mu_{M_1 - M_2} = 0$. Obviously, subtracting zero from the value $M_1 - M_2$ does not change the numerator's value at all. Nevertheless, it is still a good idea to think about the zero, at least at first, because it teaches an important lesson about the real meaning of the *t* ratio. It also allows us to view the *t* ratio in the context of something with which, by now, we do have some familiarity—the *z* score.

Independent t Ratio with Samples of Equal Size

Example Calculate an independent *t* ratio for the following sets of scores from two randomly and independently selected samples.

Group 1: 13, 12, 12, 9, 8, 8

Group 2: 8, 8, 5, 3, 3, 2

Group 1		Group 2	
X_1	X_1^2	X_2	X_2^2
13	169	8	64
12	144	8	64
12	144	5	25
9	81	3	9
8	64	3	9
8	64	2	4
$\Sigma X_1 = 62$	$\Sigma X_1^2 = 666$	$\Sigma X_2 = 29$	$\Sigma X_2^2 = 175$

If the use of the estimated standard deviation is preferred, the procedure is as follows:

1. Find the mean for each group.

$$M_1 = \frac{\Sigma X_1}{N_1} \qquad M_2 = \frac{\Sigma X_2}{N_2}$$

$$M_1 = \frac{62}{6} = 10.333 \qquad M_2 = \frac{29}{6} = 4.833$$

2. Find the standard deviation for each group.

$$s_1 = \sqrt{\frac{\Sigma X_1^2 - (\Sigma X_1)^2/N_1}{N_1 - 1}}$$

$$s_1 = \sqrt{\frac{666 - 62^2/6}{6-1}} = \sqrt{\frac{666 - 3844/6}{5}}$$

$$s_1 = \sqrt{\frac{666 - 640.667}{5}} = \sqrt{\frac{25.33}{5}} = \sqrt{5.067} = 2.251$$

$$s_2 = \sqrt{\frac{\Sigma X_2^2 - (\Sigma X_2)^2/N}{N_2 - 1}}$$

$$s_2 = \sqrt{\frac{175 - 29^2/6}{6-1}} = \sqrt{\frac{175 - 841/6}{5}} = \sqrt{\frac{175 - 140.167}{5}}$$

$$s_2 = \sqrt{\frac{34.833}{5}} = \sqrt{6.967} = 2.640$$

3. Find the estimated standard error of the mean for each group.

$$SE_{M_1} = \frac{s_1}{\sqrt{N}} = \frac{2.251}{\sqrt{6}} = \frac{2.251}{2.449} = .919 \quad SE_{M_2} = \frac{s_2}{\sqrt{N}} = \frac{2.640}{\sqrt{6}} = \frac{2.640}{2.449} = 1.078$$

4. Find the estimated standard error of difference for both groups combined.

$$SE_D = \sqrt{SE_{M_1}^2 + SE_{M_2}^2}$$

$$= \sqrt{.919^2 + 1.078^2} = \sqrt{.845 + 1.162} = \sqrt{2.007}$$

$$= 1.417$$

5. Find the t ratio.

$$t = \frac{10.333 - 4.833}{1.417} = \frac{5.500}{1.417} = 3.881$$

If the use of the actual standard deviation of the sample is preferred, then

1. Find the mean for each group.

$$M_1 = \frac{\Sigma X_1}{N_1} \qquad M_2 = \frac{\Sigma X_2}{N_2}$$

$$M_1 = \frac{62}{6} = 10.333 \qquad M_2 = \frac{29}{6} = 4.833$$

2. Find the standard deviation for each group.

$$SD_1 = \sqrt{\frac{\Sigma X_1^2}{N_1} - M_1^2} \qquad SD_2 = \sqrt{\frac{\Sigma X_2^2}{N_2} - M_2^2}$$

$$= \sqrt{\frac{666}{6} - 10.333^2} \qquad = \sqrt{\frac{175}{6} - 4.833^2}$$

$$= \sqrt{111 - 106.771} \qquad = \sqrt{29.167 - 23.358}$$

$$= \sqrt{4.229} = 2.056 \qquad = \sqrt{5.809} = 2.410$$

3. Find the estimated standard error of the mean for each group.

$$SE_{M_1} = \frac{SD_1}{\sqrt{N_1 - 1}} \qquad SE_{M_2} = \frac{SD_2}{\sqrt{N_2 - 1}}$$

$$= \frac{2.056}{\sqrt{6-1}} = \frac{2.056}{2.236} = .919 \qquad = \frac{2.410}{\sqrt{6-1}} = \frac{2.410}{2.236} = 1.078$$

4. Find the estimated standard error of difference for both groups combined.

$$SE_D = \sqrt{SE_{M_1}^2 + SE_{M_2}^2}$$

$$= \sqrt{.919^2 + 1.078^2} = \sqrt{.845 + 1.162} = \sqrt{2.007}$$

$$= 1.417$$

5. Find the t ratio.

$$t = \frac{10.333 - 4.833}{1.417} = \frac{5.500}{1.417} = 3.881$$

Independent t Ratio with Samples of Unequal Size

The technique used in the preceding example for finding the t ratio assumes that the two samples are of equal size. When conducting research, it is best to come as close to this ideal as possible. A researcher obviously should not place 100 persons in the first group and only 1 person in the second group. However, there are times when it is simply not possible to achieve identical sample sizes. When this does occur, the following variation should be used for calculating the standard error of difference.*

$$SE_D = \sqrt{\left[\frac{(N_1 - 1)s_1^2 + (N_2 - 1)s_2^2}{(N_1 + N_2 - 2)}\right]\left(\frac{1}{N_1} + \frac{1}{N_2}\right)}$$

Or, if the use of the actual standard deviation of the sample is preferred, the procedure is

$$SE_D = \sqrt{\left[\frac{N_1 SD_1^2 + N_2 SD_2^2}{(N_1 + N_2 - 2)}\right]\left(\frac{1}{N_1} + \frac{1}{N_2}\right)}$$

SIGNIFICANCE

As was the case with the single-sample t ratio, the independent t for two samples will also be evaluated for **significance.** As we learned in the previous chapter, a significant difference is one for which the chance explanation has been rejected. Again, don't read the word *significant* as being synonymous with *profound.* Significant differences are not always especially meaningful, even though the probability is small that, for example, they could have occurred by chance.

For example, a study of personality differences among married couples indicated that "nonpossessive wives" are significantly less likely to feel their relationship has much of a future, when compared to a control group of wives judged as "possessive" (Blumstein & Schwartz, 1983). Could it be, as implied, that the nonpossessive wife doesn't care if her husband is unfaithful, since she feels the marriage isn't going to last anyway? Let's look at the numbers.

Among the possessive wives, 87% say that their marriages will last, whereas of the nonpossessive wives, only 73% make the same forecast. Now, this difference may be significant, but it may not tell us much. After all, almost three out of four of these so-called nonpossessive wives look forward to a successful marriage.

Or, suppose it were discovered (in a study on college sophomores) that there is a link between eating fast-food hamburgers and developing brain tumors. Assume further that the results of the study indicate that the chance of getting a tumor from eating one hamburger every month is only 1 in 10 million, but by eating one every day the chances double to 2 in 10 million. This difference might be statistically significant,

*Although this technique may also be used with equal sample sizes, it does tend to obscure the logic of the relationship between the standard error of difference and the standard errors of the means.

WILLIAM SEALY GOSSETT (1876–1937)

Creator of the famous *t* test, Gossett modestly published under the pseudonym "Student." In order to develop the *t* test, he revised the concept of the standard error so that it could be used for generating population parameters on the basis of measures taken on small samples.

Gossett was born in 1876 into a wealthy and socially important family. He was descended from an illustrious Huguenot family that was forced to leave France due to what they perceived to be religious persecution. The French Huguenots were Protestants and because of the Edict of Nantes, enacted in the 1500s, they were given the freedom to both practice their religion and enjoy the same citizenship privileges as the Catholics. However, in 1685, King Louis XIV revoked the edict and Huguenots by the thousands left France, including the Gossetts. The family crossed the English Channel and settled in England. In the late 1880s, when William was growing up, his father, Colonel Frederick Gossett, became a full member of the prestigious Royal Engineers. Colonel Gossett wanted the best for his son and sent him to the Winchester School, an exclusive boy's prep school. William's early intellectual development was brilliantly shown when he took a competitive examination and was awarded the coveted Winchester Scholarship, an honor, incidentally, which paid for his entire prep school tuition. After graduating from Winchester, he was accepted by Oxford University and was again given a full-tuition scholarship, despite not being an athlete. In 1899 he received his Oxford degree in mathematics and natural science. After graduation, he immediately went to Dublin, Ireland, and took a job at the famous Guinness Brewery where he remained for the rest of his working life. What fascinated him was not so much the art of brewing beer but the sampling techniques used to taste-test new and improved beers and ales. In 1906, Arthur Guinness, CEO of the brewery, sent Gossett to London for one year to study with Karl Pearson. The sabbatical certainly paid off, for in 1908, he published "The Probable Error of the Mean" (Student, 1908), the article that brought him instant fame as a world-class statistician (and led to the formulation of the *t* test).

In 1934, while driving home after having worked late at the Brewery in Ireland, Gossett had a serious auto accident. He drove his car straight into a lamp post, an accident which he said was caused by his looking down to adjust some data he was carrying in a brief case. It is clear, however, that no sobriety test was ever administered. It took him a full year to recover, and for the rest of his life he walked with a decided limp.

No greater praise for a lifetime of contributions to the field of statistics could be earned than that from the pen of Sir Ronald Fisher, who said, "it is the Student of Student's test of significance who has won and deserved to win a unique place in the history of scientific methods" (Fisher, 1939).

but would you forever avoid Big Macs on the basis of it? You'd have a far greater chance of dying in an auto accident on the way to those golden arches.

Or take the following IQ data based on sample sizes of 10,000 persons each:

Men		*Women*
$M = 100$		$M = 101$
$SD = 15$		$SD = 15$
$SE_M = .150$		$SE_M = .150$
	$SE_D = .214$	
	$t = -4.673$	

As we will soon see, given that huge sample size, a t ratio greater than ±4 would undoubtedly be statistically significant and most probably would reflect a true population difference (assuming no test error). However, a one-point edge in IQ points, especially since there must have been some measurement error, is so trivial as to have little *practical* significance in bettering the quality of life as we now know it. As we will see later in the chapter, a significant t ratio may often be produced by a very small effect size.

Null Hypothesis: No Real Difference—Just Coincidence

As we have seen, statisticians use the term **null hypothesis,** symbolized as H_0, to refer to the idea that the events in question are due only to chance. A significant difference is, therefore, one for which we have excluded chance as the explanation, or one for which we have *rejected the null hypothesis.* Again, be careful here of the connotations of the English language. The word *reject* has a negative connotation; it sounds as though it might be associated with failure. In statistics, however, the word *reject* refers only to the researcher's *decision* regarding chance, or the null hypothesis. If the null hypothesis is rejected, the event is significant. If the null hypothesis is accepted, the event is due to chance and is not significant.

Suppose that we conduct a research project in which we are endeavoring to prove that a certain carefully phrased communication will change political attitudes toward a certain candidate. We test this by selecting two random samples from a population of individuals who are opposed to the candidate. Group 1 receives the special message, while Group 2 does not. Both groups are then given a questionnaire designed to measure attitudes toward the candidate. We compare the mean attitude score for each group. On the basis of the results of the t test, we decide to reject the null hypothesis. This is anything but failure! By rejecting chance, we have established a significant difference between the groups and, furthermore, have established that our special message does indeed work. The decision to reject, then, should not necessarily be viewed as a discouraging development. It is often the very decision the researcher is hoping to be able to make. After all, not many Nobel prizes have been awarded for accepting the null hypothesis, that is, concluding that all the findings are simply due to chance.

Null Hypothesis for the *t* Test. For the two-sample independent *t* test, the null hypothesis is written as

$$H_0: \mu_1 = \mu_2$$

which we read as "mu one is equal to mu two." The samples thus provide us with *two* point estimates, and the null hypothesis insists that these two point estimates of the parameter mean are equal. In other words, H_0 states that since the two samples represent populations with the same parameter mean, μ, then the populations are really identical. The null hypothesis alleges that there is, in fact, only one population and that both samples are part of that same population. Even though the sample means may differ, they do not differ by an amount large enough to reject the possibility that the samples really represent a *single* population.

The null hypothesis does *not* state that the sample means are the same, rather that they simply are not different enough to reject the conclusion that they represent a common population mean. After all, as we saw in Chapter 8 during the discussion of the sampling distribution of means, we must expect sample means selected from a single population to differ somewhat from each other. The null hypothesis is always made with reference to the population parameters and does not refer to the sample statistics. Also, the statistical decision as to whether to accept or reject is always based on the null hypothesis.

The Alternative Hypothesis

The opposite of the null hypothesis is the **alternative hypothesis** (sometimes called the research hypothesis), symbolized as H_a. Whereas the null hypothesis states that the parameter means are equal, the alternative hypothesis states that they are different.

$$H_a: \mu_1 \neq \mu_2$$

H_a says that Sample 1 is representative of a population whose mean is not equal to the mean of the population being represented by Sample 2. The alternative hypothesis theorizes that since each of these samples generates a *different point estimate* of the population mean, the population means themselves must therefore be different. In short, H_a states that the two samples represent truly different populations.

Thus, whenever we reject H_0, we are betting that H_a is true, and whenever we accept H_0, we are betting that H_a is false. The researcher, though always basing the statistical decision to accept or reject on the null hypothesis, in doing so, makes inferences about the alternative hypothesis. The null and alternative hypotheses can never both be true or false. Either μ_1 is equal to μ_2 or it is not. There is no in-between choice. This is not a "shades-of-gray" issue. H_0 and H_a are qualitatively different positions.

Since the alternative hypothesis for the *t* test states that the parameter means are different, it is also called the *hypothesis of difference*. The *t* test is, therefore, said to be designed to test the hypothesis of difference, even though the statistical decision uses only the null hypothesis. This is because rejection of the null hypothesis indicates that the samples represent different populations, whereas acceptance of the null hypothesis means that the samples represent a single population.

H_0, H_a, *and the Distribution of Differences*

Earlier in this chapter, we discovered that the mean of the distribution of differences is equal to zero whenever all pairs of random samples are selected from a single population. The statistical decision regarding the null hypothesis, then, goes right to the heart of this earlier discussion. Accepting H_0 tells us that the two samples in question are both taken from a single population and that the difference between the sample means is part of a normal distribution of differences with a mean value of zero. Conversely, rejecting H_0 indicates that the two samples represent different populations and that the difference between the obtained sample means is part of a larger distribution of differences whose mean is not equal to zero. The size of the t ratio is the determining factor in deciding whether the samples represent the same population or different populations. The larger the t ratio, the more likely it is that the mean of the distribution of differences is not equal to zero and that the null hypothesis can be rejected.

Degrees of Freedom for the Independent t *Test*

In Chapter 8 we learned that when finding the mean from a set of sample scores, there are $N - 1$ degrees of freedom. Since the independent t always compares the means from two sets of sample scores, the **degrees of freedom** in this case must equal the size of the first sample minus one, plus the size of the second sample minus one. Thus, for the t ratio,

$$\text{df} = N_1 - 1 + N_2 - 1$$

or, perhaps more conveniently,

$$\text{df} = N_1 + N_2 - 2$$

This rule holds true for both equations for the t ratio—for equal sample sizes and for unequal sample sizes. Again, we see that the larger the sample sizes, the more degrees of freedom allowed, and the smaller the sample sizes, the fewer the degrees of freedom. For other statistical tests, the calculation of the degrees of freedom will be different. Thus, the equation is valid only for the independent t test.

In general, degrees of freedom are based on the number of values in any set of scores that is free to vary once certain restrictions are set in place.

TWO-TAIL *t* TABLE

As we saw in Chapter 8, the two-tail t table, Appendix Table C, presents the critical values of t for alpha errors of either .05 or .01 and for the various degrees of freedom.

To use this table, we must first calculate the t ratio and then compare our obtained value with the table value of t listed beside the appropriate degrees of freedom. In an example earlier in this chapter, the calculated value of t was 3.881. Since the sample sizes were six for each group in that example, the degrees of freedom equal 10.

$$N_1 + N_2 - 2 = 6 + 6 - 2 = 10$$

The rule for the statistical decision is that if our calculated value of t is equal to or greater than the table value of t, we reject the null hypothesis. We look down the df (degrees of freedom) column until we find 10. Then we compare the obtained value of t, which is 3.881, with the table value of t in that row.

Meet It or Beat It to Reject H_0

Once you have found the appropriate table value, that's the number you have to reach in order to reject null. You have to "meet it or beat it" in order to reject H_0. And keep in mind that rejecting H_0 is usually the researcher's goal and is the decision that typically triggers the victory dance.

Sign of the t *Ratio*

When making the comparison on this two-tail table, we have already learned (in Chapter 8) to disregard the sign of the t ratio. The importance of both the calculated t and the table value of t lies in their *absolute* values, since it is an arbitrary decision whether to place the larger of the two mean values first in the numerator. The table values for our example are 2.228 in the .05 column and 3.169 in the .01 column. Our calculated value is certainly higher than the .05 value of 2.228, but it is also higher than the .01 value of 3.169. We thus reject the null hypothesis and state that our calculated value of t, 3.881, is significant at the .01 level. Since the absolute values are the ones being compared, if the calculated t ratio had been –3.881 we would still reject the null hypothesis.

Writing the t *Comparison*

When the table value of t was taken from the .01 column, we found that it was equal to ±3.169. This told us that with 10 degrees of freedom, to reject the null hypothesis we must get a t ratio equal to or greater than that critical value. Our t value was 3.881, which is greater than the critical value and which allows us to reject the null hypothesis. This would be written as: ($t(10) = 3.881, p < .01$). This indicates that our t ratio was significant at an alpha level of < (less than) .01.

If our calculated value of t had been say 2.255, we would then shift to the .05 column where we would find the critical value to be ±2.228. This would then be written as ($t(10) = 2.255, p < .05$) and is interpreted as having rejected the null hypothesis at an alpha level of .05.

Suppose, however, that our calculated t ratio had been only 1.288. Looking at the table values of t for 10 df, we see that our value is less than either one. In that case we would write: ($t(10) = 1.228, p > .05$, ns). This would be interpreted as showing that we failed to reject the null hypothesis, that the probability of alpha error was > (greater than) .05 and was ns (not significant).

Stop and check yourself by doing the following problem and checking your answers with those shown below. Take a moment to write out your conclusion. Assume that a researcher wishes to test the hypothesis that promising a reward will affect the motivation and hence the scores of persons taking the Block Design subtest of the WAIS-III. Two groups of subjects are randomly selected from a population of college students. The subjects in Group A are promised that doing well will excuse them from their next statistics exam, whereas the subjects in Group B are told nothing. Their scores follow:

Group A: 15, 12, 11, 10, 10, 10, 10, 9, 8, 5.

Group B: 10, 9, 8, 8, 8, 8, 8, 8, 7, 6.

The means are 10.000 for group A and 8.000 for group B.

The estimated standard errors of the mean are .816 for A and .333 for B.

The estimated standard error of difference is .881.

And the t ratio is 2.270.

In this example, there are 18 degrees of freedom and Table C tells us that at the .05 alpha level, a t ratio of 2.101 is needed to reject H_0.

Conclusion: A two-sample independent t ratio was computed on two groups, A and B. The difference was found to be significant ($t(18) = 2.270, p < .05$). The mean of Group A, the group that was promised a reward, $m = 10.000$, was significantly higher than the mean of Group B, the control group $m = 8.000$.

ALPHA LEVELS AND CONFIDENCE LEVELS

Since alpha sets the probability of committing the Type 1 error (being wrong when the null hypothesis is rejected), then alpha must be inversely related to the level of confidence. That is, if alpha is set at .05, then we have a confidence level of .95. In other words if there are only 5 chances in 100 of being wrong, then there are 95 chances in 100 of being right. Similarly, if alpha is set at .01, then the confidence level is .99.

Statistical inference does not produce eternal truth, but the statistician's ability to limit the probability of being wrong and thereby to maximize the probability of being right is far more accurate than random guessing. If, however, we conduct 100 experiments and in each one reject the null hypothesis with an alpha error of .05, the cold realization will eventually dawn that in 5 of those experiments we probably are making the wrong decision. What is even more frustrating is that we don't know which 5!

Confidence Interval for the Difference Between Two Independent Samples

If one did not wish to test the null hypothesis for the independent t test shown above, an interval estimate could be used. At the .95 level of confidence, we would look up the t value at the .05 alpha level (and, of course, at the .99 confidence level we would use the alpha of .01). This is because the probability of a given portion of the distribution falling within a certain interval is equal to 1 – alpha. The equation follows (remember that the t value in this equation is the Table C value of t—not the calculated value of t shown above) for the appropriate df. That is, we must use the table value of 2.101 that is shown for 18 degrees of freedom when alpha is set at .05.

$$\text{CI } .95 = (\pm t)\ (\text{SE}_{\text{D}}) + (M_1 - M_2)$$

Using the above data, we substitute as follows:

$$\begin{aligned}\text{CI } .95 &= (\pm 2.101)(.881) + (10.00 - 8.00)\\ &= (\pm 1.851) + 2.00\\ &= +1.851 + 2.000 = 3.851\\ &= -1.851 + 2.000 = 0.149\end{aligned}$$

Thus the researcher may conclude that the true population difference between these two samples falls between 0.149 and 3.851, or that the mean of the population represented by the first sample is no more than 3.851 points higher nor .149 points lower than the mean of the population represented by the second group.

Normality of the t Distribution

As we saw in Chapter 8, when sample sizes are sufficiently large, *the t distribution approaches normality,* as illustrated in Figure 10.2. Since the t ratios of ±1.96 *exclude* the extreme 5% of all possible t ratios, any t ratio as extreme or more extreme than ±1.96 can occur by chance at a probability level of .05 or less. This is why we say that rejecting the null hypothesis at the .05 level means that the decision to reject will be wrong 5% of the time. Why? Because t values of this magnitude can occur by chance, infrequently perhaps, but they can occur.

Similarly, with t ratios of ±2.58, the extreme 1% of the curve is now being excluded, and the middlemost 99% is being included. Figure 10.3 illustrates this situation.

FIGURE 10.2 The t distribution with infinite degrees of freedom; t ratios of ±1.96 exclude the extreme 5%.

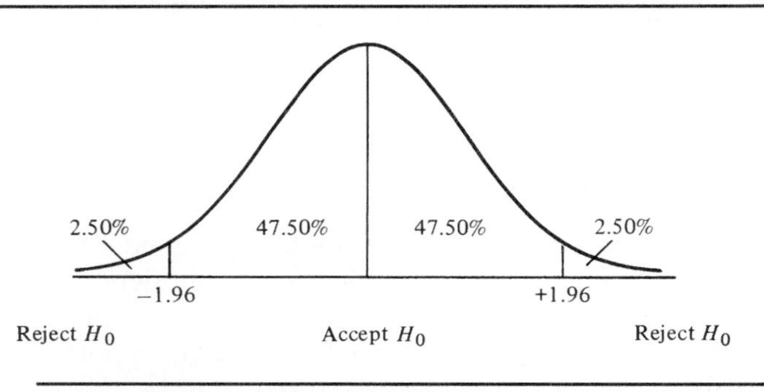

"I'm sorry, but you've been rejected at the .05 level!"

Effect of the Degrees of Freedom

Recall from Chapter 8 that as the degrees of freedom are reduced, the *t* ratios needed to exclude the extreme percentages of the curve get higher and higher. For example,

FIGURE 10.3 The *t* distribution with infinite degrees of freedom; *t* ratios of ±2.58 exclude the extreme 1%.

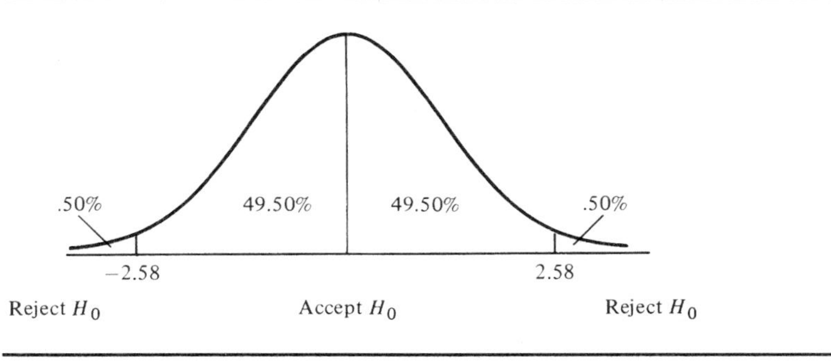

at 10 degrees of freedom, we need a t value of at least ±2.228 to exclude the extreme 5% and ±3.169 to exclude the extreme 1%. The most inflated example is found for t values at one degree of freedom. At this point, a t value of at least ±12.706 is needed to exclude the extreme 5% and ±63.657 is needed to exclude the extreme 1%.

Every degree of freedom value is specifying a different t *distribution.* As the degrees of freedom are reduced, each of these t distributions is deviating more and more from normality. As sample sizes decrease, relatively fewer cases fall around the middle of the distribution, and relatively more cases fall out in the tails.

The more degrees of freedom there are, the more the t distribution approaches the z distribution and the easier it becomes to reject the null hypothesis. Rejection of null depends, remember, on the calculated t being equal to or greater than the table value of t. A t ratio of 1.96 (which rejects the null hypothesis at .05 with an infinite number of degrees of freedom) is easier to obtain for a given difference between the sample means than a t ratio of 12.71 (needed to reject at .05 with only one degree of freedom). Other things being equal, then, the larger the sample sizes, the more likely is the prospect of obtaining a t value that will turn out to be significant. The issue will be addressed again later during the discussion of "power."

THE MINIMUM DIFFERENCE

With the two-sample t test we can determine the minimum difference needed for significance, once we know the value of the t ratio (for a given sample size and level of significance) and the standard error of the difference. For example, assume that we are using a two-group design, with six subjects in each group, and that we are setting alpha at .05. Assume further that we have found the standard error of difference to be equal to 1.417. With t at .05 and 10 degrees of freedom, Appendix Table C tells us we need a t ratio of 2.228.

$$\text{Minimum Difference} = (t \text{ from Table C})(SE_D)$$
$$= (2.228)(1.417) = 3.157$$

Thus any difference between the two sample means of 3.157 OR GREATER will be significant at .05.

OUTLIERS

As pointed out in Chapter 7, outliers are extreme scores or, as Cohen and Cohen call them, "far out observations" (Cohen & Cohen, 1983, p. 28). They are those rare scores that are so far removed from the mean of the distribution and so far out into the tails of the distribution that they seem not to represent the population the sample was selected from. For major statistical programs, such as SPSS, any value more than three standard deviations from the mean is considered to be an outlier (and is the default setting). What to do about them depends on what produced them. Most seem to result from errors, such as data-entry errors, or scoring errors, or a subject not understanding the question (such as a young man indicating his height as 12 feet), or a subject who simply responds in a noncooperative fashion. Under these circumstances, of course, the researcher has every right to simply eliminate the outliers from the analysis. If, however, the outlier truly represents one of those extremely rare events, and rare events do occur on occasion, then the researcher is faced with a dilemma. There is no magic answer to this problem. Sophisticated techniques such as Trimming, Winsorizing, and data transformations are discussed in more advanced texts, such as Howell, 1992. In general, however, if the scores can't be dropped, the safest and most conservative method of dealing with the problem is to drop down from interval to ordinal tests of significance. There will, as we shall see, be a loss of power, but violations of restrictions are less bothersome as one goes down to the nonparametric tests.

ONE-TAIL *t* TEST

The *t* test completed earlier in this chapter was based on the assumption that the Type 1 error was occurring in both tails of the *t* distribution; hence, it is called a **two-tail *t* test.** The alternative hypothesis states that $\mu_1 \neq \mu_2$, that is, that the population means are *different*—not that one population mean is greater than the other, just that they are different. This is why we disregard the sign of the *t* ratio when using the two-tail *t* table, since it is irrelevant if the first mean is greater than the second mean or less.

Sometimes, however, a researcher not only assumes that a mean difference between samples will occur but also predicts *the direction of the difference.* When this happens, the statistical decision is not based on both tails of the distribution, but only on one. In this instance, the sign of the *t* ratio is crucial. When conducted in this way, the *t* test is called a **one-tail *t* test.** *The calculation of a one-tail t test is identical to that of the two-tail t.* The only differences are in the way the alternative hypothesis is written and in the method used for looking up the table value of *t*.

One-Tail t *Test and the Alternative Hypothesis*

Since the one-tail *t* test requires that the researcher predict *beforehand* the direction of the difference between sample means, the alternative hypothesis, which simply states that a difference exists ($\mu_1 \neq \mu_2$), is no longer the full story. The researcher is now defining how the samples differ, for example, that Sample 2 is greater than Sample 1. The alternative hypothesis is still stated in terms of parameter means, but now the direction of the difference must be shown. The researcher states that $\mu_1 > \mu_2$, meaning that the first sample reflects a population whose mean is greater than the mean of the population reflected by the second sample. The statement still implies that the samples represent different populations, but now the way the populations differ is also being predicted. Of course, depending on the logic behind the research, the alternative hypothesis can instead be written as $\mu_1 < \mu_2$.

Using the One-Tail t *Table*

When the direction of the mean difference is predicted, we can turn to the one-tail *t* table, Appendix Table D, to make the statistical decision whether or not to reject H_0.

FIGURE 10.4 Percentage of cases falling above a z score of 1.65.

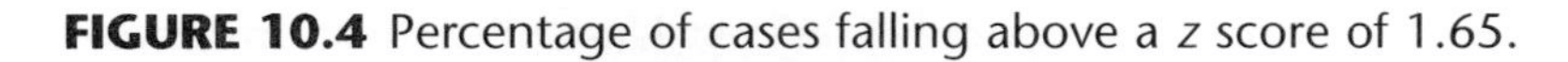

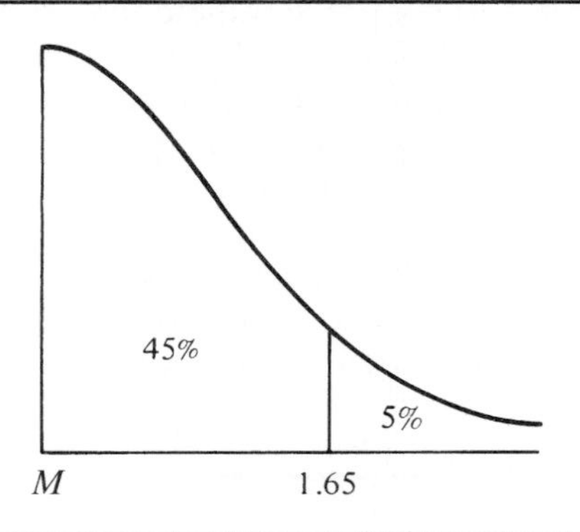

On the bottom of this table, where degrees of freedom are infinite, note the value of 1.645. We know that with infinite degrees of freedom the t distribution approaches the z distribution. A z score of 1.65 is exceeded by only 5% of the cases. This is pictured in Figure 10.4.

The probability, therefore, of receiving a z score of 1.65 or higher is .05 or less. This is also true for the t ratio when the distribution is normal. But as the t distributions depart from normality—that is, as sample sizes decrease—we must compensate for this by utilizing the degrees of freedom found in the one-tail t table. Again, as degrees of freedom decrease, the value of t necessary to reject H_0 increases.

In the one-tail t table, instead of splitting the extreme 5% of the distribution into the lowest 2½% and the highest 2½% (as was the case with the two-tail t), we are concerned here with only the most extreme 5% *on one tail of the curve.* That is, we are now dealing with only half of the distribution.

Example Assume that we are researching the theory that phonics training increases reading achievement among learning disabled children. We randomly select 20 learning disabled children, placing 10 of them in the training group and the other 10 in a group receiving no training. Since our hypothesis stipulates an *increase* in reading achievement, the one-tail t test can be used in this instance.

Assume that the mean for Sample 1 (the group receiving the training) is 55 and that the mean for Sample 2 (the group not receiving training) is 50. Assume further that the t ratio is +1.870. (The t ratio has to be plus since M_1 is greater than M_2.) The degrees of freedom are found from the equation $N_1 + N_2 - 2 = 18$. Looking up 18 degrees of freedom on the one-tail t table, we find that the table value of t is 1.734 at the .05 alpha level. By comparing the obtained t of 1.870 with the table value of 1.734, the null hypothesis is rejected. We can conclude that the two samples originally selected from a single population now represent two separate populations.

There is indeed a significant difference. Therefore, training has been shown to increase achievement.

Sign of the *t* Ratio and the One-Tail *t* Table. Since using the one-tail *t* table requires that we predict the direction of the mean difference, in the preceding example, we must accept the null hypothesis for all *t* ratios less than 1.734, including in this case all negative *t* ratios. Even a *t* ratio of –3.00, which with the 18 degrees of freedom is more than enough to reject H_0 with the two-tail table, produces an accept of H_0 with the one-tail table.

If, however, we predict a negative difference between the sample means, then the one-tail *t* table is read as though all values are negative. Returning to the previous example, suppose that we had labeled the group not receiving the training as Sample 1 and the group receiving the training as Sample 2. We would then predict a lower mean value for Sample 1, or $M_1 < M_2$. The numerator of the *t* ratio would now be written as $M_1(50) - M_2(55) = -5$. This negative numerator, of course, produces a negative *t* ratio. (The denominator of the *t* ratio can never be negative, since the estimated standard error of difference is a variability estimate, and there can never be less than zero variability.)

We again assume a resulting *t* ratio of 1.870, but now it must be written as –1.870. With 18 degrees of freedom, the table value for the one-tail *t*, as already shown, is ±1.734 at the .05 level. We therefore reject the null hypothesis. Because we are now concerned with only the negative tail of the curve, all positive values of *t*, no matter how high, must necessarily lack significance (cannot yield a reject of H_0).

Advantage and Disadvantage of Using the One-Tail *t* Table: Good News and Bad News. Using the one-tail *t* table has one decided advantage, but also an important disadvantage. The advantage is that we do not have to obtain as high a *t* value to reject the null hypothesis as we do when using the two-tail *t* table. However, the other side of the coin is that *t* values that would have been significant on the two-tail *t* table, where the direction of difference is not important, are not significant on the one-tail *t* table—simply because they do not vary in the predicted direction.

The bottom line on the one-tail *t* test is as follows: When predicting a positive mean difference, only plus *t* ratios can possibly be significant, and when predicting a negative mean difference, only minus *t* ratios can possibly be significant.

Deciding Which *t* Table to Use. When deciding which *t* table to use, one-tail or two-tail, we must be consistent. If we can logically predict the direction of the mean difference and decide to use a one-tail *t*, we must then stick with the one-tail table, no matter how the *t* ratio comes out.

For example, if we state the alternative hypothesis as $\mu_1 > \mu_2$, and then obtain a highly negative *t* ratio, we cannot restate the alternative hypothesis as $\mu_1 \neq \mu_2$ just to get a reject of H_0. Similarly, if we start out doing a two-tail *t* test with H_a as $\mu_1 \neq \mu_2$, and then obtain a *t* ratio that does not reach significance on the two-tail *t* table, we cannot suddenly switch to the one-tail t table just to get the rejection of the null hypothesis.

Because of the possibility of statistical sleight of hand, some statisticians believe that the one-tail t test should never be used. They insist that the only legitimate alternative hypothesis for the t test is the tried-and-true $\mu_1 \neq \mu_2$.

After all, if a t ratio is significant, but not in the predicted direction, it is still considered to be a nonchance event and should be interpreted as such. It might very well be important for other researchers to know about. For example, in 1994 a major drug company completed clinical trials of a new medication designed to reduce the death rate among victims of head injuries. When it turned out that there were significantly more deaths among patients who took the drug than with those who received a placebo, further testing of the new drug was discontinued (*New York Times*, December 29, 1994). Thus, as has been stated by a number of researchers, the two-tailed analysis is always seen as more conservative (Ruffolo et al 2001). It might also be argued that the one-tail t opens the door to the ethical problem of exactly when the researcher decides to make the one-tail prediction, before or after discovering that the two-tail t was not significant.

That this is not a new controversy can be seen in the words of the English psychologist and psychometrician, Hans J. Eysenck, who told us over half a century ago, "In the writer's opinion, it would be better to drop statements of one-tailed probability altogether and rely entirely on appropriate argumentation to establish the meaning of the observed two-tail probabilities" (Eysenck, 1960, p. 270).

If your instructor happens to be one of those holding this opinion, then forget about the one-tail t test, and do all t problems as though they are two-tail, nondirectional t ratios. Remember, in terms of technique, the only difference between a one-tail and a two-tail t is in how we look up the significance level.

The Independent t Test as a Research Tool

The statistical theory behind the use of the t test in analyzing experimental research (where the IV has been actively manipulated) is simple and straightforward. Since the two samples are assumed to represent a single population when originally selected, if they are later found to differ significantly, the assumption is made that they now represent different populations. And the cause of this difference is further assumed to result from the manipulation of the independent variable.

The independent t can also be used for post-facto analyses. When samples are independently selected from population groups that are known to differ on some measure, then the measure on which they differ is defined as a subject IV. For example, a researcher hypothesizes that boys and girls differ with respect to the personality trait of "need gregariousness." Random samples of boys and girls are selected, and each subject is given the Edwards Personal Preference Schedule (EPPS). The scores on the scale for need gregariousness are then compared via the independent t. In this case, if the two sample means were to differ significantly, we could not, of course, make a cause-and-effect inference. We could not say that the two samples, originally selected from a single population, now represent different populations, for the very obvious reason that these two samples were selected from two different populations to begin with. We could say, however, that the two samples, which are known to differ in gender, have also been found to differ in need gregariousness.

IMPORTANCE OF HAVING AT LEAST TWO SAMPLES

The full significance of Henny Youngman's question "Compared to what?" can now be more fully appreciated. When doing research on questions of difference, there must be at least two groups of subjects. If the claim is made that persons using Brand X toothpaste have 27% fewer cavities, the logical question is, "Compared to what?" Compared to persons who do not brush at all? Compared to persons who brush less frequently? Compared to the same group's cavity record before switching to Brand X?

"t for Two"

In Chapter 9 we discussed in some detail the techniques researchers use to create equivalent groups of subjects, and for now the main message when testing the hypothesis of difference is that there must be at least two sets of sample scores. Many years ago there was a popular song-and-dance hit titled "Tea for Two." Keep that title in mind. Whenever we want to establish whether there are differences between the sample means of two independent sets of interval scores, the t test is for two. As we saw in Chapter 8, even the one-sample t ratio was comparing two means, the sample mean with a population mean.

As a research example, a group of elementary school children was randomly selected and then randomly divided into two groups. Both groups were given the same math unit to learn, but later, during the word-problem testing, one group was given a more varied pictorial representation of the items to be added and subtracted than was the other group. A two-sample, independent t test was used to determine whether there was a significant difference between the abilities displayed by the two groups (Sprinthall & Nolan, 1991).

Some Limitations on the Use of the t Test

To calculate the t ratio, both samples must have been measured *on the same trait*. We can use the t test to find out if Group 1 is taller than Group 2 or smarter than Group 2. We cannot use the t test for comparing qualitatively different measures—for comparing apples and oranges. As a blatant example (haven't they all been?), it makes no sense to compare the mean height of one group with the mean weight of another group. A statement such as Americans are heavier than they are tall is absurd. Also, we cannot use the t test to compare different measures taken on the same group. We cannot say that a sample of Republicans has more dollars than IQ points.

Requirements for Using the t Test

To use the *t* test, the following requirements must be met:

1. The samples have been randomly selected.
2. The traits being measured do not depart significantly from normality within the population.
3. The standard deviations of the two samples must be fairly similar.
4. The two samples are independent of each other.
5. Comparisons are made only between measures of the same trait.
6. The sample scores provide at least interval data. (Any test that utilizes interval data can also be used with ratio data.)

POWER

Up to this point our interest in the statistical decision has been riveted on the alpha level, or the probability of being wrong when rejecting the null hypothesis. This is indeed as it should be because in rejecting null you are, in effect, going out on a statistical limb. By rejecting chance, you're saying that you have discovered a significant result, a public fact that may be of interest to a rather wide audience. When null is accepted, on the other hand, it may be more of a private decision. The public isn't always as interested in studies whose results can be explained on the basis of chance. But when null is accepted, another possible source of difficulty arises.

Whenever a statistical decision is made, be it an accept or a reject, there is a possibility of being wrong. Since all statistical decisions are based on a probability model, there must always be the probability of being right or of being wrong when making the decision. If the null hypothesis is erroneously rejected, we are committing the Type 1 error; but if the null hypothesis is erroneously accepted, we are committing the Type 2 error. That is, whenever the null hypothesis is accepted when it really should be rejected, the Type 2 error is committed. The Type 2 error is truly important because it means that perfectly valid research hypotheses may have been needlessly thrown away whenever it is committed. And this is too bad, since good research hypotheses are not always that easy to come by. In fact, it has recently been stated that there is a definite "bias among editors and reviewers for publishing almost exclusively studies that reject the null hypothesis" (Kupfersmid, 1988). Couple this with the fact that most research studies in which null has been accepted are never sent to journals in the first place and we suspect that only a tiny fraction of the possible Type 2 errors ever see the light of day—most of them simply languishing in someone's file drawer. As one researcher asks plaintively, "where are reports of *p* values greater than .05 going to be published, given the editorial bias against nonsignificant findings?" (Sohn, 2000).

Now, just as alpha sets the level of the Type 1 error, so too **beta** sets the level of the Type 2 error. The **power** of a statistical test is based on the beta concept and is equal to 1 minus beta ($1 - \beta$). Thus, anything that helps to reduce beta also helps to increase the power of the test. For example, if beta were equal to .10, then the power of the test is equal to (1 – .10), or .90. However, if beta were reduced to .05, then the power increases to (1 – .05), or .95.

According to one noted statistician (Wilcox, 1998), the main data problems threatening the power of a statistical analysis, such as the t test, are

1. Skewness
2. Heteroscedasticity (unequal variances within the sample groups)
3. Outliers

EFFECT SIZE

An important factor influencing power is the size of the treatment effect. The larger the effect size in the population, which in the case of the t test is estimated on the basis of the size of the t ratio for a given number of degrees of freedom, the higher the likelihood of rejecting null when it should be rejected (Cohen, 1988). Technically, the **effect size** is the difference between two population means in units of the population standard deviations. Thus, if we were to raise population scores on the verbal SAT by, say, 30 points (from 500 to 530), and since we know that the population standard deviation for this test is 100, then the effect size would be $530 - 500/100 = .30$. The larger the effect size, the higher is the likelihood of detecting the population differences through the use of inferential statistical techniques. Thus, power is a direct function of the magnitude of the effect in the population, since anything that drives the means further apart increases the likelihood of rejecting H_0.

The t ratio can be looked at as having three main components: an estimate of the effect size, the sample size, and the amount of variability (Rosnow & Rosenthal, 1993). The predicted effect size is shown in the numerator of the t ratio, $M_1 - M_2$, and the sample size is reflected in the denominator, the estimated standard error of difference. As we have seen, the standard error of difference is a combination of the two standard errors of the mean, each of which is decreased in size as the sample sizes are increased. Variability is also reflected in the denominator, since the standard error of difference is an estimate of the standard deviation of the sampling distribution of mean differences. Thus, the value of the t ratio can be maximized by increasing the numerator (the estimated effect size) and decreasing the denominator (the estimated variability).

A powerful test, therefore, is one that is less likely to commit the Type 2 error and is more likely to reject the null hypothesis when null should be rejected (Keppel, 1991); that is, when there is significance, a powerful test detects that significance. The more powerful the test, therefore, the higher is the likelihood of rejecting null when null should be rejected. What kinds of techniques can be used to increase a test's power?

1. *Increasing Sample Sizes.* A quick look at the tables of significance confirms the fact that the likelihood of rejecting null increases as sample sizes increase. For example, with a t ratio, when the df are equal to 5, a t of 2.571 is needed to reject null at the .05 level. However, with 120 df, a t of only 1.98 is needed to reject null. The message here is always to use the largest sample available to you, of course, under the constraints of practicality.

2. *Increasing the Alpha Level.* A test's power is also increased by increasing the alpha level from .01 to .05. The significance tables again confirm the fact that with a given number of degrees of freedom, the statistical value needed to reject H_0 at .05 is smaller than the value needed at .01. The researcher should rarely slavishly insist on .01 significance, for by so doing, beta has to be increased. As we have seen, statisticians do not increase the alpha level above .05. A line has to be drawn somewhere, and, since the days of Sir Ronald Fisher, that's the bottom line.

3. *Using All the Information the Data Provide.* When you have the opportunity to use interval-ratio data, use those data, and also use the statistical tests that have been designed to take advantage of that level of data. Interval measures provide more information than do ordinal or nominal measures, and the more information contained in the measurement, the more sensitive can be the statistical analysis. In Chapter 11, we will find that both the Pearson r and the rank-order Spearman r can be used to assess the possibility of a linear correlation between two sets of measures. Now, although the r is a direct, ordinal derivation of the Pearson r, it is still a less powerful test.

4. *Using a One-Tail Versus a Two-Tail Test.* As we saw earlier, using a one-tail test, as with the t test, increases the likelihood of rejecting the null hypothesis. Comparing significance values for the t ratio, you will note that lower values of t become significant values when using Appendix Table D, the one-tail table, than when using Appendix Table C, the two-tail table. As previously pointed out, however, there is legitimate controversy over the use of one-tail t tests, meaning that your conscience should guide your table of choice.

5. *Fitting the Statistical Test to the Research Design.* Your chance of rejecting null when it should be rejected, and thereby increasing test power, can be greatly increased by a very careful fitting of the statistical analysis to the research design. In fact, this will be a key issue in Chapter 15 when we focus on the relationship between the statistical analysis and the appropriate research design.

6. *Reliability of the Measures.* The more reliable the measures being analyzed, the greater is the likelihood that a true difference or correlation will be determined to be significant when it should be found to be significant. Reliable scores are dependable, consistent, and not overly laden with measurement error. Much more will be said on this important topic in Chapters 11 and 17.

In summary, then, the power of a statistical test is a measure of its sensitivity in detecting a significant result. As the value of a test's power increases, the possibility decreases of accepting the null hypothesis and thereby perhaps of committing the Type 2 error. If the statistical test has been appropriately matched to the research design, then, as we have seen, power will be a function of the alpha level, the sample size, the choice of test (one- or two-tailed), the magnitude of the true effect in the population, and the reliability of the measures. However, as a practical matter the main technique for raising power is to INCREASE the sample. It is axiomatic that large samples produce higher power than do smaller samples. However, power is still a function of the research design and the theoretical basis for the design. One researcher has said that doubling one's thinking is likely to be more productive than simply doubling one's sample size (McClelland, 2000).

Evaluating the Analysis

The evaluation of a statistical analysis depends on the interdependence among four important values: power, sample size, effect size, and level of significance. Whenever three of these values are known the last one is fixed. For example, power will be increased as increases occur in the sample size, effect size, and level of significance. For this reason, some researchers set the power, level of significance, and effect size first and then use those values to predict how large the sample should be. Although researchers tend to want to keep the power of the test at a maximum, some statisticians argue that there are times when too much power can actually be a disadvantage. Thus, there are some correlations or differences that are so small, so trivial, that even though they do occur in the population, they are of little or no practical value. Thus some researchers even elect a small effect size since it is assumed that any significant difference no matter how small, is important. Others may choose a larger effect size since it is believed that a small difference has no real-world importance.

Calculating the Effect Size

As we have seen, significant differences may be trivial differences. Thus, it is important to examine a significant difference and determine if it is meaningful, important, or even useful. An effect size, when reported as "d," is considered small at .20, medium at .50, and strong at .80 or greater. Later we will find that some effect sizes are calculated as eta squared, and a separate table has been constructed for making d to eta square conversions.

Following a two-sample t test, the effect size can be calculated as

$$d = t\sqrt{\frac{n_1 + n_2}{(n_1)\,(n_2)}}$$

where n_1 = the sample size for the first group and n_2 the sample size for the second group (Dunlap et al., 1996). For purposes of effect size, treat all t ratios as positive. Suppose the t ratio is equal to 2.70 and there are 20 subjects in each group.

$$d = 2.70\sqrt{\frac{20 + 20}{(20)(20)}} = 2.70\sqrt{\frac{40}{400}} = 2.70\sqrt{.10}$$

$$d = (2.70)(.316) = .853 \text{ (a strong effect)}$$

Calculate the effect size for the following:

$t = 2.61$, $n_1 = 30$, $n_2 = 30$

ANS: $d = .674$

$t = 3.10, n_1 = 15, n_2 = 15$
ANS: $d = 1.132$

$t = 2.13, n_1 = 18, n_2 = 18$
ANS: $d = .71$

Although seeming to state the obvious, remember that when the effect size is large it is more apt to be significant, since the likelihood of detecting a difference does depend, after all, on how big the difference actually is.

Accepting H_0

Although most studies that conclude by accepting chance as the explanation of the findings never see the light of day, there are exceptions. Sometimes researchers can capture headlines by *accepting* the null hypothesis, especially when the results are used to refute some bit of ancient folklore—you know, the kind of advice that usually begins with "they say." For example, conventional wisdom has warned teenagers to avoid chocolate or face the possibility of a disfiguring case of acne. The studies that have tested that hypothesis have come up empty. Null was accepted, and the conclusions were that chocolate has not been proved to cause acne. Of course, accepting null isn't convincing proof either. After all, there is that thing called the type 2 error. Similarly, health scientists have shown that eating just before going swimming failed to result in stomach cramps or other discomforts. In fact, marathon swimmers even improved their performances after eating high-carbohydrate meals just before entering the pool. Other attempts at disproving alleged myths, however, have not always led to an acceptance of H_0. There's the one that usually begins with, "They say that eating fish makes you smart." Well, guess what? In this case, "they" are a group of respected scientists and they concur. Researchers now say that fish really can be brain food, and their argument is based partly on the fact that one of the reasons that humans evolved in areas bordering oceans and lakes was because the fish diet provided just the right material for human brain development. It has also been found that, among other things, fish contain zinc, and studies have found that a lack of zinc can impair mental functioning and memory in human subjects (Carper, 1993). So maybe the fish-brain myth is not really a fish story after all.

And what about the case of "Jewish penicillin," or the use of chicken soup to alleviate the symptoms of the common cold? The Mount Sinai Medical Center in Miami Beach published the results of a controlled study showing that chicken soup really did work better than either hot or cold water in lessening congestion. Since then it has been found that chicken contains an amino acid called cysteine, which is chemically almost identical to a drug called acetylcysteine, a medication that is known to thin the mucus in the lungs and is currently prescribed by physicians for persons suffering from respiratory infections (Carper, 1993).

Finally, we've all heard the one about garlic being a miracle food that can be used to prevent the onset of colds, flu, and any number of other maladies. Some have even suggested that, along with eating garlic, wearing cloves of garlic offers a double protection. Cynics have scoffed that wearing garlic may prevent colds, not because of any special medical powers but simply because it keeps other people at a safe distance, especially in the winter when close contact indoors may of course lead to shared infections. It's like saying that eating Limburger cheese makes you lose weight, or at least you look thinner from a distance. However, James North, chief of microbiology at Brigham Young University in Utah, says that garlic extract killed nearly 100% of both a human rhinovirus, which causes colds, and parainfluenza 3, a flu and respiratory virus. In fact, North tells us to load up on garlic whenever we feel a cold coming on (Carper, 1993).

Type 1 Versus Type 2 Errors

Aside from the theoretical issues, the difference between committing Type 1 and Type 2 errors may also carry important practical considerations. For example, the research that showed the relationship between aspirins and the reduction of the risk of heart attacks may at first have seemed so low as to be trivial, the risk being reduced from 2.17% in the control group to 1.26% in the experimental group, for a total reduction of only .91% (Steering Committee of the Physicians Health Study Research Group, 1989). However, as pointed out by Boyd Spencer, the risk-reduction rate between the two groups produced by the aspirin was a whopping 41.9% (Spencer, 1995). Because of the huge sample size, the statistical analysis of these data yielded a significant difference (a rejection of H_0 at less than .0001). However, had the analysis committed the Type 1 error (that there really was no significant difference, even though a difference was inadvertently detected), it would have led to persons needlessly taking aspirin in the false hope that they were reducing their risks of cardiac problems. However, if the Type 2 error had been committed (that there really was a significant difference and it wasn't detected), then perhaps people may have needlessly died of heart attacks. "In this case, it is a good bet that society would much rather make a Type 1 error than a Type 2 error" (Spencer, 1995, p. 472).

With regard to testing the null hypothesis, there are four possible outcomes.

1. The researcher rejects Null when it should have been rejected, because in the population it is false—a correct decision.
2. The researcher accepts Null when it should have been accepted, because in the population it is true—a correct decision.
3. The researcher rejects Null when it should have been accepted, because in the population it is true—an incorrect decision leading to the Type 1 error.
4. The researcher accepts Null when it should have been rejected, because in the populations it is false—an incorrect decision leading to the Type 2 error.

As we have seen, the probability level for a Type 1 error is called alpha, and the probability of a Type 2 error is called beta. However, whereas the alpha level can be arbitrarily set at either .05 or .01, beta has to be found through the use of a complex series of equations and tables (Mertler & Vannatta, 2001). The issue with regard to the Type 2 error is that when H_0 is false and we incorrectly call it true, we really never know what the population distribution (from which our sample was selected) actually looks like. We do know the distribution under H_0, but not for H_a. Finally, we must keep in mind that when alpha is reduced, beta is increased and power is lost. When alpha is increased, beta is reduced and power is increased.

Hypothesis Testing and the Guarantee

Both Type 1 and Type 2 errors remind us that hypothesis testing does not offer a foolproof, full-service, absolute guarantee. Neither rejecting nor accepting Null will forever seal the deal, but the statistical decision, within the context of the Types 1 and 2 errors, does give us a far better shot than the mere flip of a coin. An error probability of .05 or .01 does guarantee that most of the time, we'll be right more often than wrong.

A Significant Difference

Even when H_0 is rejected and a statistically significant difference is established, it does not mean that all members of one group are different from all members of the other group. Though the groups differ in the aggregate, there will almost always be individual exceptions. For example, although it is a well-known fact that adult males differ significantly in height from adult females, it is fallacious to conclude that all men are taller than all women. It is also an empty rebuttal to try to debate the difference by pointing to a particular woman who happens to be taller than a particular man.

SUMMARY

When successive pairs of random samples are selected from a single population and the differences between the paired means are calculated, the mean of the resulting distribution of differences is assumed to be zero. Furthermore, if the mean differs significantly (a nonchance difference) from zero, it is then assumed that the sample pairs were not drawn from a single population. Probability statements can be made as to how large this difference must be in order to assume significance, if the variability of the distribution of differences is known. Even when not known, this variability can be inferred through the use of a statistic called the estimated standard error of difference, which is an estimate of the standard deviation of the entire distribution of differences. This estimate is made on the basis of the information contained in just two samples. All that needs to be known are the estimated standard errors of the mean for each sample.

Using the estimated standard error of difference, we can calculate an independent t ratio, which is the ratio of the difference between sample means minus the mean of the distribution of differences to the estimated standard error of difference. The value we get tells us by how much the difference between the means deviates from zero (the assumed mean of the distribution of differences) in units of estimated standard error of difference. The two-sample t test, therefore, compares the difference between two separate and independently selected samples. The two groups must differ on the independent or grouping variable, and then the *t* test determines if they also differ on some test value, called the dependent variable. For example, the samples may be grouped according to gender (the independent variable) and then compared on the basis of IQ (which in this case would be the dependent variable).

The size of the two samples determines the degrees of freedom, and, with the appropriate degrees of freedom, the calculated value of *t* can be checked for significance by comparing it with the table values of *t*. When the obtained value of *t* is equal to or greater than the table value, the null hypothesis (H_0) is rejected.

When the null hypothesis is rejected, however, there is still some possibility that the wrong statistical decision has been made. Incorrectly rejecting the null hypothesis is called the Type 1 error, and alpha expresses the probability that the error has occurred. The alpha level should be kept to a value of .05 or lower.

Whenever the researcher is testing the straight hypothesis of difference (that one sample mean is simply different from the other), the obtained *t* must be compared to the critical table values of *t* for the two-tail distribution. If, on the other hand, the researcher predicts beforehand the direction of the difference between the sample means, a one-tail check of significance may be made. Power refers to a statistical test's ability to detect significant differences among samples when there really are true differences in the population. Power is based on the Type 2 error concept, or the error of making the wrong decision when accepting null. The beta level establishes the probability of committing the Type 2 error. Thus, power equals 1 minus the beta error (the smaller the beta error, the higher the power). An important contribution to power is the size of the true effect in the population. The larger the population effect, the greater is the power. Several methods are available to the researcher for increasing power. Among them are increasing the sample sizes, increasing the alpha level (not usually above .05, however), using all the information the data provide, when appropriate, shifting from a two-tail to a one-tail analysis, and fitting the statistical test to the research design. If one did not wish to test the null hypothesis for the independent *t* test shown above, an interval estimate could be used. With this technique a range of difference values are computed, and, on a probability basis, it is assumed that the true population difference falls somewhere within the interval.

Key Terms

- alpha level
- alternative hypothesis
- beta level
- degrees of freedom
- effect size
- estimated standard error of difference
- null hypothesis
- one-tail (directional) *t* test
- power
- significance
- standard error of difference
- *t* ratio
- two-tail (nondirectional) *t* test

PROBLEMS

1. A researcher wants to determine if there is a significant difference between the Picture Arrangement scores (a subtest of the WAIS III that some feel may tap right-brain processing powers) between groups of right- and left-handed college students. Calculate a two-tail t ratio between their scores:

   Left-Handed: 12, 10, 12, 14, 12, 10, 8.

   Right-Handed: 8, 10, 10, 12, 11, 6, 7.

   a. Do you accept or reject the null hypothesis?
   b. Is there a significant difference in the Picture Arrangement scores between the right- and left-handed students?
   c. What type of research is this, post facto or experimental?
   d. Identify both the IV and DV.

2. Hypothesis: Democrats and Republicans differ with respect to the personality trait of "need dependence." Random samples of 10 Republicans and 10 Democrats are selected, and each subject is given the Edwards Personal Preference Schedule (EPPS). The scores on the scale for "need dependence" are then compared.

   Democrats: 14, 8, 10, 12, 3, 10, 12, 10, 9, 8.

   Republicans: 13, 10, 11, 14, 8, 12, 14, 12, 10, 10.

   a. Do you accept or reject the null hypothesis?
   b. Is there a significant difference in the "need dependence" scores between the Democrats and Republicans?
   c. What type of research is this, post facto or experimental?
   d. Identify both the IV and DV.

3. Hypothesis: Increasing illumination increases speed of productivity among piecework employees. Two groups are randomly selected from among the piecework employees at the Hawthorne Electric Company. Each subject is given 100 transistors to solder to the connecting relays. The time taken (in hours) is recorded for each subject to complete the task. Group A works under normal plant lighting conditions, while for Group B, the illumination level is increased by 50%. Their scores are as follows:

   Group A: 6.50, 5.00, 3.90, 4.20, 4.50, 6.20, 5.30

   Group B: 5.00, 4.00, 3.00, 3.50, 3.70, 4.20, 3.70

   a. Do you accept or reject the null hypothesis?
   b. Is there a significant difference in the speed of productivity scores between Groups A and B?
   c. What type of research is this, post facto or experimental?
   d. Identify both the IV and DV.

4. A researcher is interested in whether a certain hour-long film that portrays the insidious effects of racial prejudice will affect attitudes toward a minority group. Two

groups of 31 subjects each were randomly selected and randomly assigned to one of two conditions: Group A watched the movie, and Group B spent the hour playing cards. Both groups were then given a racial-attitude test, wherein high scores represented a higher level of prejudice. The data are as follows:

Group A	Group B
$N = 31$	$N = 31$
$M = 39.62$	$M = 42.60$

The estimated standard error of difference between the means was 1.360.

a. Do you accept or reject the null hypothesis?
b. Is there a significant difference in the racial attitude scores between the two groups?
c. What type of research is this, post facto or experimental?
d. Identify both the IV and DV.

5. A random sample of 122 delinquent boys was selected and randomly divided into two groups. The researcher was interested in discovering whether a six-week, nondirective, individual therapy program would affect levels of measured anxiety. The boys in Group A all received the therapy, whereas those in Group B did not. Both groups were then given an anxiety-level test (high scores indicating more anxiety). The data were as follows:

Group A	Group B
$N = 61$	$N = 61$
$M = 98.06$	$M = 102.35$
$SE_M = 1.98$	$SE_M = 2.02$

a. Do you accept or reject the null hypothesis?
b. Is there a significant difference in the anxiety test scores between the two groups?
c. What type of research is this, post facto or experimental?
d. Identify both the IV and DV.

6. Hypothesis: The use of full-spectrum (*FS*) lighting, also called "phototherapy," reduces symptoms of depression during the winter months. A researcher conducts an experiment with two groups of randomly selected subjects from the same population of college freshmen, with 15 subjects in each group. Group A is exposed to *FS* lighting (which has the same balance of color and the same amount of ultraviolet light as sunlight does) for a period of two hours every day. Group B is exposed to distorted-spectrum (*DS*) light, the kind most of us have in our homes, offices, and schoolrooms, for the same period of two hours a day. *DS* lighting differs from sunlight in that it doesn't have as much red; has more orange, yellow, and green; and has no ultraviolet light. At the end of two months (from the beginning of January until the end of February), all the subjects in both groups are given the Adolescent Depression Test for detecting an emotional condition known as SAD (seasonal affective disorder). The higher the scores on this test, the more depressed is the subject. In Group A, the mean score is 55 with an an SE_m of 2.004. In Group B, the mean score is 60 with an an SE_m of 2.093.

a. Do you accept or reject the null hypothesis?
b. Is there a significant difference in the Adolescent Depression Test scores between the two groups of subjects?
c. What type of research is this, post facto or experimental?
d. Identify both the IV and DV.

7. A group of 20 obese adult males was selected and randomly divided into two groups of 10 subjects each. In Group A, the experimental group, the men were involved in a vigorous, 20-minute-per-day exercise program and also given instructions on the importance of diet. Members of Group B, the control group, received the diet instructions but were not involved in the exercise program. The reduction in their cholesterol levels was as follows:

   Group A: 12, 10, 15, 11, 12, 8, 17, 11, 12, 12

   Group B: 5, 3, 7, 2, 3, 5, 4, 5, 1, 5

   a. Do you accept or reject the null hypothesis?
   b. Is there a significant difference in the cholesterol levels between the two groups of subjects?
   c. What type of research is this, post facto or experimental?
   d. Identify both the IV and DV.
   e. If the t ratio was significant, calculate the effect size.

8. A group of freshman football players were told that they had to bulk up in order to have any chance of playing on the varsity the following year. The group was randomly divided into two groups, and for three months the young men were subjected to a strict weight training regimen. Group A, however, was also given a daily food supplement alleged to increase muscle mass, while Group B was given an identical-appearing placebo.

   The data were as follows (all values given in pounds of weight increase):

   Group A: 10, 26, 34, 20, 27, 29, 23, 26, 26, 24

   Group B: 22, 24, 24, 21, 27, 20, 32, 11, 24, 20

   a. Do you accept or reject the null hypothesis?
   b. Is there a significant difference in the weight scores between the two groups of subjects?
   c. What type of research is this, post facto or experimental?
   d. Identify both the IV and DV.
   e. If the t ratio was significant, calculate the effect size.

9. A researcher is interested in whether a difference can occur between scores on one of the MAQ (Maryland Addiction Questionnaire) subtests, namely the MOT or (Motivation for Continued Treatment) among inmates incarcerated for OUI (operating a

vehicle under the influence). One group of 20 subjects was assigned to a treatment facility for dealing with drug-abuse issues, and the other group of 20 subjects continued in the main jail with no treatment being provided. The scores follow:

Treatment Group 47, 50, 41, 52, 48, 46, 50, 44, 60, 48, 46, 47, 41, 39, 58, 45, 62, 50, 39, 47.

Non Treated Group 41, 45, 40, 40, 33, 41, 40, 43, 37, 33, 53, 40, 38, 43, 55, 31, 37, 53, 38, 40.

a. Test the research hypothesis that there will be a difference between the two sets of scores.
b. Calculate a confidence interval at the .95 level for establishing the true population difference between the groups.
c. Indicate the type of research and, if appropriate, the experimental design.

10. A researcher is interested in discovering whether a record of past violence could be used to predict differences on the LSI scores (Level of Service Index) among a presentenced group of male residents at a certain jail facility. (The LSI is used to indicate how serious a security threat a given resident might be while incarcerated and is given and scored during intake). The LSI is scored for Security Risk as follows:

Scoring 0–2 Low risk, 3–5 Medium risk, 6–8 High risk

Two groups of felons were selected, Group A in which all had committed crimes which involved violence toward another person and Group B in which all had committed nonviolent felonies.

Their scores were as follows:

Group A: 3, 5, 4, 5, 6, 3, 2, 5, 6, 7, 4, 8, 3, 4, 3, 2, 5, 5, 4, 5, 5, 6, 2, 3, 1, 2, 4, 1, 3, 4.

Group B: 2, 4, 3, 5, 5, 4, 3, 2, 3, 2, 4, 2, 1, 3, 4, 2, 2, 3, 4, 4, 2, 4, 3, 5, 3, 2, 2, 3, 1, 3.

a. Test the hypothesis that inmates who had committed violent crimes were greater security risks while incarcerated than those who had not committed crimes of violence.
b. Identify the IV and DV.
c. Indicate whether the research was post-facto or experimental.

11. With extremely small samples, does the opportunity for rejecting H_0 become more or less likely?

12. When all pairs of samples come from a single population, the mean of the distribution of differences assumes what numerical value?

13. To calculate a t ratio, the data must come from which scale of measurement?

14. Does setting alpha at .01, rather than at .05, increase or decrease the likelihood of rejecting H_0?

15. The t ratio tells us by how much the difference between the sample means deviates from a mean difference of zero in units of what?

16. The standard error of difference is literally the standard deviation of which important sampling distribution?

Indicate what term is being defined by each of problems 17 through 21.

17. An estimate of the standard deviation of the entire distribution of differences.
18. The probability of committing the Type 1 error, or of being wrong when the null hypothesis is rejected.
19. $\mu_1 = \mu_2$.
20. The two samples represent a single population.
21. $\mu_1 \neq \mu_2$.

True or False. Indicate either T or F for problems 22 through 28.

22. When the two sample means differ, no matter by how much, we always reject H_0.
23. The effect size may only be calculated when H_0 has been rejected.
24. Alpha level becomes zero only when both samples reflect different populations.
25. The mean of the distribution of differences becomes zero only when all pairs of samples represent a single population.
26. The alternative hypothesis for the *t* test always specifies the direction of the difference.
27. Reducing the alpha level from .05 to .01 increases the power of a statistical test.
28. If null is accepted when it should have been rejected, the Type 2 error has been committed.

COMPUTER PROBLEMS

C1. A researcher wondered whether there was any difference in pulse rates for smokers versus nonsmokers among female college sophomores. Random samples of 15 students each were selected, and the recorded pulse rates for each group were as follows:

Smokers:	64	68	92	94	68	72	80	99	90	64	76	78	80	78	82
Nonsmokers:	66	60	66	74	84	68	62	76	80	70	62	72	76	74	78

Is there a significant difference between the two groups?

C2. Some psychologists theorize that there is a link between personality type and body type. Body types, called somatotypes, are sorted into three general categories: ectomorphs (thin and wiry), mesomorphs (stocky and muscular), and endomorphs (obese). Further, it has been assumed that mesomorphs are typically more extraverted than either ectomorphs or endomorphs. A group of 21 children, all of whom were rated above average in mesomorphy, was selected, along with another group of 21 children who had been rated below average in mesomorphy. All the selected children were 13-year-old males. Both groups were given the EPIE test, the "extraversion" scale taken from the Eysenck Personality Inventory. On this test, higher scores indicate a higher degree of extraversion. The scores obtained by the mesomorphs were

12 13 11 12 14 10 9 15 20 4 12 12 10 14 11 13 12 13 12 13 10

The scores achieved by the nonmesomorphs were

14 12 10 10 10 11 7 12 14 2 11 10 12 11 10 14 18 12 11 15 5

Do the two groups represent the same population with regard to extraversion?

C3. The Hooper Visual Organization Test (VOT) is composed of a series of pictures of common objects that have been cut into two or more parts and illogically arranged. The subject is asked to name each object. A study was conducted to discover if a difference existed on the VOT between normals and neuropsychological patients who had right-hemisphere cortical lesions. The data are as follows:

Normals: 26 26 12 26 27 36 26 26 40 33 27 19 25
29 23 26 26 24 30 26 27 20 34 26 10

Brain damaged: 17 21 38 25 24 24 29 10 24 24 24 23 24
24 24 18 32 32 28 24 25 22 25 31 27

C4. The Shaywitzs, husband and wife psychologists at Yale University, are now saying that the brains of men and women differ for certain language activities. Women appear to be physiologically better at using words and in the speed of their verbal reactions, whereas men are better at visualizing three-dimensional objects in space. Thus, women seem to be constitutionally better able to use and define words, whereas men are better at imagining what objects would look like if they were rotated. When reading a map in a car, for instance, men are less likely to have to turn the map in order to point it in the direction they are going in. Let's assume that a researcher wishes to test a part of this hypothesis on young six-year-old boys and girls using the Three-Dimensional Block Construction (3-D) test. The 3-D test has been used to assess visuoconstructional ability indicated by how well constructions in three-dimensional space are copied. Subjects reproduce models of increasing complexity using 6, 9, and 15 blocks from an assortment of blocks on a tray. If the Shaywitzs are right, then boys should outperform girls on this test.

Boys: 149 149 135 149 150 159 149 149 163 156 150 142 148
152 146 149 149 147 153 149 150 143 157 149 133

Girls: 153 149 132 145 146 146 141 160 146 146 146 147 146
146 146 162 138 147 142 146 145 148 145 139 143

C5. Researchers believe that total or partial sleep deprivation has antidepressant benefits in clinically depressed psychiatric patients (Gillin, 1994). Two groups of clinically depressed patients, 31 subjects per group, were randomly selected. Group A was sleep deprived during the second half of the night beginning at 3 A.M. Group B was allowed to sleep through the night and was not awakened until 7 A.M. At 12 noon, both groups were tested for depression. Test the hypothesis that sleep deprivation affected depression test scores. The scaled scores were as follows (higher scores indicating more depression):

Group A: 15 13 17 14 12 16 15 14 17 13 15 15 16 14 15 15 12 18
14 14 15 14 13 16 16 15 15 16 14 14 14

Group B: 29 14 18 15 16 16 17 15 16 16 28 29 17 18 17 16 17 17
17 17 18 16 28 19 17 18 19 15 17 17 17

C6. Researchers at M.I.T. suspected that the hormone melatonin, a natural hormone secreted by the pineal gland at the base of the brain, when given in small doses may be used to combat insomnia. Two groups of subjects were randomly selected; Group 1 received a placebo and Group 2 a small dose of melatonin. The subjects were then put to bed in a darkened room, and the length of time (in minutes) it took each subject to fall asleep was recorded. (The data presented here are simulated, but the overall averages are close to the original data.)

Group 1: 15.00 13.20 17.50 14.70 12.20 16.50 15.30 14.80 17.00 13.00
14.90 15.10 16.00 14.00 15.07 14.96 12.00 18.00 13.90 14.60
15.30 14.70 13.90 16.20 16.00 15.80 15.00 16.00 14.80 14.60
14.96

Group 2: 6.00 3.80 7.90 4.60 5.30 5.40 6.20 3.80 4.90 5.20 7.00 6.00
6.20 7.60 6.20 5.50 6.50 6.05 5.97 5.50 6.60 4.90 8.00 8.19
6.50 7.30 8.20 3.90 5.50 5.50 5.80

a. Calculate the independent t ratio.
b. Indicate the number of degrees of freedom in this study.
c. Setting the alpha error at .01, test the null hypothesis that the two groups represent a common population.

As a sidebar to this sleep research, physiologists have long known that changes in the natural production of melatonin act as an important factor among the hormonal effects that accompany the development of adolescent sexuality. During early childhood, melatonin inhibits the development of sexual maturity, and yet later, when melatonin production begins to naturally decrease, the stage is set for the sex hormones of adolescence to flourish. Also, melatonin production is inhibited by daylight. In fact, there is little if any difference between children and young adults in the amount of melatonin produced during daylight hours, but at night younger children produced large amounts of melatonin, whereas older children produced less and young adults even less. In fact, some researchers have suggested that humans living in areas where light exposure is greater, for example, the Mediterranean countries, tend to mature at an earlier age than those in the northern climates because of the sun's influence on melatonin production. So perhaps melatonin will be found to induce sleep, but it may also act to restrain some of the physiological aspects of sexuality. Thus, with increased levels of melatonin you may sleep well, but you may wind up sleeping alone.

C7. A researcher is interested in discovering whether scores on the Social Symptomatology Scale of the Holden Psychological Screening Inventory (HPSI) can differentiate between psychopaths and non-psychopaths. A total of 20 psychopaths (diagnosed using DSM-IV criteria during a clinical interview) and 20 non-psychopaths (similarly diagnosed) were randomly selected from a Federal Correctional facility (all inmates were serving a minimum of two years). The scores follow (although these scores are fictitious, they are modeled on the research results of Book, Knap & Holden, 2001).

Psychopaths: 17, 12, 22, 13, 21, 17, 19, 10, 17, 25, 17, 10, 27, 15, 19, 16, 15, 16, 19, 17.

Non-Psychopaths: 12, 14, 11, 10, 11, 14, 10, 21, 5, 12, 19, 12, 5, 14, 12, 16, 8, 17, 7, 12.

a. Test the research hypothesis that there will be a difference between the two sets of scores.

b. Calculate a confidence interval at the .95 level for establishing the true population difference between the groups.

Chapter

11

The Hypothesis of Association: Correlation

We now enter the treacherous and murky waters of correlation. Perhaps no other area of research demands more caution or is fraught with more danger. Too many of us assume that because a correlation is easy to say, that it's likewise easy to understand. In some respects, it is. Mathematically, the correlation coefficient is rather straightforward and easily calculated. What can legitimately be inferred from its numerical value, however, is quite another story.

The meaning of the word correlation comes literally from its parts: "co" means with, together, or jointly, and "relation" means association. Thus, when two events regularly occur together, then they are said to be correlated, as with blond hair and blue eyes. Also, when changes in one set of events are regularly accompanied by changes in another set of events, correlation is said to exist; for example, as children get taller, they also tend to get heavier. As we shall soon see, the correlation coefficient provides us with a numerical value that states the extent to which two events, or two sets of measurements, tend to occur or to change together—the extent to which they covary.

CAUSE AND EFFECT

So far, correlation seems straightforward, easy, and obvious. Why then the warning in the first paragraph about "treacherous and murky waters"? Consider the following findings. A great number of studies have shown that there is a large, positive correlation between reading speed and IQ. People who read quickly tend to have higher IQs than do people who read slowly. Therefore, enrolling in a speed-reading course will lead to an increase in IQ. Right? Wrong! Because the same factors that make a person a fast reader (verbal ability, alertness, quick reactions) probably also enable that person to perform well on IQ tests. Another study reported a positive correlation between coffee consumption and heart attacks among men. Men who had heart attacks were asked how much coffee they drank. This rate was compared to the rate of coffee consumption among a comparable group of men (matched on such characteristics as age and weight) who had never had heart attacks. The results tallied. Therefore, drinking coffee causes heart attacks. Right? Wrong! It may only be that elements of the preheart attack condition, such as fatigue or stress, cause the person to seek relief by turning to the stimulating effects produced by coffee. It's like saying that just because your alarm clock went off at 7:30, it must have been the sound of the alarm that caused it to be 7:30 (or like the rooster taking credit for the dawn).

Another example of a post-facto correlation being erroneously used to establish causation is the correlation between the amount of ice cream sold and the number of drownings. Again, the strong possibility of a third variable, warm weather, being related to both the sales of ice cream and the number of drownings should be taken into account. One should not seek to reduce the number of drownings by banning the sale of ice cream. Some post-facto research studies, called *epidemiological* studies, are based on observing groups of people and then showing statistical correlations between their lifestyles or behavior and what happens to them later in life. Many of the accusatory allegations found in studies of this type must be taken with a large grain of salt, since third (and even fourth and fifth) variables are so often overlooked. Studies showing the relationship between a child's TV viewing habits and various later outcomes, usually some sort of aggressive behavior problem, are of the epidemiological type and often omit other possibly more important variables, such as drug use, the easy availability of guns, or the loss of hope due to relentless economic deprivation.

In the foregoing examples, legitimate correlation statements were made, but illegitimate inferences were drawn. It's not that the existence of cause and effect has been ruled out in these cases; however, the given correlations, in and of themselves, do not prove it. It may be true that drinking coffee causes heart attacks or that increased reading speed causes a better IQ test performance. The point is that correlation does not directly address the cause-and-effect issue.

Problem of Isolating the Cause

Suppose that a researcher is interested in finding out whether or not parental rejection causes juvenile delinquency among teenage boys. Furthermore, the researcher has developed an accurate method for assessing perceived parental rejection; that is, the researcher has a tool for determining a given adolescent's perception of his parents' rejection-acceptance attitudes toward him. After gathering the data, the researcher checks police records and finds that a relationship does indeed exist. The more a boy perceives himself the victim of parental rejection, the more frequent have been his confrontations with those in blue uniforms. But has a cause-and-effect relationship been established? We can use A to symbolize parental rejection, B to indicate juvenile delinquency, and an arrow to indicate the causal direction. (Using letter symbols when analyzing correlational studies is usually a good idea, since such symbols do not have any of the literary overtones that are often inherent in word descriptions of the variables.) Three hypotheses are possible to account for the findings:

1. $A \rightarrow B$. It is possible, though not proven by this study, that A (parental rejection) does cause B (juvenile delinquency).
2. $B \rightarrow A$. It is possible that B (juvenile delinquency) causes A (parental rejection). A parent may not project warm feelings of affection toward a son who is brought home night after night in a patrol car.
3. $X \rightarrow A + B$. It is further possible that X (some unknown variable) is the real cause of both A and B. For instance, X could be an atmosphere of frustration and despair that permeates a given neighborhood, leading both parents and children to generate feelings of hostility.

All three of these explanations are possible; the point is that the correlation alone is not enough to identify which is the real explanation.

The problem with using correlation as a tool for establishing causation is due to the difficulty in predicting *the direction* of the relationship. For example, research on married couples has found that persons who find their partners attractive have happier sex lives. There is no guarantee in this finding, however, that personal attractiveness increases sexual pleasure, for it is just as likely that persons with happier sex lives tend to view their partners as being more attractive. Beauty may indeed reside in the eye of the beholder. Or perhaps a third variable—say, an optimistic approach to life in general—causes people both to find their partners more attractive and to view their sex lives as more complete. In short, perhaps both variables are the results of a person's rosy, self-fulfilling prophecy.

Sometimes isolating the third variable is easy as when reading a newspaper article stating that John Doe died from smoking in bed. Later in the article it points out that John was smoking in bed when a burglar broke into his bedroom and shot him. At other times it's not so blatant. Fred Mosteller found that bombing accuracy in World War II correlated positively with the number of fighter planes sent up to attack the bombers. Thus, the more enemy fighter planes sent up to attack the bombers, the more accurate the bombing. To unlock the puzzle, a third variable was sought, and it was discovered that when there was heavy cloud cover the fighter planes couldn't find the bombers and the bombers couldn't find their targets.

In the field of psychology, the literature on the condition known as agoraphobia (the fear of having a panic attack away from a safe place) shows high and significant correlations between agoraphobic symptoms and passive-dependent personality types. Perhaps conventional wisdom might interpret this relationship to mean that the passive-dependent personality tends to produce agoraphobic symptoms—even to the point of being rendered house bound. On further analysis, however, it should be pointed out that the data may just as likely be saying that a person with agoraphobic symptoms might then become a passive dependent—that is, that the symptoms might be producing the personality type. Or it might be that some unknown third factor (perhaps even a specific gene) is causing the individual both to have panic attacks and to be a passive dependent.

Using Correlation

If correlational research is so limited as to cause and effect, then why do we bother? The reason is that correlational research can yield better-than-chance predictions. If two events are correlated, then a knowledge of one of those events allows a researcher to predict the occurrence of the other, regardless of what might have caused what. This is helpful because if we can predict an event, we may later be able to control it.

A number of studies have found a significant correlation between the grades given to students by a given professor, and that professor's student evaluations (Greenwald, 1997). Some faculty members have concluded that this is proof that a self-serving professor can in effect buy an outstanding student evaluation by generously handing out high grades. However, since this is strictly correlational evidence, the reverse may be true—that students who evaluate their professors highly are more willing to study longer, work harder, and therefore learn more and as a result earn higher grades. That the latter explanation may be valid has been shown by a number of studies where students in sectioned classes of the same course all have to take the same final. The research shows that the sections whose faculty members had the highest ratings contained the students with the highest grades on the common final (Marsh & Dunkin, 1992).

THE PEARSON *r*

During the latter half of the nineteenth century, an English scholar and mathematician, **Karl Pearson,** became impressed with the fact that individuals vary so widely in such characteristics as height, weight, and reaction time. At the time Pearson was working with **Sir Francis Galton,** considered to be the father of the very important concept of individual differences. Pearson reasoned that since there was such wide variety to human characteristics, it would be extremely useful if the measurement of these characteristics could be expressed in relational terms rather than in absolute units of measurement. After all, it is just as useful to know that a woman's height places her in the *relative* position of exceeding 50% of the adult female population as it is to know that, in *absolute* terms, she is 65 inches tall.

The idea of relative position also makes it possible to create a common ground of comparison between qualitatively different measurements. Even though measurements of height and weight cannot be compared directly, because the two are conceptually different and are expressed in different units, such measurements can be compared in terms of how much they both vary *from their own average.* In effect, this means that the proverbial apples and oranges can be compared on the basis of whether a given orange and a given apple are both larger, smaller, juicier, or riper than the average orange and apple. Thus, the relationship between two qualitatively different objects can be expressed in quantitative terms. Pearson called this expression the **correlation coefficient.** Pearson's correlation coefficient, then, is a numerical statement of the linear relationship between two variables. A single value can be calculated to express this relationship.

INTERCLASS VERSUS INTRACLASS

Correlation coefficients may be either interclass or intraclass (Fisher, 1938). Interclass correlations are those between measures sharing the same measurement class, such as correlating IQ scores with IQ scores between twins, or correlating height measures with height measures for individuals at different ages, or test scores with the same test's scores taken at different times. In this case, the same metric and variance are found in both measures. Intraclass correlations, on the other hand, share neither the same metric nor the same variance, as when measuring the relationship between IQ points (a class of measures representing intelligence) and grade-point average (a class of measures representing actual achievement). That is, correlation may be used to compare apples with apples (interclass), as well as apples with oranges (intraclass). A variety of procedures have been developed for doing both significance tests and confidence intervals for intraclass correlations (McGraw & Wong, 1996).

Although correlation does not necessarily imply causation, it is still a useful tool for making predictions. The amount of correlation tells us the extent to which two or more variables are associated, or the extent to which they covary or occur together. Although the previous sentence mentions two *or more* variables, the focus in this chapter is on two-variable correlations, or bivariate analyses. Later in the book, we will present a technique for handling multivariable correlations.

Positive, Negative, and Zero Correlations

Correlations take three general forms: positive, negative, and zero. Positive correlations are produced when persons or things having high scores on one variable also have high scores on a second variable and those with low scores on the first variable are also low on the second. For example, a positive correlation between height and weight means that those individuals who are above average in height are also above average in weight and that those who are below average in height are correspondingly below average in weight.

Negative correlations are produced when high scores on one variable are associated with low scores on a second variable and low scores on the first correspond with high scores on the second. A negative correlation between college grades and absenteeism means that those who are above average in college grades tend to have fewer than the average number of absences, while those who are below average in grades tend to have more than the average number of absences. An example of a negative correlation familiar to every homeowner is the one between temperature and the amount of heating oil consumed each year. Colder weather is usually associated with higher fuel bills. Among professional golfers, a significant negative correlation exists between the amount of money earned on the tour and the scores achieved by the golfers

KARL PEARSON (1857–1936)

Called by many the father of the science of statistics, Pearson's credits are too numerous to cover completely in a short biography. However, among his major contributions are the following: the product moment correlation (which we call the Pearson r), the chi square, the normal curve (previously called the error curve by Gauss), the Multiple R, the partial correlation, and the bi serial correlation. And if that were not enough, he also introduced the *standard deviation* and, in fact, coined the term and its Greek symbol back in 1894 (Pearson, 1894).

Pearson was the consummate scientist and empiricist and, according to one of his biographers, Helen Walker, was so almost from birth (Walker, 1978). At the tender age of four, he was sitting in a highchair sucking his thumb, when his mother ordered him to stop such infantile behavior, with the admonition that if he kept it up, his thumb would wither away. He immediately compared his two thumbs, studied them for a while, and then said, "They look alike to me. I can't see that the thumb I suck is any smaller than the other. I wonder if you are lying to me" (Walker, 1978). Thus, his strong belief in observation as the source of knowledge and his confrontational manner were part of him from his very earliest years. In fact his early education was marked by his strong intellect, his wide range of interests, his almost perverse love of controversy, and his total unwillingness to accept unproven authority, traits with which his mother was all too familiar. In 1875 he was accepted, on full academic scholarship, at Cambridge University, and he received his bachelor's degree in mathematics in 1879. Later he would say that his undergraduate years were the happiest of his life. He said that he took great pleasure in being with friends, working with his coaches, having fights, and searching for new insights in religion, mathematics, and philosophy (Pearson, 1936). His college years were not without controversy, however. Although deeply religious, Pearson got into trouble at Cambridge because of his refusal to attend compulsory chapel. Although this created a significant campus stir, the Dean finally backed down and the university bent the rule to accommodate Pearson's demands. His first publication, at the age of 23, was a biographical account of his search for truth and knowledge. He says that he would "rush from science to philosophy, and from philosophy to my old friends, the poets, and then overwearied by too much idealism, I fancy, I become practical and return to science. Have you ever attempted to conceive all there is in the world worth knowing—that no one subject in the universe is unworthy of study?" (Pearson, 1880). After graduation from Cambridge, he went to law school and passed the bar exam in 1881. After spending some time traveling and studying in Germany, he returned to England, and in 1884 he was appointed as a professor and chairman of the graduate department of applied mathematics at the University of London. A few years later he wrote his famous "Grammar of Science." In 1889 he befriended and came under the influence of the eminent scholar and researcher, Sir Francis Galton. He was able to quantify Galton's twin concepts of correlation and regression. Later, while using these new tools, he discovered a strong correlation between alcoholism and various mental defects. So even Pearson, himself, tried to sneak causation through correlation's back door. His interpretation was that alcoholism was the result, not the cause, of the mental defect. Also, while trying to find a method for fitting observed data to a known probability curve, he created the enormously useful chi square statistic, which in the words of Helen Walker has a range of applications far greater than the specific problem for which it was first designed. The chi square still occupies a crucial position in modern statistical theory (Walker, 1978, p. 695). In 1903 he started a new journal, called *Biometrika,* which remains even to this day as perhaps the leading periodical in statistics. The first issue contained a picture of Charles Darwin (who was Galton's cousin), with the inscription, "*Ignoramus in hoc signo laboremus,*" which translates to—we are ignorant so let's get to work. When Galton died (1906), he left a sizable portion of his estate to the University of London to establish a new chair in the field of applied statistics, aptly named the Galton chair. The first recipient of this honor—Karl Pearson. In 1915 the department admitted its FIRST undergraduates, a fact that did not totally please Pearson. His fear was that the department would lower its standards and turn its attention to teaching rather than research, which of course was his first love. During his latter years, he engaged in a rather heated controversy with R. A. Fisher, creator of the ANOVA. Pearson was vehemently opposed to Fisher's reliance on small samples rather than population values. He also spent a great deal of his time fighting for what were then considered radical causes, such as the liberation of women and the pursuit of free thought and speech. His son, Egon Pearson (also a world-class statistician), said that his father fought for the use of solid empirical data as the basis of public policy and reform, not the untrained arguments of the social do-gooders.

In 1933, at the age of 76 he finally resigned his professorship and ironically had to turn the Galton chair over to R. A. Fisher. Three years later, Pearson was dead.

(remember that in the game of golf, as anyone who has ever played the game sadly knows, lower scores are better scores).

Finally, zero correlations are produced when high scores on one variable are as likely to correspond with high scores on a second variable as they are with low scores on the second or when low scores on the first are just as apt to be associated with either high or low scores on the second.

Interpreting Correlation Values

To express the degree to which variables are associated, or correlated, a single number is used. This number may vary from +1.00 through zero to –1.00. A value of +1.00

indicates the maximum positive correlation. A maximum positive correlation occurs when two measures associate perfectly. For example, if there were a perfect positive correlation between height and weight, a person taller than another person would always be heavier too, with no exceptions. Needless to say, perfect positive correlations are extremely rare in the social sciences.

In math and physics, there are numerous examples of perfect, 1.00, correlations. The radius of a circle bears an exact relationship to the circle's circumference. A greater radius will always coincide with a greater circumference. Perfect negative correlations are also easily found. The greater the distance an object is from its light source, the less illumination will fall on the object, hence, a correlation of –1.00. In psychology and education, where the variables are almost always empirically found, things just aren't that simple.

A value of zero indicates no relationship at all, a zero correlation. A value of –1.00 indicates a maximum negative correlation, meaning every single individual in a group who is higher than another on one measure is lower than that other on a second measure. The closer the correlation value is to either ±1.00, the more accurate is the resulting prediction, and prediction, after all, is the name of the game in correlational research. The predictive efficiency of a correlation, then, increases as the correlation value differs from zero, *regardless of sign*. A correlation of –.95 predicts more accurately than does a correlation of +.85. Remember, for correlation research, that a negative value does not mean a lack of correlation; it simply refers to an inverse relationship.

Deterministic Versus Stochastic

Using the radius of a circle to predict its circumference is based on a deterministic model. The correlation (being unity) contains no error and the prediction is perfect. Using weight to predict health, however, is based on a stochastic model, where the less-than-perfect correlation is used to account for the error.

Graphs and Correlation

For a visual representation of how much two distributions of scores correlate, statisticians use a graphic format known as a **scatter plot.** The scatter plot allows the visual representation of two separate distributions on a single diagram.

A Positive Correlation. Table 11.1 lists some data (some of which might be propaganda) from seven students who have been measured for both hours per week spent studying and grade-point average (GPA).

The data from the table are plotted in Figure 11.1. Note that each point represents two scores. For Student A, the point is directly over 40 on the X axis and directly across from 3.75 on the Y axis. Also, note that the points drift upward, from lower left to upper right. This configuration represents a positive correlation between the variables.

A Negative Correlation. Scatter plots can also portray negative correlations. For example, suppose that we select a sample of seven students and measure them on both

TABLE 11.1 Data on hours per week spent studying and GPA for seven students.

Student	Hours/Week Studying, X	GPA, Y
A	40	3.75
B	30	3.00
C	35	3.25
D	5	1.75
E	10	2.00
F	15	2.25
G	25	3.00

hours per week of Frisbee playing and grade-point average. Our data are summarized in Table 11.2.

We again plot the data, using a single point to represent the pair of scores for each student. (The point directly above 2 on the X axis and across from 3.50 on the Y axis represents the pair of scores for Student A.) Our plot is shown in Figure 11.2. Note that high scores on one measure correspond with low scores on the other. This situation constitutes, as stated earlier, a negative correlation. The scatter plot allows us to see this relationship. When the correlation is negative, the general slope of the points in the scatter plot falls from upper left to lower right.

A Reminder about Cause and Effect. No direct statement regarding cause and effect can or should be made about either of the preceding relationships. For the data on the relationship between study time and grades, although it is certainly a viable hypothesis that hours spent studying affect grades ($A \rightarrow B$), other hypotheses are at least possible. Perhaps the student who receives high grades is encouraged by them to

FIGURE 11.1 Scatter plot of study hours versus GPA showing positive correlation.

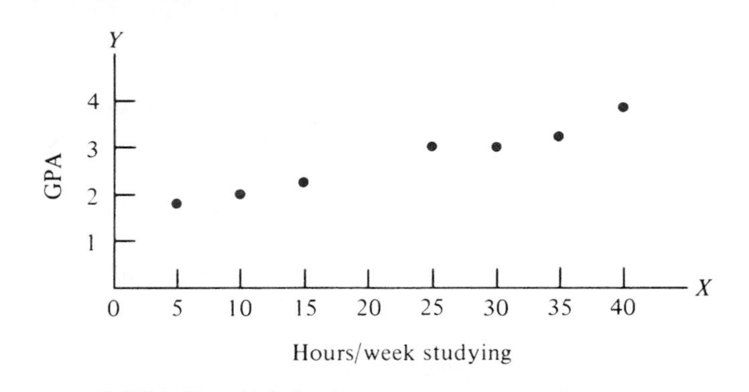

TABLE 11.2 Data on hours per week spent playing Frisbee and GPA for seven students.

Student	*Frisbee Hours/Week, X*	*GPA, Y*
A	2	3.50
B	8	2.50
C	10	2.25
D	12	1.75
E	16	.50
F	6	3.00
G	4	3.25

spend more time studying ($B \rightarrow A$). Finally, it may be a third variable, perhaps old-fashioned ambition, that drives a student to want to study and also to want high grades ($X \rightarrow A + B$). The negative correlation between hours spent playing Frisbee and grade-point average does not mean that playing Frisbee causes a lower GPA (although that is one possible explanation: $A \rightarrow B$). Perhaps, instead, students who are doing poorly in school use Frisbee tossing as an outlet for pent-up frustrations ($B \rightarrow A$). Or, perhaps, a personality factor of a gregarious nature exists in some students, who need constant social stimulation and so seek the companionship of Frisbee partners *and* avoid the lonely hours involved in academic preparation ($X \rightarrow A + B$).

Scatter Plot Configurations. Figure 11.3 shows how the three general types of correlations are revealed by the scatter plots formed when large numbers of subjects are measured on two variables and the pairs of scores graphed. The scatter plot on the left portrays a positive correlation; the array of points tilts from lower left to upper right,

FIGURE 11.2 Scatter plot of Frisbee-playing hours versus GPA showing negative correlation.

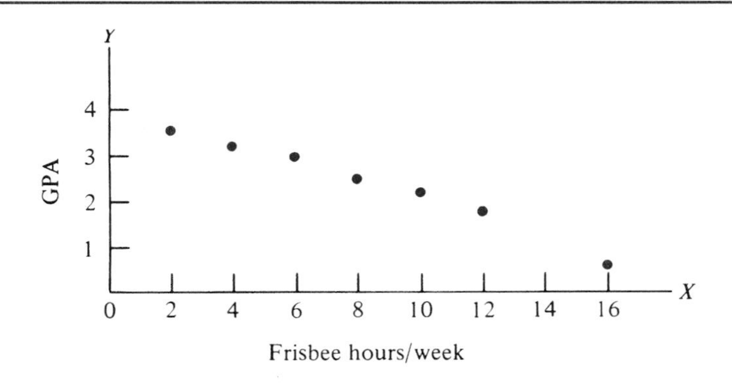

FIGURE 11.3 The three types of correlation.

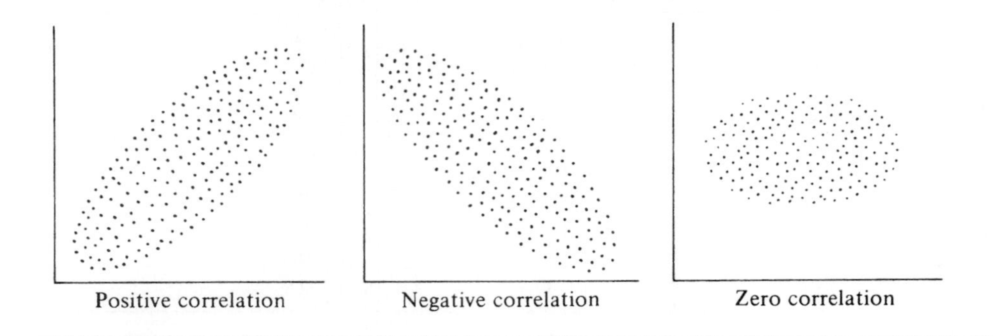

telling us that as one variable increases, so does the other. The scatter plot in the middle portrays a negative correlation; the array of points falls from upper left to lower right, telling us that as one variable increases, the other decreases. Finally, the scatter plot on the right portrays a zero correlation; as one variable changes, there is no related change in the other.

It is important to note that the array of points on a scatter plot generally tends to form an oval shape. This is due to central tendency. Since most scores in any distribution tend to cluster around the middle of that distribution, the configuration points in a scatter plot, which reflects two distributions, typically show a bulge of scores in the middle and a tapering off at the extremes.

The z Score Method for Calculating the Pearson r

Karl Pearson called his equation the product-moment correlation coefficient, but in his honor it is more familiarly known as the **Pearson *r*.** Pearson based his equation on the concept of z scores, defining r as the mean of the z score products for the X and Y variables, as follows:

$$r = \frac{\Sigma z_x z_y}{N} \qquad z_X = \frac{X - M_X}{SD_X} \qquad \text{and} \qquad z_Y = \frac{Y - M_Y}{SD_Y}$$

To calculate r, each raw score for both distributions must be converted into a z score. The z score pairs are then multiplied and the z products are added. To get the mean, this product sum is divided by the number of *pairs* of z scores. This is admittedly a rather complicated process, but, fortunately, an easier, more user-friendly equation has been derived.

Computational Method for Calculating the Pearson r

Although working a few correlations via the z score method may be instructive, it is also extremely time-consuming. Getting all those z scores is laborious indeed, especially when there are long lists of paired scores. The computational equation that follows makes the calculations far easier. If you can calculate a mean and standard deviation, you can easily calculate the Pearson *r*. If the estimated population standard deviation is used, the computation is as follows:

$$r = \frac{\dfrac{\Sigma XY - (M_x)(M_y)(N)}{N-1}}{(s_x)(s_y)}$$

Follow the steps for the estimated population standard deviation as shown on page 164.

Example Calculate the Pearson *r* value for the data from Table 11.1.

Student	Hours of Study, X	X^2	GPA, Y	Y^2	XY
A	40	1600	3.75	14.063	150.000
B	30	900	3.00	9.000	90.000
C	35	1225	3.25	10.563	113.750
D	5	25	1.75	3.063	8.750
E	10	100	2.00	4.000	20.000
F	15	225	2.25	5.063	33.750
G	25	625	3.00	9.000	75.000
	$\Sigma X = 160$	$\Sigma X^2 = 4700$	$\Sigma Y = 19.00$	$\Sigma Y^2 = 54.752$	$\Sigma XY = 491.250$

1. Calculate the means.

$$M_x = \frac{\Sigma X}{N} = \frac{160}{7} = 22.857 \qquad M_y = \frac{\Sigma Y}{N} = \frac{19.00}{7} = 2.714$$

2. Calculate the two estimated standard deviations as follows.

$$s_x = \sqrt{\frac{\Sigma X^2 - (\Sigma X)^2 / N}{N-1}} = 13.184$$

$$s_y = \sqrt{\frac{\Sigma Y^2 - (\Sigma Y)^2 / N}{N-1}} = .728$$

3. The equation: Plug all values into the Pearson r equation. Subtract the product of the means and *N* from the ΣXY term and divide by $N - 1$, or 6. Then divide this result by the product of the two estimated standard deviations (13.184)(.728) and get the Pearson *r*. Note that by using this equation, you can still do a reality check on both the means and standard deviations.

$$r = \frac{\frac{491.250 - (22.857)(2.714)(7)}{6}}{(13.184)(.728)}$$

$$r = \frac{\frac{491.250 - 434.237}{6}}{9.598} = \frac{\frac{57.013}{6}}{9.598} = \frac{9.502}{9.598} = .989$$

If the actual sample standard deviation is used, then the equation becomes

$$r = \frac{\frac{\Sigma XY}{N} - (M_x)(M_y)}{SD_X SD_Y}$$

1. Calculate the means.

$$M_x = \frac{\Sigma X}{N} = \frac{160}{7} = 22.857 \qquad M_y = \frac{\Sigma Y}{N} = \frac{19.00}{7} = 2.714$$

2. Calculate the two actual sample standard deviations as follows:

$$SD_x = \sqrt{\frac{4700}{7} - 22.857^2} = 12.206 \qquad SD_y = \sqrt{\frac{54.752}{7} - 2.714^2} = .675$$

3. The equation—plug all values into the Pearson r equation. Divide the ΣXY value by N, which is the number of paired scores, and then subtract the product of the means. Be careful of the sign in this step; if the $(\Sigma XY)/N$ term is smaller than the product of the means, the numerator will be negative and a negative correlation will result. Divide by the product of the standard deviations to get the Pearson r. Notice that the denominator in this final fraction is composed of the product of the standard deviations, and since, as has been shown many times, the standard deviations can *never* be negative, then the denominator can never be negative. As already mentioned, the only way to arrive at a negative correlation is when the *numerator* is negative.

$$r = \frac{\Sigma XY/N - (M_x)(M_y)}{SD_X SD_Y} = \frac{491.250/7 - 62.034}{(12.206)(.675)} = \frac{8.145}{8.239}$$

$$r = .989$$

Testing the Pearson r for Significance

Although the focus in this chapter will be on correlation as an inferential procedure, the correlation coefficient can also be viewed simply as another descriptive statistic. As a descriptive measure, the correlation is used to portray the relationship between two sets of scores, and the findings are then concerned only with the observed data and the variables being represented by these data. In most research situations, especially in the behavioral sciences, the importance of the correlation lies in its generalizability, not in its straight description. Thus, our concern throughout this chapter will be aimed at discovering whether the observed correlation, as shown by the sample, can be extrapolated to the population being represented by that sample.

Therefore, as we did with the t test, we must determine whether or not the correlation, which was obtained from sample data, can be legitimately generalized to the entire population. When applying the t test, differences between sample means may or may not reflect population differences. To find out, we have to test for the significance of our obtained value. Similarly, with the Pearson r, an association between sets of sample scores may or may not reflect an association that truly exists in the population. We must also, therefore, test r for significance. If the results of any statistical test are going to be used to infer population characteristics, those results must be subjected to a significance test.

The Null Hypothesis. We have learned that the null hypothesis is the hypothesis of chance. No matter what results have been obtained on a sample, the null hypothesis insists that these results are strictly due to the vagaries of chance and, therefore, cannot be generalized to the population. For the Pearson r, the null hypothesis states that $\rho = 0$; this means that there is no correlation in the population, regardless of the value that has been obtained for the sample. The alternative hypothesis, on the other hand, is that there is indeed a correlation in the population, or that $\rho \neq 0$. The Greek letter ρ (rho) is used to refer to the correlation that may or may not exist in a population, whereas r always refers to a sample.

Using the Pearson r Table

To make the statistical decision, we use Appendix Table E, the Pearson r table. This table is set up like the t table, with degrees of freedom in the far left column and the critical values of r in two columns, one headed .05 and the other headed .01. As in the t table, .05 and .01 represent the probability of alpha error—the probability of rejecting the null hypothesis when we should have accepted it. Degrees of freedom are assigned on the basis of $N - 2$, the number of *pairs* of scores minus 2. We compare the obtained value of r with the table value of r for the appropriate degrees of freedom. If the absolute value of the obtained r is equal to or greater than the critical table value, we reject the null hypothesis. If the absolute value calculated is less than the table value, we accept the null hypothesis.

Example In the last example (found on page 282), the obtained value of r was .989. Is that a significant correlation?

Since we have 7 pairs of scores, the degrees of freedom ($N - 2$) are $7 - 2$, which equals 5. At 5 degrees of freedom, the table values of r are .754 for .05 and .874 for .01. Our obtained r is higher than the .05 value of .754, but it is also higher than the .01 value of .874. We thus reject the null hypothesis and state that our calculated value of r is significant at the .01 level.

Conclusion: A Pearson correlation was computed between Hours of Study and Grade-Point-Average. A strong, positive correlation was established and found to be significant ($r(5) = .989$, $p < .01$).

The statistical reasoning behind this procedure is that since the obtained correlation coefficient equals .989, its probability is less than .01 of occurring by chance when the assumed population correlation is zero. Since the population distribution of correlations under the null hypothesis is assumed to center on zero, the obtained correlation of .989 has a probability of occurring by chance less than one time in a hundred.

In making the table comparison, it is the absolute value of r that determines significance. Although the sign of the correlation is crucial in terms of the interpretation of the results, it is totally irrelevant to the significance check. With five degrees of freedom, and if the obtained r were equal to .770, we would have dropped down to the .05 level and written: A Pearson correlation was computed between Hours of Study and Grade-Point-Average. A strong, positive correlation was established and found to be significant ($r(5) = .770$, $p < .05$).

Finally, again with 5*df*, if the obtained r were equal to .5000 we would have to conclude: A Pearson correlation was computed between Hours of Study and Grade-Point-Average. A correlation of .500 was established and found not to be significant ($r(5) = .500$, $p > .05$, ns).

When determining significance, there are only three possible outcomes:

1. Reject H_0 at the .01 level.
2. Reject H_0 at the .05 level.
3. Accept H_0.

Meaning of a Significant Correlation. Even though a correlation is significant, it does not necessarily mean that it is communicating an especially profound message. A significant correlation is one that is *not* likely to be a result of chance. Scanning the df column of the r table reveals that as sample size increases, extremely small correlations attain significance. With 400 degrees of freedom, for example, an r of only .10 may be significant. An r of .10 is indicative of only a very slight associational trend; when it is found to be significant, this slight trend does indeed exist in the population.

Studies show that there is a small but significant correlation between hair color and frequency of temper tantrums among children. Yes, it's true; redheads do have slightly more temper outbursts than do their blonde and brunette siblings. However, we cannot leap to the conclusion that this association is genetic; it could just as easily be environmental. Perhaps some parents permit only their redheaded children to act out in this fashion. Also, we cannot assume that red hair causes temper tantrums. Although this conclusion might increase hair color sales in other shades, it is simply not justified on the basis of the evidence.

For reasons of space, all possible df values for the significance of r could not be included in Appendix Table E. For a rough and, as we shall see, conservative estimate, however, you may evaluate the r by using the next smaller df value. For example, if the number of pairs of scores were 122, which produces 120 degrees of freedom, you might safely use the table value for 100 degrees of freedom, or a significance value of .195. If an exact value is required, use the equation

$$t = \frac{r\sqrt{(N-2)}}{\sqrt{1-r^2}}$$

and then evaluate the t ratio using the two-tail t table and, of course, the appropriate degrees of freedom for the Pearson r.

For example, assume that you obtained an r of .16, with 122 pairs of scores.

$$t = \frac{.16\sqrt{122-2}}{\sqrt{1-.16^2}} = \frac{.16\sqrt{120}}{\sqrt{1-.026}} = \frac{(.16)(10.954)}{\sqrt{.974}}$$

$$= \frac{1.753}{.987} = 1.776$$

Since a table value of 1.980 is needed at the .05 level for 120 degrees of freedom, we would have to accept H_0 and conclude that the correlation was not significant, or that the two variables are really independent of each other.

Example Calculate the t ratio, and make a statistical decision at the .05 level for the following combinations of r and N.

a. $r = .600$ and $N = 20$
b. $r = .150$ and $N = 52$
c. $r = .500$ and $N = 102$
d. $r = .085$ and $N = 507$

Answers

a. $t = 3.182$. Reject H_0. Correlation is significant.
b. $t = 1.073$. Accept H_0. Correlation is *not* significant.
c. $t = 5.774$. Reject H_0. Correlation is significant.
d. $t = 1.917$. Accept H_0. Correlation is *not* significant.

Restricted Range

The Pearson r has the potential for yielding its highest value when both sampling distributions, X and Y, represent the entire range of normally distributed values. If the range of either or both sampling distributions is in any way restricted, the Pearson r may seriously underestimate the true population correlation. As stated, the Pearson r was designed to show the strength of the relationship between high and low scores on one variable with the high and low scores on the other. If either of the variables fails to contain the high or low end of its distribution, the resulting correlation will tend to be closer to zero than it otherwise would be. For example, suppose a researcher were interested in correlating IQ and SAT scores among students at a special private school for gifted adolescents. In this case, both the IQ and SAT distributions might show severely restricted ranges, especially if the school's admission policy were to select students largely on the basis of IQ. The correlation, because this group was so homogeneous, would tend to be considerably lower than it would for unspecified cases taken from the secondary school population at large.

As another example of how a restricted range can affect the correlation, it has been reported that the correlation between IQ scores and grades in school drops dramatically as the size of the group being tested becomes smaller and more homogeneous. The correlations average in the .60s for elementary school children (the largest and most heterogeneous of the groups tested), to the .50s for high school students, to the .40s for college students, to only the .30s for graduate students (the smallest and most homogeneous group) (Fancher, 1985). The groups tested became smaller and obviously more homogeneous, since the lower IQs are consistently and systematically weeded out as the students progress toward more intellectually demanding experiences.

Outliers

A possible trap awaiting the unwary investigator is the damage that can be done to a Pearson correlation by just one or two outliers.

For example, take the following distributions of SAT scores and college GPAs

Subj #	*SAT*	*GPA*
1.	450	2.00
2.	400	2.50
3.	510	3.10
4.	550	3.20
5.	480	2.90
6.	500	2.80
7.	570	3.30
8.	460	2.60
9.	510	2.90
10.	500	3.00

If you calculate this correlation, you'll get a Pearson r of +.811, very high and very significant. However, by changing the last student's pairing (subject #10) from 500 and 3.00 to the outliers 800 on the SAT and a GPA of .10, the correlation reverses to a –.670, also significant at .05, but very misleading. Could those outliers have been real values, or is it most likely an entry error on someone's part? Either is possible. It is conceivable that a very bright student could hit the SAT for an 800 and then either never attend class or willfully do anything in his/her power to flunk out (maybe as an aggressive response to overbearing parents, with the implicit message "show these grades around the country club"). In any case these outliers have completely changed the nature and meaning of the correlation.

Interpreting the Pearson r

As we have seen, the value of the Pearson r varies between –1.00 and +1.00. Since r gives a precise numerical value within this range, it can express an enormous variety of associational meanings. Extremely subtle degrees of associational strength are communicated with the Pearson r. Despite this, certain broad categories should be kept in mind when attempting to explain any given correlation. One such set of verbal tags has been supplied by the distinguished American statistician J. P. Guilford (see Table 11.3).

Guilford points out that these interpretations may be used only when the correlation coefficient is significant. Also, the same interpretations "apply alike to negative and positive r's of the same numerical size" (Guilford, 1956, p. 145). An r of –.90 indicates the same degree of relationship as an r of +.90 does.

Coefficient of Determination. Although Table 11.3 gives a few correlational benchmarks with which to interpret the value of r, far more precise methods have been devised. In Chapter 14, for example, we will utilize the Pearson r as a means for making specific predictions and will also master techniques for assessing the accuracy of those predictions. For now, however, an easily calculated statistic, called the **coefficient of determination,** can help bridge the gap. The coefficient of determination is simply the square of the Pearson r.

$$\text{coefficient of determination} = r^2$$

The coefficient of determination is used to establish the proportion of the variability among the Y scores that can be accounted for by the variability among the X scores. That is, anytime there is a correlation between X and Y, then X contains some information about Y. When the correlation is a maximum of 1.00, then the coefficient of determination is also 1.00 (since one squared equals one). It tells us that X is carrying

TABLE 11.3 Guilford's suggested interpretations for values of r.

r *Value*	*Interpretation*
Less than .20	Slight; almost negligible relationship
.20–.40	Low correlation; definite but small relationship
.40–.70	Moderate correlation; substantial relationship
.70–.90	High correlation; marked relationship
.90–1.00	Very high correlation; very dependable relationship

all there is to know about Y. In fact, when $r = 1.00$, then X and Y may be measuring a single underlying trait, and knowing the X score automatically gives the Y score. By multiplying the coefficient of determination by 100, we get the *percentage* of information about Y that is contained in X. For example, a correlation value of .70 yields a coefficient of determination of .49, meaning that 49% of the information about Y is contained in X. Therefore, although a correlation value of .70 does *not* mean 70% accuracy, the square of the correlation, or .49, can be used to give us a percentage estimate of accuracy.

It is important to keep in mind that the Pearson r is not a proportion, and the distances between successive points on an r scale are not equal. Thus an r of .50 doesn't mean we have half the strength of an r of 1.00. The strength of various correlations, from .10 to 1.00, breaks out like this:

0.10.20.30 .40 .50 .60 .70 .80 .90 1.00

The square of r, however, the coefficient of determination, does create equal intervals. So that for r^2

0 .10 .20 .30 .40 .50 .60 .70 .80 .90 1.00

Note how r^2 lines up. That is, notice that .10 aligns almost under .30, whose square is .09, and .50 aligns almost under .70, whose square is .49 and so on. Thus r^2 is a proportion, meaning that an r^2 of .60 has twice the predictive power of an r^2 of .30. The only problem with r^2 is that it can never show the sign of the correlation, so from r^2 alone we can't tell if the correlation is positive or negative.

Suppose that we were attempting to predict adult male height, the Y variable, on the basis of weight, the X variable, and we obtained an r of .50. Squaring that correlation yields .25, and multiplying that by 100 indicates how much, in percentage terms, the variation in height would *decrease* if everyone were the same weight. In this case, then, the variation in height would decrease by 25% for all persons of the same weight. Look at it another way. Suppose that the correlation between height and weight were 1.00 (which it obviously isn't), and we then squared the correlation and obtained 1.00. This would mean that 100% of the variability in height would be "explained" on the basis of weight, or that if everyone had the same weight, the variation in height would be *reduced by 100%*. That would mean that *everyone* of a given weight would be exactly the same height. Finally, if the correlation were zero, then none of the variability in height could be explained (reduced) on the basis of weight, and the variation in height would be the same regardless of how much a person weighed. When you hear statisticians talking about "explaining" a certain percentage of the variability, it simply means the amount of variability that can be *reduced*. Why is this important? Because the more the variability is reduced, the more error is being taken out of the prediction and the more accurate and precise the resulting prediction becomes. A prediction that the temperature tomorrow is going to be between 50 and 60 degrees is a lot more functional than if the variability were so large that the prediction went from 0 degrees to 100 degrees. To take another example, since education has been found to explain 20% of the variance in income ($r = .45$), then if everyone had exactly the same number of years of education, the variation in income for a given number of years of schooling would *decrease* by 20%, meaning, of course, more uniformity of incomes for a given level of schooling. Thus, the coefficient of determination provides a value that *determines* what percentage of the information about Y is contained in X. More will be said concerning this important topic in Chapter 14 when we will use the coefficient of determination's cousin, the standard error of estimate, to assess the confidence level of our correlational predictions.

Comparing Two Pearson Correlations: Sample Independent

There are times when the researcher has calculated two Pearson r correlations from separate and unmatched samples and wishes to establish whether they differ significantly from each other. This can be easily accomplished by converting each of the Pearson r values into a variable called Fisher's z_r and then applying the Z test. Table 11.4 begins with the r value for .20 since for all r values of less than .25, z_r is equal to r (as shown for r values of .24 on down to .20). For example, suppose that we have discovered a significant correlation of .70 between self-esteem and physical attractiveness among ninth-grade female students. Among ninth-grade male students, however, the correlation was found to be lower, yet still significant at .55. Both samples contain 100

TABLE 11.4 Converting Pearson r values into Fisher's Z_r

r	Z_r	r	Z_r	r	Z_r	r	Z_r
.20	.203	.40	.424	.60	.693	.80	1.099
.21	.213	.41	.436	.61	.709	.81	1.127
.22	.224	.42	.448	.62	.725	.82	1.157
.23	.234	.43	.460	.63	.741	.83	1.188
.24	.245	.44	.472	.64	.758	.84	1.221
.25	.255	.45	.485	.65	.775	.85	1.256
.26	.266	.46	.497	.66	.793	.86	1.293
.27	.277	.47	.510	.67	.811	.87	1.333
.28	.288	.48	.523	.68	.829	.88	1.376
.29	.299	.49	.536	.69	.848	.89	1.422
.30	.310	.50	.549	.70	.867	.90	1.472
.31	.321	.51	.563	.71	.887	.91	1.528
.32	.332	.52	.576	.72	.908	.92	1.589
.33	.343	.53	.590	.73	.929	.93	1.658
.34	.354	.54	.604	.74	.950	.94	1.738
.35	.365	.55	.618	.75	.973	.95	1.832
.36	.377	.56	.633	.76	.996	.96	1.946
.37	.388	.57	.648	.77	1.020	.97	2.092
.38	.400	.58	.662	.78	1.045	.98	2.298
.39	.412	.59	.678	.79	1.071	.99	2.647

Based on Fisher's equation: the table was constructed using Fisher's original formula: $z_r = .50[\log e(1 + r) - \log e(1 - r)] = [\ln(1 + r) - \ln(1 - r)]$.

students. Do the two correlations represent a common population, or do they differ significantly as a function of gender?

Since the sampling distribution of Fisher's z_r is normal, we can test the null hypothesis by using the Z test, with Z values of ±1.96 showing significance at the .05 level and ±2.58 for the .01 level.

$$Z = \frac{z_{r_1} - z_{r_2}}{\sqrt{\frac{1}{N_1 - 3} + \frac{1}{N_2 - 3}}}$$

For the numerator, find the two z_r values from Table 11.4. The r of .70 translates to a z_r of .87 and the r of .55 to a z_r of .62. For the denominator, the only variables are the two sample sizes. Thus,

$$Z = \frac{.87 - .62}{\sqrt{\frac{1}{100 - 3} + \frac{1}{100 - 3}}} = \frac{.25}{\sqrt{\frac{1}{97} + \frac{1}{97}}} = \frac{.25}{\sqrt{.010 + .010}}$$

$$= \frac{.25}{\sqrt{.02}} = \frac{.25}{.141}$$

$$= 1.773$$

Accept H_0. The difference is not significant.

Therefore, since ±1.96 is needed to reject H_0 at the .05 level, we conclude that the correlations between physical attractiveness and self-esteem found from the samples of males and females actually represent a common population correlation. There is no significant difference between the two correlations.

Example A researcher has computed two significant Pearson r values from independent random samples. Sample A, 57 subjects, produced an r of .67 and Sample B, 62 subjects, produced an r of .41. Do the two sample correlations represent a common population correlation?

$Z = 1.96$: Reject H_0 at .05 and conclude that the two correlations are significantly different and represent separate populations.

MISSING DATA

Let's say we are conducting a survey, and among other things we ask each respondent to indicate both their height and weight. To do a Pearson r correlation we obviously need two scores for each individual selected. Suppose that a few subjects filled out their height items but failed to indicate their weights or vice versa. Must these subjects be eliminated from the analysis and thereby reduce the degrees of freedom? One method for handling the missing data problem is to insert the mean value for the entire distribution. This is a conservative approach, since the mean has not been changed, and also by including a mid-range value like the mean, the correlation is not artificially raised. Another more sophisticated solution is to calculate the correlation that was obtained without using any of the subjects with missing values, and then by the use of a regression analysis (as shown in Chapter 14) placing the resulting predicted values in the empty slots. This second method is less conservative because it might tend to increase the value of the correlation if there is one to begin with, which is a more powerful approach. Using either of these methods, however, demands that the researcher is using at least ordinal or interval data. You can assign a mean rank or a mean score, but with nominal data you can't average discreet categories. For instance, if 1 is used for identifying a female and 2 for a male, you can't fill in an empty slot by inserting a 1.5.

CORRELATION MATRIX

Correlations among several variables are conveniently shown in what is called a correlation matrix. It is a square, symmetrical array where each row and each column represent a different variable. The point where the row and column intersect is called a cell, and this is where the correlation between the variables is shown. The following is a correlation matrix taken from data supplied by the author and collected at the Western Massachusetts Correctional Center. A total of 60 incarcerated men were tested on five variables:

1. Anger—as shown on the Novaco Provocation Inventory, a test of the range and intensity of a person's anger (Novaco, 1975).
2. Alcohol—as measured on the Alcohol Scale that is taken from the SASSI—Substance Abuse Subtle Screening Inventory (Miller, 1990).
3. Drug—as measured on the SASSI Drug Scale.
4. Age.
5. Observed Anger—as reported on a 10-point behavioral scale filled out by the case workers and counselors.

Notice that the correlation between Anger and Alcohol (.315) appears at the intersection of the first row and second column as well as the second row and first column. This is true of all the other correlations so presented. They all appear twice. The pair of dotted lines, which have been drawn in for illustrative purposes, identifies what is called the main diagonal, a series of 1.00 correlations resulting from the redundant correlations of each variable with itself. The matrix is therefore said to be symmetrical around the main diagonal and for this reason only the top or bottom half of the matrix is usually shown (meaning that each correlation will then only appear once). On this matrix a single asterisk signifies significance at .05 or less and the double asterisk at .01 or less.

Correlations

		ANG	ALC	DRU	AGE	OBS
ANG	Pearson Correlation	1.000	.315*	.338**	.115	.550**
	Sig. (2-tailed)	.	.014	.008	.383	.000
	N	60	60	60	60	60
ALC	Pearson Correlation	.315*	1.000	.566**	−.167	.056
	Sig. (2-tailed)	.014	.	.000	.201	.669
	N	60	60	60	60	60
DRU	Pearson Correlation	.338**	.566**	1.000	−.081	−.004
	Sig. (2-tailed)	.008	.000	.	.538	.975
	N	60	60	60	60	60
AGE	Pearson Correlation	.115	−.167	−.081	1.000	.212
	Sig. (2-tailed)	.383	.201	.538	.	.105
	N	60	60	60	60	60
OBS	Pearson Correlation	.550**	.056	−.004	.212	1.000
	Sig. (2-tailed)	.000	.669	.975	.105	.
	N	60	60	60	60	60

* Correlation is significant at the 0.05 level (2-tailed).
** Correlation is significant at the 0.01 level (2-tailed).

Requirements for Using the Pearson r

To use the Pearson *r*, the following requirements must be met.

1. The sample has been randomly selected from the population.

2. The traits being measured do not depart significantly from normality. This is not as severe a limitation as it might appear. The Pearson *r* is a "robust" test, meaning that rather large departures from normality still allow for its use. The distribution forms, however, must be unimodal and fairly symmetrical.

3. Measurements on both distributions are in the form of, at least, interval data. Of course, any test that can handle interval data can also be used with ratio data.

4. The variation in scores in both the X and Y distributions must be similar. This property, known as **homoscedasticity,** may be assumed unless either of the distributions is markedly skewed.

5. The association between X and Y is linear. That is, the Pearson *r* cannot be used unless the relationship forms a straight line. Curvilinear relationships (as when an increase in X is accompanied by an increase in Y up to a point and is then accompanied by a decrease in Y) should not be assessed by the Pearson *r*.

Linearity can usually be identified by examining the scatter plot (which can easily be constructed by using the SPSS program found at the back of the book). If the scatter plot is roughly oval shaped, then one can assume that the correlation is basically linear (see page 280). You do need a fairly decent-sized sample for this oval shape to appear. It won't show itself on problems containing 9 or 10 pairs of scores. Nonlinear relationships can be observed if the scatter plot shows a horseshoe shape or has a series of strange little twists and turns.

Effect Size. The Pearson *r* is itself a measure of effect size, so no further calculations are needed (Cohen, 1988). In the behavioral sciences, significant correlations of .10, .30, and .50, regardless of sign, are sometimes interpreted as small, medium, or large respectively (Green, Salkind, & Akey, 1997).

Limitations on the Use of the Pearson *r*. The Pearson *r*, as we have seen, assesses the strength of a linear relationship between two sets of sample scores in interval form and representing normal distributions. Sometimes the data do not meet those requirements.

In Chapter 9, it was pointed out that interval data demand that the distances between successive scale points be equal. Therefore, with interval data, we know not only that one observation is greater than another but also how much greater. Ordinal data, on the other hand, which come in the form of ranks, present no information regarding the distances between scale points. With ordinal data, we know that if one observation has a higher rank than another, then the first is greater than the second, but we do not know how much greater.

Therefore, whenever the distribution of interval scores is definitely not normal, or whenever the data come to us in ordinal form, then we should not use the Pearson *r*. For example, the Pearson *r* can almost never be used on income data, since the income distribution in the population is often skewed. Similarly, the Pearson *r* should not be used on a distribution of rank-ordered scores, for example, the order of finish of a group of race horses.

THE SPEARMAN r_S

When the Pearson *r* is inapplicable, the correlation value may be obtained using an equation derived by **Charles Spearman.** The resulting value is called the *Spearman r*, or, the **Spearman r_S**. (In some texts, this statistical test is called the Spearman rho.) Since the Spearman r_S does not make any assumptions regarding either the parameter mean or the parameter standard deviation, it is called a nonparametric test. The Pearson *r*, on the other hand, is a parametric test. Much more will be said later, especially in Chapters 16 and 19, concerning the difference between parametric and nonparametric statistical tests, especially with regard to power.

Three examples of calculating the r_S are to follow: one in which both sets of scores are originally given in ordinal form (rank order); one in which the two sets of scores are given in different forms, one interval and the other ordinal; and one in which both of the distributions of interval scores are known not to be normal.

Calculating r_S with Data Originally Given in Ordinal Form

In this first example, both sets of raw scores are in the form of ordinal data.

Example Suppose that a researcher wishes to establish whether there is a significant correlation between aggressiveness and leadership ability among female military recruits. A random sample of 10 recruits is selected. The soldiers are watched closely by a trained observer who rank-orders the women on the basis of aggressiveness exhibited. Independently, another trained observer watches the women and rank-orders them on the basis of leadership abilities. (Independent observers are required in this study to prevent a type of research error known as the halo effect; see Chapter 9.) The data, the R_1 being the rank on aggressiveness and R_2 the rank on leadership, are as follows:

Subject	R_1	R_2	$d = \|R_1 - R_2\|$	d^2
1	2	3	1	1
2	3	1	2	4
3	7	5	2	4
4	6	9	3	9
5	1	2	1	1
6	5	6	1	1
7	10	8	2	4
8	8	10	2	4
9	9	7	2	4
10	4	4	0	0
				$\Sigma d^2 = 32$

We use the following steps for calculating the value of r_S:

1. Pair off the two ranks for each subject. In this case, Subject 1 ranks second in aggressiveness and third in leadership.
2. Obtain the absolute difference, d, between each subject's pair of ranks. In the case of Subject 1, d is equal to 1, the difference between 2 and 3.
3. Square each difference.
4. Calculate Σd^2 by adding the squared differences.
5. Substitute this value, Σd^2, into the r_S equation, along with N, the number of paired ranks. These are the only variables in the equation the numerical terms are constants.

$$r_S = 1 - \frac{6\Sigma d^2}{N(N^2 - 1)}$$

$$= 1 - \frac{6(32)}{10(100 - 1)} = 1 - \frac{192}{10(99)} = 1 - \frac{192}{990}$$

$$= 1 - .194 = .806$$

The calculated correlation is .806. It must next be evaluated for significance. The null hypothesis is $\rho_S = 0$ (that there is no correlation in the population from which the sample was drawn), and the alternative hypothesis is $\rho_S \neq 0$ (that there is a correlation in the population).

With the r_S equation, there is no need to establish the degrees of freedom. To check for significance, all we need to know is the value of N, the number of paired ranks. We turn to Appendix Table F. For an N of 10, we find correlations of .649 for an alpha error level of .05 and .794 for an alpha error level of .01. We compare our obtained value with the table values. If the absolute value of r_S is equal to or greater than the table value, we reject the null hypothesis.

Conclusion: A Spearman correlation was computed between aggressiveness and leadership among female military recruits. A strong positive correlation was found ($r(10) = .806, p < .01$). The correlation was clearly significant and indicates that the more aggressive recruits were assessed as having higher leadership abilities. (We would have to be very careful in interpreting this finding, however. Despite the use of different observers, the same behaviors may have been seen both as aggressive and as signs of leadership.)

Calculating r_S with Data Originally Given in Interval (or Ratio) and Ordinal Form

For our second example, we take the quite common situation in which the data are given in different measurement scales, one interval and one ordinal. The solution to this apparent dilemma is really quite simple—we convert the interval data to ordinal data and then calculate the r_S. Since interval data give information as to greater than or less than status, they can therefore be rank-ordered. Since ordinal data contain no information regarding how much greater or less, they cannot be converted into interval data. This is, of course, also true of ratio data, and in the example to follow, the "years served" distribution is actually composed of ratio values.

Example Assume that a researcher wishes to discover whether there is a significant correlation between a congressional member's physical attractiveness and years in office. A random sample of eight members from throughout the country is selected and rank-ordered on the basis of physical attractiveness. These rankings are paired with the number of years each has served in the House of Representatives.

These data are measures on two different scales. The rank order for physical attractiveness is clearly an example of ordinal scaling, whereas the number of years served represents at least an interval scale. Before calculating the r_S, these measures must be rank-ordered. We assign a rank of 1 for the most years served, 24, and a rank of 8 for the fewest years, 6.

Member	*Rank for Physical Attractiveness*	*Years Served*
A	1	6
B	2	8
C	3	12
D	4	14
E	5	12
F	6	20
G	7	20
H	8	24

The Case of Ties. Note that there are two ties, one at 12 years served and the other at 20. When ranking these data, we handle ties by adding the ranks at the given positions and dividing by the number of tied scores. The tied scores are then assigned the same average rank. For example, Member H, with 24 years, receives the rank of 1, first place, but Members F and G, each with 20 years, are tied for second and third place. Add the ranks 2 + 3; divide by the number of tied scores, 2; and assign both F and G the rank of 2.5. Also, Members C and E are tied for fifth and sixth place, so each is given the average rank of (5 + 6)/2, or 5.5. The ranks, then, are as follows:

Member	*Years Served*	*Ranks for Years Served*
A	6	8.0
B	8	7.0
C	12	5.5
D	14	4.0
E	12	5.5
F	20	2.5
G	20	2.5
H	24	1.0

Note that Member D is assigned the rank of 4. This is because ranks 1, 2, and 3 have already been filled. Similarly, Member B is assigned the rank of 7.

Now that both measures are clearly ordinal (R_1 is the rank for attractiveness and R_2 is the rank for length of service), the r_S can be calculated as before.

Member	R_1	R_2	$d = \|R_1 - R_2\|$	d_2
A	1	8.0	7.0	49.00
B	2	7.0	5.0	25.00
C	3	5.5	2.5	6.25
D	4	4.0	0.0	0.00
E	5	5.5	0.5	0.25
F	6	2.5	3.5	12.25
G	7	2.5	4.5	20.25
H	8	1.0	7.0	49.00
				$\Sigma d^2 = 162.00$

$$r_S = 1 - \frac{6\Sigma d^2}{N(N^2 - 1)}$$

$$= 1 - \frac{(6)(162)}{8(64 - 1)} = \frac{972}{(8)(63)} = \frac{972}{504}$$

$$= 1 - 1.929 = -.929$$

$$= .05(8) = .881$$

$$= -.929 \qquad \text{Reject } H_0;\ \text{significant at } p < .01$$

As with the Pearson r, the null hypothesis is rejected whenever the *absolute* value of r_S is equal to or greater than the table value. This r_S is therefore significant, and we reject the null hypothesis.

Conclusion: A Spearman correlation was computed between the physical attractiveness of congressional members and their number of years of service. A strong negative correlation was found ($r(8) = -.929, p < .01$). The correlation was clearly significant and indicates that the members of congress with the longest service are considered to be the least attractive physically.

Thus, it seems that more attractive congressional members are less likely to be reelected. We must be very careful of this interpretation, however. In this study, another extremely important variable, age, may be influencing or being influenced by the other two variables. Probably the members of Congress with the longest service are also the oldest and therefore, alas, perhaps considered the least attractive.

Calculating r_S for Nonnormal Distributions of Interval Data

In the next example we present a situation that, on the surface, may appear to be appropriate to the Pearson r rather than the r_S. That is, scores on two distributions, both of which are made up of at least interval data, are being assessed for the possibility of correlation. In this case, however, the interval distributions deviate markedly from normality, thus violating one of the Pearson r's important assumptions. The rule is that if one or both sets of interval scores lack normality, then all the interval scores *must be converted into ordinal ranks,* and the r_S, not the Pearson r, must be chosen as the correlation equation.

Example A university researcher wishes to assess whether a significant correlation exists between the amount of money contributed to the school by an alumnus and the age of the alumnus. It is obvious to the researcher that the distribution of alumni donations is skewed. The mean donation last year was more than $100, but the mode was less than $20 and the median less than $40. Thus, a few large donors jacked up the mean and skewed the entire distribution toward the high end. The age distribution is also skewed, with most of the alumni less than 40 years old but a few in their 80s and even 90s. Both sets of interval scores, then, have to be translated into ordinal data before an analysis can take place. A random sample of 12 alumni is selected, and both the size of their donations and their ages are noted.

Alumnus	*Amount of Donation*	*Age (Years)*
A	$ 0	25
B	10	30
C	400	80
D	200	90
E	25	23
F	20	35
G	5	22
H	35	28
I	40	55
J	15	26
K	28	40
L	8	24

The rankings of the two sets of scores (R_1 is the rank for the size of donation, and R_2 is the rank for the age of the donor) are as follows:

Alumnus	R_1	R_2	$d = \|R_1 - R_2\|$	d^2
A	12.0	9.0	3.0	9.00
B	9.0	6.0	3.0	9.00
C	1.0	2.0	1.0	1.00
D	2.0	1.0	1.0	1.00
E	6.0	11.0	5.0	25.00
F	7.0	5.0	2.0	4.00
G	11.0	12.0	1.0	1.00

H	4.0	7.0	3.0	9.00
I	3.0	3.0	.0	.00
J	8.0	8.0	.0	.00
K	5.0	4.0	1.0	1.00
L	10.0	10.0	.0	.00
				60.00

$$r_S = 1 - \frac{6\Sigma d^2}{N(N^2-1)} = 1 - \frac{(6)(60)}{12(144-1)} = 1 - \frac{360}{1716} = 1 - .210$$

$$r_S = .790$$

$$r_{S.01(12)} = .735$$

$r_S = .790$ Reject H_0; significant at $p < .01$.

Thus, there is a significant positive correlation between alumni age and amount donated. A correlation of .790 is not perfect, but predicting from it is better than chance. Now, since a Spearman correlation of .735 is needed for significance at .01 with an N of 12, we can conclude: A Spearman correlation was computed between the amount of donation and the age of a group of college alumni. A positive correlation was found ($r(12) = .790, p < .01$). The correlation was significant and indicated that donation amounts are positively correlated with age.

Requirements for Using the r_S

To use the r_S, the following requirements must be met:

1. The sample has been randomly selected from the population.
2. Both distributions of scores must be in ordinal form.
3. The relationship between the two measures must be linear.

Deciding Which r to Use

The r_S was derived directly from the Pearson r. However, whenever the requirements for the Pearson r are met, it should definitely be used. Converting interval scores into ordinal ranks because the r_S is easier to calculate is a mistake. Interval data contain more information than do ordinal data, and one should not throw away perfectly good information.

The Pearson r is a more sensitive test than the r_S. For a given sample size, an r of a certain value is more apt to be significant than is an r_S of the same value. That is, if there is in fact a correlation existing in the population, the Pearson r is more likely than the r_S to detect that correlation.

A quick look at the tables for r and r_S (Appendix Tables E and F) shows that for a given sample size, the Pearson r value can be consistently lower than the r_S value *and still attain significance.* With a sample of, say, seven pairs of scores (five degrees of freedom for the Pearson r), a Pearson r value of .75 reaches significance at the .05 level. For that same size sample, the r_S value needed for significance at .05 is a larger .786. Thus, the Pearson r value reaches significance, whereas the same r_S does not.

In Chapter 10 we discussed the concept of power. The Pearson r, a parametric test of correlation, is a more powerful test than is the nonparametric r_S, since the Pearson r reduces the likelihood of committing the type 2 error (which then reduces the beta probability).

AN IMPORTANT DIFFERENCE BETWEEN THE CORRELATION COEFFICIENT AND THE *t* Test

During the discussion of the t test in the preceding chapter, it was emphasized that both of the distributions being compared have to be composed of measures of the same trait. For instance, we cannot do a t test if the data are weight versus height scores, since the t test is only concerned with comparisons on a single measured variable. With correlation, however, where the focus is on the association rather than the difference between measures, qualitatively different measures can be compared. A correlation coefficient can be used to compare the same types of measurements, as in correlating heights of fathers with heights of sons. It can also be used to compare different measures, as in correlating heights and weights. Correlation is used to assess whether measures change together in some systematic fashion. Correlation does not ferret out a causal factor, but it can be utilized to make better-than-chance predictions.

SUMMARY

Correlation is defined as the degree to which two or more variables are associated. This chapter focused on establishing the linear relationship between two variables, or what are called bivariate associations. When the association is direct—that is, high scores on the first measure link up with high scores on the second *and* low scores on the first measure pair with low scores on the second—the correlation is said to be positive. When the association is inverse—that is, high scores on the first measure pair off with low scores on the second or low scores on the first pair off with high scores on the second—the correlation is said to be negative. When there is no consistent pairing of the measured scores, a zero correlation is said to exist. Through the use of correlation, even qualitatively different measures can be compared and stated as a precise quantitative value.

The major hazard when interpreting correlational studies lies in assuming that merely because two events are associated, one of the events must necessarily be the cause of the other. Although correlation does not imply causation, it is still an extremely useful tool for establishing better-than-chance predictions.

Correlations can be detected on a scatter plot, in which two distributions are graphed simultaneously. Each point on a scatter plot represents a pair of scores. When the array of points slopes from lower left to upper right, a positive correlation is portrayed. When the array slopes from upper left to lower right, a negative correlation is shown.

The procedure for calculating the correlation coefficient for interval data was introduced by Karl Pearson, who called it the product-moment correlation coefficient. It is now generally known as the Pearson r. The Pearson r ranges in value from +1.00 (a perfect positive correlation) through zero (no correlation) to –1.00 (a perfect negative correlation). The more the value of r deviates from zero, the greater is its predictive accuracy. The value of the r calculated on the basis of sample scores must be assessed for significance before being extrapolated to the population.

The coefficient of determination, r^2, provides specific information about a given correlation's predictive accuracy. By multiplying r^2 by 100, the approximate percentage of information about one variable that is supplied by the other variable can be determined.

If the data are in ordinal form or the interval scores are skewed, a different correlation coefficient, the r_S, is used. It was developed by Charles Spearman. The r_S, however, is a nonparametric test and is not as powerful as the Pearson r. Therefore, the r_S should not be used unless the requirements for the use of the Pearson r are not met.

There are times when the researcher has calculated two Pearson r correlations from separate and unmatched samples and wishes to establish whether they differ significantly from each other. This can be easily accomplished by converting each of the Pearson r values into a variable called Fisher's z_r and then applying the Z test.

Key Terms and Names

coefficient of determination
correlation coefficient
Galton, Sir Francis
homoscedasticity
linearity
Pearson, Karl
Pearson r
scatter plot
Spearman, Charles
Spearman r_S

PROBLEMS

1. Hypothesis: Among Republicans, there is a dependable relationship between party loyalty and number of hours per week spent working for their candidate. A 20-item "Party Loyalty Test" is devised and given to a random sample of 10 Republicans (high scores indicate greater loyalty). These scores are then compared with the average number of hours per week being volunteered to the candidate.

Subject	Party Loyalty Score	Volunteer Hours/Week
1	18	30
2	3	5
3	7	6
4	10	11
5	9	9
6	10	7
7	12	10
8	8	6
9	16	25
10	5	6

Test the hypothesis and state conclusions.

2. Hypothesis: Among elementary school children, there is a significant association between height and running speed. A random sample of eight children is selected. Each child is measured for height (in inches) and is timed in the 40-yard dash (in seconds).

Subject	Height (Inches)	Running Speed (Seconds)
A	60	8
B	55	11
C	56	10
D	52	12
E	48	14
F	44	16
G	47	13
H	52	12

Test the hypothesis and state conclusions. Does being taller cause one to run faster? If not, what other factor might account for the relationship?

3. An investigator studying the relationship between anxiety and school achievement selects a random sample of 15 fifth-grade students, all aged 10 years. Each student is given an anxiety test (high scores signifying high anxiety), and then these measures are paired with the student's score on an academic achievement test. The data are as follows:

Subject	Anxiety Score	Achievement Score
1	10	2
2	2	10
3	8	7
4	6	5
5	9	1

Continued

Note: Scores continued from previous page.

Subject	*Anxiety Score*	*Achievement Score*
6	5	6
7	2	10
8	6	5
9	4	6
10	9	3
11	5	6
12	6	5
13	8	2
14	5	4
15	1	9

a. Calculate the Pearson r correlation.
b. Is it significant?
c. Interpret these results. Does anxiety prevent some children from performing academically? Does a poor academic performance tend to produce anxiety in some children?

4. A group of 12 third-grade students is randomly selected and given a standardized arithmetic test. A trained observer watches the children for a period of one week and then rank-orders them in terms of the amount of extraversion each child displays. The data are as follows:

Student	*Arithmetic Score*	*Extraversion Rank*
1	100	12
2	50	8
3	90	7
4	65	6
5	87	5
6	75	9
7	80	10
8	95	11
9	80	3
10	75	4
11	60	1
12	75	2

a. Find the rank-order correlation between arithmetic ability and extraversion.
b. Is it significant?

5. A random sample of seven junior high students is selected, and each student is given both a math and a spelling test. Their scores are as follows:

Math	Spelling
15	13
5	6
16	14
10	13
11	11
3	5
12	10

Is math ability related to spelling ability? Test at .05.

6. A researcher wishes to test part of the Addictive Personality theory, that individuals addicted in one area tend to be addicted in other areas. Specifically the researcher wants to find out if there is a dependable relationship between alcohol abuse and drug abuse among persons incarcerated for substance abuse. A random sample of 10 incarcerated male inmates from a county house of correction is selected from the population of inmates serving time for substance abuse. Each man is tested on the two major SASSI scales (Substance Abuse Subtle Screening Inventory), Alcohol Abuse and Drug Abuse. The scores (higher scores indicating greater addiction) were as follows:

Subject	Alcohol Score	Drug Score
1	16	12
2	6	10
3	5	8
4	12	10
5	8	7
6	15	12
7	10	8
8	8	10
9	11	8
10	9	10

a. Is the correlation significant?
b. What conclusions can you draw?

7. The following represent distributions of both verbal SAT scores and college rank in class for a random sample of 10 recent college graduates.

Subject	SAT	Rank in Class
1	600	1
2	550	3
3	540	5
4	320	4
5	480	7
6	600	6
7	510	8
8	480	9

Continued

Note: Scores continued from previous page.

Subject	*SAT*	*Rank in Class*
9	470	10
10	700	2

Test the hypothesis that there is a significant association between verbal SAT scores and rank in college class.

8. Hypothesis: There is a significant correlation in the population between income and health. A random sample of 10 men, all 30 years old, is selected and given complete physical examinations. The men are then rank-ordered by the physician in terms of their overall health. Each man's yearly income is also ascertained. The data are as follows:

Subject	*Rank for Health*	*Income*	*Subject*	*Rank for Health*	*Income*
A	2	$70,000	F	10	$20,000
B	7	18,000	G	6	25,000
C	9	20,000	H	3	29,000
D	8	20,000	I	5	27,000
E	4	31,000	J	1	40,000

Test the hypothesis and state conclusions. Do men tend to earn more income because they are healthy, or are they healthy because they can afford better medical care, or might there be another explanation?

9. Some physiologists believe that language skills are controlled by one brain hemisphere and musical skills by the other. To test this, a researcher is studying whether musical ability correlates inversely with reading ability among six-year-old children. A random sample of six-year-olds is selected and given a reading comprehension test. Each child is also evaluated by a musicologist (for pitch, rhythm, etc.) and then rank-ordered on musical ability. The data are as follows:

Subject	*Reading Scores*	*Rank for Musical Ability*
A	120	10
B	100	12
C	100	11
D	100	13
E	95	9
F	90	6
G	80	8
H	80	7
I	78	3
J	76	5
K	70	4
L	62	1
M	41	2

Test the hypothesis and state the conclusions.

10. Audience sizes (in millions of listeners) were announced for 1999. A random sample of college students was selected and asked to rank-order these talk-show hosts in terms of their own personal preferences. The audience size and median rank (adjusted for ties) for each radio show were as follows:

	In Millions	Preference Rank
1. Dr. Laura Schlessinger	18.00	1.0
2. Howard Stern	17.50	3.5
3. Rush Limbaugh	17.25	5.0
4. Art Bell	8.75	3.5
5. Dr. Joy Browne	8.70	2.0
6. Don Imus	7.50	6.0
7. Jim Bohannon	6.50	7.0
8. Michael Reagan	3.75	8.5
9. Dr. Dean Edell	3.70	8.5
10. Bob Grant	2.00	10.0

a. Determine the correlation between audience size and student preference.
b. Test the correlation for significance.

11. The following data represent the salaries *and* benefits (total compensations) for United Way executives in 10 cities with the highest compensations for one year during the early 1990s (as reported in various newspapers during March of 1992). Also included are the per-capita contributions to the United Way in each of these cities. Is there a significant correlation between the *total* earnings of the executives and the contributions received? Do high compensations for executives correlate with increased contributions by the city residents?

City	Executive's Total Compensation (Salary and Benefits)	Per-Capita Contributions
New York	$263,986	$10.64
Washington, D.C.	221,137	17.94
Detroit	215,790	16.74
St. Louis	206,387	22.44
San Francisco	199,688	15.09
Philadelphia	195,211	21.20
Chicago	193,771	15.81
Cleveland	189,541	31.49
Minneapolis	178,033	29.84
Atlanta	174,042	17.35

12. A researcher has computed two significant Pearson r values from independent random samples. Sample A, 20 subjects, produced an r of .77, and Sample B, 22 subjects, produced an r of .51. Do the two sample correlations represent a common population correlation?

13. A researcher has computed two significant Pearson r values from independent random samples, and the correlations are the same as those shown in problem 10. Sample A produced an r of .77 and Sample B, an r of .51. However, this time the samples are considerably larger. Sample A contains 100 subjects and Sample B has 80 subjects. Do the two sample correlations represent a common population correlation?

14. When the array of data points in a scatter plot slopes from lower left to upper right, is the correlation indicating a positive sign or is it indicating a negative sign?

15. The correlation between state spending on education and the IQs of its schoolchildren is .70.
    a. Does this mean that increased state spending on education leads to an increase in the IQs of the children in the state?
    b. What other factors might produce this correlation?

16. Why does the array of points in a scatter plot usually form an oval shape?

17. When the data are in interval form and the distributions are normal, which correlation coefficient should be used?

18. When the data are in ordinal form, which correlation coefficient should be used?

19. When the data are in interval form, but the distribution is known to be heavily skewed, which correlation coefficient should be used?

20. When possible (that is, all of its requirements are met), why should a researcher choose the Pearson r over the r_S?

True or False. Indicate either T or F for problems 21 through 31.

21. If the correlation between X and Y is high and the correlation between X and Z is high, then the correlation between Y and Z must be high.

22. Significant correlations always predict better than chance.

23. Negative correlations, even when significant, never predict better than chance.

24. The higher the correlation between X and Y, the more information about Y is contained in X.

25. Whenever a correlation is significant, the possibility of a cause-and-effect relationship is totally ruled out.

26. The higher the Pearson r, the higher is the coefficient of determination.

27. A Pearson r of .90 means that the percentage of information about Y contained in X is roughly 81%.

28. If sample sizes are equal, the Pearson r and the r_S predict with exactly the same degree of accuracy.

29. If the range of either set of sample scores is in any way restricted, the Pearson r will overestimate the degree of correlation.

30. To use the Pearson r, both sets of paired scores must be composed of at least interval data.

31. The Pearson r assumes that the association between X and Y is always curvilinear.

COMPUTER PROBLEMS

C1. According to Milton Rokeach, there is a positive correlation between dogmatism and anxiety. Dogmatism is defined as a rigidity of attitude that produces a closed belief system (or a closed mind) and a general attitude of intolerance. In the following study, dogmatism was measured on the basis of the Rokeach D Scale (Rokeach, 1960), and anxiety is here measured on the 30-item Welch Anxiety Scale, an adaptation taken from the MMPI (Welch, 1952). A random sample of 30 undergraduate students from a large western university was selected and given both the D Scale and the Welch Anxiety test. The results were as follows:

Subject	D Scale	Anxiety Test	Subject	D Scale	Anxiety Test
1	180	60	16	193	73
2	174	64	17	188	68
3	150	45	18	205	68
4	222	102	19	185	80
5	120	40	20	197	75
6	195	75	21	193	73
7	165	45	22	196	76
8	150	30	23	194	60
9	245	145	24	180	81
10	200	125	25	186	50
11	194	80	26	201	66
12	285	80	27	194	75
13	200	74	28	165	73
14	180	60	29	195	74
15	210	74	30	170	50

Find the linear correlation between dogmatism and anxiety. As a research consumer, what conclusions might you draw?

C2. Hypothesis: The use of illegal substances (cocaine, marijuana, or hashish) is a function of, among other things, shyness and a low level of sociability. A group of over 600 adolescent boys (from ninth to twelfth grades) was selected from a representative sample of high schools in a western state. The average age of the respondents was 15.48 years.

Subjects were asked how frequently they had used the previously mentioned substances during the past month. An index of substance use was formed by summing the number of times a subject had used an illicit drug during that time period. All subjects were also given the Cheek–Buss Shyness scale, and their scores were recorded. The researchers wanted to establish whether shyness was a risk factor in illicit substance use among adolescent males. The researchers suggested that shy males turn to proactive substances in an effort to alleviate their social discomfort and inhibition. If the correlation turns out to be positive and significant, what conclusions might you draw from these data? Could it be that shy people are drawn to the use of drugs, or could it be that the use of drugs causes persons to become more withdrawn and shy? Or could some other variable(s) be responsible for the relationship? Your sample will consist of 20 adolescent boys.

Subject	*Shy*	*Drug Use*	*Subject*	*Shy*	*Drug Use*
	X	Y		X	X
1	36	30	11	25	16
2	29	18	12	25	16
3	27	24	13	24	5
4	27	18	14	24	7
5	26	17	15	24	6
6	26	10	16	24	0
7	26	12	17	23	10
8	25	15	18	23	0
9	25	15	19	22	5
10	25	16	20	14	0

C3. *Presidents of Colleges and How They Survive.* The following lists presidential salaries and fringe benefits but does *not include* presidential housing or house expenses. Also included is the average faculty salary for each institution.

College or University	*Presidential Compensation*	*Faculty Average Salary*
Vanderbilt	$410,900	$73,300
Columbia	409,800	87,400
Duke	348,800	83,800
Yale	314,000	82,700
Tulane	311,300	69,100
Chicago	303,300	88,900
Princeton	295,500	85,500
Barnard	291,200	61,800
Cornell	265,600	77,500
Harvard	249,100	93,300
Brown	233,600	76,100
Mount Holyoke	209,800	64,300

Data continued on next page

College or University	Presidential Compensation	Faculty Average Salary
Oberlin	197,800	64,400
Reed	194,700	58,100
Amherst	189,900	77,300
Haverford	189,400	68,800
Smith	160,600	71,300
Carleton	159,700	63,200
Grinnell	139,000	60,000
Middlebury	135,600	63,700

Source: Chronicle of Higher Education, 1994.
Note: Enter the numbers only. Do not enter the commas into the computer.

Does presidential compensation correlate with average faculty salary?

C4. A study was conducted to test the hypothesis that birth order impacts intelligence. One theorist suggests that the more children there are in a family, the lower the overall IQs of the children, especially the later-borns (Zajonc, 1986). A random sample of 25 convicted felons was selected from the County House of Correction and their birth orders were compared with their respective IQs.
a. Test the hypothesis that birth order correlates significantly with IQ.
b. What other variables might be playing a role?
c. What conclusions might you draw? (Be very careful.)

Subject #	Birth Order	IQ	Subject #	Birth Order	IQ
1.	3	100	14.	3	90
2.	2	120	15.	2	100
3.	7	95	16.	3	80
4.	2	110	17.	5	90
5.	2	100	18.	1	95
6.	3	125	19.	3	90
7.	1	115	20.	2	105
8.	4	105	21.	1	115
9.	2	105	22.	6	60
10.	3	100	23.	3	95
11.	6	80	24.	4	100
12.	2	135	25.	3	90
13.	2	110			

C5. The following is the list of total team salaries as of opening day, April 2000, as well as the wins, losses, and percentages of wins for all the Major League teams for the 2000 season, both American and National League.
a. Find the mean and median salaries.
b. Find the correlation between salary and winning percentage.

League	Team	Win%	Win	Loss	Total	Salary in dollars
AL	New York	54.0	87	74	161	92,538,260
NL	Los Angeles	53.1	86	76	162	88,124,286
NL	Atlanta	58.6	95	67	162	84,537,836
AL	Baltimore	45.7	74	88	162	81,447,435
NL	Arizona	52.5	85	77	162	81,027,833
NL	New York	58.0	94	68	162	79,509,776
AL	Boston	52.5	85	77	162	77,940,333
AL	Cleveland	55.6	90	72	162	75,880,871
AL	Texas	43.8	71	91	162	70,795,921
AL	Tampa Bay	42.9	69	92	161	62,765,129
NL	St. Louis	58.6	95	67	162	61,453,863
NL	Colorado	50.6	82	80	162	61,111,190
NL	Chicago	40.1	65	97	162	60,539,333
AL	Seattle	56.2	91	71	162	58,915,000
AL	Detroit	48.8	79	83	162	58,265,167
NL	San Diego	46.9	76	86	162	54,821,000
NL	San Francisco	59.9	97	65	162	53,737,826
AL	Anaheim	50.6	82	80	162	51,464,187
NL	Houston	44.4	72	90	162	51,289,111
NL	Philadelphia	40.1	65	97	162	47,308,000
NL	Cincinnati	52.5	85	77	162	46,867,200
AL	Toronto	51.2	83	79	162	46,238,333
NL	Milwaukee	45.1	73	89	162	36,505,333
NL	Montreal	41.4	67	95	162	34,807,833
AL	Oakland	56.5	91	70	161	31,971,333
AL	Chicago	58.6	95	67	162	31,133,500
NL	Pittsburgh	42.6	69	93	162	28,928,333
AL	Kansas City	47.5	77	85	162	23,433,000
NL	Florida	49.1	79	82	161	20,072,000
AL	Minnesota	42.6	69	93	162	16,519,500

C6. A researcher in health care is interested in discovering if there is a correlation between government health spending and the longevity of its citizenry. The following is a list of countries and both the percentages of GDP (gross domestic production) and the life expectancy for that country (source Pocket World in Figures, 2001 Edition). London: The Economist 2001

	Health Spending % of GDP	Life Expectancy in years
United States	14.00	77
Germany	10.50	77
Switzerland	9.80	79
France	9.70	78
Canada	9.20	79

Data continued on next page

	Health Spending % of GDP	*Life Expectancy in years*
Czech Republic	9.10	74
Netherlands	8.60	78
Costa Rica	8.50	76
Portugal	8.20	75
Austria	8.00	77
Iceland	8.00	79
Belgium	7.90	77
Norway	7.90	78
Spain	7.70	78
Italy	7.60	78

a. Compute the correlation and state your conclusions.

The correlation found may be somewhat misleading, since both these distributions are for the countries who were both the highest spenders and had the longest longevity, thus severely restricting the range. Had the list continued it would have contained the lowest-spending countries such as:

Cote d'Ivoire	3.50	46
Togo	3.40	49
Zambia	3.30	40
Ethiopia	2.60	43
Kenya	2.60	51
Bangladesh	2.40	58
Ghana	1.50	60
Cameroon	1.40	55
Nigeria	1.30	49
Niger	1.00	49
Sierra Leone	.04	39
Sudan	.03	55

Note: The above includes only the countries where both health spending AND longevity could be found.

b. Has adding to the range changed the correlation? If so, what conclusions might you now draw? What other factors might influence both health spending and longevity? Remember that when evaluating world data, it's important to consider cultural and ethnic differences. This is especially true when examining the health data, which can be distorted on the basis of different methods of tracking and reporting.

Chapter

12

Analysis of Variance

Since the time around the turn of the century that William Sealy Gossett first created the t *test, another statistical test of the hypothesis of difference has been developed. During the 1920s,* **Sir Ronald Fisher** *introduced a technique called the* ***analysis of variance,*** *and the resulting statistic, the* F *ratio, was named in his honor. The procedure is now known by the acronym* **ANOVA,** *for ANalysis Of VAriance. ANOVA is a powerful and versatile statistical technique. It allows us to do virtually everything we use the* t *test for, plus a lot more. Why, then, did we bother with the* t *test? Why not skip directly to* F*? There are two reasons: First, the concepts on which the* t *ratio is built provide an excellent background for appreciating the overall logic of statistical reasoning, from samples to sampling error and on to parameter predictions; second, one form of the* t *test, the one-tail* t*, allows the researcher to predict the direction of a difference, which is a feature ANOVA simply does not possess.*

ADVANTAGES OF ANOVA

The main advantage of ANOVA is that it allows the researcher to compare differences among many sample groups. Whereas t is "for two," the F ratio can theoretically handle any number of group comparisons. This is a big plus; it means that we can design experiments in which the independent variable is manipulated through a whole range of values. Analysis using the t test means that the independent variable can have only two levels, one for the experimental group and one for the control group. With ANOVA, we may set up a number of experimental groups to compare with the control group.

For example, in problem 3 at the end of Chapter 10 a t test was performed for a study in which increased illumination was hypothesized to cause an increase in worker productivity. Two groups of subjects were selected, one working under normal light conditions (control group) and the other working under an illumination level that was increased by 50% (experimental group). The independent variable (illumination) was manipulated at two levels. With ANOVA, we can, by adding more experimental groups, manipulate the independent variable over a range of lighting conditions, for instance, as in the following:

Control Group A: Normal illumination

Experimental Group B: Illumination increased by 50%

Experimental Group C: Illumination increased by 75%

Experimental Group D: Illumination increased by 100%

Drawbacks of Doing Successive t *Tests*

It may immediately be obvious that comparisons among more than 2 groups can be made by doing successive *t* tests, for example, comparing Group A with Group B, then A with C, A with D, B with C, B with D, and finally C with D. That means calculating six *t* tests for only a 4-group experimental design. For a 6-group design, it would take 15 *t* tests, and for a 10-group design, it would take 45 *t* tests. One reason, then, for not performing successive *t* tests is the enormous amount of work involved to calculate them. Since computers can calculate them so quickly, however, a far more important reason concerns the alpha level, the probability of being wrong when the null hypothesis is rejected. If H_0 is rejected several times, as it might be when performing successive *t* tests, the alpha level for each decision combine to produce a dangerously high overall level. The probability of committing the type 1 error in this circumstance is called the *familywise* error rate because it is addressing the issue of a group (or family) of comparisons, not just the single comparison for which the *t* test was originally designed. Table 12.1 illustrates how the alpha level first set at the .05 level inflates as

TABLE 12.1 Alpha levels for successive rejections of H_0.*

Number of Decisions to Reject	*Alpha Error Level*
1	.05
2	.10
3	.14
4	.19
5	.23
6	.26
7	.30
8	.34
9	.37
10	.40

*Constructed from the general formula p of $\alpha = 1 - (1 - \alpha)^d$, where d equals the number of decisions. Thus, with alpha set at .05 and three decisions to reject, p of $.05 = 1 - (1 - .05)^3 = 1 - .95^3 = 1 - .86 = .14$.

the number of decisions to reject increases. Setting the alpha level at .05 and making only one decision to reject H_0 keeps the alpha error at precisely .05. But by the tenth decision to reject, the alpha error has skyrocketed to .40. Were the alpha error to have a value of .50, which it would reach after 13 decisions, we may as well not have done the experiment because the probability of significant results would be equal merely to flipping a coin.

BONFERRONI TEST

The use of multiple t tests has, however, become somewhat more fashionable recently due to something called the Bonferroni correction (which appears on several of the popular statistical software programs). This is used as an attempt to even out the level of significance among those t comparisons that were *planned from the outset of the study.* As with the one-tail t, the researcher using this correction has to be trusted, in this case to only test those differences that had been chosen for analysis before the data were in. The Bonferroni test is based on setting a lower alpha level for all the t tests combined and is usually accomplished by taking the original alpha, say of .05, and dividing it by the number of t tests performed. If the study called for for three t tests, then .05 divided by 3 would yield an overall alpha of .016 or rounded to .02. This is certainly a conservative adjustment since it makes it less likely to get significance than had each decision remained at .05. The theory behind the Bonferroni is that the probability of a set of events can never be greater than the sum of the separate probabilities. Thus, in the above example .016 + .016 + .016 = .048 (which keeps the sum under the original .05).

ANOVA and the Null Hypothesis

Using ANOVA, the overall hypothesis of difference among more than two groups can be tested while making *only one statistical decision.* This means that if the alpha error is originally set at .05, it remains .05, no matter how many groups are being compared. For a four-group experimental design, for example, the null hypothesis states that $\mu_1 = \mu_2 = \mu_3 = \mu_4$, that is, that the means of the populations being represented by the four sample groups are all identical (or that the four samples represent the same population). If the statistical decision for the study is to reject H_0, then H_a, the alternative hypothesis (that the same groups do not all represent the same population), is accepted. Thus, a *single* decision to reject encompasses *all* sample groups. In short, then, if we begin an ANOVA with the alpha error level set at .05 and reject H_0, the alpha error level stays at .05, no matter how many sample groups are involved.

Therefore the F ratio can really accomplish all that the t ratio can plus having the advantage of being able to detect differences among three, four, or more means, not just the two demanded by the t ratio. Also, as we will see later in the chapter, the F ratio can handle more than just one independent variable, giving the researcher the opportunity to examine the possible effects of the interactions among IVs.

RONALD AYLMER FISHER (1890–1962)

During the 1930s and 1940s R. A. Fisher was without question the most famous statistician in the world. He carried on his work at the University of London in the tradition of his predecessor, the great Karl Pearson. Fisher was the originator of the ANOVA technique, called F ratio in honor of Fisher. He also introduced hypothesis testing and multivariate statistical techniques, especially the use of several variables in determining discrimination or regression.

Fisher was born just outside London, the youngest of seven children. His father was an auctioneer, not a scientist or mathematician, although it is said that he was exceptionally good with numbers. Also, his father made more than just a decent living. He was able to send young Fisher to Harrow and then on to Cambridge where he graduated as a mathematics major in 1912. From there he went into the investment business and tried to apply mathematical theory to the world of finance. He found that tiresome and boring and soon began teaching in a succession of English public schools. Because of very poor eyesight, he was not able to serve his country during World War I. In 1917 he married Ruth Guiness and went on to father eight children. In 1919 he took a job at the Rothamsted Experimental Station, a leading agricultural research center. It was there that he had some of his most brilliant insights, including the creation of the analysis of variance. Up to that time, experiments were carried out varying only one factor at a time, and the statistical test of choice was the two-sample t test. It was Fisher who saw the importance of multifactor designs, especially because they could be used to determine the possibility of interactions. In 1933 Fisher joined the department of applied mathematics at the University of London and upon Pearson's retirement was awarded the Galton chair. Although he followed in Pearson's footsteps, Fisher and Pearson had a stormy relationship, and when Fisher joined the University of London faculty, this only fanned the flames of their feud. One of the controversies that flared between Pearson and Fisher concerned large versus small samples. Pearson's procedures were always based on theoretically infinite samples, whereas Fisher was able to set exact probabilities for smaller, more realistic samples. It was Fisher who introduced the seemingly paradoxical concept of testing the logical contradiction of the hypothesis being tested, called the null hypothesis. Thus, the null hypothesis states that there is no difference (or relationship) greater than that which could be reasonably accounted for on the basis of chance. That means that when the null hypothesis is disproved, then a significant difference has been proved. Fisher was unbending in his insistence on significance testing, or that statistical proof could only come through the testing of the null hypothesis. Incidentally it was Fisher who set the maximum alpha level at .05. He also was clear in his warning that correlation should not be used to imply causation, an error that he thought Pearson may have committed on occasion. His most important works were *Statistical Methods for Research Workers*, 1925, and *The Design of Experiments*, published in 1935. Perhaps more than any other statistician, Fisher set modern statistical theory on its present course.

ANALYZING THE VARIANCE

In Chapter 3, we saw that variance, like the standard deviation, was a measure of the amount that all scores in a distribution vary from the mean of that distribution. A large variance indicates a large spread in the distribution—many of the scores are deviating widely from the mean. A large variance, like a large standard deviation, is characteristic of a platykurtic distribution. A small variance indicates that there is little dispersion of the scores from the mean and that the distribution of scores is leptokurtic. The basic fact to remember is that when scores are very similar to one another, or homogeneous, the variance is small, but when scores are dissimilar, or heterogeneous, the variance is large.

When several groups of scores are being compared, however, the question arises as to which mean a given score should be compared to, the mean of its own sample group or the mean of all of the scores from all of the sample groups. Furthermore, the means of the separate sample groups are almost certain to differ; this could be yet another source of variability. Sir Ronald Fisher's solution to this apparent dilemma is, in fact, rather simple—do it all! Analyze (sort out) the different sources of variance into separate components, and then compare these variance components—hence his term analysis of variance.

Sum of Squares

Before calculating the actual variance components, we must look at a necessary ANOVA concept called the **sum of squares,** symbolized by SS. The sum of squares is our route to the calculation of the variance. It is equal to the sum of all of the squared deviations of scores from the mean.

$$SS = \Sigma x^2$$

$$x = X - M$$

Therefore, for the sets of scores given in Table 12.2 the sum of squares equals 80 for Sample A and only 10 for Sample B. These two samples demonstrate that the sum

TABLE 12.2 The sum of squares for two samples.

Sample A			Sample B		
X	x	x_A^2	X	x	x_B^2
14	+6	36	10	2	4
10	+2	4	9	1	1
8	0	0	8	0	0
6	−2	4	7	−1	1
2	−6	36	6	−2	4
$\Sigma X_A = 40$		$\Sigma x_A^2 = 80$	$\Sigma X_B = 40$		$\Sigma x_B^2 = 10$

$$M_A = \frac{\Sigma X_A}{N} = \frac{40}{5} = 8 \qquad M_B = \frac{\Sigma X_B}{N} = \frac{40}{5} = 8$$

of squares, *like the standard deviation and the variance,* is large when the scores are spread out and small when the scores are bunched together. The sum of squares is, in fact, another measure of variability. Note, therefore, that the sum of squares can *never be negative.* The smallest possible sum of squares is zero, and that value only occurs when every score in a distribution is the same.

The procedure in Table 12.2, although it nicely conveys the relationship between the definition of the sum of squares and its calculation, can be exceedingly cumbersome in practice. The samples given there have scores that are nice whole numbers, and the means come out as nice whole numbers. As we know by now, that is not the way it usually happens out there in research land. To lighten the mathematical burden, another equation has been derived for the sum of squares.

Computational Method for Calculating the Sum of Squares. The equation for the computational method for calculating the sum of squares is

$$SS = \Sigma X^2 - \frac{(\Sigma X)^2}{N}$$

That is, the sum of squares is equal to the summation of the squared raw scores minus the squared summation of raw scores divided by the number of scores.

As we saw in Chapter 7, those parentheses are important. ΣX^2 does not equal $(\Sigma X)^2$. In fact, ΣX^2 tells us to square the scores and then add, whereas $(\Sigma X)^2$ tells us to add the scores and then square.

The value $(\Sigma X)^2/N$ is also labeled C, a correction factor for using raw scores rather than deviation scores.

$$C = \frac{(\Sigma X)^2}{N}$$

Table 12.3 shows the computational method for calculating the sum of squares using the same data as in Table 12.2.

This method results in the same sum of squares values, 80 and 10. It is important to realize that with this computational method the mean is never calculated directly. However, the resulting sum of squares still affords information regarding the amount by which the scores vary around the mean. Thus, we gain information about the mean without having calculated it.

Components of Variability. Depending on which mean the scores are compared with, three different sums of squares can be calculated.

1. *The total sum of squares.* When several groups are being compared, the scores from all the groups can be added to compute a total mean. A total sum of squares can then be calculated; it is based on how far each score in each group differs from this total mean. SS_t is therefore composed of all values of $X - M_t$, the difference between each score and the total mean. Figure 12.1 illustrates one such value in a comparison of four sample groups.

TABLE 12.3 The computational method for calculating the sum of squares.

Sample A		Sample B	
X_A	X_A^2	X_B	X_B^2
14	196	10	100
10	100	9	81
8	64	8	64
6	36	7	49
2	4	6	36
$\Sigma X_A = 40$	$\Sigma X_A^2 = 400$	$\Sigma X_B = 40$	$\Sigma X_B^2 = 330$

$$SS_A = \Sigma X_A^2 - \frac{(\Sigma X_A)^2}{N} = \Sigma X_A^2 - C$$
$$= 400 - \frac{40^2}{N} = 400 - \frac{1600}{5}$$
$$= 400 - 320 = 80$$

$$SS_B = \Sigma X_A^2 - \frac{(\Sigma X_B)^2}{N} = \Sigma X_B^2 - C$$
$$= 330 - \frac{40^2}{N} = 330 - \frac{1600}{5}$$
$$= 330 - 320 = 10$$

2. *The variability between groups.* When several groups are being compared, the difference between each group mean and the total mean can be calculated. This is called the sum of squares between groups SS_b. Figure 12.2 illustrates this value again as part of a comparison of four groups.

3. *The variability within groups.* Finally, when several groups are being compared, the difference between each individual score and the mean of the group from which that score comes can be calculated. This is called the sum of squares within groups, SS_w. Figure 12.3 is an illustration of this value.

Now, by combining each of the three sum of squares components on a single graph, we get the relationship shown in Figure 12.4. This shows that the total sum of

FIGURE 12.1 The total sum of squares.

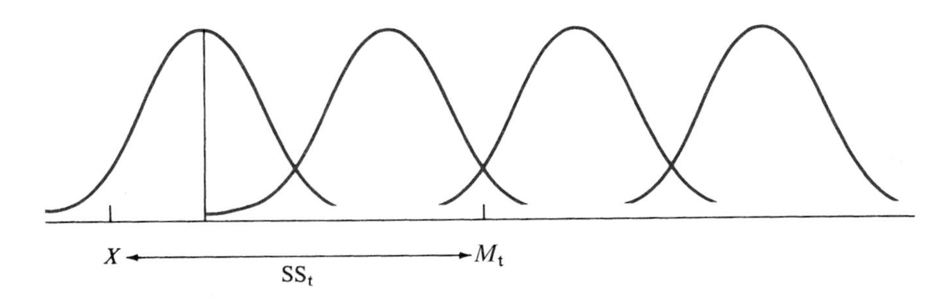

FIGURE 12.2 The sum of squares between groups.

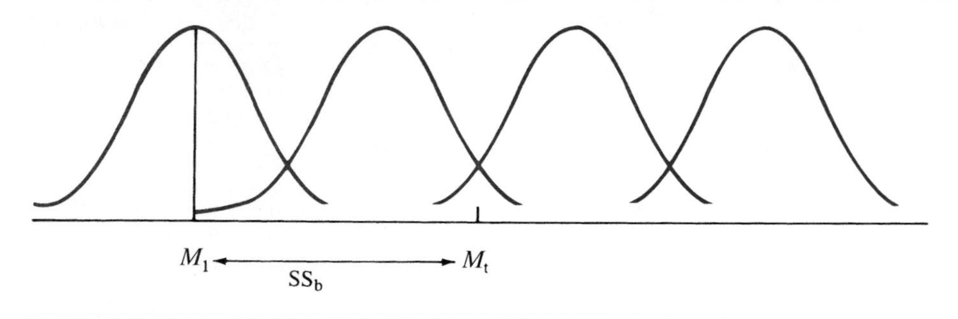

squares is equal to the sum of squares between groups *plus* the sum of squares within groups, or $SS_t = SS_b + SS_w$. Knowing any two of these three values, then, automatically gives us the third. Since

$$SS_t = SS_b + SS_w$$

then

$$SS_b = SS_t - SS_w$$

or

$$SS_w = SS_t - SS_b$$

The total sum of squares can, therefore, be partitioned into two components, the sum of squares between groups and the sum of squares within groups. In other words, the total variability is composed of between-group variability and within-group variability.

FIGURE 12.3 The sum of squares within a group.

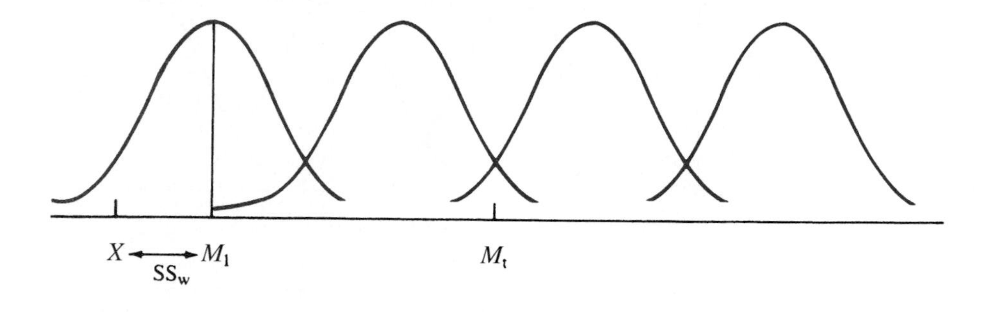

FIGURE 12.4 The relationship between total sum of squares, sum of squares between groups, and sum of squares within groups.

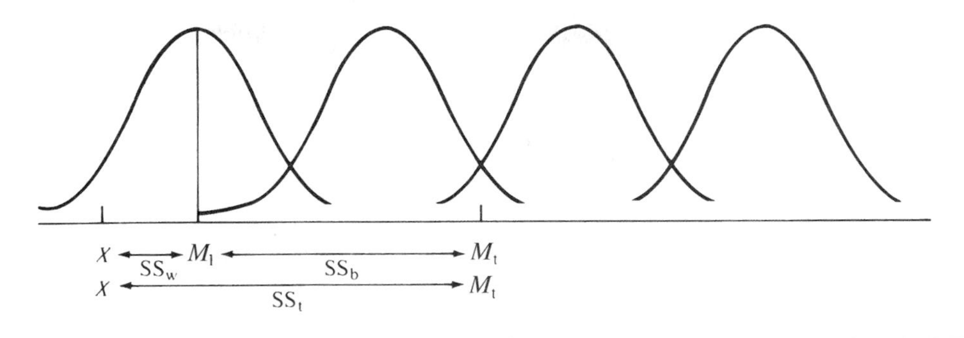

Interpreting the Variability Components. A close look at these variability components should now shed some light on the reason for this type of analysis.

The sum of squares between groups gives us a measure of how far each group mean is from the total mean *and of how far the group means are from each other.* Figure 12.5a shows a situation in which the SS_b is relatively large, meaning that the sample groups are spread apart from one another. In this case, the spread is so great that there is no overlap among the distributions.

Figure 12.5b, on the other hand, illustrates a smaller SS_b, with much less spread among the samples, in fact, considerable overlap. Intuitively, it seems likely that the sample groups in Figure 12.5a represent different populations and those in Figure 12.5b represent the same population. But what about the sum of squares within groups? Both parts of Figure 12.5 have the same SS_w.

The value of the SS_w reveals how far each individual score varies from the mean of its own sample group. The more leptokurtic the sample distribution is, the smaller the SS_w is; the more platykurtic the sample distribution, the larger is the SS_w. Figure 12.6 illustrates this difference in the size of the SS_w, holding SS_b constant. Figure 12.6a shows four very leptokurtic distributions, samples whose SS_w is quite small. Figure 12.6b, however, illustrates a larger value of SS_w, hence the platykurtic shape of the four distributions. Since leptokurtic distributions are less apt to overlap, *the smaller the* SS_w, *the more likely it is that the samples are representing different populations.*

Putting together what we now know of the two variability components, SS_b and SS_w, sample groups are most likely to reflect different populations when SS_b is relatively large and SS_w relatively small. Figure 12.7 illustrates this relationship. In this illustration, the sample means are spread apart, meaning a large SS_b, but each sample distribution is itself tightly leptokurtic, meaning a small SS_w. Analysis of these data would undoubtedly lead to a reject of H_0 and a conclusion that the samples represent different populations. Figure 12.8, on the other hand, shows four groups whose means lie close to each other, meaning a small SS_b, but whose distributions are platykurtic, meaning a large SS_w. Analysis of these data would lead to an acceptance of H_0 and a conclusion that the sample groups represent the same population.

FIGURE 12.5 Difference in the values of the sum of squares between groups (a) when the distributions are spread out and (b) when the distributions overlap.

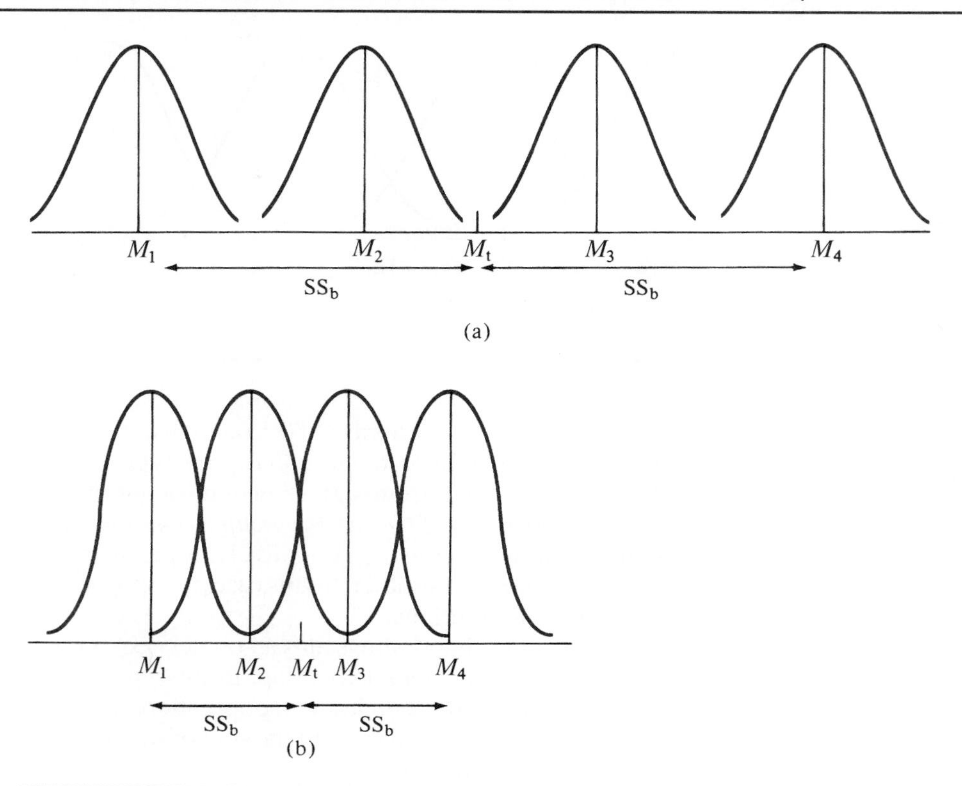

Therefore, when comparing the variability components, if the SS_b is large relative to the SS_w, we reject the null hypothesis. The less the sample groups overlap each other, the more sure we are that they represent different populations. The *F* ratio will tell us precisely how much overlap can be tolerated while remaining confident that the samples represent different populations.

Converting Sums of Squares to Variance Estimates

Using generalities, such as "the SS_b is large relative to the SS_w," is fine for openers, but a far more precise method must be used to calculate a testable inferential statistic. The sums of squares—total, between groups, and within groups—must be converted into variance estimates, since the sum of squares is greatly affected by the sample size. Adding scores to a distribution increases the value of the sum of squares, unless the additional scores happen to fall right on the mean. Therefore, to make accurate variability comparisons regardless of sample size, a method must be used that will, in effect, "average out" the variability. (We place "average out" in quotation marks because

FIGURE 12.6 Difference in the value of the sum of squares within a group (a) when the distribution of sample scores is leptokurtic and (b) when the distribution of sample scores is platykurtic.

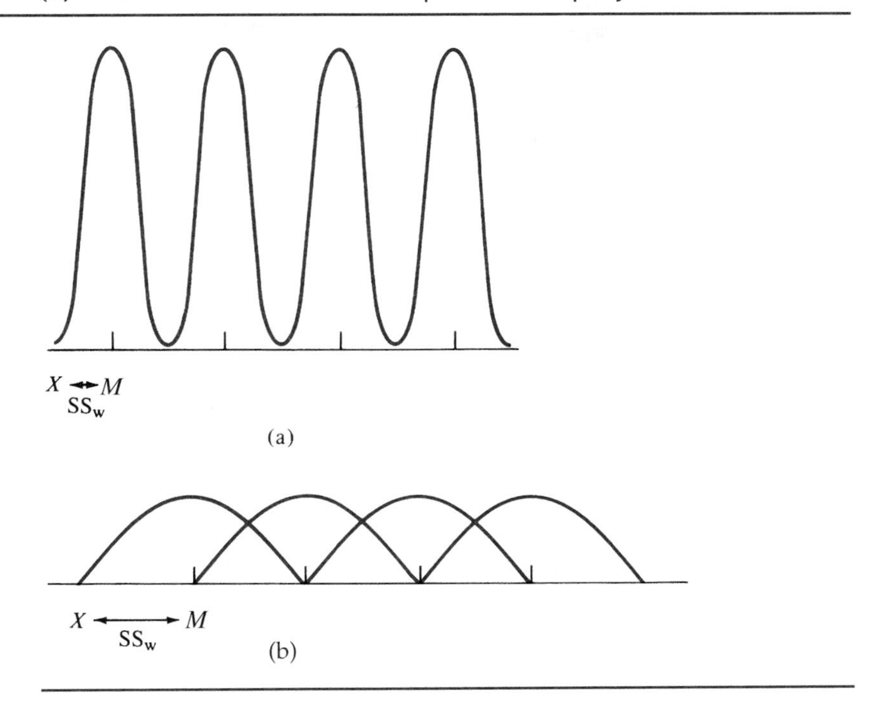

FIGURE 12.7 Sample groups that represent different populations; SS_b is relatively large and SS_w is relatively small.

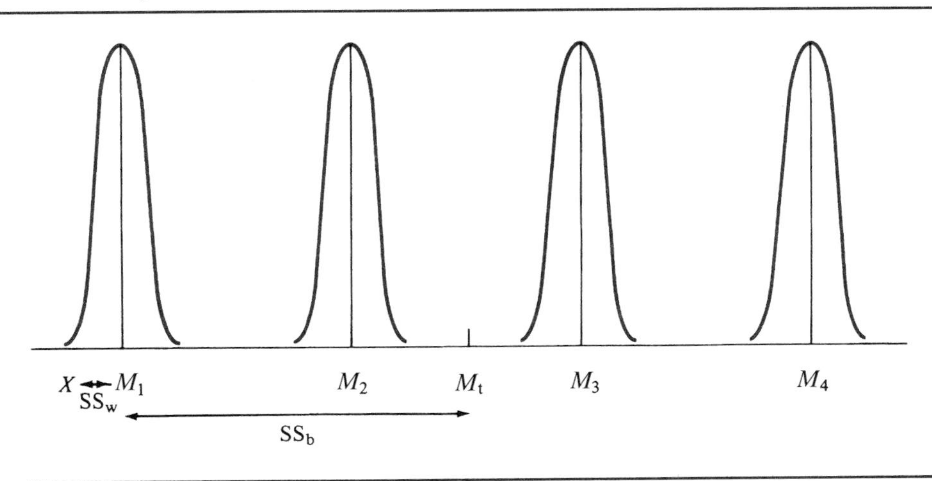

FIGURE 12.8 Sample groups that represent the same population; SS_b is relatively small and SS_w is relatively large.

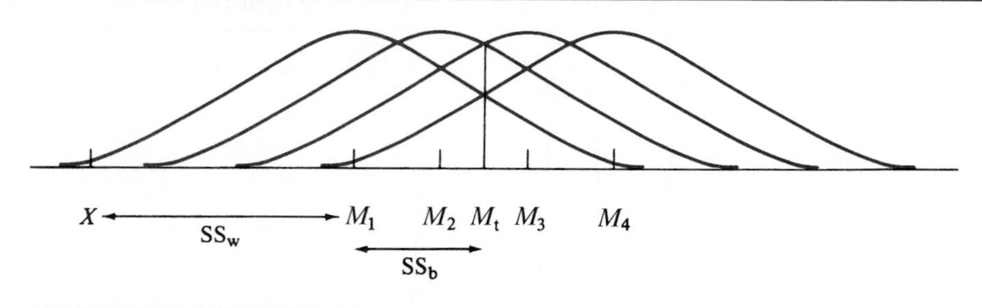

the method used does not achieve an exact arithmetic average, but it does come very close to it.) The sums of squares must be divided by some value that reflects the sample size. The resulting value is called the variance, or the mean square.

Degrees of Freedom. The conversion of a sum of squares to a variance is simple and direct—we divide the sum of squares by the appropriate degrees of freedom. The total degrees of freedom are equal to the total number of scores in all the groups combined, N, minus one.

$$df_t = N - 1$$

The between degrees of freedom are equal to the actual number of sample groups, k, minus one.

$$df_b = k - 1$$

Finally, the within degrees of freedom are equal to the total number of scores, N, minus the number of sample groups, k.

$$df_w = N - k$$

Suppose that we use a three-group design: Groups A, B, and C have three subjects each. The df_t equals 8 (the 9 subjects less 1). The df_b equals 2 (the 3 groups less 1). The df_w equals 6 (the 9 subjects less the 3 groups). As a check, df_b and df_w must add up to the df_t. In this case, df_b plus df_w equals df_t, or $2 + 6 = 8$.

The *F* Ratio. After the sums of squares are converted into variance components, the F ratio can be computed; the F ratio is equal to the variance between groups, V_b, divided by the variance within groups, V_w.

$$F = \frac{V_b}{V_w}$$

The Mean Square. When the variance is calculated by using the sum of squares, it is called the mean square, or MS. The mean square is the mean of the squared deviations, *not* the square of the mean. Throughout the following calculations, then, variance will be symbolized as MS, and

$$F = \frac{MS_b}{MS_w}$$

If the value of the F ratio is large, it tells us that the mean square between is larger than the mean square within, or that the variability between the groups is larger than the variability occurring within the groups. Thus, the larger the F ratio, the more likely it is that the null hypothesis will be rejected. An F ratio of 12, for example, indicates that the variability between the groups is exactly 12 times greater than the variability within groups. The smaller the F ratio is, on the other hand, the greater the likelihood of accepting the null hypothesis. An F ratio of .50, for example, indicates that the variability between the groups is only half as large as the variability within groups.

Despite its name, the final goal of analysis of variance is to detect differences among MEANS. Analyzing and comparing variances are techniques used en route to determining mean differences.

Calculation of the One-Way F Ratio: Between Subjects

The following procedure is for a *one-way* ANOVA, that is, where there is only one independent variable set at various levels and where each of the sample groups is independent of the others. To use this analysis, each level of the independent variable must consist of a different group of subjects. A given subject thus may *serve in only one condition,* and each score represents a separate subject. Hence, the number of scores will always equal the number of subjects.

Using the following data, we will now calculate an F ratio.

X_1	X_1^2	X_2	X_2^2	X_3	X_3^2
1	1	2	4	4	16
2	4	3	9	5	25
3	9	4	16	6	36
$\Sigma = 6$	14	9	29	15	77

1. *Calculate C, the correction factor:* Add all scores in all the groups combined, square the total, and divide by the total number of scores.

$$C = \frac{(\Sigma X)^2}{N} = \frac{(6 + 9 + 15)^2}{9} = \frac{30^2}{9} = 100$$

2. *Find the total sum of squares:* Square all the scores for each group, add the squares, and then subtract C from the total of the squares.

$$SS_t = \Sigma X^2 - C = 14 + 29 + 77 - 100 = 20$$

3. *Find the sum of squares between groups:* Add the scores for each group, square each total, and divide it by the number of subjects *within each group,* or n_1, n_2, and so on. Add the resulting values, and again subtract the C term.

$$SS_b = \frac{(\Sigma X_1)^2}{n_1} + \frac{(\Sigma X_2)^2}{n_2} + \frac{(\Sigma X_3)^2}{n_3} - C = \frac{6^2}{3} + \frac{9^2}{3} + \frac{15^2}{3} - C$$

$$= 12 + 27 + 75 - 100 = 14$$

4. *Find the sum of squares within groups:* From the SS_t value, subtract the SS_b value. Remember that the total sum of squares is composed of the sum of squares between groups plus the sum of squares within groups.

$$SS_w = SS_t - SS_b = 20 - 14 = 6$$

5. *Find the mean square between groups:* Divide the sum of squares between groups by its degrees of freedom.

$$MS_b = \frac{SS_b}{df_b} = \frac{14}{2} = 7$$

6. *Find the mean square within groups:* Divide the sum of squares within groups by its degrees of freedom ($df_w = N - k$).

$$MS_w = \frac{SS_w}{df_w} = \frac{6}{6} = 1.00$$

7. *Obtain the F ratio:* Divide the mean square between groups by the mean square within groups.

$$F = \frac{MS_b}{MS_w} = \frac{7}{1} = 7.000$$

Summarizing the Results. Statisticians then set up a summary ANOVA table of the results of the F ratio calculations, with the between-group values in the first row and the within-group values just below.

	Source of Variance			
	SS	*df*	*MS*	F
Between groups	14	2	7	7.000
Within groups	6	6	1	

The MS column of the table contains the numerator and denominator of the F ratio, in this case 7 and 1. F equals 7/1, or 7.000. You are urged to use this summary technique, not only because it's standard operating procedure but also because it provides a convenient recap of the ANOVA results.

Using the *F* Table. With a three-group design, the H_0 being tested is that $\mu_1 = \mu_2 = \mu_3$. We turn now to the table of critical values of *F*, Appendix Table G, and compare our obtained value of *F* with the critical value of *F* for the appropriate degrees of freedom. The row across the top indicates the between degrees of freedom, and the column on the far left indicates the within degrees of freedom. For the calculations just done, then, the column for 2 df and the row for 6 df intersect on Appendix Table G at two *F* values: 5.14 for an alpha level of .05 and, in boldface, 10.92, for an alpha level of .01. As with the *t* test, the null hypothesis is rejected when the obtained value of *F* is equal to, or greater than, the critical, or table, value of *F*. Thus, a one-way ANOVA was computed comparing scores of subjects who were tested under three conditions. A significant difference was found between the groups ($F(2,6) = 7.000$, $p < .05$). The null hypothesis was rejected.

However, if our obtained *F* had been say, 3.000, then we would conclude: a one-way ANOVA was computed comparing scores of subjects who were tested under three conditions. The difference in their scores was not significant and the null hypothesis was accepted ($F(2,6) = 3.000$, $p > .05$ ns). Finally, if our obtained *F* had been equal to, say, 12.000, then the conclusion would state that one-way ANOVA was computed comparing scores of subjects who were tested under three conditions. A significant difference was found between the groups ($F(2,6) = 12.000$, $p < .01$). The null hypothesis was rejected.

Effect Size

For the one-way ANOVA, effect size (ES) is calculated as eta square η^2 and is equal to

$$\frac{SS_b}{SS_t}$$

which is the sum of squares between divided by the sum of squares total.

For eta square, values of .01, .06, and .14 represent small, medium, and large effects.

For the previous problem, where the *F* ratio was a significant 7.000, the effect size would be 14 divided by 20 (the SS_b from step 3 divided by the SS_t from step 2).

$$\eta^2 = 14/20 = .700$$

which is considered to be a strong effect.

Requirements for Using the *F* Ratio. To use the *F* ratio, the following requirements must be met.

1. The sample groups have been randomly and independently selected.

2. There is a normal distribution in the population from which the samples are selected.

3. The data are in interval form (or, of course, ratio).

4. The within-group variances of the samples should be fairly similar. This property, called **homogeneity of variance,** simply means that ANOVA demands sample groups that do not differ too much with regard to their internal variabilities. For

example, you should not do an ANOVA on three sample groups, one of which has a leptokurtic shape, the second a mesokurtic shape, and the third a platykurtic shape.

Tukey's HSD: A Post-Hoc Multiple Comparison Test

If the *F* ratio has been found to be significant, it is important to ferret out precisely where these sample differences came from. The reason for this step is that a significant *F* ratio can be the result of differing patterns of group differences.

For example, the results of a three-group design graphed in Figure 12.9 produce a significant *F* ratio, but the actual pattern of the group differences makes it clear that the effects are not spread out evenly. The between-group variance is certainly large, but most of it is due to the fact that Group C is so far away from both Groups A and B. In fact, perhaps Groups A and B do not differ significantly from each other. The alternative hypothesis for the *F* ratio is not, therefore, a straightforward $\mu_1 \neq \mu_2 \neq \mu_3 \neq \mu_4$, and so on but instead is any one of a series of choices, $\mu_1 \neq \mu_2 = \mu_3 = \mu_4$ or $\mu_1 \neq \mu_2 \neq \mu_3 = \mu_4$ or any other such combination.

To determine where the significant differences might be, multiple comparison tests can be used. Although several statistical tests have been created to perform this type of analysis, one of the most popular is **Tukey's HSD** (honestly significant difference) test. Sometimes these comparisons have been planned ahead of time, called a priori comparisons, and sometimes they come after the fact and are then called post-hoc comparisons. In either case, the Tukey HSD test can be the appropriate tool. This test is admittedly conservative and not as powerful as some of the others, but when all the assumptions of ANOVA are met, the Tukey test is considered to be the safest (Stoline, 1981). "You'll find that the Tukey HSD is generally regarded as the best procedure for controlling the error rate when you are making all pairwise comparisons among many group means" (Howell, 1992, p. 353).

Calculating Tukey's HSD. The first step in calculating Tukey's HSD is to set the alpha level, typically .05, and then turn to Appendix Table H. This table has the within

FIGURE 12.9 One possible pattern for the between-group differences with three sample groups.

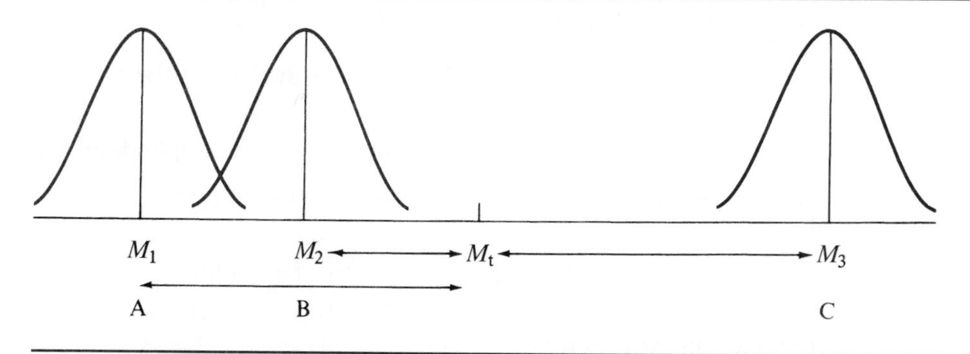

degrees of freedom in the left column and k, the number of sample groups, across the top row. Since the F ratio obtained in our previous problem was found to be significant, we will now use those data to illustrate the HSD calculation.

$$\text{HSD} = \alpha_{.05}\sqrt{\frac{\text{MS}_\text{w}}{n}}$$

The data are from a three-group design, so we look across the top row of Appendix Table H until we find 3. Then we follow that column down until it intersects the row led by 6, which is within our degrees of freedom. The value found there, in this case 4.34 (or 6.33 if the alpha level had been set at .01), is then plugged into the equation in place of $\alpha_{.05}$. Then we plug in the mean square within groups, 1, and divide by the number of subjects in any single sample, 3.

$$\text{HSD} = 4.34\sqrt{\tfrac{1}{3}} = 4.34\sqrt{.333} = (4.34)(.577) = 2.504$$

Interpreting the HSD Values. The calculated value for HSD just obtained indicates that in order to be statistically significant at an alpha probability of .05, the difference between any pair of means must be at least 2.504. From our data, examine the means:

1. $M_1 = 6/3 = 2$
2. $M_2 = 9/3 = 3$
3. $M_3 = 15/3 = 5$

All the mean differences equal to or greater than the calculated HSD of 2.504 are significant at the .05 level. When the calculated mean differences are placed in matrix form, it is customary to mark the significant differences with an asterisk. Sometimes a double asterisk is used if the differences reach significance at .01 (which in this case would have required an HSD of 3.652).

	$M_1 = 2$	$M_2 = 3$	$M_3 = 5$
$M_1 = 2$	—	1	3*
$M_2 = 3$	1	—	2
$M_3 = 5$	3*	2	—

We therefore conclude that the only significant mean difference in this study is between the first and third groups. The difference between Groups 1 and 2 failed to reach significance, as did the difference between Groups 2 and 3.

Applications of Tukey's HSD. The procedure just outlined can be used only when comparing pairs of means taken from sample groups of equal size, and this, of course, is the way ANOVAs should be planned. Sometimes, however, fate intervenes and the researcher's best-laid plans go awry. One of your human subjects may not show up on test day, or, in an animal study, one of the rats might even die. Fortunately, Tukey's HSD may also be used on data from samples of unequal sizes by making the slight adjustment shown next, called the Tukey–Kramer correction. This correction for unequal numbers of subjects is for a specific group comparison (two groups at a time).

$$\text{HSD} = \alpha_{.05}\sqrt{\frac{MS_{\text{w}}}{2}\left(\frac{1}{n_1} + \frac{1}{n_2}\right)}$$

When the sample sizes are equal, the correction shown here reduces to the previously shown Tukey critical value.

APPLICATIONS OF ANOVA

Obtaining the Value of t

ANOVA sometimes is used for testing the hypothesis of difference between two independent sample means. In this situation, the between degree of freedom, $k - 1$, equals $2 - 1$, or 1. The resulting F ratio is equal to t^2.

$$F = t^2 \text{ or } t = \sqrt{F}$$

The interpretation of a t ratio calculated in this way must be consistent with the two-tail t format, since the alternative hypothesis for F can never be directional. That is, if t is computed by taking the square root of F, then its significance must be evaluated against the critical values in the two-tail t table (Appendix Table C).

ANOVA can be used to test the hypothesis of difference in both experimental and post-facto research.

Analysis of Experimental Research

The action of the independent variable in experimental research is always reflected in the variability between groups, while the variability within groups simply reflects random error and/or individual differences among the subjects. (Between-group variability also reflects individual differences to a small degree, since individuals may differ in their receptivity to the independent variable.) The effects of the independent variable tend to separate the groups, to cause the means of the samples to fall farther and farther apart. This produces an increase in the size of the variability between the groups and, therefore, an increased probability of rejecting the null hypothesis. When the independent variable "takes"—that is, has an effect—then the following properties are observed:

1. Between-group variance is increased.
2. The F ratio is larger.
3. The probability of rejecting the null hypothesis is increased.

Analysis of Post-Facto Research

In the analysis of post-facto differences, the F ratio, of course, is not used for establishing whether an independent variable has any effect, since in such studies the independent variable is assigned, not manipulated. Therefore, in post-facto research, an increase in the variance between the groups relative to the variance within groups reflects the fact that the samples represent different populations. Why this difference exists cannot be directly determined. But since it does exist, predictions can be made using it.

Robustness and ANOVA

Robustness indicates how sensitive a statistical test is to violations of its basic assumptions. That is, it is the degree to which a statistical test is still appropriate to apply even though some of its assumptions are not met (Mertler and Vannata, 2001). ANOVA is considered to be a very robust test and when its assumptions are not fully met little harm results.

THE FACTORIAL ANOVA

All the experimental research studies covered so far in this book have been interested in the influence one independent variable might have on a dependent variable, with all other variables held constant. That is, we have created equivalent groups of subjects and exposed the groups to different levels of the independent variable. However, in all other ways the groups have been treated equally (for example, testing all groups at the same time of day, under identical conditions of illumination, etc.). Sometimes, however, researchers are interested in the effects of more than one independent variable. In these situations, the sample groups must be treated differently on a number of different dimensions.

For example, suppose that we are interested in conditions that might affect the height to which grass grows. One research technique would be to manipulate several independent variables separately, holding all other factors constant. We could treat two lawn patches with different amounts of nitrogen but keep the patches equal with regard to amount of sunlight, moisture, lime, temperature, and so on. Then, in another study, the amount of moisture could be manipulated, and nitrogen content (as well as all other factors) would be held constant. However, studies of this nature are often somewhat misleading because the different conditions involved often have an interactive effect. Perhaps nitrogen is most effective under certain moisture conditions and is even damaging under others. The point is that when several independent variables are manipulated *simultaneously,* it often leads to the discovery of cumulative effects acting above and beyond the effects of any of the independent variables working separately. We all know that phenobarbitol has a depressing effect on the human physiology, as does alcohol. But put the two together and you get that big dependent variable in the sky.

Factors

Any variation in an independent variable can also be called a *factor.* For example, if groups of subjects are treated differently with regard to, say, amount of illumination, that would be one factor. Similarly, if the groups are also treated differently on something else, such as noise level, that would be another factor. Whenever a study focuses on more than one factor, the data analysis is called factorial. If the data scale is at least interval, the statistic used for this analysis is called the **factorial ANOVA.**

One of the reasons for the popularity of factorial ANOVA is that behavior is often the result of several motives working together. The human behavioral repertoire, especially, is complex and multi-motivated. Examining one IV as a causal factor can often be unrealistic, since very few behavioral causes actually exist in splendid isolation. In fact, it is often the case that a given treatment may affect subjects in several ways, that is, impact more than just a single DV. Researchers using advanced statistical techniques are now turning to multivariate statistical analysis, where several IVs can be looked at in conjunction with several DVs. An excellent coverage of this important topic can be found in Mertler and Vannata's "Advanced and Multivariate Statistical Methods" (2001).

Between-Group Variability

We know that the between-group variability is that portion of the total variability separating the sample groups and that it results from the action of the independent variable. When more than one independent variable is involved, we must factor out the portion of the between-group variability that is attributable to each independent variable. Furthermore, since the independent variables may interact to produce a cumulative effect, we also need to know what part of the between-group variability is being created or increased by such interaction among the independent variables. Just as we have previously taken the total variability and partitioned out its between-group and within-group components, we can now take the between-group variability and separate its components.

The actions of the independent-variables taken separately are called the **main effects,** as opposed to their combination, or **interaction effects.**

Calculating the Two-Way ANOVA

In the two-independent-variable research design, which calls for the two-way ANOVA (rest assured, this is as far as we will go in this chapter), one of the independent variables is set up in columns and the other in rows to form a block of cells. A cell is a combination of treatment conditions that is unique to one group of subjects, or *where a row intersects a column.* In the following example, we examine a four-cell, completely randomized design. Two independent variables, diet and exercise, are each being manipulated at two levels.

Example A researcher is interested in the effects of both diet and exercise on percentage of body fat. A sample of 20 male, high school sophomores is randomly selected. The subjects are randomly assigned to four different treatment conditions. Since the two independent variables, diet and exercise, are each manipulated in two ways, the four cells shown represent all possible combinations of treatment conditions.

		Factor A	
		Diet below 2000 calories	***Diet above 2000 calories***
Factor B	*Exercise*	a	b
	No exercise	c	d

Cell a contains only those subjects who follow the exercise program and are also on the low-calorie diet. Cell b contains only those subjects who follow the exercise program and are on the high-calorie diet. Cell c contains only those subjects who follow the no-exercise program and are on the low-calorie diet. Cell d contains those subjects who follow the no-exercise program and are on the high-calorie diet.

After a three-month period, each subject is measured for percentage of body fat compared to total weight, and their scores are recorded as follows:

	Diet below 2000 calories		***Diet above 2000 calories***		
	a		b		
	X_a	X_a^2	X_b	X_b^2	
	10	100	12	144	
	12	144	14	196	
Exercise	14	196	15	225	
	12	144	16	256	
	10	100	14	196	$\Sigma X_a + \Sigma X_b = 129$
	$\Sigma X_a = 58$	$\Sigma X_a^2 = 684$	$\Sigma X_b = 71$	$\Sigma X_b^2 = 1017$	
	c		d		
	X_c	X_c^2	X_d	X_d^2	
	16	256	22	484	
	18	324	24	576	
No exercise	20	400	24	576	
	22	484	22	484	
	20	400	24	576	$\Sigma X_c + \Sigma X_d = 212$
	$\Sigma X_c = 96$	$\Sigma X_c^2 = 1864$	$\Sigma X_d = 116$	$\Sigma X_d^2 = 2696$	
	$\Sigma X_a + \Sigma X_c = 154$		$\Sigma X_b + \Sigma X_d = 187$		$\Sigma X = 341$

To perform a two-way ANOVA on these data, we go through the following steps:

1. Calculate the correction factor, *C*.

$$C = \frac{(\Sigma X)^2}{N} = \frac{341^2}{20} = 5814.050$$

2. Calculate the total sum of squares, just as was done earlier in this chapter for the one-way ANOVA.

$$SS_t = \Sigma X^2 - C$$
$$= (684 + 1017 + 1864 + 2696) - 5814.050$$
$$= 6261 - 5814.050 = 446.950$$

3. Calculate the sum of squares between groups, again as was done for the one-way ANOVA.

$$SS_b = \frac{(\Sigma X_a)^2}{n_a} + \frac{(\Sigma X_b)^2}{n_b} + \frac{(\Sigma X_c)^2}{n_c} + \frac{(\Sigma X_d)^2}{n_d} - C$$
$$= \frac{58^2}{5} + \frac{71^2}{5} + \frac{96^2}{5} + \frac{116^2}{5} - 5814.050$$
$$= \frac{3364}{5} + \frac{5041}{5} + \frac{9216}{5} + \frac{13,456}{5} - 5814.050$$
$$= 672.800 + 1008.200 + 1843.200 + 2691.200 - 5814.050$$
$$= 6215.400 - 5814.050 = 401.350$$

4. From the values just found for SS_t and SS_b, calculate the sum of squares within groups.

$$SS_w = SS_t - SS_b$$
$$SS_w = 446.950 - 401.350 = 45.600$$

5. Calculate the sum of squares for rows, SS_{row} for the first main effect. We add all the scores for subjects on the exercise program, square the total, and divide by the number of scores in that row. We do the same for the row of subjects on the no-exercise program. Then we add the values for the two rows and subtract the correction factor.

$$SS_{row} = \frac{(\Sigma X_a + \Sigma X_b)^2}{n_a + n_b} + \frac{(\Sigma X_c + \Sigma X_d)^2}{n_c + n_d} - C$$

$$= \frac{(58 + 71)^2}{5 + 5} + \frac{(96 + 116)^2}{5 + 5} - 5814.050$$

$$= \frac{129^2}{10} + \frac{212^2}{10} - 5814.050$$

$$= \frac{16,641}{10} + \frac{44,944}{10} - 5814.050$$

$$= 1664.100 + 4494.400 - 5814.050 = 344.450$$

6. Calculate the sum of squares for columns, SS_{col} for the second main effect. We add all the scores for subjects on the diet below 2000 calories, square the total, and divide it by the number of scores in that column. We do the same for the column of subjects on the diet above 2000 calories. We then add the values for the two columns and subtract the correction factor.

$$SS_{col} = \frac{(\Sigma X_a + \Sigma X_c)^2}{n_a + n_c} + \frac{(\Sigma X_b + \Sigma X_d)^2}{n_b + n_d} - C$$

$$= \frac{(58 + 96)^2}{5 + 5} + \frac{(71 + 116)^2}{5 + 5} - 5814.050$$

$$= \frac{154^2}{10} + \frac{187^2}{10} - 5814.050$$

$$= \frac{23,716}{10} + \frac{34,969}{10} - 5814.050$$

$$= 2371.600 + 3496.900 - 5814.050 = 54.450$$

7. Calculate the interaction sum of squares, $SS_{r \times c}$. Since the sum of squares between groups is composed of the sum of squares for columns plus the sum of squares for rows plus the interaction sum of squares, we can get the interaction value by finding the difference between values we already have.

$$SS_{r \times c} = SS_b - SS_{col} - SS_{row}$$
$$= 401.350 - 54.450 - 344.450 = 2.450$$

8. Set up a summary table of our calculations for the factorial (two-way) ANOVA. The degrees of freedom for the table are allocated as follows. The degrees of freedom for rows equal the number of rows minus one.

$$df_{row} = n_{row} - 1$$
$$= 2 - 1 = 1$$

The degrees of freedom for columns equal the number of columns minus one.

$$\begin{aligned} df_{col} &= n_{col} - 1 \\ &= 2 - 1 = 1 \end{aligned}$$

The interaction degrees of freedom equal the degrees of freedom for rows multiplied by the degrees of freedom for columns.

$$\begin{aligned} df_{r \times c} &= (df_{row})(df_{col}) = (n_{row} - 1)(n_{col} - 1) \\ &= (1)(1) = 1 \end{aligned}$$

The degrees of freedom within groups are equal to the total sample size, N, minus the number of sample groups, k.

$$\begin{aligned} df_w &= N - k \\ &= 20 - 4 = 16 \end{aligned}$$

As a check, we can add the various degrees of freedom, $1 + 1 + 1 + 16 = 19$. This should equal $N - 1$; $20 - 1 = 19$.

	Source of Variability			
	SS	*df*	*MS*	F
Rows (exercise)	344.450	1	344.450	120.860
Columns (diet)	54.450	1	54.450	19.105
r × c (interaction)	2.450	1	2.450	.860
Within groups	45.600	16	2.850	

The mean square values for the table are calculated as for the one-way ANOVA; that is, the mean square equals the sum of squares divided by the appropriate degrees of freedom.

$$MS_{row} = \frac{SS_{row}}{df_{row}} = \frac{344.450}{1} = 344.450$$

$$MS_{col} = \frac{SS_{col}}{df_{col}} = \frac{54.450}{1} = 54.450$$

$$MS_{r \times c} = \frac{SS_{r \times c}}{df_{r \times c}} = \frac{2.450}{1} = 2.450$$

$$MS_w = \frac{SS_w}{df_w} = \frac{45.600}{16} = 2.850$$

The F ratios are determined exactly as in a one-way ANOVA. There we divided the mean square between groups by the mean square within groups to get the value of F. Here the mean square between groups is broken down into three values, so we get three F ratios.

$$F_{\text{row}} = \frac{\text{MS}_{\text{row}}}{\text{MS}_{\text{w}}} = \frac{344.450}{2.850} = 120.860$$

$$F_{\text{col}} = \frac{\text{MS}_{\text{col}}}{\text{MS}_{\text{w}}} = \frac{54.450}{2.850} = 19.105$$

$$F_{\text{r}\times\text{c}} = \frac{\text{MS}_{\text{r}\times\text{c}}}{\text{MS}_{\text{w}}} = \frac{2.450}{2.850} = .860$$

9. Each of the three F ratios obtained is now compared with the critical value of F in Appendix Table G. The degrees of freedom are 1 in the numerator and 16 in the denominator (each of the mean squares between groups has 1 df, and the mean square within groups has 16 df). Again, the rule is that the null hypothesis is rejected whenever the calculated value of F equals or exceeds the table value.

$F_{.01(1.16)} = 8.53$

$F_{\text{row}} = 120.860$ Reject H_0; significant at $p < .01$.

$F_{\text{col}} = 19.105$ Reject H_0; significant at $p < .01$.

$F_{\text{r}\times\text{c}} = .860$ Accept H_0; not significant.

Conclusions: A 2 × 2 between-subjects factorial ANOVA was conducted in an attempt to discover possible differences in the percentage of body fat as a function of the two independent variables, diet and exercise. A significant main effect for exercise was found ($F(1,16) = 120.860$, $p < .01$), as well as a significant main effect for diet ($F(1,16) = 19.105$, $p < .01$). Subjects had lower body fat percentages the more they exercised. They also had lower body fat percentages when they were on the low calory diet rather than the high calory diet. However, there was no significant interaction effect ($F(1,16) = -.860$, $p > .05$, ns.). Neither of the main effects was significantly influenced by the other.

The interaction F, therefore, indicates whether the effect of one of the IVs is influenced by differing levels of the second IV. Thus, when significant differences occur that are not explained by the main effects, then an interaction between the IVs has occurred.

Because each IV in the study had only two levels, Tukey's HSD is not needed.

If one or both of the IVs had been set at more than two levels, then HSD could be performed on the main effects using the same equation shown earlier. When analyzing factorial data for main effects, however, it is important that the means being compared are the marginal means (column means and row means), not the individual cell means. This is probably obvious to you, since the main-effect IVs only show up as column and row data. Also, the N now stands for the number of subjects in each row or column, *not* the number in the individual cell.

Effect Size

For the factorial ANOVA, **effect size** may be determined as a partial eta square. An ES is found for each of the factorial components, the main effects as well as the interaction.

For the rows, divide the sum of squares for rows by the sum of squares for rows PLUS the sum of squares within

$$\text{Partial } \eta^2 = \frac{SS_r}{SS_r + SS_w}$$

For the columns, divide the sum of squares for columns by the sum of squares for columns PLUS the sum of squares within

$$\text{Partial } \eta^2 = \frac{SS_c}{SS_c + SS_w}$$

And for the interaction, divide the sum of squares for the interaction by the sum of squares for interaction PLUS the sum of squares within

$$\text{Partial } \eta^2 = \frac{SS_{r \times c}}{SS_{r \times c} + SS_w}$$

Using the previous problem, where we had significant F ratios for the columns and for the rows, but not for the interaction, the effect sizes would be

$$\text{Partial } \eta^2 \text{ Rows} = \frac{344.450}{344.450 + 45.600} = .883$$

$$\text{Partial } \eta^2 \text{ Columns} = \frac{54.450}{54.450 + 45.600} = .544$$

Both of these are considered to be strong effects.

We cannot calculate a partial eta square for the interaction because in this problem the interaction F ratio was not significant.

ETA SQUARE AND *d*

Since effect sizes for ANOVA are typically reported as "eta square" values, and others as "*d*" values (see Chapter 10), a conversion table has been constructed to allow for quick and easy comparisons. Notice that the eta square values are always LOWER than the *d* values. Whereas *d* values of .20, .50, and .80 represent small, medium, and large values respectively, eta square values of .01, .06, and .14 represent small, medium, and large effects (Green, Salkind, & Akey, 1997).

eta sq	d	*eta sq*	d	*eta sq*	d	*eta sq*	d
0.01	0.20	0.26	1.19	0.51	2.04	0.76	3.56
0.02	0.29	0.27	1.22	0.52	2.08	0.77	3.66
0.03	0.35	0.28	1.25	0.53	2.12	0.78	3.77
0.04	0.41	0.29	1.28	0.54	2.17	0.79	3.88
0.05	0.46	0.30	1.31	0.55	2.21	0.80	4.00
0.06	0.51	0.31	1.34	0.56	2.26	0.81	4.13
0.07	0.55	0.32	1.37	0.57	2.30	0.82	4.27
0.08	0.59	0.33	1.40	0.58	2.35	0.83	4.42
0.09	0.63	0.34	1.44	0.59	2.40	0.84	4.58
0.10	0.67	0.35	1.47	0.60	2.45	0.85	4.76
0.11	0.70	0.36	1.50	0.61	2.50	0.86	4.96
0.12	0.74	0.37	1.53	0.62	2.55	0.87	5.17
0.13	0.77	0.38	1.57	0.63	2.61	0.88	5.42
0.14	0.81	0.39	1.60	0.64	2.67	0.89	5.69
0.15	0.84	0.40	1.63	0.65	2.73	0.90	6.00
0.16	0.87	0.41	1.67	0.66	2.79	0.91	6.36
0.17	0.91	0.42	1.70	0.67	2.85	0.92	6.78
0.18	0.94	0.43	1.74	0.68	2.92	0.93	7.29
0.19	0.97	0.44	1.77	0.69	2.98	0.94	7.92
0.20	1.00	0.45	1.81	0.70	3.06	0.95	8.72
0.21	1.03	0.46	1.85	0.71	3.13	0.96	9.80
0.22	1.06	0.47	1.88	0.72	3.21	0.97	11.37
0.23	1.09	0.48	1.92	0.73	3.29	0.98	14.00
0.24	1.12	0.49	1.96	0.74	3.37	0.99	19.90
0.25	1.15	0.50	2.00	0.75	3.46		

*Eta squares can be calculated for *t* ratios, since $\eta^2 = \dfrac{t^2}{t^2 + df}$.

Although in this era of powerful computer programs, it is easy to produce interaction *F* ratios among 3, 10, or even 30 variables, the researcher should proceed with great caution when contemplating pursuing these higher-order interactions. First, they can become very difficult to interpret. Second, Cohen and Cohen tell us that three-way or higher-order interactions are rarely of central interest to the serious researcher and are used as only ancillary backups to the original intent of the design. Finally, again from Cohen and Cohen, "higher order interactions increase dramatically the risks of the spurious occurrence of significance (Type 1 errors) *and* of the failure to detect real effects (Type 2 errors)" (Cohen & Cohen, 1983 p. 347). They go on to say that with 31 interactions being tested, the probability of 1 or more computing out as significant is in the range of .90, even when the null hypothesis is actually true for ALL OF THEM.

GRAPHING THE INTERACTION

After completing a factorial ANOVA, it is often instructive to graph the means, looking for the directions of change and the possibility of an interaction. The graph does not have to be elaborate, and the few minutes it takes may show important trends that might easily remain obscured by just looking at the *F* ratios themselves. Using the data from the previous example, first compute the means for each cell.

The results are as follows:

		Factor A (diet)	
		Below 2000 calories	Above 2000 calories
Factor B (exercise)	Exercise	a M = 11.600	b M = 14.200
	No exercise	c M = 19.200	d M = 23.200

The graph would appear as shown in Figure 12.10. Notice that the trend is clearly in evidence. Both the low-calorie diet and exercise distinctly show lower body fat. Also notice that the two graph lines remain fairly parallel, a pictorial indication that there

FIGURE 12.10 Graphing the results with no significant interaction.

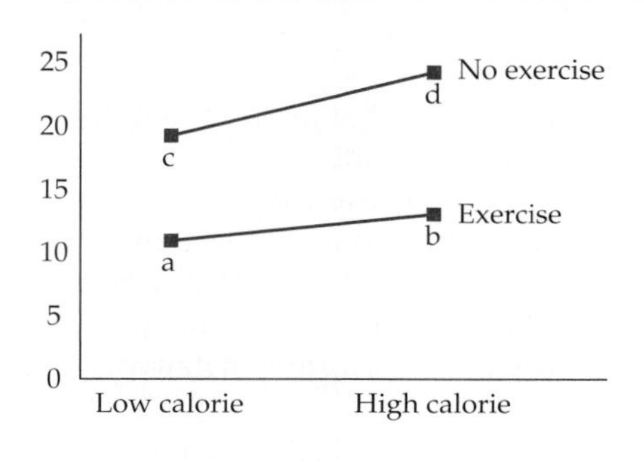

was no interaction. As long as the graph lines do not differ significantly from being parallel, there is no interaction effect. Had the graph lines crossed, however, the interaction would have been significant.

When looking at the graph of the interaction, when the two lines run somewhat parallel, as they do in Figure 12.10, there is no interaction since the differences are the same for both calorie groups; the difference between low-calorie subjects is essentially the same as the difference between high-calorie subjects. This would be true in this case even if the graph lines were both raised or lowered. That is, even if the main effect of exercise had changed, there would still be no change in the interaction (as long as the lines do not differ significantly from parallellity). In the next example, we will examine a significant interaction.

A Significant Interaction: The Paradox of Ritalin

As we just saw, the analysis of the interaction term in the factorial ANOVA can be extremely instructive and becomes especially so when neither of the main effect IVs appears to be significant. As an example, assume that a researcher wishes to discover whether the drug Ritalin has any effect on reducing the activity levels of hyperactive boys. Further, the researcher is concerned as to whether the drug's effect is similar despite age differences among the subjects. A random sample of 40 subjects is selected from the population of hyperactive boys; 20 are over age 16 and 20 are under age 16. Each of the two age groupings is further divided into two groups of 10 subjects each and randomly assigned to either the Ritalin condition or the placebo condition.

The boys are then measured on their activity levels on a scale of 0 to 50, with high scores indicating higher levels of activity. The results are as follows:

		Factor A (drug)	
		Placebo	*Ritalin*
Factor B (age)	*Under 16*	a Mean = 35	b Mean = 27
	Over 16	c Mean = 33	d Mean = 37

Calculation of the F ratios showed that

Age (in rows):	$F = 3.900$	Accept H_0.
Drug (in columns):	$F = 3.800$	Accept H_0.
Interaction ($r \times c$):	$F = 15.200$	Reject H_0: significant at $p < .01$.

$F_{.05}$ with 1 and 36 degrees of freedom = 4.11. $F_{.01}$ with 1 and 36 degrees of freedom = 7.39.

Conclusion: A 2 × 2 between-subjects factorial ANOVA was conducted in an attempt to discover possible differences in activity levels as a function of the two independent variables, age and ritalin. The main effect for age was not significant and the null hypothesis was accepted ($F(1,36) = 3.900$, $p > .05$, ns.). Nor was there a significant main effect for ritalin ($F(1,36) = 3.800$, $p > .05$, ns.). However, there was a significant interaction effect ($F(1,36) = 15.200$, $p < .01$). Thus, the drug ritalin appeared to increase the activity levels of older children while decreasing the activity levels of younger children. (This finding is called a paradoxical effect.)

Thus, the F ratios tell us that the drug and the age variable both appear to have no effect, but that the interaction between the two factors is clearly significant. To more fully understand this situation, a graph of these results can be drawn by connecting the means from the same age groupings, as shown in Figure 12.11.

Thus, it becomes clear that since the over-16 boys increased their activity levels and the under-16 boys decreased their activity levels, the effects of the drug were seemingly canceled out. That is why the F ratio for drug (in the columns) is only 3.800 and produced an accept of the null hypothesis. This finding, that Ritalin appears to increase the activity levels of older children and yet at the same time to decrease the activity levels of younger children, is known as a paradoxical effect.

Theory of the Two-Way ANOVA

Whenever there are significant differences among several sample groups, these differences are related to the between-group variability. When only one independent variable is involved, we have no problem identifying the source of these differences, so only one F ratio is calculated (a one-way ANOVA). But when significant differences result from the effects of several independent variables, a number of explanations of the sources of difference are possible; each of these possible accountings requires a separate F ratio. Since the action of the independent variables is reflected in the between-group variability, it is this *between-group variability* that must be broken up

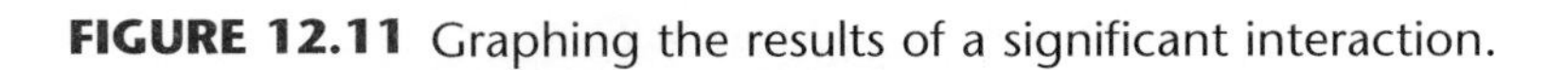

FIGURE 12.11 Graphing the results of a significant interaction.

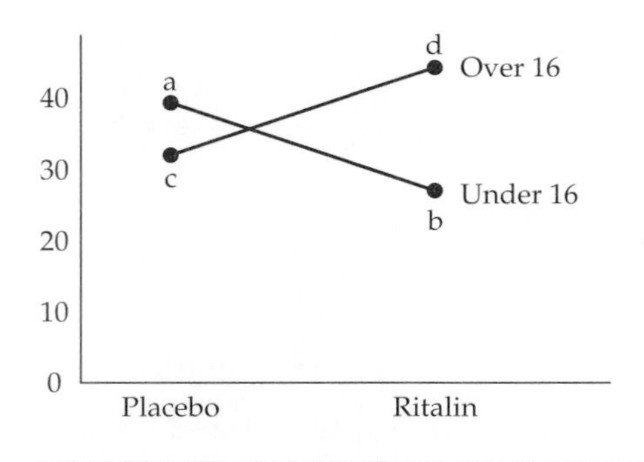

into its various parts—one part for each independent variable and the remaining part for the interaction.*

To accomplish the two-way ANOVA, each component of the between-group variability is analyzed separately. For one of the independent variables, we look only at the column scores, holding the row comparisons in abeyance. This is why in our example we added the scores of all of the subjects on the low-calorie diet, regardless of whether they exercised. Next, we look only at the row scores, holding column variability constant. In our example, that meant adding all the scores for an exercise level, regardless of which diet the subjects followed. Finally, by subtraction, we factor out the interaction effect, that is, whether the independent variables are in any way acting in concert to produce still another effect. The effects of each independent variable separately are called *main effects,* as opposed to the interaction effect produced by the combination of variables.

Limitation on the Use of ANOVA Techniques. All the ANOVA techniques outlined in this chapter can be applied only if the groups of subjects are randomly selected and are *independent* of each other. Among the various experimental designs, then, our ANOVAs are restricted to to studies based on between-group comparisons, where each subject has been randomly assigned to a specific combination of treatment conditions. Later, in Chapter 15, other ANOVA techniques will be presented that do allow for the analysis of repeated-measures and matched-subject designs.

Finally, before leaving the ANOVA, two other variations should be mentioned. You will undoubtedly run across these procedures, so although you won't have to learn to compute them at this point, at least you will have heard the terms and hopefully hold them in mind when reading research articles.

MANOVA

MANOVA is a type of analysis of variance that tests whether two or more IVs can significantly impact two or more DVs. This is a variation of the straight factorial design just shown where several IVs were used to establish a possible difference on a *single* DV.

Discriminant Analysis. Discriminant Analysis (DA) is a procedure that can be thought of as a reverse ANOVA or MANOVA. Whereas in an ANOVA the researcher takes two or more groups of subjects and then compares their scores on the DV to discover whether there are significant DV differences, with discriminant analysis the researcher reverses the process. That is, the scores on the DV are used to predict which group they came from. If there were several DVs, then a reverse MANOVA would be performed. For example, one could have groups of high-, medium-, and low-income persons who have also been measured on a series of such variables as age, IQ, various

*Note that if there are three or more independent variables, then more than just one possible interaction effect must be included in the analysis.

personality traits, and amount of education. The discriminant analysis would be based on the identification of the variables or combinations of variables that produced the most accurate prediction of which of the three income groups the person represented.

SUMMARY

ANOVA, an acronym for ANalysis Of VAriance, is used on interval data when more than two sample means are to be compared for differences. The resulting statistic, the *F* ratio, determines the ratio between the variability occurring between the sample groups and the variability occurring within each of the sample groups. The higher the *F* ratio, the greater is the likelihood that the samples represent different populations; a high *F* ratio indicates that there is a great deal of between-group variability and little within-group variability occurring (the sample distributions show little or no overlap). The lower the *F* ratio, the greater is the chance that the samples represent a single population; a low *F* ratio indicates that there is little between-group variability compared to the amount of within-group variability occurring (the sample distributions show a great deal of overlap).

Variability can be determined through the use of the sum of squares, the sum of the squared deviations of all the scores from the mean. The variance, *V*, or mean square, MS, is calculated by dividing the sum of squares by the appropriate degrees of freedom, df. This is a necessary step; although adding new scores to a distribution has a dramatic effect on the sum of squares, this effect is minimized through the averaging process used to obtain the mean square. The *F* ratio, then, is equal to MS_b/MS_w, the mean square between the groups divided by the mean square within the groups.

The calculated *F* ratio must then be checked for significance by comparing it against the critical values of *F* for the appropriate degrees of freedom. When the calculated *F* is equal to or greater than the table value, the null hypothesis is rejected. As with the *t* test, alpha should be kept at a probability level of .05 or less.

Although a significant *F* ratio tells us that there are significant differences among the several group means, it does not, in and of itself, specify precisely where those differences are occurring. For this analysis, Tukey's HSD (honestly significant difference) can be used.

The *F* ratio so far described is called a one-way ANOVA in that, although several sample groups are being compared, only one independent variable is used. When the experimental design calls for more than one independent variable, the factorial ANOVA must be used. This technique allows us not only to discover whether the independent variables taken separately are having any effect but *also* whether the independent variables are interacting and producing a cumulative effect. The actions of the separate independent variables are called main effects, and the cumulative effects are called interaction effects (produced by the combination of independent variables). When an *F* ratio has been found to be significant, the strength or magnitude of the effect (effect size) can be determined by the use of eta squares for the one-way ANOVA and partial eta squares for the rows, columns, and interaction for the factorial ANOVA.

Key Terms and Names

analysis of variance (ANOVA)
effect size
eta squared
factorial ANOVA
Fisher, Sir Ronald
homogeneity of variance
interaction effect
main effect
sum of squares
Tukey's HSD

PROBLEMS

1. Calculate an F ratio for the following four-group design. Do you accept or reject the null hypothesis?

A	*B*	*C*	*D*
3	6	7	9
4	7	8	9
5	6	7	8
4	7	8	10
5	6	6	12

2. If the results in problem 1 make it appropriate, compare the four sample means using Tukey's HSD.

3. For the following two-group design, calculate both a t and an F ratio. Do you accept or reject the null hypothesis? (Be sure that $t = \sqrt{F}$ and $F = t^2$, within rounding error.)

A	*B*
12	8
10	6
8	4
6	2
4	0

4. A researcher is interested in determining the possible effect of shock on memory of verbal material. Subjects are randomly selected and then randomly assigned to one of three treatment conditions. Group A receives no shock, Group B receives

medium shock, and Group C gets high-intensity shock levels. The various shock levels are introduced during the learning period. All subjects then take a verbal retention test, with high scores indicating more retention. Their scores are as follows:

A	*B*	*C*
24	20	16
30	18	14
25	22	12
24	20	15
20	17	16

a. Calculate the *F* ratio.
b. Do you accept or reject the null hypothesis?
c. If appropriate, calculate Tukey's HSD.

5. A researcher is interested in proving that ingestion of the drug magnesium pemoline (MgPe) increases retention of learned material. A group of 16 subjects is randomly selected from the population of students at a large university. The subjects are then randomly assigned to one of four conditions: A receives placebo, B receives 10 cc of MgPe, C receives 20 cc of MgPe, and D receives 30 cc of MgPe. All subjects are then given some material to read and four hours later are tested for retention (high scores indicating high retention).

A	*B*	*C*	*D*
8	10	11	10
6	7	6	8
7	8	8	7
5	6	9	9

a. Calculate the *F* ratio.
b. Do you accept or reject the null hypothesis?
c. If appropriate, calculate Tukey's HSD.
d. State the kind of research this represents, experimental or post-facto.
e. Identify the independent and dependent variables.
f. What conclusions can you legitimately draw from this study?

6. Hypothesis: Increasing movie violence increases viewer aggression. Three groups of subjects are randomly selected. All groups view a 30-minute motion picture, but each version of the movie has a different number of violent scenes. Following the movie, all subjects are given the TAT (Thematic Apperception Test) and scored on the amount of fantasy aggression displayed (high scores indicating more aggression).

Group A (No Violent Scenes)	Group B (5 Violent Scenes)	Group C (10 Violent Scenes)
2	4	6
4	6	8
6	8	10
8	10	12

a. Calculate the *F* ratio.
b. Should H_0 be accepted or rejected?
c. If appropriate, calculate Tukey's HSD.

7. Hypothesis: Increasing temperature and physical exercise increases the body's skin conductivity (as measured by a galvanometer). A sample of 24 males, all 21 years old, is randomly selected, and the subjects are randomly assigned to four treatment conditions. The high temperature was set at 80 degrees F and the low temperature was 60 degrees F. Exercise consisted of jogging in place for 10 minutes and no exercise meant sitting quietly for 10 minutes. Their GSR scores (Galvanic Skin Response) were measured in decivolts. High scores indicating greater skin conductivity. The data follow:

	High Temperature	Low Temperature
	9	7
	8	6
Exercise	8	5
	10	5
	7	4
	8	5
	5	3
	3	2
No Exercise	3	2
	4	1
	3	2
	4	1

Test the hypothesis.

8. A psychologist is interested in discovering the possible effect of both a certain drug and psychotherapy on the anxiety levels of anxious patients. Six groups of anxious patients, 10 per group, were randomly selected and independently assigned to the various treatment conditions.

(Each of the scores represents the measure of anxiety taken on the patient after the treatments were completed, with higher scores indicating higher levels of anxiety.)
a. Was there a significant effect for the drug?

b. Was there a significant effect for the psychotherapy?
c. Was there a significant interaction between the drug and the psychotherapy?

Factor B		Factor A: Placebo A1	Drug (low dose) A2	Drug (high dose) A3
	B1 Psychotherapy	9	6	7
		7	4	1
		8	3	1
		4	6	2
		7	5	4
		9	6	5
		10	2	4
		5	4	2
		6	8	1
		9	2	1
	B2 No psychotherapy	9	7	6
		6	5	1
		9	7	1
		10	6	5
		14	5	4
		6	3	3
		8	2	2
		5	4	1
		4	5	2
		5	9	3

9. A researcher wants to know if length of time and type of training affects one form of arithmetic ability (dealing with simple fractions). Six groups of elementary students (all from the fifth grade) were randomly selected and assigned to different time (from 1 to 3 hours) and different teaching conditions (old drill method vs. the reach, meaningful method). All the students were then tested on the "fractions" subtest of a standard arithmetic achievement test. Their scores (higher scores indicating greater achievement) were as follows:

	Time: One Hour	Two Hours	Three Hours
Old Drill	1	2	4
	2	3	5
	3	4	6
Meaningful	6	8	10
	7	9	12
	8	10	14

Test the hypothesis.

10. Using the following summary table, based on a design that had 3 subjects in each group, answer the questions that follow.

	SS	df	MS	F
Between rows	910.00	2	455	65.000
Between Columns	378.00	2	189	27.000
Interaction	64.00	4	16	2.285
Within	126.00	18	7	

a. The main effect for rows showed the IV to be set at how many levels?
b. Was the main effect for rows significant at .05?
c. Was the main effect for rows significant at .01?
d. The main effect for columns showed the IV to be set at how many levels?
e. Was the main effect for columns significant at .05?
f. Was the main effect for columns significant at .01?
g. Was the interaction significant at .05?
h. Was the interaction significant at .01?
i. How many subjects were there in all the groups combined?
j. How many groups were there?

Fill in the blanks in problems 11 through 20.

11. Total variability results from the accumulated differences between each individual score and ________.

12. Between-group variability results from the accumulated differences between each sample mean and ________.

13. Within-group variability results from the accumulated differences between each individual score and ________.

14. When a calculated F ratio has a large value, it indicates that the variability between groups is ________ (larger or smaller) than the variability within groups.

15. The variance, or mean square, results from dividing the sum of squares by ________.

16. The total sum of squares is made up of two major components, the ________ and the ________.

17. When an ANOVA results in the rejection of the null hypothesis, then the ________ variability must be larger than the ________ variability.

18. If the within-group variability is small, then the separate sample groups are most likely to have ________ (platykurtic or leptokurtic) distributions.

19. A five-group research design with six subjects in each group has ________ between degrees of freedom and ________ within degrees of freedom.

20. The greater the spread among the various sample means, the larger is the ________ (between or within) variability.

True or False. Indicate either T or F for problems 21 through 29.

21. The F ratio is a nondirectional, two-tail test of differences among sample groups used whenever the data are in interval form.
22. ANOVA demands that *at least* four sample groups must be compared.
23. On a four-group design, the between degrees of freedom for a one-way ANOVA must equal 4.
24. ANOVA assumes that the data are at least interval.
25. An F ratio of 5.00 indicates that the variance between groups is five times greater than the variance within groups.
26. The use of the factorial ANOVA is required whenever there is more than one independent variable and the data are in interval form.
27. On a factorial ANOVA, the interaction effect will always be significant if the main effects are themselves significant.
28. To do a factorial ANOVA, there must be a minimum of at least four different treatment conditions.
29. When the obtained value of F is larger than the table value of F for a given number of degrees of freedom, the null hypothesis must be accepted.

COMPUTER PROBLEMS

C1. The Novaco Provocation Inventory (NPI) is either an 80- or 90-item instrument for assessing anger responsiveness. (In order to maximize the variability, the 90-item version was used in this study.) The test consists of brief descriptions of situations of provocation, for which the respondent notes the degree of anger that he/she might experience if that event should occur in his/her life. The ratings are on a 5-point Likert-type scale. The main purpose of the NPI is to gauge the range and intensity of a person's anger. Three groups of inmates from a state prison were selected and grouped on the basis of whether this was a first incarceration, a second incarceration or a third incarceration. Each man took the Novaco and the results were as follows:

First Incarceration	Second Incarceration	Third Incarceration
223	347	297
202	319	310
241	310	290
234	320	335
266	330	320
240	300	290
292	290	315
237	337	310
176	310	337
201	315	290

Continued

First Incarceration	Second Incarceration	Third Incarceration
229	290	300
225	295	330
147	327	310
149	290	300
210	320	317
319	335	320
278	290	300
266	310	247
176	287	310

Test the hypothesis that anger is a function of the number of incarcerations. If the *F* is significant, follow it with a Tukey.

C2. A researcher is interested in whether the phonics method of teaching reading is more or less effective than the sight method, depending on what grade the child is in. Twenty children were randomly selected from each of three grades: kindergarten (K), first grade (1), and second grade (2). Achievement was measured in terms of reading comprehension, higher scores indicating more comprehension. Within each grade, 10 children were assigned to each of two methods of teaching reading, phonics or sight. The data are as follows:

	Grades K	Grades 1	Grades 2
Sight	14	25	49
	20	29	49
	16	27	46
	21	31	46
	20	27	44
	14	34	43
	21	32	50
	23	34	43
	14	35	48
	15	28	52

	Grades K	Grades 1	Grades 2
Phonics	17	35	34
	22	36	33
	19	40	34
	20	37	39
	26	37	38
	18	41	33
	26	42	35
	18	33	42
	25	34	42
	23	43	38

a. Test the hypothesis for both main effects and interactions.
b. Specify all variables.
c. Write a statement as to your conclusions.

C3. In a study of hyperactivity among elementary school boys, nine groups of subjects were randomly selected from a school population of ADHD, seven-year-old boys. (ADHD is attention deficits with hyperactivity, and left untreated, it can prevent a child from attending to incoming learning stimuli and may also create major disruptions in the classroom.) The researcher wanted to study the classroom effects of both the drug Ritalin and a behavior modification program on the activity levels of the subjects. The drug was varied from no dosage (in the form of a placebo), through 10 mg of Ritalin to 20 mg of Ritalin per day. The behavior modification program consisted of giving the child 10 tokens to start the day and then taking away a token for each hyperactive infraction. The tokens that were saved could then be exchanged for some valued prize. The behavior mod program was varied from no program to the program being utilized every other day, to the program being in force every day. After four weeks, all the children were evaluated for hyperactivity and were assigned scaled scores ranging from a possible low of zero (no indication of hyperactivity) to a high of 40 (extreme hyperactivity). The nine groups and their hyperactivity scores are as follows:

		Drug dose	
	Ritalin, 20 mg	*Ritalin, 10 mg*	*Placebo*
No behavior mod	12	18	21
	10	16	22
	19	27	33
	19	23	35
	15	26	27
	18	22	34
	22	22	32
Behavior mod every other day	28	30	38
	15	15	20
	20	20	23
	12	14	20
	14	15	14
	14	20	21
	17	19	21
Behavior mod every day	3	7	13
	10	14	18
	11	11	12
	8	12	13
	11	17	19
	5	10	9
	13	15	17

a. Conduct the overall analysis. Calculate the means for all nine cells, and write a verbal description of the results.
b. Compute the main effects for the drug.
c. Compute the main effects for behavior mod.
d. Compute the interaction and write a paragraph explaining these results.

C4. A researcher is interested in whether a certain amphetamine drug will affect retention of learned behavior among rats as a function of how much time is allowed to elapse between the training and the retention test. Four groups, 10 subjects in each group, were randomly selected from a colony of Sprague–Dawley rats. All the rats were given the same amount of training on a simple maze. Groups A and C were given the amphetamine, whereas Groups B and D were given a placebo. Groups A and B were then tested for retention immediately following the training, whereas Groups C and D were tested four hours later. The retention measures consisted of scaled scores based on how many correct responses the rats would exhibit while running the maze; the higher the score, the greater is the retention. Test the hypotheses. The data were as follows:

	Amphetamine Group A	*Placebo Group B*
Immediate testing	14	12
	8	8
	8	11
	10	6
	9	6
	5	7
	9	9
	11	8
	13	4
	6	10
	Group C	*Group D*
Delayed testing	2	9
	6	7
	3	10
	2	5
	5	5
	6	6
	7	7
	6	7
	3	7
	7	8

Chapter

13

Nominal Data and the Chi Square

By far the most popular test for ***nominal data*** *is the chi square. Although nominal data were described in Chapter 9, it is important to repeat here that the nominal case only gives us information regarding the frequency of occurrence within categories. From the nominal scale we get nose-counting data. It tells us in how* many *cases a certain trait occurs. From nominal data we get no information as to how much of the given trait any individual possesses, only information as to whether that individual has the trait at all. A nominal scale can be constructed for political party affiliation by grouping individuals according to party loyalties—so many Democrats, so many Republicans, so many Independents, and so on. We cannot characterize the strength of any individual's party affiliation from this nominal scale; we can only know what that affiliation is. With nominal data, there are no shades of gray; an observation either has the trait or not. In short, then, nominal data are generated by sorting and counting—sorting the data into discrete, mutually exclusive categories and then counting the frequency of occurrence within each category. The statistical analysis of nominal data is sometimes called* categorical data analysis *(Agresti, 1990).*

CHI SQUARE AND INDEPENDENT SAMPLES

When we are setting up a nominal scale of measurement, the categories must be totally independent of each other—that is, an individual that is counted as a Democrat cannot also be counted as a Republican. All the cases in any given category must share a common trait, but they cannot share the same trait with cases from any other category. This is what is meant by the equality versus nonequality rule for the nominal scale that was mentioned in Chapter 9. All observations in a given category are equal to all other observations in that category and not equal to observations in any other category.

The 1 × k Chi Square (Goodness of Fit)

Suppose that a market researcher wishes to find out if one radio station is more popular than others among teenagers. A random sample of 100 teenagers is selected, and they are categorized on the basis of their radio station preference. The data are as follows: Station A, 40; Station B, 30; Station C, 20; and Station D, 10. These data—40, 30, 20, and 10—make up the frequency *observed*, or f_o, which is to be compared to the frequency expected, f_e. The frequency expected can be calculated either on the basis of pure chance or on the basis of some a priori hypothesis. (An a priori hypothesis is one that either is known or is presumed to be true, either because of previous research findings or, more commonly, because it is consistent with some scientific theory.) We then compare the observed data with an expected data set to determine how well these observations "fit" the expectations.

Frequency Expected Due to Chance. We first consider the effect of chance. If 100 teenagers are selected and sorted into four categories, and if only chance determines the sorting, then the result should be a frequency of 25 individuals per category. The frequency expected due to chance, then, is the total number in our sample, N, divided by the number of categories, k.

$$\text{chance } f_e = \frac{N}{k}$$

Calculating the Chi Square Value. The value of **chi square** is obtained from the following equation:

$$\chi^2 = \Sigma \frac{(f_o - f_e)^2}{f_e}$$

To illustrate the method for calculating chi square with this equation, we can use the data on radio station preference among teenagers. We construct a table of four columns (one for each category, that is, station preference) and five rows (see Table 13.1). We fill in this table according to the following steps:

1. In the first row, f_o, we place our observed data on radio station preference, that is, 40 for Station A, 30 for Station B, 20 for Station C, and 10 for Station D.
2. Our calculated value for frequency expected, f_e, is placed in each category in the second row. For the radio station survey, this value is 25.
3. The third row of the table represents the difference between the observed and expected frequency in each category, that is, $f_o - f_e$. Subtracting in each column of Table 13.1, we get 15, 5, –5, and –15 across the third row.
4. We square each difference and place these values in the fourth row. The squared differences are 225, 25, 25, and 225.

TABLE 13.1 Chi square table of radio station preference among 100 teenagers.

	Station A	Station B	Station C	Station D	
f_o	40	30	20	10	
f_e	25	25	25	25	
$f_o - f_e$	15	5	−5	−15	
$(f_o - f_e)^2$	225	25	25	225	
$(f_o - f_e)^2/f_e$	9	1	1	9	$\Sigma \frac{(f_o - f_e)^2}{f_e} = 20$

5. For the fifth row, each squared difference is divided by the value for f_e for that category. For Table 13.1, this step yields 9, 1, 1, and 9.
6. Adding the values across the fifth row gives us the summation result for chi square. Our chi square value for the radio station survey is therefore 20.

Interpreting the Chi Square Value. The chi square value tells us whether the frequency observed differs significantly from that expected. The null, or chance, hypothesis, of course, says that there is no difference.

$$H_0: f_o = f_e$$

The alternative hypothesis says there is a difference.

$$H_a: f_o \neq f_e$$

To make the statistical decision, we compare our obtained value with the critical chi square values in Appendix Table I. The table is set up with the degrees of freedom in the column at the far left and the critical chi square values for the .05 and .01 alpha error levels in the next two columns. Note that in this table as the degrees of freedom increase, the critical chi square values also increase. Unlike the *t* or *r* tables, here, as the degrees of freedom increase, it becomes more, rather than less, difficult to reject the null hypothesis.

The degrees of freedom for 1 × *k* chi square are based on $k - 1$, the number of categories minus one. Note that for the chi square the degrees of freedom are *not* based on the size of the sample.

Degrees of Freedom for the 1 × k Chi Square (1 by k)

For the 1 × *k* chi square, degrees of freedom are found on the basis of how many cell entries, not number of subjects, are free to vary after Σf_o or N is fixed. For example, if we were to place a total of 100 people into four categories, once the first three categories were filled

	A	B	C	D	
f_o	30	20	40	?	$\Sigma f_o = N = 100$

the fourth now must contain 10 people. The value allowed in that fourth category is therefore *not free to vary*. With four categories, then, only three of the values are free to vary, giving us $k - 1$, or $4 - 1 = 3$ degrees of freedom. This remains true whether N is equal to 100, 1000, or whatever. For the chi square, as for both t and r, if our obtained value is equal to or greater than the tabled value, the null hypothesis is rejected. For our radio station survey the degrees of freedom are equal to $4 - 1$, or 3. Looking this up in Table I, we find a value of 11.34 at the .01 alpha level. Since our calculated value is clearly greater than the tabled value, we reject the null hypothesis and conclude that Station A is indeed more popular among the population of teenagers from which the sample was selected. The observed frequency does differ significantly from that predicted by chance. In this case, the chi square value is large enough that the probability of a type 1 error is less than one in a hundred.

Conclusion: A 1 × k (goodness-of-fit) chi square was computed comparing the frequency of occurrence of a sample of TV viewers according to four radio station categories. A significant difference was found between the observed and expected values (chi square(3) = 20.000, $p < .01$.).

Frequency Expected Due to an A Priori Hypothesis. Sometimes a certain theory, or perhaps an existing piece of research, is used to generate the values for the expected frequency of occurrence of a trait being studied. For example, geneticists know that of the genes controlling eye color, those for brown eyes are dominant over those for blue. This means that blue-eyed parents will only have children with blue eyes. It also means that a brown-eyed parent who carries *only* genes for brown eyes will only have children with brown eyes, even if the other parent's eyes are blue. However, if a known hybrid—that is, a man with brown eyes who carries a gene for blue eyes because one of his parents had blue eyes—has children with a brown-eyed hybrid woman, the chances are 3 to 1 that each resulting child will have brown eyes. This 3-to-1 ratio, generated from Mendel's genetic theory of dominants and recessives, can be used as an a priori hypothesis.

To test this a priori hypothesis, a researcher might select a sample of parents and observe the eye color of their children. These observations could be compared with the theoretical frequencies in an attempt to assess how well the observed frequencies fit with those expected—in short, a goodness-of-fit test.

Example A sample of 40 couples is selected. In each couple both of the parents are brown-eyed hybrids. The eye color of each couple's first-born child is noted. The data are as follows: brown-eyed children, 23; blue-eyed children, 17. These values for observed frequency are then compared with the frequency of eye colors expected on the basis of the a priori, or 3-to-1, hypothesis. Of the total of 40, then, the frequency expected is that 30 of the children will have brown eyes and 10 will have blue. Now we have values for f_o and f_e and can complete the chi square table as follows:

	Brown Eyes	Blue Eyes	
f_o	23	17	
f_e	30	10	
$f_o - f_e$	−7	7	$\Sigma \frac{(f_o - f_e)^2}{f_e} = 6.533$
$(f_o - f_e)^2$	49	49	
$(f_o - f_e)^2/f_e$	1.633	4.900	

With degrees of freedom of 1, Table I shows that a value of 3.84 is needed to reject H_0 at an alpha level of .05. Since 6.533 is greater than 3.84, we can reject H_0, and state our conclusions as follows:

Conclusion: A 1 by k (goodness-of-fit) chi square was computed comparing the frequency of occurrence of eye color among children of brown-eyed hybrid parents. The frequencies observed were compared to an a priori, or 3-to-1, hypothesis. A significant difference was found between the observed and expected values (chi square(1) = 6.533, $p < .05$.).

Do the results in this example disprove the genetic theory? They certainly do not validate it, but neither do they refute it. Remember, this was post-facto research. The subjects are assigned to categories on the basis of a trait that they already possessed. It may be that the theory is wrong, or perhaps people do not always know who their real fathers are.

The r × k *Chi Square (r by k)*

The chi squares shown so far are called $1 \times k$ chi squares, meaning that one sample group of subjects has been assigned to any number, k, of categories. Sometimes, however, a researcher wishes to select more than one group—for example, an experimental and a control group—and then to compare these groups with respect to some observed frequency. For this, the $r \times k$ chi square is used.

Although the basic equation remains the same, there are two differences in the steps required to complete the $r \times k$ chi square. First, the frequency expected on the basis of chance cannot be computed in the same way. Second, the degrees of freedom are assigned in a slightly different manner.

A researcher is interested in discovering whether vitamin C aids in the prevention of influenza. Two groups are randomly selected, with 30 subjects in each group. Group A is given 250 mg of vitamin C daily for a period of three months, while Group B is given a placebo. In Group A, 10 subjects report having caught influenza during that time, while in Group B, 15 subjects report having the flu.

Contingency Table. The data are set up in what is called a contingency table, shown in Table 13.2. The cells in the contingency table are lettered a, b, c, and d and represent specific and unique categories. For example, cell a contains only those subjects who took vitamin C and also caught influenza, cell b contains only those who took vitamin

TABLE 13.2 Contingency table for vitamin C/influenza data.

	Had influenza	*Did Not Have influenza*	
Group A: vitamin C	a 10	b 20	30 = a + b
Group B: placebo	c 15	d 15	30 = c + d
	a + c = 25	b + d = 35	60 = a + b + c = *N*

C and did not catch influenza, and so on. Note that this particular contingency table has two rows and two columns; it is called a 2 × 2 contingency table. Also note that the two groups define the rows, whereas the categories on which the subjects are being measured (flu versus no flu) are heading the columns. Another way to remember this is that the independent variable (whether or not the subjects took the vitamin) is set in the rows and the dependent variable (whether or not they caught the flu) in the columns.

To the right of each row, and at the bottom of each column, we place the marginal totals: 30 and 30 for the rows and 25 and 35 for the columns. This is an important step; these marginal totals are used for calculating the values of the frequency expected. Also, as a check, we verify that the row total (30 + 30) equals the column total (25 + 35) and that *each adds up to the total* N, in this case 60.

Calculating the Chi Square Value. As we did for the 1 × *k*, we set up a table. Here, each column is headed by a particular cell. As we fill in this table, we must keep the contingency table (Table 13.2) clearly in view because some of the values are taken from it.

The value of the *r* by *k* chi square is obtained by the following steps (see Table 13.3):

1. We take the values for f_o directly from the contingency table and put them in the first row: 10 for cell a, 20 for cell b, and so on.

TABLE 13.3 An *r* × *k* chi square with data from Table 13.2.

	a	*b*	*c*	*d*	
f_o	10	20	15	15	
f_e	12.500	17.500	12.500	17.500	
$f_o - f_e$	−2.500	2.500	2.500	−2.500	
$(f_o - f_e)^2$	6.250	6.250	6.250	6.250	
$(f_o - f_e)^2/f_e$	.500	.357	.500	.357	$\chi^2 = 1.714$

2. To fill the second row, we calculate the values for f_e for all cells of the contingency table. For a given cell, f_e equals the product of its column total and its row total divided by N, the total number of cases. Cell a is in the column whose total is 25 and in the row whose total is 30. Therefore, f_e for cell a equals $(25 \times 30) \div 60$, or 12.500. By this same formula, we obtain values of f_e for b, c, and d of 17.500, 12.500, and 17.500, respectively.

3. In the third row, we place the difference between f_o and f_e. For cell a, this value is –2.500, from 10 – 12.500, and so on.

4. We square the values in the third row and place the squared differences in the fourth row, that is, 6.250 for all cells.

5. We divide each squared difference by the f_e for that cell and put the resulting values in the fifth row: .500, .357, .500, and .357.

6. We add across all the values in the fifth row to obtain the value for chi square, that is, $\chi^2 = 1.714$.

Interpreting the Chi Square Value. The degrees of freedom for the $r \times k$ chi square are found by the following equation: $\text{df} = (r - 1)(k - 1)$. The numbers of rows and columns are taken *from the contingency table.* Thus, in the vitamin C study, we had two rows (vitamin C versus no vitamin C) and two columns (influenza versus no influenza). The degrees of freedom, then, are equal to $(2 - 1)(2 - 1)$, or 1. Then

$$\chi^2_{.05(1)} = 3.84$$

$$\chi^2 = 1.714 \qquad \text{Accept } H_0\text{; not significant.}$$

Checking our obtained chi square value of 1.714 against the critical value of 3.84 for an alpha level of .05 shows that the statistical decision must be an acceptance of the null hypothesis. There are no significant differences between the two groups with respect to their frequency of catching flu. Since this was experimental methodology between-subjects, we conclude that the independent variable (vitamin C) did not affect the dependent variable (influenza). We would write the conclusions as follows: A 2 × 2 chi square was computed comparing the frequency of influenza between groups who had either received vitamin C or a placebo. The difference was found not to be significant (chi square(1) = 1.714, $p > .05$, ns.).

Degrees of Freedom and the r × k Chi Square

The $r \times k$ chi square allots degrees of freedom on the basis of how many cell entries are free to vary, *once the marginal totals are fixed.* For example, a 2 × 2 chi square, with marginal totals of 30, 70, 60, and 40, is set up as follows:

a 10	b ?	40
c ?	d ?	60
30	70	100

Once the value of cell a is entered, in this case 10, all the other cell entries are fixed. Cell b can now only have a value of 40 – 10, or 30. Similarly, cell c now must contain 20 and cell d, 40. Thus, with a 2 × 2 chi square, only the value for one cell is free to vary—producing, of course, one degree of freedom.

Variations of the Chi Square Design. The $r \times k$ chi square just shown, because the contingency table has two rows and two columns, is called a 2 × 2 chi square. Without question, this is a very popular chi square design. However, many other variations are possible. For example, in the vitamin C study, the researcher might have wished to compare three sample groups with respect to influenza. The contingency table then would appear as in Table 13.4. This is a 3 × 2 chi square, with degrees of freedom equal to (3 – 1)(2 – 1), or 2. The researcher might even have wished to make a finer discrimination on the dependent variable by setting up a 3 × 3 chi square, as shown in Table 13.5. In this case, the degrees of freedom are equal to (3 – 1)(3 – 1), or 4.

Checking the Calculated Degrees of Freedom. To be sure that the degrees of freedom for a chi square have been calculated correctly, we can perform a simple manual check using the contingency table. We merely cross out one row and one column of the contingency table and then count the cells remaining. A contingency table for a 3 × 3 chi square yields a value of 4 by this process, as shown.

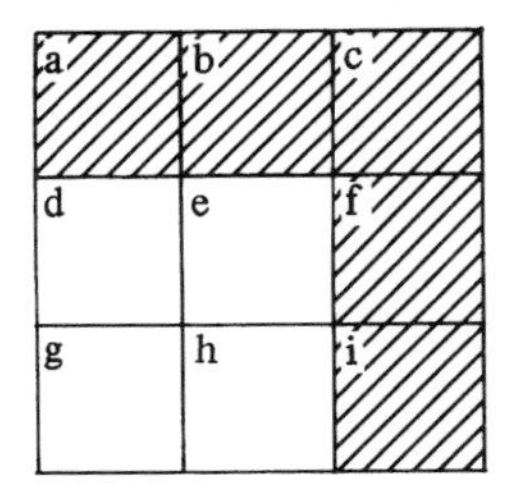

TABLE 13.4 Contingency table for a 3 × 2 chi square.

	Had influenza	*Did not have influenza*
Group A: placebo	a	b
Group B: vitamin C (low dose)	c	d
Group C: vitamin C (high dose)	e	f

TABLE 13.5 Contingency table for a 3 × 3 chi square.

	Had severe influenza	*Had mild influenza*	*Did not have influenza*
Group A: placebo	a	b	c
Group B: vitamin C (low dose)	d	e	f
Group C: vitamin C (high dose)	g	h	i

A contingency table for a 2 × 2 chi square handled in the same manner yields a value of 1 for the degrees of freedom.

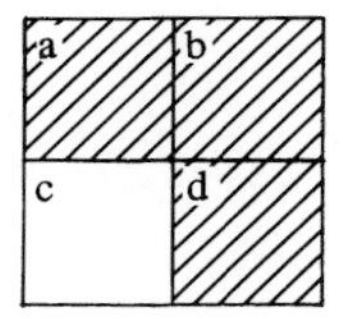

Yates Correction for the 2 × 2 Chi Square

Whenever the expected frequency within a given cell is small—that is, less than 10—the 2 × 2 chi square value may become slightly inflated. Since higher chi square values make it easier to reject the null hypothesis, then any such artificial increase in chi square may increase the alpha error. To prevent this, a correction formula has been developed that slightly lowers the chi square value. The **Yates correction** for continuity is simply this: Whenever any *expected frequency* is less than 10, subtract .50 from the absolute size of all the $|f_o - f_e|$ values. Even if only one value of f_e is less than 10, all of the differences must be reduced. This correction is only necessary when df = 1, as for a 2 × 2 chi square.

There is considerable debate concerning the value of the Yates correction (Conover, 1974; Camilli & Hopkins, 1979; Overall, 1980). One of the arguments rests on the fact that the uncorrected chi square may provide a better approximation of the true probabilities than does the value produced by the Yates. Because of this a growing number of statisticians, including David Howell, have suggested that Yates be avoided (Howell, 1992). In this instance, let your class instructor be your guide.

Example Hypothesis: Persons with high incomes are more apt to watch the 11 P.M. TV news than the 6 P.M. news. A random sample of 40 subjects is selected. The subjects are separated on the basis of whether they earn $60,000 per year; 23 report income below that level and 17 above. Of the high-income subjects, 7 indicate watching the 6 P.M. news, whereas 10 watch the 11 P.M. news. Of the low-income subjects, 15 watch the news at 6 P.M. and 8 at 11 P.M. We set up the 2 × 2 contingency table as follows:

	6 P.M. news	*11 P.M. news*	
High income	a 7	b 10	17 = a + b
Low income	c 15	d 8	23 = c + d
	a + c = 22	b + d = 18	40 = a + b + c + d = *N*

Then we calculate the chi square value, using a table format as before. The circled values of f_e trigger the Yates correction. The Yates correction stipulates the subtraction of the value .50 from the *absolute* value of each difference between f_o and f_e. Thus, when the Yates correction is applied, all these absolute difference values are *decreased*.

	a	*b*	*c*	*d*	
f_o	7	10	15	8	
f_e	(9.35)	(7.65)	12.65	10.35	
$f_o - f_e$	−2.35	2.35	2.35	−2.35	
|Yates correction| − .50	1.85	1.85	1.85	1.85	
$(f_o - f_e)^2$	3.423	3.423	3.423	3.423	
$(f_o - f_e)^2/f_e$	.366	.447	.271	.331	$\chi^2 = 1.415$

We compare our obtained value for chi square, 1.415, with the critical value from Appendix Table I, which for a 2 × 2 chi square at the .05 alpha level is equal to 3.84. Since our calculated value did not meet or beat the tabled value, we had to accept the null hypothesis. The two income groups did not differ significantly with respect to TV news viewing. This, of course, is post-facto research; the IV (income) was an assigned subject variable, rather than having been manipulated.

The conclusions would be stated as follows: A 2 × 2 chi square, using the Yates correction, was computed comparing the frequency of persons watching the 6 P.M. news with those watching the 11 P.M. news between high and low income groups, The difference was found not to be significant (chi square(1) = 1.415, $p > .05$, ns.).

Some Checkpoints When Calculating a Chi Square Value

1. The cell entries must be independent of one another.
2. The sum total of the values for f_o must equal N, the total number of observations ($\Sigma f_o = N$).
3. All the values of f_o must be whole numbers, as opposed to fractions or decimals. (With nominal data, either an event is placed in a given category or it is not. No observation can be halfway in or three-quarters of the way in a given category.)

LOCATING THE DIFFERENCE

—**Exercise 13a** A social psychologist is interested in testing the "mere presence" phenomenon, that is, that just having another person around you can affect your performance—even if that other person is neither watching nor judging you. Hiding a movie camera beside the university's jogging path, the psychologist filmed passing male joggers under three conditions: (1) a woman, seated on the grass and facing the joggers; (2) a woman, seated on the grass but with her back to the joggers; and (3) no observer at all. A sample of 300 joggers was filmed, 100 under each condition. The joggers were then categorized as to whether they (a) increased their pace, (b) remained at the same pace, or (c) decreased their pace.

	Increase	*Same*	*Decrease*	
Woman facing	a 80	b 10	c 10	100
Woman's back	d 30	e 40	f 30	100
No observer	g 33	h 33	i 34	100
	143	83	74	300

	a	*b*	*c*	*d*	*e*	*f*	*g*	*h*	*i*
f_o	80	10	10	30	40	30	33	33	34
f_e	47.667	27.667	24.667	47.667	27.667	24.667	47.667	27.667	24.667
$f_o - f_e$	32.333	−17.667	−14.667	−17.667	12.333	5.333	−14.667	5.333	9.333
$(f_o - f_e)^2$	1045.423	312.123	215.121	312.123	152.103	28.441	215.121	28.441	87.105
$(f_o - f_e)^2/f_e$	21.932	11.281	8.721	6.548	5.498	1.153	4.513	1.028	3.531

$$\chi^2 = \Sigma \frac{(f_o - f_e)^2}{f_e} = 64.205$$

With 4 degrees of freedom, we needed a tabled value of 13.28 to reject H_0 at the .01 alpha level. Our obtained chi square of 64.205 was clearly enough to reject the null hypothesis. The conclusions would be stated as follows: A 3 × 3 chi square was computed comparing the frequency of male joggers who increased the pace, kept the same pace, or decreased the pace as a function of whether a woman was watching. It was found that the joggers increased the pace when the woman was watching them. The difference was found to be significant (chi square(4) = 64.205, $p < .01$).

The decision to reject the null hypothesis means that the observed frequencies found among the three categories did not occur by chance under the assumption that all frequencies were the result of independent selection from the same population. On the basis of the chi square value, we have decided that significant differences did occur, *but the chi square value itself has not defined their precise location.* We must, therefore, go back and inspect the contingency table. We are then able to find that the major source of the difference occurs in the first row, that is, among joggers being watched. In the other two conditions, the frequencies, roughly a third in each category, are not showing divergence from the chance hypothesis. We may conclude, then, that male joggers do pick up the pace when being observed (by a woman) but that the "mere presence" of another person seems to have no apparent effect—a finding that challenges the "mere presence" hypothesis.

The 2 × 2 Chi Square: A Special Equation

A special equation has been derived to use for the 2 × 2 chi square. With this equation the tedious process of calculating values of f_e is avoided.

$$\chi^2 = \frac{N(\text{ad} - \text{bc})^2}{(\text{a} + \text{b})(\text{c} + \text{d})(\text{a} + \text{c})(\text{b} + \text{d})}$$

Using the data from the vitamin C study in Table 13.2, we have

	Had influenza	*Did not have influenza*	
Group A: vitamin C	a 10	b 20	30 = a + b
Group B: placebo	c 15	d 15	30 = c + d
	a + c = 25	b + d = 35	60 = a + b + c + d = N

$$\chi^2 = \frac{(60)[(10)(15) - (15)(20)]^2}{(30)(30)(25)(35)^*}$$

$$= \frac{(60)(150 - 300)^2}{787,500} = \frac{(60)(-150)^2}{787,500} = \frac{(60)(22,500)}{787,500}$$

$$= \frac{1,350,000}{787,500} = 1.714$$

We can make the Yates correction with a form of this equation. We correct by subtracting $N/2$ from the *absolute* value of the difference between ad and bc.

$$\chi^2 = \frac{N\left(|\text{ad} - \text{bc}| - \frac{N}{2}\right)^2}{(\text{a} + \text{b})(\text{c} + \text{d})(\text{a} + \text{c})(\text{b} + \text{d})}$$

Using the data from the TV news study earlier in this chapter, we have

	6 P.M. news	*11 P.M. news*	
High income	a 7	b 10	17 = a + b
Low income	c 15	d 8	23 = c + d
	a + c = 22	b + d = 18	40

$$\chi^2 = \frac{(40)\left[|(7)(8) - (10)(15)| - \frac{40}{2}\right]^2}{(17)(23)(22)(18)}$$

$$= \frac{(40)(|56 - 150| - 20)^2}{154,836} = \frac{(40)(94 - 20)^2}{154,836}$$

$$= \frac{(40)(74)^2}{154,836} = \frac{(40)(5476)}{154,836} = \frac{219,040}{154,836} = 1.415$$

Chi Square: A Test of Independence

Since the $r \times k$ tests whether the variables in the contingency table are independent, it can be used for testing the hypothesis of difference. Therefore, chi square can be applied to both experimental data, which are always aimed at the hypothesis of difference, and post-facto data, which sometimes test for difference.

---

*To obtain the denominator value, use your calculator to chain-multiply; that is, key in 30 times 30 times 25 times 35 and then hit the "equals" key.

CHI SQUARE AND PERCENTAGES

If the chi square has been calculated on the basis of percentages, it must be corrected to reflect the sample size. The same percentages taken from different sample sizes yield very different chi square values: the larger the sample, the larger the chi square.

To make this correction, you must take the percentage-based chi square and multiply it by $N/100$, where N is equal to the sample size. (A percentage-based chi square, of course, may have cell entries that are not whole numbers.)

For example, suppose that a $1 \times k$ chi square has been calculated based on the following percentages:

	a	*b*	*c*	*d*	
f_o	10%	20%	30%	40%	
f_e	25	25	25	25	
$f_o - f_e$	–15	–5	5	15	
$(f_o - f_e)^2$	225	25	25	225	
$(f_o - f_e)^2/f_e$	9	1	1	1	$\chi^2 = 20$

This value must then *be corrected for sample size before being checked for significance.* If the sample size had been 200, then

$$\chi^2 = \frac{20(200)}{(100)} = 40$$

Or, if the sample size had been 1000, then

$$\chi^2 = \frac{20(1000)}{(100)} = 200$$

The same procedure is used on the $r \times k$ chi square. Assume that a 2×2 chi square has been calculated on the following percentages:

a 5%	b 45%	50%
c 15%	d 35%	50%
20%	80%	100%

	a	*b*	*c*	*d*	
f_o	5	45	15	35	
f_e	10	40	10	40	
$f_o - f_e$	–5	5	5	–5	
$(f_o - f_e)^2$	25	25	25	25	
$(f_o - f_e)^2/f_e$	2.50	0.625	2.50	0.625	$\chi^2 = 6.25$

Had N been equal to 2000, then

$$\chi^2 = \frac{6.25(2000)}{(100)} = 125.00$$

Or, if the sample size had been 200, then

$$\chi^2 = \frac{6.25(200)}{(100)} = 12.50$$

Both of these values clearly lead to a reject of the null hypothesis, since $\chi^2_{.05(1)} = 3.84$. But if the same percentages had been obtained when N was equal to 60 (still a fairly large sample), then

$$\chi^2 = \frac{6.25(60)}{(100)} = 3.75$$

which, of course, would then lead to an accept of the null hypothesis. With this test, then, although sample size does not enter into the allocation of the degrees of freedom, it is absolutely crucial in determining the actual chi square value.

CHI SQUARE AND *z* SCORES

The chi square test is another Karl Pearson creation. Notice that the chi square significance values, Appendix Table I, for one degree of freedom are based on the squares of the z distribution. At the .05 significance level, the chi square value of 3.84 is equal to 1.96 squared, ±1.96 being the z score that excludes the extreme 5% of the normal curve. Also, at the .01 level, the chi square value of 6.66 is equal to the square of 2.58 and, again, z's of ±2.58 exclude the extreme 1%.

CHI SQUARE AND DEPENDENT SAMPLES

The chi square test has very few requirements; therefore, it is safe and extremely versatile. However, the *chi square does demand independent cell entries.* There can be no exceptions to this fundamental requirement. It would seem that this might be a rather severe limitation and that chi square might not be so versatile after all. For example, how can we possibly analyze nominal data when the sample groups are correlated, as they are when the same group of subjects is measured twice or when different groups are matched?

Suppose that a researcher is interested in determining whether a direct-mail campaign sells a certain product. A sample of 100 persons is selected, and they are asked whether they have the product in their homes—30 have the product and 70 do not.

The sample is then subjected to a heavy direct-mail campaign extolling the virtues of this product. Then the subjects are checked again. This time 70 have the product and only 30 do not. A four-cell contingency table for this is as follows:

	Had product	***Did not have product***
Before campaign	a 30	b 70
After campaign	c 70	d 30

200

At this point, the researcher realizes that something has gone awry. Adding all values for f_o should yield the total number of subjects, $N = 100$. Instead it yields 200. What went wrong? The basic independence rule for the chi square has been violated, and the data are simply not testable in this form. However, there is a solution to this dilemma—set up the categories on the basis of change scores. That is, instead of categorizing the same group of subjects, before and after, on whether they have the product, the subjects should be categorized on the basis of whether they change from not having the product to later having the product, or vice versa. The *change scores are independent of each other,* and the sum of the values of f_o now equal N, the sample size.

McNemar Test for Dependent Samples

The originator of this type of chi square analysis for dependent samples is Quinn McNemar, and the procedure is known as the **McNemar test.** The equation for the McNemar test is as follows:

$$\chi^2 = \frac{|a - d|^2}{a + d}$$

The data must be set up in a 2 × 2 contingency table as follows:

		Post –	+
Pre	+	a	b
	–	c	d

N of possible changes

Here, cell a contains *only* those subjects who changed from a + pre-condition to a – post-condition. Cell b contains only those subjects who did not change, remaining + both pre and post. Cell c contains only those subjects who did not change, remaining – both pre and post, and cell d contains only those subjects who were – in the pre-condition and changed to + in the post. The equation for the analysis, as we can see, concerns only the change cells, a and d.

Reworking the results of the direct-mail campaign study, we find that of the 30 who had the product before the campaign, 20 still had the product after the campaign, leaving 10 who did not. Also, of the 70 who did not have the product before the campaign, 50 had the product after the campaign, leaving 20 who did not.

		Post		
		No product (–)	*Had product (+)*	
Pre	*Had product (+)*	a 10	b 20	30
	No product (–)	c 20	d 50	70
		30	70	100

$$\chi^2 = \frac{|a - d|^2}{a + d} = \frac{|10 - 50|^2}{10 + 50} = \frac{40^2}{60} = \frac{1600}{60}$$

$$= 26.667$$

Checking our results for significance (Appendix Table I), we find that the tabled value for chi square at the .01 level and with 1 degree of freedom is 6.64, and our obtained chi square is 26.667 which is more than enough to reject the null hypothesis. Our results would be stated as follows: A 2 × 2 chi square, utilizing the McNemar test for dependent samples was computed comparing the frequency of persons having or not having a certain product as a result of an advertising campaign. It was found that significantly more persons had the product after the ad campaign than before (chi square(1) = 26.667, $p < .01$).

Therefore, the difference in change scores is significant between the pre- and post-conditions. The manipulated independent variable (direct-mail campaign) did cause a change in the direction of obtaining the product. The null hypothesis is rejected at an alpha probability level of .01.

McNemar Test and Matched-Subjects Designs

The same type of analysis can be done on matched-subjects designs. The cells of the contingency table in this case indicate the agreement (+) or the disagreement (–) between the nominal scores of the matched pairs. The agreement-disagreement match-ups are independent of each other.

		Group 2	
		Different (–)	*Same (+)*
Group 1	*Same (+)*	a	b
	Different (–)	c	d

N of possible changes

Yates Correction for the McNemar Test

Like any 2 × 2 chi square, the McNemar test may be adjusted for continuity when dealing with small cell entries. With one degree of freedom (as is the case for all 2 × 2 chi squares), the McNemar equation becomes

$$\chi^2 = \frac{(|a - d| - 1)^2}{a + d}$$

Remember that the vertical bars in this equation indicate that the value for the difference between a and d is inserted without regard to its sign. We subtract one from the absolute value of the difference between a and d, the resulting value is squared, and then it is divided by the sum of a and d.

Correlation and Nominal Data: The Coefficient of Contingency

There are a variety of nominal tests for correlation, but the most popular among social researchers is the **coefficient of contingency,** symbolized by *C*. It is a versatile test that can be used with any size of contingency table, 2 × 2, 2 × 3, 4 × 4, or whatever. Also, *C* is very easy to calculate. When the chi square value is known, *C* is only a few seconds away via this equation:

$$C = \sqrt{\frac{\chi^2}{N + \chi^2}}$$

Finally, *C* is easy to check for significance. If the calculated chi square is significant, so too is the resulting value of *C* (Appendix Table I).

Example Hypothesis: There is a significant correlation among male children between watching violent TV shows and actual overt aggression on the part of the viewer. A random sample of 100 fifth-grade, male children is selected from a certain school, and the children are asked to indicate their favorite TV show. Observers then visit the children during recess and record, for each child, whether overtly aggressive acts are exhibited. Of the 65 children who prefer violent TV shows, 50 are recorded as being overtly aggressive. Of the 35 children who prefer nonviolent TV, 10 are recorded as overtly aggressive.

	Did commit aggressive acts	***Did not commit aggressive acts***	
Preferred violent TV	a 50	b 15	65 = a + b
Preferred nonviolent TV	c 10	d 25	35 = c + d
	a + c = 60	b + d = 40	100 = a + b + c + d = *N*

We calculate the value of chi square.

	a	*b*	*c*	*d*	
f_o	50	15	10	25	
f_e	39	26	21	14	
$f_o - f_e$	11	−11	−11	11	
$(f_o - f_e)^2$	121	121	121	121	
$(f_o - f_e)^2/f_e$	3.103	4.654	5.762	8.643	$\chi^2 = 22.162$

We compare our calculated value with the critical value from Appendix Table I:

$\chi^2_{.01(1)} = 6.64$

$\chi^2 = 22.162$ Reject H_0; significant at $p < .01$.

Since the chi square value is significant, we can use that value in the *C* equation. No matter what value we now get for *C*, it must also be significant.

$$C = \sqrt{\frac{\chi^2}{N + \chi^2}}$$

$$= \sqrt{\frac{22.162}{122.162}} = \sqrt{.181} = .425$$

$= .425$ Reject H_0; significant at $p < .01$.

Therefore, there is a significant correlation between watching violent TV and actual overt aggression on the part of the young male viewer. Our conclusions would be stated as follows: A 2 × 2 chi square was computed comparing the frequency of aggressive acts among children who preferred watching violent or non-violent TV. The chi square was found to be significant (chi square(1) = 22.162, $p < .01$). The chi-square was then used to test for correlation and the C (Contingency Coefficient) was found to be equal to .425.

This is an example of correlational, post-facto research. No conclusions as to causation are permissible from the results of this study. Perhaps watching violent TV does cause aggressive behavior, or perhaps aggressive behavior causes one to watch violence on TV. Finally, there may be a third factor, variable X, that causes both overt aggression and preference for TV violence.

Limitations of the Coefficient of Contingency. The resulting correlation of .425 in the last example may seem to be somewhat low, especially in light of the rather dramatic differences between the two groups of TV viewers. The value of C was low, far lower than a Pearson r would have been had we been able to obtain interval measures on the subjects. The fact is that the C correlation calculated for a 2×2 chi square can only, at best, reach a value of .87, not 1.00. Because it is a nonparametric method of determining correlation, C is a much less powerful test than the parametric test, r. All nonparametric tests underestimate the degree of differences and correlations, and the C test is no exception.

The maximum attainable values for C are determined by the number of cells in the contingency table. (See Table 13.6.) With a 2×2 chi square (four cells), the maximum value of C is .87. Therefore, the previously calculated C of .425 is reflecting a stronger association than might have appeared at first.

Effect Size

For the chi square, one measure of effect size is based on the coefficient of contingency, which, as a measure of correlation, may also be interpreted as an effect size. Thus, if the chi square value were 22.162, as it was in the previous example, the effect size would be the C value of .425, a fairly strong effect. C values of .10, .25, and .40 are said to be small, medium, and large effect sizes, respectively.

Example A researcher is interested in discovering if a prison treatment program for substance abuse has any impact on recidivism. A total of 200 men were selected and

TABLE 13.6 Maximum values of the coefficient of contingency.

Number of cells	4	5	6	7	8
Maximum value of C	.87	.89	.91	.93	.94

randomly assigned to treatment or nontreatment conditions. Half the men went into treatment and the other half did not. After their release, the men were checked to find out if they were reincarcerated within a three-year period. The results follow:

	Incarcerated Again			
	Yes	***No***	***Total***	
Treatment program	30	70	100	
No treatment	62	38	100	Chi Square = 20.612

$$C = \sqrt{\frac{\chi^2}{N + \chi^2}} = \sqrt{\frac{20.612}{200 + 20.612}} = \sqrt{\frac{20.612}{220.612}} = \sqrt{.093}$$

C = .305 (a medium-to-strong effect)

For the following, calculate the effect size (ES)

Chi Square = 15, $N = 50$

ANS: ES or $C = .480$ (strong effect)

Chi Square = 3.84, $N = 55$

ANS: ES or $C = .255$ (medium effect)

Chi Square = 4.95, $N = 500$

ANS: ES or $C = .10$ (small effect)

REQUIREMENTS FOR USING THE CHI SQUARE

The requirements for using the chi square are as follows:

1. The samples must have been randomly selected.
2. The data must be in nominal form.
3. There must be independent cell entries.
4. No value for expected frequency should be less than 5.

The last requirement is only of interest when the contingency table has just four cells. With larger tables, as in a 4×5 table, a few values of less than 5 are permissible (Siegel & Castellan, 1988). Some statisticians even argue that the *5-rule* may be too stringent in smaller contingency tables (Everitt, 1977).

There's Always Chi Square. Although as a nonparametric test, chi square is not the world's most powerful statistical test, it certainly is one of the most popular. It is an extremely safe test to use because it requires so few assumptions. It really is a workhorse test for researchers. Distributions can be skewed, variability can be dramatically different for two samples, the intervals between successive scale points can vary, and the researcher can relax—because if nothing else works, one can usually use chi square.

SUMMARY

Nominal data are those that have been identified on the basis of equality versus nonequality. Observations are sorted into mutually exclusive categories and then counted to get the frequency of occurrence within each category. The categories must be completely independent of one another, that is, an observation sorted into one category cannot also be placed in another category.

The statistical test spotlighted in this chapter is the chi square, a test of whether the observed frequency of occurrence differs significantly from the frequency expected on the basis of chance or on the basis of some a priori hypothesis. The $1 \times k$ chi square is applicable if one sample group is sorted into any number of categories. The $r \times k$ chi square is applied when two or more sample groups are sorted into any number of categories.

An extremely common form of the $r \times k$ is the 2×2 chi square (two sample groups sorted into two categories), and a special computational equation has been derived for this particular situation. Whenever any of the values of f_e for the 2×2 chi square are less than 10, the Yates correction is typically applied to the basic chi square equation.

Since one of the limitations of the chi square is that it only can be used when all cell entries are independent of each other, the testing of correlated samples requires a special technique. This is the McNemar test and to apply it, only the change scores are used in the analysis.

Although the chi square itself only tests the hypothesis of difference, variations on the chi square theme have been developed for assessing the possibility of a correlation. The test of correlation covered in this chapter is the coefficient of contingency, C, which can be applied to any number of independent cells, 2×2, 3×4, and so on. The value of C may also be used to assess effect size.

Chi square, since it makes no assumptions regarding either the population mean, μ, or the shape of the underlying distribution, is called a nonparametric test. Also, for calculating the degrees of freedom for the chi square, the size of the sample is irrelevant; the only information that is required is the number of sample groups and the number of categories.

Key Terms chi square
coefficient of contingency
McNemar test
nominal data
Yates correction

PROBLEMS

1. A researcher is interested in whether a significant trend exists regarding the popularity of certain work shifts among police officers. A random sample of 60 uniformed police officers is selected from a large metropolitan police force. The officers are asked to indicate which of three work shifts they prefer. The resulting data show that 40 officers prefer the first shift, 10 the second shift, and 10 the third. Do the results deviate significantly from what would be expected due to chance?
2. A sample of 48 college students is randomly selected, and the students are asked to indicate their attitudes toward the statement, "The United States military should invade the island of Bermuda." Of the students polled, 12 agreed with the statement, 12 had no opinion, and 24 disagreed. Do these results deviate significantly from what would be expected due to chance?
3. A group of 90 economists are randomly chosen from members of the profession in this country. They are asked to predict whether the prime lending rate (the interest charged by major banks to their best customers) six months from now will be higher, the same, or lower than it is today. Of 90 economists, 30 said "higher," 35 said "the same," and 25 said "lower." Do these results deviate significantly from what would be expected due to chance?
4. A dog-racing enthusiast wonders whether there is any systematic bias connected with the lane in which the dogs run. Results of the first race on a given night are checked at 108 dog tracks selected randomly from throughout the country. The data indicate that the number of winning dogs is 26 for lane one, 17 for lane two, 16 for lane three, 15 for lane four, 17 for lane five, and 17 for lane six. Check the hypothesis that dog-track wins are a function of the lane.
5. Two random samples have been selected. One sample consists of 220 Catholics and the other contains 200 Protestants. All subjects were then asked to indicate their attitudes toward voluntary birth control. The data are as follows:

	Favor	*Oppose*
Catholics	70	150
Protestants	120	80

Do the two groups differ regarding their attitudes toward birth control?

6. A commonly heard, and perhaps misleading, statistic is that 50% of Americans contracting lung cancer are nonsmokers, which on the surface seems to indicate that lung-cancer chances are 50-50, regardless of whether one smokes (*New York Times*, March 16, 1996). However, breaking the numbers down, we find that only 30% of Americans are smokers. Thus, out of 100 persons, 70 don't smoke and 30 do. Assume that the same absolute number from each group contract the disease, let's say 15 from each group. Is the risk equal for smokers and non-smokers? As a chi square,

	Cancer	*No Cancer*	*Total*
Smokers	15	15	30
Nonsmokers	15	55	70

7. Some researchers, such as Michael Eysenck (son of the famous English psychologist, Hans Eysenck), have concluded that there are cognitive biases among anxiety-disordered patients. In order to test part of this theory, researchers made the prediction that anxious individuals have more memories of defeat than of victory. A group of senior college varsity football players was selected and both given a test of trait anxiety and asked to recall the most vivid memories of their high school football careers. The responses were categorized as to whether they represented moments of personal triumph or defeat. A random sample of 200 players was selected from a specific southern conference, and their responses were as follows:

	Personal Triumph	*Personal Defeat*	*Total*
High anxious	45	55	100
Low anxious	60	40	100

Test the hypothesis that high- and low-anxious athletes will differ in their memories of triumph and defeat.

8. A random sample of 150 persons was selected from the membership rolls of the National Rifle Association, and another random sample of 150 persons was selected from the list of contributors to the "Save the Whales" foundation. The two populations do not overlap. All subjects were asked to indicate their attitudes toward handgun control. The data are as follows:

	Oppose	*Favor*
NRA group	90	60
"Save the Whales" group	50	100

Do the two groups differ regarding their attitudes toward handgun control?

9. Hypothesis: Union membership differs on the basis of sex. A random sample of 100 adult workers, 60 men and 40 women, is selected in a large midwestern city. Of the group, 42 men and 17 women are union members.

	Union members	Not union members	
Men	42	18	60
Women	17	23	40
			100

a. Test the hypothesis.
b. Indicate the type of research (experimental or post-facto).
c. State the independent and dependent variables.

10. A researcher suspects there are differences in reading problems as a function of gender among first-grade children. A random sample of 100 first-graders is selected—50 boys and 50 girls. Of the boys, 25 are found to have reading problems and 25 are found to have no reading problems. Of the girls, 10 are found to have reading problems and 40 to have no reading problems. The contingency table is as follows:

	Reading problems	No reading problems	
Boys	a 25	b 25	50
Girls	c 10	d 40	50
	35	65	100

Test the hypothesis.

11. A researcher hypothesizes that there are differences in personality type as a function of body type among adult males. A random sample of 200 men is selected and categorized as to body type. Of the sample selected, 50 are found to be ectomorphs (underweight and small-boned), 60 are found to be mesomorphs (muscular and normal weight), and 90 are found to be endomorphs (overweight). Of the ectomorphs, 20 are diagnosed as having extraverted personalities and 30 as introverted. Of the mesomorphs, 32 are diagnosed as extraverted and 28 as introverted. Of the endomorphs, 70 are diagnosed as extraverted and 20 as introverted. The contingency table is as follows:

	Extravert	Introvert	
Ectomorph	a 20	b 30	50
Mesomorph	c 32	d 28	60
Endomorph	e 70	f 20	90
	122	78	200

Test the hypothesis.

12. A major pharmaceutical company is testing the possible effectiveness of a certain new antiarthritic drug. A sample of 50 arthritic patients is randomly selected and divided into two groups of 25 each. Group A is treated with pills containing the new drug, and Group B is given a placebo (a pill that appears identical to the drug but is nonactive). After three months of treatment, the two groups are compared. Of the patients in Group A, 15 report symptom "improvement," and 10 report "no improvement." Of the patients in Group B, 12 report "improvement," and 13 report "no improvement." Test the hypothesis that the new drug produces significantly higher rates of reported symptom improvement.
13. Hypothesis: More Republicans than either Democrats or Independents own their own homes. A random sample of 71 city residents is selected in a large eastern city. Of these, 23 indicate that they are Republicans, 25 that they are Independents, and 23 that they are Democrats. Of the Republicans, 15 are homeowners and 8 are not. Of the Independents, 10 are homeowners and 15 are not. Of the Democrats, 8 are homeowners and 15 are not. Test the hypothesis.
14. Hypothesis: There is a significant difference between students living in dorms and students who commute regarding whether or not they receive financial aid. A random sample of 55 university students is selected—15 commuters and 40 dorm students. Of the commuters, 5 are receiving university aid and 10 are not. Of the dorm students, 30 are receiving financial aid, and 10 are not. Test the hypothesis.
15. Hypothesis: A weekend seminar in assertiveness training increases the assertiveness of the female participants. A random sample of 90 female university students is selected. Each woman is first evaluated for assertiveness by specially trained interviewers. In that pre-test, 25 are judged to be assertive and 65 nonassertive. All the women then attend a weekend seminar in assertiveness training. At the end of the weekend, each woman is again interviewed and evaluated. In the post-condition, 60 of the women are judged to be assertive and 30 nonassertive. The data table, with marginal totals, follows:

		Post: Nonassertive	Post: Assertive	
Pre	Assertive	10	15	25
	Nonassertive	20	45	65
		30	60	90

Test the hypothesis.

16. A researcher is interested in establishing whether a correlation exists between a student's choice of college major and qualification for the academic honor roll. A random sample of 80 university seniors is selected and categorized on the basis of college major. Of the students selected, 30 are majoring in liberal arts, 30 are in business, and 20 are in education. Of the liberal arts majors, 10 are on the honor roll; of the business majors, 15 are on the honor roll; of the education majors, 10 are on the honor roll. The data table follows:

	Honors	No honors	
Liberal arts	10	20	30
Business	15	15	30
Education	10	10	20
			80

a. Calculate the chi square.
b. If appropriate, calculate the correlation.

17. A researcher is examining the possibility of a correlation existing between participation in varsity athletics and the incidence of personal and/or emotional problems among male university students. Random samples of 50 varsity athletes and 50 nonathletes are selected from among the male population of several large midwestern universities. The incidence of personal problems is gauged by asking each student whether he had ever visited the university counseling center. Of the athletes, 10 acknowledged that they had paid at least one visit to the counseling center. Among the nonathletes, 22 said they had gone to the counseling center at least once. Test the correlation.

18. A social psychologist, in an attempt to test part of Bandura's social-learning hypothesis, wanted to find out whether helping behavior could be affected by modeling. The suggestion was that motorists might be more likely to help a woman in distress if they had just passed a helping scene on the highway than if they hadn't. A total of 177 motorists were observed, 96 of whom drove past an arranged scene where a male driver had stopped to change a tire for a female driver, and 81 of whom had not witnessed such a helping scene. A few miles down the road another scene was arranged where a second female would be found on the side of the road attempting to change a tire. The researcher counted the number of drivers who attempted to help the second woman.

	Stopped to help	***Did not help***
Saw the helping scene	58	38
Did not see the helping scene	37	44

a. Is there a significant difference in the numbers who stopped to help?
b. What other variables or controls might the researcher consider before drawing any conclusions?

19. A researcher wished to find out whether there was a significant difference between younger and older subjects with respect to whether they would return valuable "lost letters" (Gabor & Tonia, 1989). As part of the study, 56 stamped, self-addressed envelopes were set up so that each contained a penny in an authentic coin holder and a letter stating the coin was worth over $150. The envelopes were unsealed, and the letters were typed on what seemed to be the official letterhead of "The Coin Collector's Association of Canada." The letters were then placed under the windshield wipers of cars that had been seen arriving and parking at various shopping malls throughout the city of Ottawa. Only solitary drivers were selected, half of them appearing to the researchers to be under 30 years of age and half over 30. Also, for each age group the sample was equally divided between men and women. A scrawled note that read "Found near your car" was also placed with the envelope on the windshield. The data were as follows:

	Returned	***Not returned***	***Total***
Younger	19	9	28
Older	20	8	28
			56

Test the hypothesis that there was an age difference regarding the return of the "lost letters."

20. A study was done to determine if suicide attempts are more likely among patients with bipolar disorder comorbid with alcoholism than among patients who were bipolar alone (Potash et al., 2000). A sample of 337 psychiatric patients was selected, all of whom had been diagnosed as having a bipolar disorder. The patients were divided into two groups. Group A was both bipolar and alcoholic, while Group B was bipolar only. All the patients were checked for lifetime suicide attempts. The data follow:

	Attempted Suicide	*Did Not Attempt Suicide*
Group A	40	86
Group B	37	174

Is there a significant difference in suicide attempts between the two groups?

21. A large number of comparable cases tried in criminal court were examined and categorized on the basis of the jury size and jury verdict. The data were as follows:

	Acquittals	*Convictions*	*Hung Juries*
12 member juries	40	60	10
6 member juries	30	55	8

Are there differences in jury outcome as a function of jury size? Test the hypothesis.

22. A researcher is interested in discovering if a prison treatment program for substance abuse has any impact on recidivism. A total sample of 100 men was selected and randomly assigned to either a treatment or no-treatment group, 50 men in each group. After their release, the men were checked to find out if they were reincarcerated within a three-year time frame. The results follow:

	Incarcerated Again?	
	Yes	*No*
Treatment Program	20	30
No Treatment	40	10

a. Is this experimental or post-facto research?
b. Identify both the IV and DV
c. Test the hypothesis and state your conclusions.

23. Researchers have long been wondering about the deterrent effect of various forms of treatments for crimes of violence. In one study, conducted in Minneapolis, police officers were instructed to use one of three different methods for handling the violence of domestic disputes that did not involve life-threatening injuries (Sherman & Berk, 1984). In each case the wife or the female live-in companion contacted the police. The methods used were (1) arrest, (2) advice and counseling, or (3) ordering the offender off the premises for at least eight hours. The choice of method was randomly assigned, since it was felt that if the tactics used had been optional to the officers, they might be more likely to use arrest in the more severe cases, or they might be characteristically more prone to using one technique rather than another, thus biasing the results. All the offenders were then followed through police reports for the next six months, and data were gathered regarding repeated offenses. The data were as follows:

	Repeat	*Did Not Repeat*	*Totals*
Arrest	25	59	84
Counseling	40	43	83
Order off Premises	55	28	83

Test the hypothesis that when a man has a few hours of real incarceration, it lowers a woman's risk of being battered again. It might seem on the surface that even if the results are significant, then of course the arrested offenders were less likely to repeat, simply because they were in jail and did not have the opportunity for further harassment. However, if significance is obtained, then the deterrent effect of arrest was in this instance very real, since of the 84 arrested men the arrest was accompanied by only a very brief incarceration, typically lasting only a few hours.

24. When data are in the form of frequencies of occurrence within mutually exclusive categories, which scale of measurement is being used?
25. How is the null hypothesis for the chi square test stated?
26. Under what condition(s) should the Yates correction be used?
27. For which experimental design(s) is the McNemar test an appropriate analysis of nominal data?

True or False. Indicate either T or F for problems 28 through 32.

28. Chi square may be used only with nominal data.
29. Chi square can be used only if the frequency expected value of all cell entries is at least 5.
30. An a priori hypothesis always states that the frequency expected must be the same in every cell.
31. Chi square is a nonparametric statistical test.
32. To test for "goodness of fit," a $1 \times k$ chi square may be performed.

COMPUTER PROBLEMS

C1. Sports physiologists have hypothesized that among skiers a lack of knee strength is a major factor in predicting knee injuries. A random sample of 623 skiers was selected and tested for knee strength on an isokinetic testing device. Of these, 491 showed enough knee strength as to be judged adequate for skiing, while 132 demonstrated deficient strength. At the end of the ski season, 14 of the skiers who had passed the test did have knee injuries, while of those who failed the test, 45 had knee injuries. Test the hypothesis and state all conclusions.

	Injured	Not injured	
Passed the test	14	477	491
Failed the test	45	87	132
	59	564	623

C2. Researchers at the Veterans Studies Center in New Haven, Connecticut, wondered whether there was any difference in the rate of later heart attacks among heart patients based on whether they received bypass surgery or were treated medically with drug therapy. A total of 803 patients were recruited, all of whom suffered from stable angina (chest pains that result when narrowed arteries fail to provide enough blood to the heart). One group of patients had bypass surgery (where the surgeon uses veins from elsewhere in the body to replace the narrowed arteries), and the other group was treated with drugs. All the patients were then monitored and the number who had subsequent heart attacks was tallied. The results were as follows (adapted from the American Heart Association's journal, *Circulation,* March 1991):

	Heart attack	No heart attack
Surgery	141	202
Drug therapy	117	343

Test the hypothesis that there is a difference in the number of later heart attacks based on the mode of therapy.

C3. Continuing computer problem 2, the researchers then wanted to find whether, among those patients who did have heart attacks, there was a difference in the number of *fatal* attacks, depending on which mode of therapy had been provided. Of the surgery group, 25 of the patients suffered fatal heart attacks, whereas in the drug-therapy group, 33 had fatal attacks.

	Fatal attacks	Nonfatal attacks
Surgery	25	116
Drug therapy	33	84

Test the hypothesis that there is a difference in the risk of death from heart attacks, depending on the treatment provided.

C4. To assess whether time of day might be a factor in assaults on law enforcement officers, data were collected from the state of Utah for the time period 1981–1990. The number of police officers assaulted during that time was 2370 (*Utah Law Enforcement Statistics*, 1992). By time of day, the data were as follows:

Time of day	*10 P.M.–2 A.M.*	*2 A.M.–6 A.M.*	*6 A.M.–10 A.M.*	*10 A.M.–2 P.M.*	*2 P.M.–6 P.M.*	*6 P.M.–10 P.M.*
N of assaults	965	320	123	171	273	518

Test the hypothesis that assaults on officers differ as a function of time of day.

C5. A study was done to determine if the number of murders is in any way connected to the day of the week on which the crime was committed. Data were collected, again from the state of Utah, and the following results were recorded. For the period 1978–1990, there was a total of 665 murders, categorized into the seven separate days (*Utah Law Enforcement Statistics*, 1992).

Day of Week	*Monday*	*Tuesday*	*Wednesday*	*Thursday*	*Friday*	*Saturday*	*Sunday*
N of murders	74	97	94	83	106	101	110

Test the hypothesis that the number of murders differs as a function of day of the week.

C6. Researchers wished to find out if early referral to the juvenile justice system in any way relates to the odds of that juvenile later being sent to adult prison. A random sample of 476 cases of juveniles adjudicated delinquent in a certain Pennsylvania county was selected from the court records. Of the 243 juveniles adjudicated delinquent on their first referral to juvenile justice, a total of 51 went on to adult prison after the age of 18. Of the 233 juveniles not taken to juvenile court on their first referral, 99 were imprisoned in adult life after the age of 18 (Brown et al., 1991).

At First Referral	*No Prison*	*Prison*	*Totals*
Case adjudicated	192	51	243
Case not adjudicated	134	99	233
Total	326	150	476

a. Determine if the difference is significant.
b. State all conclusions.

C7. The Greenwich Laboratories wished to test their then new antiarthritic drug Therafectin, and their researchers selected a sample of 200 arthritis sufferers and randomly divided them into two groups. Group A was treated with Therafectin and Group B was given a placebo. Twelve weeks later, all subjects were tested for possible symptom improvement. The data were as follows (from the September 1989 issue of *Internal Medicine*):

	Improvement	No improvement
Therafectin	41	59
Placebo	21	79

Test the hypothesis that Therafectin leads to arthritic symptom relief.

UNIT

III

Advanced Topics in Inferential Statistics

Chapter

14

Regression Analysis

One of the major themes of this book, recurring again and again like the cadence of a marching drumbeat, is that correlation in and of itself cannot be used to assign causation. Correlation does not rule out causation; however, it cannot guarantee it. What a correlational technique can do, however, is allow the researcher to make better-than-chance predictions. Regression analysis provides one method for making such predictions and even, at times, for pointing a finger of suspicion in the direction of possible causation.

When using the correlation for purposes of prediction, the goal is to estimate one variable on the basis of the information contained in one or more other variables. The variable being predicted is called the criterion variable, and the variable(s) from which the prediction is being made is (are) called the predictor variable(s).

REGRESSION OF *Y* ON *X*

Throughout this chapter the focus is on predicting a value of Y having been given the value of X. This is called the linear regression of Y on X, and it attempts to utilize the correlation between X and Y to make very specific predictions of the Y value. The correlation between X and Y specifies how much information about Y is contained in X. If the correlation is a perfect +1.00, then X contains all there is to know about Y. A correlation of zero, on the other hand, indicates that X tells us nothing at all about Y. Thus, the higher the correlation, the more information about Y is contained in the X score. The variable whose value is known is called the independent variable, whereas the variable whose value is to be determined from the formula is called the dependent variable.

Bivariate Scatter Plot

In Chapter 11, a discussion of the scatter plot was presented. We learned then that a scatter plot is a graphic format in which *each data point represents a pair of scores*—a measurement on X as well as a measurement on Y. A bivariate, or two-variable, scatter plot is shown in Figure 14.1. The points in the scatter plot are arrayed from lower left to upper right, indicating a positive correlation. Also, the points form the elliptical shape that is a result of the central tendency occurring within the two distributions of X and Y scores.

Regression Line. Three straight lines are drawn on the plot in Figure 14.1. Line A is close to the points along the top of the array of points but is relatively far from those along the bottom. Similarly, line C, although near the points along the bottom, is a considerable distance from those along the top. However, line B is fairly close to *all* the points in the scatter plot. For this reason, line B is called the prediction line or, more often, the **regression line** of Y on X. The regression line is therefore the single straight line that lies closest to all the data points in the scatter plot, or the line that passes through what is called the *centroid* of the scatter plot. Although there are some exceptions (Loh, 1987), the general rule is that the Pearson r typically measures how tightly the points in a scatter plot cluster around the regression line.

To make predictions, three important facts about the regression line must be known: (1) the extent of the scatter around the line, (2) the slope of the line, and (3) the point where the line crosses the Y axis.

Extent of the Scatter Around the Regression Line. The closer the points on the scatter plot cluster around the regression line, the higher is the resulting correlation between X and Y, *and the more accurate is the resulting prediction.*

Figure 14.2 shows two scatter plots, both of which illustrate positive correlations. Figure 14.2b, however, portrays a much higher correlation—the points do not deviate at all from the regression line. In this scatter plot, the correlation is a perfect 1.00; each point that has a higher value on the X variable is also higher on Y. With a correlation of 1.00, there are no exceptions, that is, no reversals. Figure 14.2a, on the other hand,

FIGURE 14.1 A bivariate scatter plot.

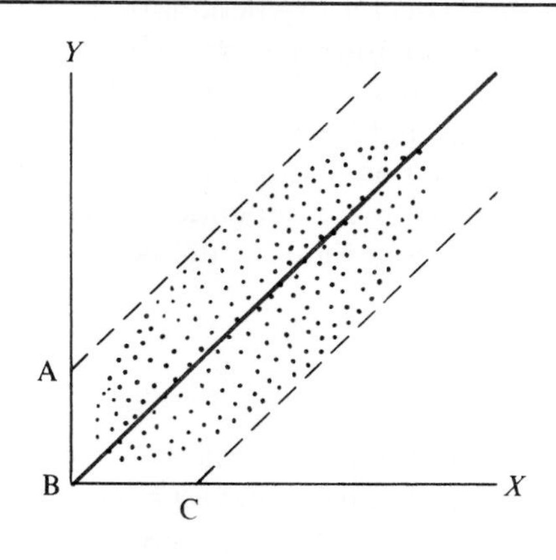

FIGURE 14.2 Scatter plots showing positive correlations.

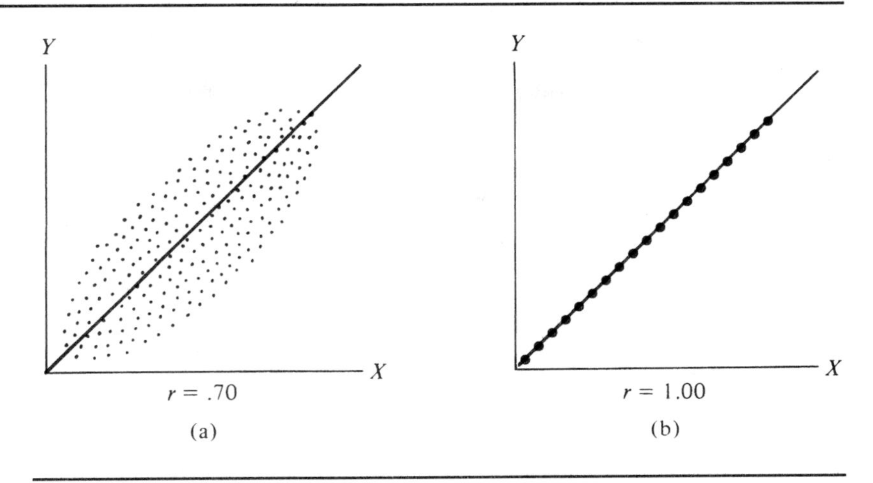

is more typical of real research in the social sciences. In this scatter plot, although there is still a positive trend to the correlation, the scattering of points around the regression line shows that the correlation is not perfect. Here, while in general the high scores match with high scores and the low scores with low scores, there are numerous exceptions. In short, then, the higher the correlation, the closer the scatter points cluster around the regression line. Also, the higher the correlation, the more accurate is the prediction because more information about Y is being carried in X.

Slope of the Regression Line. The manner in which the regression line tips, or slopes, greatly affects the prediction. For example, the regression lines in Figure 14.3a

FIGURE 14.3 Regression lines showing (a) positive and (b) negative correlations.

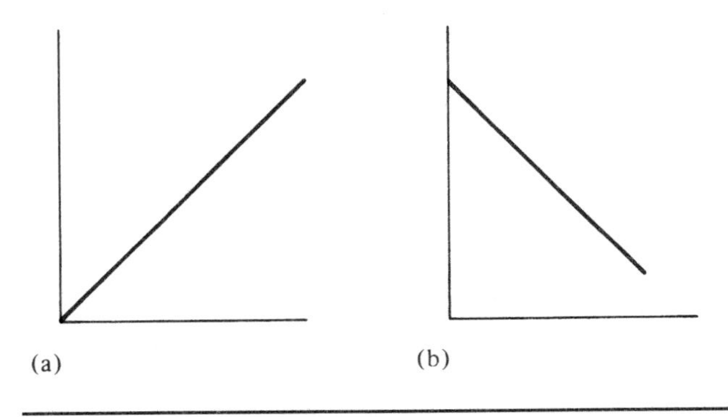

FIGURE 14.4 Regression lines with different positive slopes.

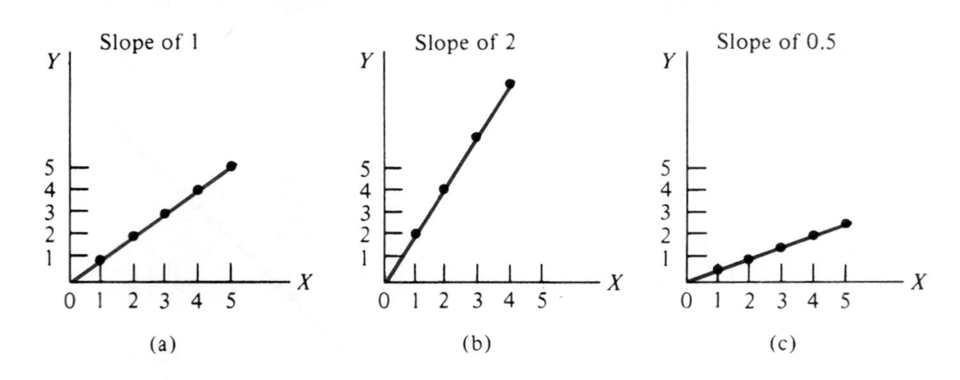

(going from lower left to upper right) and in Figure 14.3b (going from upper left to lower right) demonstrate entirely different correlations. The line in Figure 14.3a portrays a positive correlation. Here, for a high score on *X*, we would predict a similarly high score on *Y*. The line in Figure 14.3b, however, shows a negative correlation. A high score on *X* would lead to a prediction of a low score on *Y*. Therefore, the line in Figure 14.3a has a positive slope and the line in Figure 14.3b a negative slope.

Knowing the sign of the slope, however, is not enough. Figure 14.4 shows three regression lines with positive slopes. These would yield very different predictions. The degree of the slope is determined by the amount of change in *Y* that accompanies a given unit change in *X*. Figure 14.4a shows a line with a slope of 1.00; *Y* increases by one unit for each single unit increase in *X*. The line in Figure 14.4b is much steeper. The slope is 2.00; that is, *Y* increases by two units for each single unit increase in *X*. Finally, the line in Figure 14.4c has a slope of one-half, or .50. Here, *Y* only increases by half of a unit for each full unit increase in *X*.

Point Where the Regression Line Crosses the *Y* Axis. Finally, to make a prediction we must know precisely where the regression line crosses the ordinate. That is, we must establish the value of *Y* when *X* is equal to zero. This is called the **point of intercept,** or the *Y* intercept. Figure 14.5 shows two possible situations.

In Figure 14.5a, the regression line meets the ordinate at zero. In this situation, where *X* equals zero, so does *Y*. In Figure 14.5b, however, the regression line intercepts the ordinate at a value of 2 (*Y* equals 2, when *X* equals 0). Note, in these two examples, that both lines have the same slope, despite their different points of intercept.

Regression Equation

To draw the graph of the regression line, the following equation is used:

$$Y_{\text{pred}} = bX + a$$

FIGURE 14.5 Regression lines with the same positive slope but different Y intercepts.

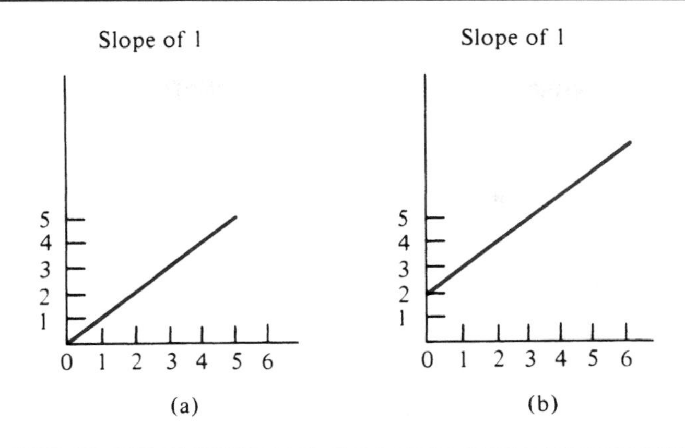

In this equation, Y_{pred} is the predicted value of Y; b is the slope of the regression line; X is a known value, from which Y_{pred} is to be calculated; and a is the point of intercept.

The equation for the b term (slope) is

$$b = \frac{rs_y}{s_x}$$

The equation for the a term (intercept) is

$$a = M_y - bM_x$$

Thus, since

$$Y_{\text{pred}} = bX + a$$

then

$$Y_{\text{pred}} = \frac{rs_y}{s_x} X + M_y - \left(\frac{rs_y}{s_x}\right)(M_x)$$

Rearranging terms, we write

$$Y_{\text{pred}} = \frac{rs_y}{s_x} X - \left(\frac{rs_y}{s_x}\right)(M_x) + M_y$$

This equation is much easier to work out than it may appear. For one thing, the rs_y/s_x, or slope, term appears twice, but we only have to calculate it once.

Example A correlation between hours per week spent studying and grade-point average was found in Chapter 11 to be a significant .989. The mean hours of study time were 22.857, with an estimated population standard deviation of 13.184. The mean grade-point average was 2.714 with an estimated population standard deviation of .728. Predict the grade-point averages for students who spend the following weekly study hours: 37, 22, and 8.

1. Set up the problem by clearly separating the X and Y values. We are given the hours of study time, making study time the X distribution, and we are asked to predict grade-point average, making grade-point average the Y distribution.

Study Hours, X	*Grade Average, Y*
$M_x = 22.857$	$M_y = 2.714$
$s_x = 13.184$	$s_y = .728$
$X_1 = 37$	$Y_{pred} = ?$
$X_2 = 22$	$Y_{pred} = ?$
$X_3 = 8$	$Y_{pred} = ?$

2. We plug all values except the specific values of X into the regression equation.

$$Y_{pred} = \frac{rs_y}{s_x} X - \frac{rs_y}{s_x} M_x M_y$$

$$= \frac{(.989)(.728)}{13.184} X - \frac{(.989)(.728)}{13.184}(22.857) + 2.714$$

$$= .055X - (.055)(22.857) + 2.714$$

$$= .055X + 1.457$$

To add two numbers with unlike signs, take the difference between the two quantities and affix the sign of the larger quantity.

$$Y_{pred} = .055X + 1.457$$

Slope $= .055$

Y intercept $= +1.457$

Stop for a moment and look at the resulting values carefully. Remembering our definitions, the slope value, which is based on how much change in Y we can expect for a given unit change in X, is now telling us that the grade average will *increase* (since the correlation is positive) by .055 for each single hour increase in study time. The intercept, which is the value of Y when X equals zero, is indicating that a grade average of 1.457 will accompany a study time value of zero hours.

3. Using the elements of the regression equation, we set up a regression table, now including the specific values of X. Multiply the X values by the slope (in this case, .055) and then add the intercept (in this case, +1.457).

X	.055X	.055X + 1.457 = Y_{pred}
37	2.035	3.492
22	1.210	2.667
8	.440	1.897

The predicted grade-point averages are 3.492 for 37 hours of study time, 2.667 for 22 hours, and 1.897 for 8 hours.

Now try making a prediction for someone who studied 9 hours (just one hour more than the student above whose GPA was solved for 8 hours). Multiply 9 by the slope value of .055 and add in the intercept of 1.457 and we get a predicted GPA of 1.952. Compare this prediction with the GPA of 1.897 that we got from the 8 hours, and you'll find that they differ by exactly .055, or by the value found for the slope. This reminds us that since the slope is telling us how much of a change in Y we can expect for a given unit change in X, when we changed X by a single unit, from 8 hours to 9 hours, the resulting prediction changed by precisely the value of the slope.

We hold the X values out of the regression equation initially so that when we set up the regression table, we can use a long list of X values to predict the corresponding Y values very quickly, without having to resolve the equation each time. These predicted Y values are point estimates of population parameters, since they are based on sample data.

When both the X and Y values are known, we can easily check how much discrepancy there is between the actual values of Y and their predicted values. This deviation, between the known values of Y and the regression line, is known in correlational analysis as the residual error.* The regression line has sometimes been called the line of best fit and it displays the best possible equation for predicting scores on the DV.

Using the Regression When $\rho = 0$

We can now use the regression equation to predict how future members of a population might score; but, to do so, the *correlation must be significant*. That is, the null hypothesis ($\rho = 0$) must be rejected. If the correlation in the population between X and Y is zero, then the only prediction we can make about the Y variable is to predict the *mean* of the Y distribution. This is because of the effect of central tendency, or the fact that most scores do cluster around the mean. Therefore, with a correlation of zero (the null hypothesis accepted) or with no knowledge of the correlation value, the best prediction we can make is simply the mean of the Y distribution.

*The term *residual error* is also used in the analysis of variance (see Chapter 15) and defines the denominator in the within-subjects F ratio.

This is clear if we look again at the regression equation. If r is equal to zero, then all the terms, since they are products, also become zero, except for M_y.

$$Y_{\text{pred}} = \frac{(0)(s_y)}{s_x}X - \frac{(0)(s_y)}{s_x}M_x + M_y$$

$$= 0 - 0 + M_y$$

$$= M_y$$

When r assumes a significant value, however, the resulting predictions of Y may deviate from the mean of the Y distribution. The higher the correlation, then, the more the predicted Y values may vary from M_y, and the lower the correlation, the more the predictions regress toward the mean. Thus, if you are told that the average rainfall in India is 58 inches and then are asked to predict the height of a specific male college student, your best estimate would be 5 feet 9 inches, the mean of the adult male height distribution.

The Beta Coefficient

The equation for the slope of the regression line is called the **beta coefficient.** As we have seen, this b value indicates the amount of change in Y (a rise or drop in the ordinate) for a given change in X (a run on the abscissa). If the correlation between X and Y is perfect ($r = 1$), then the slope is the direct relationship of s_y over s_x. When $r = 1.00$, then

$$b = \frac{(1.00)s_y}{s_x} = \frac{s_y}{s_x}$$

When r is less than 1.00, however, then the s_y over s_x relationship is a weighted proportion. An r of .50 yields half as much of a Y-on-X increase as does an r of 1.00. For example, if we assume $s_y = 2$ and $s_x = 1$, then with an r of 1.00,

$$b = \frac{(1.00)(2)}{1} = 2.00$$

That is, with a beta coefficient of 2, a rather steep slope, there will be a two-unit increase in Y for each unit increase in X. But with an r of .50,

$$b = \frac{.50(2)}{1} = 1.00$$

Here, Y increases only one unit for each unit increase in X. Thus, with an r of .50, the increase in Y is half of what it is if r is equal to 1.00.

Note that when the standard deviations for both the X and Y distributions are equal to 1.00, as when comparing two distributions of z scores, then b (the slope) becomes identical with the Pearson r.

$$b = \frac{rs_y}{s_x} = \frac{r(1)}{(1)} = r$$

When the Intercept Is Zero

To illustrate a zero intercept, take the following imaginary data. Assume that a significant Pearson r correlation of .90 has been found between IQ and grade-point average. If the mean GPA is 3.00 with an estimated population standard deviation of .50 and the mean IQ is 100 with an estimated population standard deviation of 15, predict the GPA for a student who has an IQ of 110. Since GPA is being predicted, let's make all the GPA values Y values and all the IQs X values.

IQ	GPA
$M_x = 100$	$M_y = 3.00$
$s_x = 15$	$s_y = .50$

$$\begin{aligned} Y_{\text{pred}} &= \frac{rs_y}{s_x}X - \frac{rs_y}{s_x}M_x + M_y \\ &= [(.90)(.50) / 15]X - (.90)(.50) / 15 * 100 + 3.00 \\ &= .03X - 3.00 + 3.00 \\ &= .03X + 0 \end{aligned}$$

Thus, with an intercept of zero, the value of Y when X equals zero is itself equal to zero. In this case then, an IQ of zero would project a GPA of zero (hardly a mind-bending thought). When the intercept is zero, the predicted GPA would be simply the IQ times the slope for any possible IQ value. For example, an IQ of 110 would predict a GPA of (110)(.03) or 3.30, and an IQ of 90 would predict a GPA of (90)(.03) or 2.70.

X	$.03X$	$.03X + 0$
110	3.300	3.300
90	2.700	2.700

If the reader, or your instructor, prefers to use the actual sample standard deviation, the regression equation remains exactly the same. Although the actual standard deviations have slightly lower values, the ratios remain the same.

$$Y_{\text{pred}} = \frac{rSD_y}{SD_x} X - \frac{rSD_y}{SD_x} M_x + M_y$$

Theory of Regression

When **Sir Francis Galton** (1822–1911), younger cousin of Charles Darwin, began studying heredity, he first wanted to find out if the physical height of humans had a genetic component. He compared the heights of fathers and sons, and to put both distributions on a common measuring scale, he converted all the raw score values into Z scores. The unit he then plotted for both sets of scores was equal to 1 *SD*. He next calculated the means of the *z* scores for the sons and placed them on the *y* ordinate. Under the abscissa, he placed the *z* score values for the fathers' heights. When he looked at

SIR FRANCIS GALTON (1822–1911)

Galton is considered to be the person most responsible for developing the basic ideas for the methods used for quantifying the data of today's behavioral sciences. He was born into a wealthy and highly educated English family. Among his relatives were many of England's most intelligent, accomplished, and gifted citizens, including Charles Darwin, the founder of the modern theory of evolution, and Arthur Hallam, the subject of Tennyson's "In Memoriam." Galton even published a list of his wife's "connections," indicating that her father had been headmaster at Harrow, and a brilliant and dedicated educator. In 1838 Galton began his studies in medicine, but two years later he shifted his career plans and decided to major in mathematics. Following college, Galton went on several trips to Africa, exploring and mapping many areas of that continent for the very first time. For his African exploration, Galton received the Royal Geographical Society's gold medal in 1854. After his marriage, in 1853, Galton turned to writing. His first book, *The Art of Travel,* was a practical guide for the explorer, and his second book, *On Meteorology,* was one of the first attempts to set forth precise techniques for predicting the weather.

During the 1860s Galton became impressed with his cousin Charles Darwin's book, *On the Origin of Species.* He was fascinated with Darwin's notion of the survival of the fittest to adapt to the environment, and he attempted to apply this concept to human beings, thus founding the field of eugenics, or the study of how the principles of heredity could be used to improve the human race.

In 1869 Galton published his first major work, *Hereditary Genius,* in which he postulated the enormous importance of heredity in determining intellectual eminence. He felt that "genius" ran in families, and he was able to point to his own brilliant family as "exhibit A." He also became impressed with the wide range of individual differences that he found for virtually all human traits, physical as well as psychological. It was Galton who introduced the world to fingerprinting, and some of his methods of analysis are still used today by law enforcement agencies. Assuming that intelligence was a function of the quality of a person's sensory apparatus, he devised a series of tests of reaction time and sensory acuity to measure intellectual ability. He is considered, therefore, to be the father of intelligence testing. Although Galton's tests may seem somewhat naïve by modern standards, his emphasis on the relationship between sensory ability and intellect foreshadows much of today's research on the importance of sensory stimulation during the early years in encouraging cognitive growth. In order to analyze the test data that he was collecting, Galton created what he termed the "index of co-relation" (Galton, 1888). It was left to one of his younger colleagues, Karl Pearson, however, to work out the mathematical equation for this index, which is now known as the Pearson r, or product moment correlation. In fact, much of what we know about Galton's life comes from a series of books written by Karl Pearson, entitled *The Life, Letters, and Labours of Francis Galton.* Galton also was the first to describe the concept of regression toward the mean, which he discovered while noting that although tall parents would usually have taller-than-average children, the children were typically not as tall as the parents (the same was, of course, true for short parents, whose children were still short but not as short as the parents).

Galton wrote other major works, including *Inquiries into Human Faculty and Its Development* (Galton, 1883) and *Natural Inheritance* (Galton, 1889). His range of interests was extremely wide, embracing such topics as imagery, free and controlled associations, personality testing, and, of course, the assessment of intellect. To gain understanding of the possible interaction of heredity and environment, Galton performed psychology's first research studies on twins. According to one author, Galton was a gigantic figure in the history of statistics, and "perhaps the last of the gentlemen scientists" (Stigler, 1986, p. 266).

In 1909, just two years before his death, Galton was knighted. Galton's place in the history of both psychology and statistics is ensured. He died in 1911, the same year in which the world lost Alfred Binet, and more than any other person, he set the behavioral sciences on the road to data quantification and statistical analysis.

the resulting scatter plot, allegedly the first ever constructed, he noticed that the *means* of the sons' heights, for each unit interval on the X axis, not only formed a straight line but a line that had less slope than he was expecting. That is, the mean of the sons' heights deviated less from their overall mean than did the heights of the fathers from their overall mean. He called this the *law of filial regression;* that is, the height of the next generation tended to fall back toward the general mean.* He found that the slope, incidentally, took on a value of .50, which in this case would have been identical to the Pearson r value had the Pearson r been available to him back then (since Galton had expressed both measures in z score units). In the graph in Figure 14.6, the dotted line shows the results of Galton's work, and the solid line indicates what the regression line would have been had the slope been equal to 1.00. Later, Karl Pearson worked out the equation for the Pearson r, an equation, as you have seen, whose derivation saved you from the tedium of having to convert all those raw scores into z scores. Galton also believed that in successive generations even intelligence regressed toward the mean, a fact he lamented by calling it the *regression to mediocrity.*

FIGURE 14.6

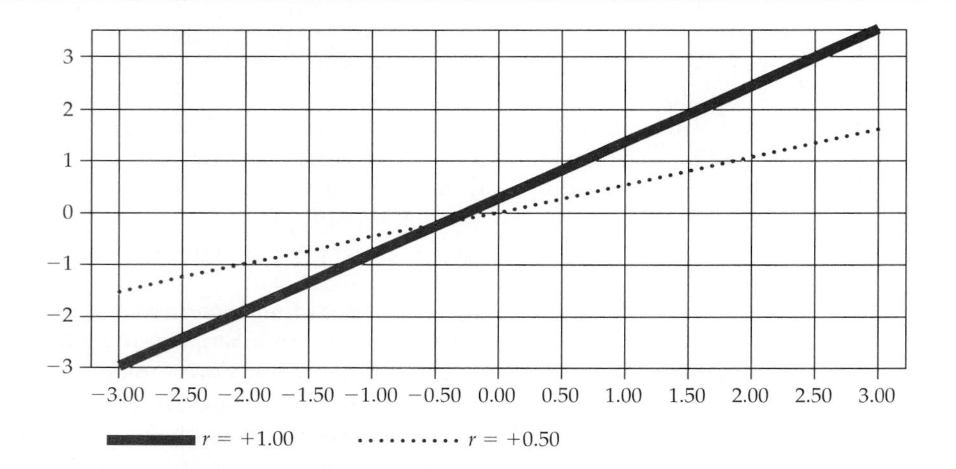

Example The correlation between the heights of fathers and their adult sons is .55 (Bloom, 1964). The mean of both height distributions is 68 inches, and the standard deviations are both 3 inches. Given two fathers with heights of 78 inches and 58 inches, predict the heights of their sons.

*Unless the correlation between X and Y is 1.00, the predicted Y value will always be closer to the mean of the Y distribution than the original X value is to the mean of the X distribution.

Fathers, X	*Sons, Y*
$M_x = 68$	$M_y = 68$
$s_x = 3.00$	$s_y = 3.00$

$$Y_{\text{pred}} = \frac{rs_y}{s_x}X - \frac{rs_y}{s_x}M_x + M_y$$

$$= \frac{(.55)(3)}{3}X - \frac{(.55)(3)}{3}(68) + 68$$

$$= .55X - (.55)(68) + 68$$

$$= .55X - 37.40 + 68$$

$$= .55X + 30.60$$

Father's Height, X	*.55X*	*Predicted Son's Height, .55X + 30.60*
78	42.90	73.50
58	31.90	62.50

Note here that the beta coefficient of +.55 is telling us that for each full inch of change on the X distribution (father's height), we can expect a corresponding change of .55 inch in the Y distribution (son's height). The sons are changing about a half inch for each full-inch change among the fathers.

The regression models are designed to minimize the average error of the predictions across a population. If the assumptions are met, then "the sum of the individual deviations of the estimate from the actual values is zero, and underestimation for one individual is compensated for by overestimation in another" (Veiel & Koopman, p. 356, 2001).

STANDARD ERROR OF ESTIMATE

Values predicted from X, the independent variable, yield the most probable values of the dependent variable. However, "most probable" does not always produce a practical or useful prediction. As we have seen, with a correlation of zero, the most probable Y value for any given X is simply the mean of Y. Our job now is to determine how accurate the predicted or most probable value is.

As we saw back in Chapter 11, whenever we square the Pearson r and multiply by 100, we get the coefficient of determination, a number that tells us how much of the variation in the DV is explained on the basis of changes in the IV. Recall from Chapter 11 that when statisticians refer to "explaining" the variability, what they really mean is reducing the variability among the predicted values (and thereby producing more precise, less varied predictions). Predicting that a given individual's IQ will fall somewhere between 120 and 130 is a lot more precise and useful than predicting that it lies between 55 and 145. A correlation of, say, .50 would explain 25% of the variability, meaning that a knowledge of one variable would decrease the variation in the predicted values by 25%. Thus, by utilizing this r^2 value, a new statistic, the **standard error of estimate,** can be generated that will allow us to predict the actual accuracy level of our predicted Y values.

Calculating the Standard Error of Estimate

With small samples, typically less than 50, the standard error of estimate can be calculated as follows:

$$SE_{est} = s_y \sqrt{\left(\frac{N-1}{N-2}\right)\left(1-r^2\right)}$$

where s_y is the estimated population standard deviation. The standard error of estimate will produce an estimate of how much all the data points in the scatter plot vary around the regression line. If the correlation were found to be a perfect 1.00, then all the data points would fall directly on the regression line and the standard error of estimate would be zero. When the standard error of estimate assumes a large value it indicates that the data points may scatter widely around the regression line.

If the actual sample standard devaition is used the equation becomes

$$SE_{est} = SD_y \sqrt{\frac{N}{N-2}\left(1-r^2\right)}$$

Establishing the Confidence Interval

Assume we have selected a random sample of 40 college students and discovered a significant correlation of .850 between scores on an introversion-extraversion personality test and hours-per-week spent in cocurricular activities. The mean personality score was 50 with an estimated population standard deviation of 7.538 (higher scores indicating more extraversion). The mean hour-per-week value was 20, with an estimated population standard deviation of 3.107. If we were to select a student whose personality score was 55 (slight extraversion), how many weekly cocurricular hours would be expected? First, solve the regression equation of Y on X, where Y equals hours per week and X equals personality scores.

$$Y_{\text{pred}} = \frac{(.850)(3.107)}{7.538} X - \frac{(.850)(3.107)}{7.538}(50) + 20$$

$$Y = .350X - (.350)(50) + 20$$

$$= .350X - 17.500 + 20$$

$$= .350X + 2.500$$

Then, using the results of this equation, set up the regression table.

X	.350X	.350X + 2.500 = Y$_{pred}$
55	19.250	21.750

We are thus predicting that a student whose personality score is 55 should spend 21.750 hours per week on cocurricular activities.

Next, we calculate the standard error of estimate.

$$SE_{\text{est}} = s_y\sqrt{\left(\frac{N-1}{N-2}\right)(1-r^2)}$$

$$= 3.107\sqrt{\left(\frac{39}{38}\right)(1-.85^2)} = 3.107\sqrt{(1.026)(1-.722)}$$

$$= 3.107\sqrt{(1.026)(.278)} = 3.107\sqrt{.285}$$

$$= (3.107)(.534) = 1.659$$

We will now use this statistic to generate a range of possible Y values within which we can feel confident that the parameter Y value could likely fall. Since the predicted Y value is a point estimate of the true parameter, we are again looking for a range of values that may lie around this estimate. To establish the confidence intervals, we will use the following equation, where the degrees of freedom for t are defined as N pairs – 2.

$$CI = (\pm t)(SE_{\text{est}}) + Y_{\text{pred}}$$

Using the previous data on personality and hours per week, the student whose personality score was 55 yielded a Y_{pred} value of 21.750 hours. The sample size was 40, giving us 38 degrees of freedom. The t ratio at .05 for 38 degrees of freedom is found on table C to be equal to 2.024.

$$CI_{.95} = (\pm 2.024)(1.659) + 21.750$$

$$+3.358 + 21.750 = 25.108$$

$$-3.358 + 21.750 = 18.392$$

Thus, we are 95% confident that a student whose personality score was 55 will work somewhere between 25.108 and 18.392 hours per week on cocurricular activities.

Or, at the .99 confidence level, we look up the t ratio at the .01 level, again for 38 degrees of freedom and find it on table C to be equal to 2.712.

$$CI_{.99} = (\pm 2.712)(1.659) + 21.75$$
$$+4.499 + 21.75 = 26.249$$
$$-4.499 + 21.75 = 17.251$$

Finally if the samples are large enough to assume normality, the z score may be substituted for the above t ratio, and .95 confidence would require a z of ± 1.96, or for .99 confidence a z of ± 2.58. This is because 95% of the cases under the normal curve are included between z's of +1.96 and −1.96, and 99% are included between z's of +2.58 and −2.58.

With large samples, where the distribution of sample scores can be assumed to be normal, the equations are as follows:

For a .95 level of confidence, the limits are

$$\mathrm{CI}_{.95} = (\pm 1.96)(SE_{\mathrm{est}}) + Y_{\mathrm{pred}}$$

And for a .99 level of confidence, the limits are defined by

$$\mathrm{CI}_{.99} = (\pm 2.58)(SE_{\mathrm{est}}) + Y_{\mathrm{pred}}$$

The procedures outlined thus far in this chapter are based on three important assumptions: (1) linearity, (2) normality, and (3) homoscedasticity.

1. *Linearity.* The two variables being correlated must bear a straight-line relationship to one another. The regression line must therefore maintain a single direction. It cannot start in an upward direction, then turn and continue in a downward one.

2. *Normality.* The two sets of sample scores must represent populations whose distributions are normal, or at least nearly so. These regression procedures will not work on badly skewed distributions.

3. *Homoscedasticity.* All sample scores on the Y axis of the scatter plot must have similar amounts of variability. There cannot be dramatic differences in the variability of the Y scores from column to column. For example, if we were predicting grade-point averages, we could not use the standard error of estimate if all of the GPAs above 2.50 had large amounts of variability, whereas those below 2.50 had little or none.

MULTIPLE R (LINEAR REGRESSION WITH MORE THAN TWO VARIABLES)

When we first discussed correlation in Chapter 11, it was defined as a numerical statement as to the relationship among two or more variables. Up to now, however, the only procedures shown have been for two-variable, or bivariate, problems. We are now going to confront multivariate problems, those in which several variables are intercorrelated with one another.

Instead of simply attempting to find out how much information about Y is being contained in X (the goal of the Pearson r), a multiple correlation could establish how much information about Y is contained in X_1, X_2, X_3, and so on. Suppose, for example, that your job is to predict the height of the very next person to enter your room. Knowing that you are to predict the height of a *person* helps you somewhat—you wouldn't guess 3 inches or 60 feet. Next, we add some information and tell you that the person in question is an adult male. You know that gender correlates to some extent with height and that the adult male height distribution tends to have a higher average value than does the female distribution. You might now guess a value of 5 feet 9 inches, the actual mean of the male distribution. Next, we provide even more information and tell you that the person weighs 230 pounds. Again, since you know that height and weight tend to correlate positively, you might now increase your estimate to 6 feet or 6 feet 1 inch. Finally, we give one last bit of information and tell you that the person plays center on a professional basketball team. You now jump your estimate to 6 feet 10 inches or 7 feet. The point is, your prediction becomes increasingly more accurate as more and more predictor variables are supplied. Now, although the example has been based partly on nominal variables (male versus female), the same sort of results can be obtained when all the correlated measures are in the form of interval/ratio data, and the appropriate statistical test then becomes the multiple R.

Although the technique we use applies to virtually any number of variables, our discussion is restricted to the three-variable correlation. This is really all that is needed by a student of elementary statistics. It is also, because of the limitations of inexpensive calculators, about all that you can realistically be asked to calculate. In any event, the reason for doing all multivariate analyses is the same—to increase predictive efficiency. After all, when it comes to correlation, prediction is the basic goal.

Recently, there has been a controversy regarding the use of Scholastic Assessment Test (SAT) scores as predictors of college success. Critics of the SAT point to the fact that high school grades are better than SAT scores as predictors of college performance. So why bother with the SAT? Defenders of the SAT, on the other hand, argue that since SAT scores do correlate significantly, though admittedly not perfectly, with college performance, why throw them out? Doesn't every little bit help?

The nationwide correlation between high school grades and college grades is .52, whereas the correlation between SAT scores (combined verbal and math) and college grades is only .41. (Also, the correlation between high school grades and SAT scores is .31.) There is no question about it, then: High school grades do predict college performance better than SAT scores do (Turnbull, 1980). But what happens when we combine them? Does combining high school grades and SAT scores increase their predictive efficiency above and beyond the performance of either factor alone? To answer this question, the **multiple R** can be used.

Multiple R Equation

Before using the multiple R equation, the following designations must be made. The subscript y always refers to the criterion variable, the variable we are trying to predict. The subscripts 1, 2, 3, 4, and so on always refer to the predictor variables, the variables we are predicting from. The multiple R equation where there are two predictors is as follows:

$$R_{y \cdot 1,2} = \sqrt{\frac{r_{y,1}^2 + r_{y,2}^2 - 2r_{y,1}\, r_{y,2}\, r_{1,2}}{1 - r_{1,2}^2}}$$

Identifying the y Variable (Keep Your Eye on the "y")

Before plugging the values into the equation, take a moment to sort out the various components. The first step, *and the most important*, is to identify and *clearly label* the y (or criterion) variable, that is, the variable whose value is being predicted. In this example, then, y must be identified as representing college grades. Once this is done, it makes no difference which variable is called "1" and which "2." For this problem, we happened to choose high school grades as the "1" variable (the first predictor) and the SAT scores as the "2" variable (the second predictor).

Label the components as follows:

y = criterion: college grades
1 = first predictor: high school grades, or X_1
2 = second predictor: SAT scores, or X_2

The second step requires that you go back to the problem and pull out the correlations as they apply to your labeled components.

$r_{y,1} = .52$ (the correlation between college grades and high school grades)
$r_{y,2} = .41$ (the correlation between college grades and SAT scores)
$r_{1,2} = .31$ (the correlation between high school grades and SAT scores)

Finally, and this is really the easy part, simply plug these values into the multiple R equation.

$$R_{y \cdot 1,2} = \sqrt{\frac{.52^2 + .41^2 - 2(.52)(.41)(.31)}{1 - .31^2}}$$

1. Square $r_{y,1}$.

   $.52^2 = .270$

   Then square $r_{y,2}$.

   $.41^2 = .168$

   Plug both values into the numerator of the equation.

2. Chain-multiply the term $2r_{y,1}r_{y,2}r_{1,2}$. That is, let your calculator be a computer. Enter 2, times .52, times .41, times .31, and then press the equals button. The result should equal .132. Plug this value into the numerator of the equation.
3. Square $r_{1,2}$.

   $.31^2 = .096$

   Subtract it from 1. This should equal .904. Plug this value into the denominator.
4. Complete the solution as follows:

$$R_{y \cdot 1,2} = \sqrt{\frac{.270 + .168 - .132}{.904}} = \sqrt{\frac{.438 - .132}{.904}} = \sqrt{\frac{.306}{.904}} = \sqrt{.338}$$

$$= .581$$

Thus, combining high school grades with SAT scores yields a multiple correlation of .581, which is higher than either separate correlation. With a correlation of .581, will predictions ever be wrong? Of course they will, sometimes. But it does beat flipping a coin!

R^2—*The Coefficient of Determination*

Just as r^2 was the coefficient of determination for the Pearson r, so too is R^2 called the coefficient of determination for the multiple R. Thus, R^2 is the proportion of the DV variance that can be explained (accounted for) by the combination of at least two IVs. Thus, R^2, when multiplied by 100, becomes the *percentage* of the explained variance. If we take the SAT data just presented, we found the r between high school grades and college grades to be .52, which was the highest of the three internal correlations. We then found the multiple R to be equal to .581. Thus, the r of .52 led to an r^2 of .27, whereas the R of .581 produced an R^2 of .338. Comparing the two, the r^2 explained 27% of the variance, whereas the R^2 explained 34%, an increase of 7%.

The Multiple R *and Significance*

Once the multiple R is obtained, we should assess whether the resulting linear relationship between this particular set of IVs (the predictors) and the Y criterion is really only zero in the population. That is, we must test the multiple R for significance, just as we previously did with the Pearson r. To do this for the case of two predictors, we use a version of the F ratio that states the following:

$$F = \frac{R^2(N - 3)}{2(1 - R^2)}$$

Using the preceding example and, for purposes of illustration, assuming a sample size of 103 students, we get

$$F = \frac{.581^2(103 - 3)}{2(1 - .581^2)}$$
$$= \frac{(.338)(100)}{2(1 - .338)}$$
$$= \frac{33.80}{2(.662)}$$
$$= \frac{33.80}{1.324}$$
$$= 25.529$$

Check this on the F table with 2 degrees of freedom in the numerator and $N - 3$ degrees of freedom in the denominator, or in this case with an F at 2 and 100. F at .01 with 2 and 100 df equals 4.82. Reject H_0 at $p < .01$. Therefore, the multiple R is significant and can be extrapolated to the population from which it was selected.

Example Assume a sample size of 58 subjects and a multiple R (based on two predictors and a single Y criterion) of .50. Is the R significant?

$F = 9.167$. Reject H_0 at $p < .01$.

The value of a significant multiple R should only be compared with those internal correlations that associate with the y (or the criterion) variable. The multiple R should then be used for making predictions only if it is higher than all the internal correlations with y. For example, suppose that $r_{y1} = .70$, $r_{y2} = .80$, and the multiple $R = .60$. Then, of course, go back and use r_{y2} for making the prediction and in this case disregard the multiple R.

Although the multiple R allows for the use of a long list of predictors, it's not always good policy to put in too many. What we should strive for is to identify those IVs that produce the highest levels of predictive power. (With some statistical programs, the culling process is part of the routine, and only significant correlation will be used.) Thus, using significant predictors means that only those IVs that explain a significant portion of the variance will be entered. Also, the major statistical programs won't enter new variables that are highly colinear.

Multiple R *and Regression*

To make predictions to the criterion variable, having been given performance measures on the predictor variables, a **multiple regression** equation must be used. The use of this equation, like that of the regression equation for Y on X, requires the calculation of a, the point of intercept. Unlike the regression of Y on X, however, the multiple regression equation also requires the calculation of at least two slopes, b_1 and b_2. The value for b_1 tells us how much change in Y will occur for a given unit change in the first predictor, *when the effects of the second predictor have been held constant.* The b_2 value tells us, on the other hand, how much change in Y will occur for a given unit change in the second predictor, *when the effects of the first predictor have been ruled out.* The multiple regression equation takes the following form, with X_1 and X_2 being specific values from the two predictor distributions.

$$Y_{\text{M pred}} = b_1X_1 + b_2X_2 + a$$

The equation for b_1 is as follows:

$$b_1 = \left(\frac{s_y}{s_1}\right)\left(\frac{r_{y,1} - r_{y,2}\, r_{1,2}}{1 - r_{1,2}^2}\right)$$

And for b_2,

$$b_2 = \left(\frac{s_y}{s_2}\right)\left(\frac{r_{y,2} - r_{y,1}\, r_{1,2}}{1 - r_{1,2}^2}\right)$$

Finally, for a we have the following:

$$a = M_y - b_1 M_{x_1} - b_2 M_{x_2}$$

Example With the multiple R we established a correlation of .581 between the criterion (college GPA) and the two predictors (high school grades and SAT scores). We can now take the data and set up a multiple regression. A college GPA will be predicted for two students, one with a high school grade average of 80 and an SAT of 500, the other with a high school grade average of 70 and an SAT score of 400. Let us assume (1) a mean college GPA of 2.710 with an estimated population standard deviation of .690, (2) a mean high school grade of 80 with an estimated population standard deviation of 6.00, and (3) a mean SAT score of 450 with an estimated population standard deviation of 69.00.

College GPA, Y	High School Grades, 1	SAT, 2
$M_y = 2.710$	$M_{x_1} = 80.00$	$M_{x_2} = 450$
$s_y = .690$	$s_1 = 6.00$	$s_2 = 69$

The correlations, as shown previously, are as follows:

$r_{y,1} = .520$

$r_{y,2} = .410$

$r_{1,2} = .310$

1. We plug in the appropriate values to solve for b_1.

$$b_1 = \left(\frac{s_y}{s_1}\right)\left(\frac{r_{y,1} - r_{y,2}\, r_{1,2}}{1 - r_{1,2}^2}\right) = \left(\frac{.69}{6}\right)\left(\frac{.52 - (.41)(.31)}{1 - (.31)^2}\right)$$

$$= (.12)\left(\frac{.52 - .13}{1 - .10}\right) = (.12)\left(\frac{.39}{.90}\right) = (.12)(.43)$$

$$= .050$$

2. We plug in the appropriate values to solve for b_2.

$$b_2 = \left(\frac{s_y}{s_2}\right)\left(\frac{r_{y,2} - r_{y,1}\, r_{1,2}}{1 - r_{1,2}^2}\right) = \left(\frac{.69}{69}\right)\left(\frac{.41 - (.52)(.31)}{1 - (.31)^2}\right)$$

$$= (.01)\left(\frac{.41 - .16}{1 - .10}\right) = (.01)\left(\frac{.25}{.90}\right) = (.01)(.28)$$

$$= .0028 = .003$$

3. We plug in the appropriate values to solve for a.

$$a = M_y - b_1 M_{x_1} - b_2 M_{x_2}$$
$$= 2.71 - (.05)(80) - (.003)(450)$$
$$= 2.71 - 4.00 - 1.35$$
$$= -2.640$$

4. Using the values obtained, we set up a multiple regression table, including the specific X_1 and X_2 values for each student.

Student	*High School Grade Average,* X_1	$.05X_1$	*SAT,* X_2	$.003X_2$	$.05X_1 + .003X_2$	*College GPA,* $Y_{Mpred} = .05X_1 + .003X_2 - 2.64$
A	80	4.00	500	1.50	5.50	2.860
B	70	3.50	400	1.20	4.70	2.060

Thus, for Student A, whose high school grade average was 80 and whose SAT score was 500, the predicted college GPA is 2.860. For Student B, the predicted GPA is 2.060.

Limits to Regression Analysis

Collinearity is a term used to describe a situation in which two variables in a matrix (see Chapter 11) are perfectly (or nearly perfectly) correlated and are intended for use in a regression analysis. (The opposite of collinearity is orthogonality, a term used to describe two variables that are totally independent.) An example would be when IQ scores on the WAIS III and on the Stanford–Binet are so highly correlated as to be collinear since they are both measuring the same underlying concept—intelligence. There are statistical methods (for example, factor analysis) where collinearity is sought after (since factor analysis tries to identify highly correlated variables in order to collect together the fewest sets of predictors). But for regression analysis, collinear pairs present a problem, since they are containing the same information and therefore probably measuring the same thing. The best way to spot collinearity is to simply look at the Pearson r values. Pearson's r's of .99 or higher will reveal high collinearity, and then, of course, the redundant variables become obvious. Tabachnick and Fidell (1996) point out that once collinearity is detected, there are several methods of dealing with the problem. For our purposes, the simplest and best method is to delete the offending variable. Because one variable is a combination of another, information is not really lost by this deletion. Other suggested methods, such as subjecting the variables to principal components analysis or using stepwise entry of the variables, go beyond the scope of this book, but the interested reader can pursue this discussion in Tabachnick and Fidell.

The Multiple R provides its greatest advantage when the predictor variables are orthogonal, or independent of each other. For example, assume for a moment that SAT scores and motivation are orthogonal. Then, if 20% of a student's GPA can be predicted on the basis of SAT scores, and 50% can be predicted on the basis of motivation, then each could be added together so that 70% of the variability in GPA could be explained on the basis of the two variables, SAT and motivation taken together. However, when the predictor variables are all correlated their combined contribution to the prediction is lowered considerably, since to the extent they are correlated, they are both measuring some aspect of the same underlying trait. With multiple regression it is especially important to have as large a sample as possible. A ratio of about 15 subjects per IV is recommended and should provide a reliable regression equation (Mertler & Vannata, 2001).

PATH ANALYSIS, THE MULTIPLE *R*, AND CAUSATION

Multiple regression analysis is also being used in a causal modeling technique called **path analysis.** Although path analysis is definitely a correlation technique, it is being used to establish the possibility of cause-and-effect relationships. The multiple regression is used to determine the interrelationships among a series of variables that are logically ordered on the basis of time (Yarenko et al., 1982). Since, logically, a causal variable must precede (in time) a variable it is supposed to influence, the multiple regression analyzes a whole series of variables, each presumed to show a causal ordering. The attempt is made to find out whether a given variable is being influenced by the variables that precede it and then, in turn, influencing the variables that follow it.

A "path" diagram is drawn that indicates the direction of the various relationships. Although not as definitive a proof of causation as when the independent variable is experimentally manipulated, this technique, proponents of path analysis tell us, has taken us a long step forward from the naïve extrapolations of causation that at one time were taken from simple bivariate correlations. Path analysis, however, is not without its share of critics. Games (1988), for example, insists that the use of correlational data as a vehicle for causative assumptions is still fallacious, despite the alleged sophistication of the modeling approach. For a more complete discussion of the use of correlational analyses as causative models, see Chapter 11 in Sprinthall, Schmutte, and Sirois (1990).

PARTIAL CORRELATION

Another correlation problem, almost the reverse of the multiple R, involves establishing an accurate correlation between two variables when both are known to be significantly influenced by a third variable. To solve this problem, a method has been developed that allows the researcher to attempt to rule out the influence of the third variable on the remaining two variables under study. The resulting correlation is called a **partial correlation.**

The equation for the partial correlation is

$$r_{y,1 \cdot 2} = \frac{r_{y,1} - r_{y,2}\, r_{1,2}}{\sqrt{\left(1 - r_{y,2}^2\right)\left(1 - r_{1,2}^2\right)}}$$

The $r_{y,1 \cdot 2}$ term is read as the correlation between y and 1, with 2 held constant. The variable held constant is called the *covariate.*

Example A researcher is interested in the relationship between reading speed and reading comprehension. However, since the subjects in this study vary considerably as to IQ, it is decided to run a partial correlation in which the influence of IQ will be statistically nullified.

Identifying the "2" variable (keep the "2" in view). The first step in this problem is to identify and clearly label the "2" variable, the variable whose influence is being partialed out. Once this is done, it doesn't matter to the equation which variable is called "y" and which "1." In this problem, IQ *must* be the "2" variable. We have chosen to call reading speed the y variable and reading comprehension the "1" variable. Label the components as follows:

2 = IQ

y = reading speed

1 = reading comprehension

From the problem, then, the correlations are as follows:

$r_{y,1} = .55$ (correlation between reading speed and comprehension)
$r_{y,2} = .70$ (correlation between reading speed and IQ)
$r_{1,2} = .72$ (correlation between comprehension and IQ)

These values are substituted into the equation for partial correlation.

$$r_{y,1\cdot 2} = \frac{r_{y,1} - r_{y,2}\, r_{1,2}}{\sqrt{\left(1 - r_{y,2}^2\right)\left(1 - r_{1,2}^2\right)}}$$

$$= \frac{.55 - (.70)(.72)}{\sqrt{(1 - .70^2)(1 - .72^2)}}$$

$$= \frac{.55 - .504}{\sqrt{(1 - .490)(1 - .518)}}$$

$$= \frac{.046}{\sqrt{(.510)(.482)}} = \frac{.046}{\sqrt{.246}}$$

$$= \frac{.046}{.496} = .093$$

With IQ partialed out, the correlation between reading speed and comprehension drops from .550 to .093.

Thus, by partialing out IQ, we have, in effect, created a statistical equivalence within the group being measured. The subjects have been statistically equated, although after the fact, through the use of the partialing process.

By partialing out IQ we were able to look at the resulting correlation between reading speed and reading comprehension as if every child in the sample had exactly the same IQ. The covariate, IQ, has thus been statistically held constant.

By partialing out the effects of one or more variables, the researcher is able to identify the correlation remaining between two of the variables, without this correlation coming under the influence of the partialed variables. That is, we can find the real correlation between two variables with the assumption that all the subjects had EXACTLY THE SAME SCORE on the partialed variables.

If the study in this example had been done by selecting subjects with identical IQs at the outset (a better procedure), there would have been no need to partial out the IQ variable.

The equation given here is called a first-order partial, that is, one variable is held constant while examining the remaining relationship between the other two variables. Higher-order partials are also calculable, in which two or more variables are simultaneously held constant. For these procedures, consult more advanced texts.

Partial Correlation and Significance

Although the internal correlations may all have been significant, we must now test the partial correlation for significance, since in many instances the resulting correlation is reduced and now has the potential for diminishing into the ethereal realm of non-significance. To do this, we can use the Pearson r table (Appendix Table E), and for a first-order partial *the degrees of freedom are assigned on the basis of* N – 3. In the problem just shown, let's assume an N of 33, and as we saw the partial correlation $r_{y,1 \cdot 2} = .093$. To test this value for significance, we assume that in a trivariate normal population the null hypothesis expects that the population correlation, ρ, will equal zero. Thus, with a three-variable partial correlation and an N of 33, we test the null hypothesis with alpha set at .05 as follows:

Null hypothesis:	H_0: $\rho_{y,1 \cdot 2} = 0$
Alternative hypothesis:	H_a: $\rho_{y,1 \cdot 2} \neq 0$
Table value:	$r_{y,1 \cdot 2}$ at .05 with 30 df = .349
Obtained value:	$r_{y,1 \cdot 2} = .093$ Accept H_0.

Conclusion: A partial correlation coefficient was computed between Reading Speed and Reading Comprehension with the IQ variable being partialled out. The resulting correlation ($r(30) = .093, p > .05$) was not significant. The two variables, Reading Speed and Reading Comprehension, are independent of each other when the IQ variable is statistically controlled.

Example Assume the following partial correlation and sample size: $r_{y,1 \cdot 2} = .24$ and $N = 43$. Using the Pearson r table, determine if this partial correlation is significant.

We find that at 40 df an r of .304 is needed at the .05 level. Thus, the partial correlation of .24 is *not* significant.

As we have now seen, when certain variables are partialed out, the correlation under investigation is often reduced, sometimes by enough to lose significance. However, sometimes it works the other way around, and the partial correlation may be found to increase the correlation. Therefore, as stated many times before, let the data talk. Don't leap to the conclusion that because it's a partial, the resulting correlation will always be lowered. For example, assume that a study was conducted on a random sample of 103 eighth-grade students, and the following significant correlations were discovered: r between IQ and study time = –.20; r between grades and study time = .20; r between grades and IQ = .55. The focus will now be on that r of .55, the correlation between grades and IQ. That is, the question will now be, what is the correlation between grades and IQ if we were to partial out the influence of study time?

First, identify the 2 variable, which in this case is study time (since it's the variable being partialed out). Next, assign the y and 1 variables, and here the order doesn't make any difference to the equation; so we will assign y to grades, since that would typically be the dependent variable, and 1 to the predictor, which in this example would be IQ.

Thus,

$$r_{y1} = .55$$
$$r_{y2} = .20$$
$$r_{1,2} = -.20$$

Working out the partial correlation, we get a final result of .615. This means that we are predicting that the true correlation between grades and IQ, which had been shown to be .55, is really an even higher .62 when the effects of study time are partialed out. Thus, if everyone in the sample of 103 selected eighth-graders were to study exactly the same amount of time, the correlation between grades and IQ would be *increased* to .615.

Testing this partial correlation for significance, we look up the r value for $N - 3$ or 100 df and get .195 at the .05 level or .254 at the .01 level. Since our partial correlation of .615 is greater than either of the table values, we reject the null hypothesis with a p of $<.01$. The partial correlation is *significant.*

Suppose a researcher found a correlation between the amount of exercise a person indulged in and the amount of money that a person had in the bank. The correlation would almost certainly be negative; that is, people who are involved in heavy exercise programs might tend to have smaller bank accounts. This would not necessarily mean that persons who exercise regularly don't have enough time left over to earn and save money, since it may well be that another variable, age has been overlooked. Elderly persons who have had a lifetime in which to acquire money are probably less apt to jog five miles a day than are high school students who are still on a family allowance. By partialling out age, the true correlation between exercise and bank accounts could be established, since when the age variable has been nullified the resulting correlation looks at the relationship as if everyone in the sample were of exactly the same age.

Example Among humans, head size is very much related to brain growth and is partly determined by age two when the sutures in the skull close. By age six, head size growth is nearly completed. A researcher wondered if head size might be used to predict verbal ability. A large random sample of children ages two to six was selected, and each child had his or her perimeter head size measured. Each child was also given a verbal ability test, and it was found that head size and verbal ability correlated at a significant .49. The researchers then wondered if this fairly impressive correlation would hold up if the age variable were partialed out. You have been called in as a consultant. Using the following significant correlations, find the correlation between head size and verbal ability with the age factor ruled out. Since age is being partialed, it becomes the 2 variable. Let's call verbal ability the y variable and head size the 1 variable.

Head size and verbal ability: $r_{y_1} = .49$

Verbal ability and age: $r_{y_2} = .68$

Head size and age: $r_{1,2} = .72$

With age partialed out, the correlation between head size and verbal ability equals zero and, of course, cannot be significant (since zero can't be significantly different from the null's hypothesized correlation of zero).

The partial correlation is logically the other side of the multiple R coin. Whereas the multiple R combines variables in order to see what their cumulative influence might be, the partial correlation focuses on peeling away variables in order to see what influences are left over.

Before leaving the topic of multiple correlations, two more techniques should at least be mentioned. You won't have to calculate either one, but at least when reading the research literature, you should be aware of what they can and cannot do.

Canonical Correlation

When groups of variables are correlated with other groups of variables, the procedure is called canonical correlation. Thus, two or more X variables are correlated with two or more Y variables, such as using several personality tests to predict several measures of leadership.

Stepwise Regression

This is a type of multiple regression in which IVs are added one at a time in what is called a stepwise process. The procedure (usually a computer program) then orders the variables on the basis of their predictive power. When the overall R stops increasing (it can never decrease), the procedure is halted. Adding new IVs to a regression can never lower the predictive power of R because the worst it can do is leave the R unchanged. This is called a forward stepwise regression. A similar procedure, called a backward stepwise regression, begins when the researcher enters all the IVs and then lets the regression procedure delete those that have the least predictive power. Stepwise regression methods are used to obtain parsimonious explanations and predictions based on as few independent variables as possible. With stepwise regression the researcher begins with a theoretical model that specifies several IVs and then attempts to find which of the IVs correlates the highest with the DV. Stepwise regression should always be based on a theoretical model, not used as a fishing expedition. In this latter case the researcher doesn't have the vaguest notion of what might be the best array of predictors so, usually because of a powerful computer program, simply tries every conceivable IV with the hope that the procedure itself will discover the best-fitting model. Of course, no backfitted model will fail to backfit. In one celebrated case, a researcher created a data set using a random number generator and set up a total of 15 IVs. He obtained a significant R of .418 (and rejected H_0 at .01). Was this a rare event? The researcher says no. "In any one test, the chance of observing a statistic as great as the one observed is quite small, but in doing the stepwise procedure the computer did so many tests that the chance of finding one or two significant results was not particularly small at all, and, of course, the significant results were the only ones the stepwise procedure bothered to tell us about" (Chalmers, 1987).

SUMMARY

Correlation allows the researcher the opportunity to make better-than-chance predictions. When the correlation between X and Y is known, the prediction of a value of Y from a given value of X is based on the linear regression of Y on X. The higher the correlation (that is, the more the correlation deviates from zero, whether positively or negatively), the greater is its predictive power.

On a scatter plot, each data point represents a pair of scores. The single straight line that comes closest to all the data points in the scatter plot is called the regression line. This line is useful for making predictions when three important facts are known: (1) the correlation between the two variables, (2) the slope of the line, and (3) where the line crosses the ordinate (its point of intercept).

Unless the correlation between the variables is perfect (±1.00), the predictions resulting from the regression equation will contain some degree of error. The predicted Y value can, however, be assessed for accuracy by calculating the standard error of estimate. This value will provide an estimate of how much the data points in the scatter plot cluster around the regression line. The standard error of estimate can then be used for calculating the interval within which the researcher feels confident that the true score would most likely be. The wider the range of this interval, the higher is the level of confidence.

Whenever more than two variables are to be correlated, the multiple R can be used. If the multiple R value is greater than the separate internal correlations with the criterion, it can be used to increase further the predictive accuracy of the correlations. The multiple R is the overall correlation between Y, the criterion variable, and the best linear combination of all the Xs, the predictors.

Just as the bivariate regression equation is used for predicting Y when X is known, so too can the multiple regression equation be used for predicting Y when several inputs, or X variables, are given. This prediction may be followed by a standard error of multiple estimate for assessing the accuracy of the result.

Finally, when it is assumed that several variables do intercorrelate, a partial correlation can be calculated to determine the degree of correlation that exists (is left over) between any two variables when the influence of the others is ruled out. This technique is sometimes even used in post-facto research in order to create an after-the-fact statistical equivalence among groups of subjects that were originally different.

Key Terms and Names

beta coefficient (slope)
collinearity
Galton, Sir Francis
homoscedasticity
multiple *R*
multiple regression
partial correlation
path analysis
point of intercept
regression line
standard error of estimate

PROBLEMS

1. The correlation between IQ and the verbal SAT score among the seniors at West High School is .90; it is significant. The mean IQ is 100 with an estimated standard deviation of 10. The mean SAT score is 500 with an estimated standard deviation of 100. Predict SAT scores for seniors with the following IQs:
   a. 105
   b. 110
   c. 120
2. A random sample of factory workers is selected. Each worker is given an IQ test and also a factual "Current Events" test. The mean IQ for the group is 110 with an estimated standard deviation of 12.85. The mean score on the "Current Events" test is 50 with an estimated standard deviation of 15.
   a. Assume a significant correlation of .95 and predict the IQ of a worker whose "Current Events" score is only 20.
   b. Assume a significant correlation of .25 and again predict the IQ of a worker whose "Current Events" score is only 20.
   c. Why is the IQ prediction in part (b) higher than the prediction in part (a)? (Hint: Note the sizes of the two correlations used and recall the meaning of regression.)
3. A researcher is interested in the possible relationship between two of the subtests on the WAIS (Wechsler Adult Intelligence Scale). A random sample of eight Army recruits is selected; they are given both the Vocabulary subtest and the Digit Span (a test of short-term memory) subtest. Their weighted, scaled scores are as follows:

Subject	*Digit Span*	*Vocabulary*
1	9	11
2	6	8
3	12	13
4	7	6
5	10	10
6	5	6
7	9	11
8	10	9

a. What is the correlation between the two sets of subtest scores?
b. If one of the subjects receives a score of 11 on the Digit Span subtest, what is your best estimate of the subtest score that person will receive on Vocabulary?

4. At a certain public high school the correlation between IQ and GPA was a *significant* .60. The mean GPA was 3.00 with an s of .16. The mean IQ is 105 with an s of 10.00
   a. Find the regression line for predicting GPA on the basis of IQ.
   b. What is the value of the slope?
   c. What is the value of the intercept?

5. Using the data from problem 4,
   a. find the GPAs for students with IQs of 80, 95, 105, and 130.
   b. what kind of research was this, post facto or experimental?
   c. identify both the IV and the DV.

6. A study conducted at the University of Munster in Germany found that shorter men (but not women) had higher diastolic blood pressures than did taller men. The research was also confirmed in a study of Rhode Island men who were tested at the Pawtucket Memorial Hospital (*New York Times,* March 16, 1996). In fact the data indicated that diastolic blood pressure increased a little over 5 points for roughly each four inches shorter the men were. Suggested explanations included the fact that shorter men have smaller blood vessels that may be more easily clogged with fatty deposits. Also being short as an adult may be the result of poor nutrition in childhood, which then might affect the heart later in life. For this problem, assume a mean height of 69 inches with an estimated standard deviation of 3 inches and assume a mean diastolic blood pressure of 80 with an estimated standard deviation of 5.50. With a significant Pearson r of –0.68, predict diastolic blood pressures for the following adult male heights:
   a. 72 inches
   b. 60 inches
   c. 58 inches
   d. 76 inches
   e. 54 inches

7. The correlation between height and IQ among the students at a certain eastern women's college is zero. The mean IQ at the college is 125 with an estimated standard deviation of 8.20. The mean height is 65 inches with an estimated standard deviation of 2.40 inches. What IQ would you predict for a student at this college who is 68 inches tall?

8. Of the men's varsity football players in the mid-America conference, there is a significant correlation between height and weight of .75. The mean height is 73 inches with an estimated standard deviation of 2.10 inches. The mean weight is 210 pounds with an estimated standard deviation of 16.25 pounds. Predict the height of each of the following players:
   a. a 175-pound halfback
   b. a 195-pound split end
   c. a 240-pound linebacker

9. For each of the height predictions in problem 8, assume a sample size of 30 and find the .95 confidence intervals.

10. Using the data (the Pearson r, the two means, and the two standard deviations) from problem 1 at the end of Chapter 11, predict the "Number of Hours Worked" from the following "Party Loyalty" scores: 11, 17, 18.

11. Using the data (the Pearson r, the two means, and the two standard deviations) from problem 3 at the end of Chapter 11, predict the achievement scores for the following anxiety scores: 3, 7, 11.

12. Assume that a large sample of adult females was selected and that the following Pearson r correlations were obtained. The correlation between weight and blood cholesterol levels was .75. The correlation between pulse rate and cholesterol levels was .70, and the correlation between weight and pulse rate was .30. Calculate the multiple R for predicting blood cholesterol levels, using both weight and pulse rate as the predictors.

13. For problem 12 and assuming a sample size of 73, test the significance of the multiple R for predicting blood cholesterol levels.

14. A large sample of adult males is selected and the following Pearson r correlations were obtained. The correlation between height and weight was .60, the correlation between weight and chest size was .70, and the correlation between height and chest size was .50. Find the partial correlation between height and chest size with the weight variable having been ruled out.

15. For problem 14, assume a sample size of 103 and test the partial correlation between height and chest size for significance.

16. A researcher selects a random sample of college sophomores and gives them three separate tests, one for reading speed (rs), one for reading comprehension (comp) and a WAIS IQ test (iq). The data are as follows:

rs	*comp*	iq
55	50	110
35	45	95
75	60	135
30	40	90
45	50	105
50	45	100
45	40	95
55	60	110
48	52	102
42	48	100

a. What is the correlation between reading speed and comprehension?
b. What is the correlation between reading speed and IQ?
c. What is the correlation between comprehension and IQ?
d. Find the correlation between reading speed and reading comprehension, with the influence of IQ being partialled out or controlled.

17. A researcher is interested in the possible intercorrelations among three of the subtest scores on the WAIS (Wechsler Adult Intelligence Scale). A random sample of ten inmates from a Federal Prison were selected: They were given the Comprehension subtest (a test of social intelligence), the Digit Span (a test of short-term memory) subtest and the Information subtest (a test of general knowledge). Their weighted scaled scores were as follows:

Subject	*Information*	*Digit Span*	*Comprehension*
1.	9	10	11
2.	6	5	8
3.	10	11	13
4.	7	8	6
5.	10	9	10
6.	5	5	6
7.	9	8	11
8.	10	8	9
9.	11	8	10
10.	5	8	6

a. What is the correlation between Digit Span and Information?
b. What is the correlation between Digit Span and Comprehension?
c. What is the correlation between Information and Comprehension?
d. What is the multiple correlation for predicting Comprehension from *both* Digit Span and Information?

18. This problem is based on the following data set:

Personality Scale			
Student	ES	S	V
1	16	5	8
2	19	6	8
3	15	7	7
4	15	6	8
5	14	5	6
6	15	6	8
7	13	6	7
8	12	6	8
9	4	6	2
10	15	8	7
11	16	4	7
12	14	5	7
13	17	6	8
14	13	7	10
15	14	2	4
16	15	7	12
17	17	5	4
18	26	14	17
19	16	5	8
20	14	6	7

The personality test scales are

V = Venturesomeness scale on the Eysenck & Eysenck Personality questionnaire
ES = Experience Seeking subscale of the Zuckerman Sensation Seeking Scale
S = Sociability Scale from the Eysenck Personality Inventory

a. Both V and ES may be measuring similar personality traits. Obtain the Pearson r between them and determine whether it is significant.

b. Find the Pearson r between ES and S. Are experience seekers typically more sociable?

c. Find the Pearson r between V and S. Are venturesome persons typically more sociable?

d. Calculate a regression line for predicting V scores on the basis of the S scale scores (that is, the line for predicting V on the basis of a single predictor, S).

e. Using the regression line obtained in part d, predict a V score for a person whose S score was 8.

19. Using the data from the above problem (18),

a. find the multiple R for predicting a V score on the basis of the two predictors, ES and S. Is this correlation significant, and is it higher than either of the internal correlations with the y variable?

b. using the multiple R results found in part a, predict a person's V score, given a score of 17 on ES and a score of 8 on S.

20. The correlation between the IQs of fathers and their children is a significant .45. The correlation between the IQs of mothers and their children is a significant .55. The IQ correlation between husbands and wives has been reported to be a significant .40 (Jensen, 1978). Find the multiple R for predicting the child's IQ as the criterion variable and the IQs of the parents as the predictor variables.

21. Assuming a sample size of 63 persons, is the multiple correlation found in problem 20 significant?
22. Assuming a sample size of 13 persons, is the multiple correlation found in problem 20 still significant?
23. Using the value of the multiple R found in problem 20, predict the IQ of a child whose father's IQ is known to be 115 and whose mother's IQ is known to be 120. For all three distributions (mother, father, and child), assume a mean IQ of 100 and an SD of 15.
24. Assume a significant correlation of .50, an estimated standard deviation on the Y variable of 15 and a sample size of 100. Find the standard error of estimate.
25. Among elementary school children, the correlation between height and strength is a significant .65. The correlation between height and age is a significant .82. The correlation between strength and age is a significant .75. What is the resulting correlation between height and strength when the age variable has been partialed out?
26. For problem 25, assume a sample size of 63 and test the partial correlation between height and strength for significance.
27. The correlation between reading speed and SAT scores is .60. The correlation between reading speed and IQ is .62. The correlation between IQ and SAT scores is .58. Find the correlation between reading speed and SAT scores with the influence of IQ partialed out.
28. Using the partial correlation between reading speed and SAT found in problem 27, and assuming a sample of 18 children, test for significance.
29. Assume a sample size of 18 subjects and a multiple R (based on two predictors and a single Y criterion) of .58. Is the R significant?
30. Assume a sample size of 33 subjects and a multiple R (based on two predictors and a single Y criterion) of .40. Is the R significant?

Indicate what term or concept is being defined in problems 31 through 35.

31. The amount of increase in Y that accompanies a given increase in X
32. The single straight line that lies closest to all the points on a scatter plot
33. The value of Y when X equals zero
34. The resulting correlation between two variables when the effects of a third variable have been statistically ruled out
35. The correlation between a criterion variable and several predictor variables

Fill in the blanks in problems 36 through 41.

36. When the correlation between X and Y is zero, what is the best prediction of a Y score that can be made from a given X score? ________
37. If the regression line has a slope of .50, then each single unit increase in X will be accompanied by how much of an increase in Y? ________

38. Each single data point on a scatter plot represents ________.

39. When the regression line slopes from upper left to lower right, then the sign of the correlation must be ________.

40. When all the points on a scatter plot lie directly on the regression line, then the value of the correlation must be ________.

41. In the regression equation, what term denotes the point where the regression line crosses the ordinate? ________

True or False. Indicate either T or F for problems 42 through 48.

42. The higher the correlation, the more a predicted *Y* value may deviate from the mean of the *Y* distribution.

43. The more a correlation deviates from zero, the better is its predictive accuracy.

44. For the Pearson *r*, degrees of freedom are assigned on the basis of the number of *pairs* of scores minus the constant 2.

45. A correlation of +.75 must be significant, regardless of the degrees of freedom.

46. No correlation is ever greater than +1 or less than –1.

47. If *X* correlates significantly with *Y*, then *X* is probably the cause of *Y*.

48. If *X* correlates .90 with *Y*, then *Y* must also correlate .90 with *X*.

COMPUTER PROBLEMS

C1. The admissions officer at a small college collected the following SAT data (V, verbal, and M, math) for 47 students who had applied to become computer science majors.

	V	*M*		*V*	*M*		*V*	*M*
1	510	640	17	400	440	33	710	780
2	470	670	18	420	550	34	710	680
3	290	420	19	420	550	35	610	710
4	380	660	20	450	600	36	290	420
5	640	730	21	420	670	37	780	730
6	240	540	22	450	470	38	370	390
7	390	410	23	490	550	39	670	670
8	330	560	24	780	700	40	460	450
9	290	310	25	660	490	41	660	650
10	430	440	26	490	500	42	440	440
11	510	500	27	610	600	43	510	380
12	460	500	28	540	590	44	490	510
13	460	480	29	510	510	45	390	400
14	650	610	30	530	530	46	610	640
15	500	430	31	450	470	47	490	470
16	780	770	32	400	460			

a. Find the linear correlation between verbal and math SAT scores for this sample.
b. Is the correlation significant?
c. Predict the math score for the following verbal SAT scores: 370, 610, 510, and 700.

C2. A random sample of 20 inmates from the county correctional center was selected and given three tests: (1) The Novaco Provocation Inventory, a test of the range and intensity of a person's anger; (2) alcohol use—as measures on the Alcohol Scale that are taken from the SASSI—Substance Abuse Subtle Screening Inventory (Miller, 1990); and (3) drug use—as measured on the SASSI Drug Scale.

Subject #	*Alcohol Score*	*Drug Score*	*Anger Score*
1.	14	12	290
2.	6	10	240
3.	5	8	180
4.	12	10	210
5.	8	7	185
6.	15	12	230
7.	10	8	225
8.	8	10	260
9.	11	8	290
10.	9	10	230
11.	14	12	290
12.	6	10	240
13.	5	8	180
14.	12	10	210
15.	8	7	185
16.	15	12	230
17.	10	8	225
18.	8	10	260
19.	11	8	290
20.	9	10	230

a. Find the Pearson r between alcohol use and drug use. Is it significant?
b. Find the Pearson r between alcohol use and anger. Is it significant?
c. Find the Pearson r between drug use and anger. Is it significant?
d. Find the multiple R for predicting anger on the basis of both alcohol use and drug use.
e. Using the multiple R results found above, predict an anger score for an inmate who had an alcohol score of 13 AND a drug score of 9.

C3. The following represent distributions of verbal SAT scores and college GPAs selected from a random sample of 20 college students, all of whom have completed their first year of study.

Subject	*SAT*	*GPA*	*Subject*	*SAT*	*GPA*
1	600	2.76	11	490	2.70
2	550	2.86	12	500	2.61
3	540	2.77	13	550	2.70
4	320	2.64	14	250	1.40
5	480	2.64	15	400	2.54
6	600	2.70	16	500	2.78
7	510	2.70	17	460	2.53
8	480	2.63	18	500	4.00
9	470	2.62	19	750	2.87
10	500	2.76	20	550	2.79

a. For the SAT distribution, find the mean and median. Is the distribution skewed? If so, which way?
b. For the SAT distribution, find the range and standard deviation.
c. For the GPA distribution, find the mean and median. Is the distribution skewed? If so, which way?
d. For the GPA distribution, find the range and standard deviation.
e. Find the linear correlation between the SAT verbal scores and grade-point average (GPA).
f. Find the regression equation for predicting GPA on the basis of SAT.
g. Predict GPAs for the following SAT scores: 470, 560, and 590.
h. Calculate the standard error of estimate.

C4. In the previous problem, the researcher measured 20 subjects on one independent variables and one dependent variable. Later the researcher wondered if the correlation between SAT and GPA might have been inflated due to a failure to control for IQ. It was decided to give each of the subjects a WAIS-III IQ test and their scores were as follows (each subject's number will be followed by the IQ score for that person):

1. 130, 2. 135, 3. 120, 4. 95, 5. 95, 6. 125, 7. 110, 8. 100, 9. 105, 10. 100, 11. 90, 12. 110, 13. 130, 14. 90, 15. 105, 16. 110, 17. 100, 18. 140, 19. 130, 20. 120.

a. For the IQ distribution shown above, find the mean, median, and range.
b. Find the correlation between IQ and SAT.
c. Find the correlation between IQ and GPA.
d. Find the partial correlation between SAT and GPA with the IQ variable being partialled out. (Remember, you already have the original correlation between SAT and GPA from problem C3 above).

Chapter

15

Repeated-Measures and Matched-Subjects Designs with Interval Data

During our earlier discussion of experimental methodology (Chapter 9), three basic designs were presented: the between-subjects (after-only), the repeated-measures (before-after), and the matched-subjects. Each of these designs is aimed at the creation of equivalent groups of subjects so that the potential effects of the manipulated independent variable can be assessed. In a between-subjects design, each subject is randomly and independently selected and independently assigned to either the control or the experimental conditions. That is, the fact that a given subject has been assigned to the experimental group has no influence on who might then be selected for the control group. Because of this principle of independence, analysis of experimental results can be accomplished with many of the tests we have already covered, such as the independent t *or* F *tests for interval data and the chi square for nominal data.*

PROBLEM OF CORRELATED OR DEPENDENT SAMPLES

The other two experimental designs, repeated-measures and matched-subjects, pose an inherent statistical problem that must be resolved before analysis of their results can be completed. This problem is that of **correlated samples,** and it arises due to the fact that whenever subjects are paired off as either subjects with themselves, as in the case of the repeated-measures design, or subjects with their matched partners, as in the matched-subjects design, a correlation almost certainly results between the paired scores. For example, if a repeated-measures design is used to conduct a weight loss study, each subject is weighed, the independent variable is manipulated, and each subject is weighed again. Even if every single subject does lose weight, however, the relative standing among the subjects might well remain very similar in both the before and after measures. (See Table 15.1.)

TABLE 15.1 Results of a before-after weight loss study.

Subject	Weight Before, X_1 (Pounds)	Weight After, X_2 (Pounds)
1	220	180
2	160	155
3	140	133
4	112	108
	$M_1 = 158$	$M_2 = 144$

We see in Table 15.1 that Subject 1, who was the heaviest in the before measure, is still heaviest in the after measure, despite the dramatic loss of 40 pounds. Also, Subject 4 is the lightest in both the before and after measures. The paired scores thus show a high degree of correlation, even though the thrust of the analysis is on the difference between the two sets of weight measures. As with all experimental designs, the appropriate statistical test for repeated-measures (R/M) data is of the hypothesis of difference, but it's to our advantage to take the correlation into account.

Similarly, in a matched-subjects (M/S) design, if our matching process has been at all effective, there should be correlation between the resulting pairs of matched scores. For example, assume that two groups of subjects are matched person for person on the basis of IQ. One group (control) is given a placebo, while the other group (experimental) is given a special math ability–enhancing drug. The two groups then take a math achievement test. (See Table 15.2.)

In this case, we expect the paired math scores to correlate—that is, if in fact matching on IQ is relevant, as it certainly appears to be. Note, too, that although the difference between the means of the two groups is not dramatically great, it is rather impressive that all subjects in the experimental group outperformed their counterparts in the control group.

TABLE 15.2 Results of a matched-subjects math achievement study.

Subject Pairs	Control Math Scores, X_1	Experimental Math Scores, X_2
1 (Both with IQs of 130)	90	95
2 (Both with IQs of 75)	55	57
3 (Both with IQs of 110)	85	88
4 (Both with IQs of 100)	70	72
	$M_1 = 75$	$M_2 = 78$

The previous discussion applies only to true repeated-measures and/or matched-subjects designs. The matched-group design, wherein whole groups are matched on the basis of average scores, does not create the potential for correlation, and the data from such a design should be treated as though the groups were independent.

REPEATED MEASURES

The researcher must be careful when using a repeated-measures design since it creates the risk of two serious errors, those due to sequence effects as well as those caused by carryover effects. That is, whenever the act of measuring subjects in the first condition produces uncontrolled influences on the measures that follow, then other designs should be considered. Obviously this is not true in all studies, such as learning studies in which the researcher is hoping to show that exposure to the first condition leads to increased levels of performance on later measures.

The repeated-measures design is a special case of what is technically called a randomized blocks design. Repeated use of the same subjects, as we have seen, leads to correlation among the measures, just as in matched-subjects designs. If we set up a three matched-subjects design, the triad of subjects, each being matched on some relevant variable should provide significant correlations among the scores.

PAIRED *t* RATIO

When two distributions of interval data are to be compared for possible differences and the data result from either an R/M or an M/S design, then the **paired *t* ratio** should be used. That is, when there is correlation between the pairs of scores, then the independent t is no longer the appropriate test.

Corrected Equation for the Standard Error of Difference

When the equation for the estimated standard error of difference (the denominator in the t ratio) was first introduced, the equation given was

$$SE_D = \sqrt{SE_{M_1}^2 + SE_{M_2}^2}$$

In a way, this equation is not theoretically correct, even though its use in calculating the independent t ratio is certainly justified. We were not lying to you—just protecting you from possible trauma.

The corrected equation for the estimated standard error of difference is

$$SE_D = \sqrt{SE_{M_1}^2 + SE_{M_2}^2 - 2r_{1,2}SE_{M_1}SE_{M_2}}$$

The element $r_{1,2}$ is read as "r sub one two." It is the correlation value between the first and second sets of measures.

Effects of the Correlation Term. The new element in the equation for the standard error of difference, the correlation term,

$$2r_{1,2}\, SE_{M_1} SE_{M_2}$$

is, in fact, a single product. Therefore, if any value in that product is equal to zero, then the whole term must be equal to zero. (Zero times any value equals zero.) When this equation is applied to independently selected samples, the correlation $r_{1,2,}$ has to be zero. There is no correlation between a pair of independently selected samples. In fact, even if we tried, we could not pair off such scores to calculate a Pearson r. So for the independent t, where no correlation between the measures is possible, the standard error of difference is, in fact,

$$SE_D = \sqrt{SE_{M_1}^2 + SE_{M_2}^2 - 0}$$

or simply

$$SE_D = \sqrt{SE_{M_1}^2 + SE_{M_2}^2}$$

Also, although the estimated standard error of difference can be calculated for unequal sample sizes, when the correlation term is included, there must be the same number of scores in each distribution.

Advantages of the Paired t *Ratio*

For the paired t, then, the full equation for the estimated standard error of difference must be employed. In fact, using the full equation provides a decided statistical advantage. Since the correlation term is subtracted from the first term, the resulting value for the estimated standard error of difference will be *less* than what it would otherwise have been. Also, since t is a ratio of mean differences divided by the estimated standard error of difference, then *reducing the estimated standard error of difference increases the size of* t. For the same numerator, then

$$\text{independent } t = \frac{2}{4-0} = \frac{2}{4} = .50$$

$$\text{paired } t = \frac{2}{4-3} = \frac{2}{1} = 2.00$$

This is important. The larger the value of the t ratio, the more likely it is that t will equal or exceed the table value and therefore be found to be significant.

Using the paired t, then, has three major effects:

1. It reduces the estimated standard error of difference by a factor related to the size of r, which in turn
2. increases the size of the t ratio, which in turn
3. increases the chances of rejecting the null hypothesis and achieving significance.

These points are valid only when the correlation has a positive value, which it certainly should have with an R/M design. If with an M/S design, we find a negative value for the correlation, then the matching variable is definitely not relevant and should be discarded. Actually, a negative correlation value *increases* the value of the standard error of difference and *decreases* the size of the t ratio. This is a somewhat academic issue, however, since the chances of getting a negative r from an M/S design are about the same as the chances of being hit by an arrow from a crossbow.

Degrees of Freedom for the Paired t *Ratio*

The degrees of freedom for the paired t ratio are equal to the number of *pairs* of scores, N, minus one:

$$\text{df} = N - 1$$

Since the paired t makes use of scores that are yoked together, the data do not include as many independent observations as there are with the independent t. As someone once said, "Going from the independent to the paired t is like getting married—you lose half your degrees of freedom." For a given number of scores, the df for the paired t are exactly half what they are if the independent t is used. As we know from our previous use of the t table, as degrees of freedom decrease, the table value of t needed to reject the null hypothesis increases. Thus, the decrease in degrees of freedom makes it more difficult to get a significant value of t.

We just discovered in the preceding section that using the correlation term decreases the value of the estimated standard error of difference and increases the t ratio.

It's all Greek to me, sir!

However, it seems now that that advantage has been thrown away into the degrees-of-freedom scrap heap. The situation is rosier than it appears. When that correlation term is large, the increase in the resulting t ratio *more than compensates for the loss of degrees of freedom.*

Example A researcher wishes to discover whether the intake of orange juice affects the potassium level in the bloodstream. A group of 12 elderly patients is selected from those in a large nursing home, where previous diet has been controlled. Potassium blood levels are measured for each subject. Then each subject is given a pint of orange juice, and, two hours later, potassium levels are again measured. The data are as follows (the scaled scores represent the potassium blood level):

Subject	*Before*, X_1	X_1^2	*After*, X_2	X_2^2	X_1X_2
1	26	676	25	625	650
2	25	625	28	784	700
3	24	576	27	729	648
4	23	529	26	676	598
5	23	529	25	625	575
6	21	441	23	529	483
7	19	361	21	441	399
8	17	289	19	361	323
9	17	289	16	256	272
10	16	256	18	324	288
11	15	225	19	361	285
12	9	81	19	361	171
	$\Sigma X_1 = 235$	$\Sigma X_1^2 = 4877$	$\Sigma X_2 = 266$	$\Sigma X_2^2 = 6072$	$\Sigma X_1X_2 = 5392$

To calculate the paired t ratio, use the following steps:

1. Calculate the means for each distribution.

$$M_1 = \frac{\Sigma X_1}{N} = 19.583 \qquad M_2 = \frac{\Sigma X_2}{N} = 22.167$$

2. Calculate the estimated population standard deviations for each distribution.

$$s_1 = \sqrt{\frac{\Sigma X_1^2 - \frac{(\Sigma X_1)^2}{N}}{N-1}} = 4.999 \qquad s_2 = \sqrt{\frac{\Sigma X_2^2 - \frac{(\Sigma X_2)^2}{N}}{N-1}} = 3.996$$

3. Calculate the Pearson r. This is the same equation shown back in Chapter 11, except that X_2 is substituted for Y.

$$r = \frac{\dfrac{(\Sigma X_1X_2) - (M_x)(M_y)(N)}{N-1}}{(s_x)(s_y)} = \frac{\dfrac{5392 - (19.583)(22.167)(12)}{11}}{(4.999)(3.996)}$$

$$r = .832$$

4. Calculate the estimated standard error of each mean.

$$SE_{M_1} = \frac{s_1}{\sqrt{N}} = 1.443 \qquad SE_{M_2} = \frac{s_2}{\sqrt{N}} = 1.153$$

5. Calculate the estimated standard error of difference. (Chain-multiply the correlation term, that is, 2 × .832 × 1.443 × 1.153 = 2.769.)

$$SE_D = \sqrt{SE^2_{M_1} + SE^2_{M_2} - 2r_{1,2}SE_{M_1}SE_{M_2}} = \sqrt{1.443^2 + 1.153^2 - 2(.832)(1.443)(1.153)}$$

$$= \sqrt{2.082 + 1.329 - 2.769} = \sqrt{3.411 - 2.769} = \sqrt{.642} = .802$$

6. Calculate the paired t ratio.

$$t = \frac{M_1 - M_2}{SE_D} = \frac{19.583 - 22.167}{.802} = -3.222$$

7. Check the paired t ratio for significance, using 11, or N pairs – 1, as the degrees of freedom. At the .01 alpha level the table tells us we need a value of ±3.106 to reject the null hypothesis. Since the absolute value of our calculated t of –3.222 is GREATER that the table value we reject H_0 and conclude that we have a significant difference. Our conclusions would be stated as follows:

Conclusions: The paired t was computed on this pre-post study. The paired t was significant ($t(11) = -3.222$, $p < .01$). The potassium blood level scores were significantly higher in the post test.

Thus, the subjects do differ significantly in their before and after potassium blood levels. The independent variable is effective, since the group, originally selected from one population, now represents a different population. If the direction of this difference had been predicted—that is, if the alternative hypothesis had been stated as $\mu_1 < \mu_2$ (or that potassium levels will increase as a result of the independent variable)—then the one-tail table (Table D), could be used for finding the critical value, which at df 11 would be 2.718

CONFIDENCE INTERVAL FOR PAIRED DIFFERENCES

Instead of testing the null hypothesis, some researchers prefer the confidence interval, and for the paired t situations, the equation is as follows:

$$CI = (\pm t)(SE_D) + (M_1 - M_2)$$

Using the above data and remembering that the value of t in this equation comes from Table C, and that the df is based on a sample size of N pairs – 1, or 11. Thus, at the .95 level the t ratio would be 2.201, or 3.106 at the .99 level. In this example, it is important to remember that the negative difference between the means simply indicates that the second mean is higher than the first.

$$\begin{aligned} CI\ .95 &= (\pm 2.201)(.802) + (19.583 - 22.167) \\ &= +1.765 + -2.584 = -0.819 \\ &= -1.765 + -2.584 = -4.349 \end{aligned}$$

Thus, we can be .95 confident that the true population difference between these means falls within the range – 0.819 and – 4.349.

To again show that the range of possible differences is extended when using the .99 confidence interval, we will again use the table value of t for 11 degrees of freedom but this time change the alpha to .01. This results in a table value of 3.106

$$\begin{aligned} CI\ .99 &= (\pm 3.106)(.802) + (19.583 - 22.167) \\ &= +2.491 + -2.584 = -0.093 \\ &= -2.491 + -2.584 = -5.075 \end{aligned}$$

Effect Size

Computing an **effect size** for a paired t first requires the calculation of the paired t ratio. If the t is significant, then the following equation can be used:

$$d = \frac{t}{\sqrt{N}} \quad \text{(where } N \text{ equals the number of } pairs \text{ of scores).}$$

Thus, if the paired t were equal to –3.222, as in the previous example, and there were 12 pairs of scores,

$$d = \frac{3.222}{\sqrt{12}} = \frac{3.222}{3.464} = .930 \text{ (a strong effect)}$$

Even though it makes no difference to the interpretation whether d is positive or negative, it's convenient here to change the sign of the t ratio and read d as a positive value.

Calculate the effect size for the following paired t ratios:

$t = 2.61$ and $N = 15$

ANS: $d = .674$

$t = 3.13$ and $N = 10$

ANS: $d = .990$

$t = 5.97$ and $N = 30$

ANS: $d = 1.090$

Paired t and Power

Previously, in Chapter 10, we discussed the concept of power, which was defined as the test's sensitivity in detecting significant results. It was mentioned then that a variety of methods could be employed for increasing power, such as *fitting the statistical test to the research design*. A discussion of the inner workings of the paired t can be especially illustrative of this point—fitting the test to the design.

Let us compare the workings of the independent and paired (or dependent) t tests. On the surface they seem to be very similar—both use interval-ratio data, both test for the possibility of a difference between two sample means, both require a normal distribution of scores in the underlying population, and so on. However, they differ dramatically in both the procedures involved and the kind of results each may be expected to give. The independent t demands that the two sample measures be absolutely independent of each other or that the selection of one sample in no way influences the selection of the second sample.

The paired or dependent t, on the other hand, demands that the two samples be somehow related to each other, where the selection of a subject for one sample group determines who will be selected for the second group. An example of an experimental design where this holds true would be the matched-subjects design.

What if, in the preceding example, the independent t had been used? That is, what if the researcher had not taken advantage of the correlation between the pairs of scores—could the decision to reject the null hypothesis still have been made? We can easily find out. The standard error of difference, without the correlation term included, would have been as follows:

$$SE_D = \sqrt{SE^2_{M_1} + SE^2_{M_2}} = \sqrt{1.443^2 + 1.153^2} = \sqrt{2.082 + 1.329} = \sqrt{3.411}$$

$$SE_D = 1.847$$

$$t = \frac{M_1 - M_2}{SE_D} = \frac{19.583 - 22.167}{1.847} = \frac{-2.584}{1.847} = -1.399$$

Now, no matter how many degrees of freedom, no matter whether a one- or two-tail test is used, this is an acceptance of H_0. The important point is that it should not be an accept. Therefore, by not taking advantage of the correlation term, the decision would have been to accept *when it should have been to reject*. Using the wrong t test caused the Type 2 error to occur, that is, accepting the null hypothesis when we should be rejecting it.

On the other hand, suppose that a researcher decides that the paired t works so beautifully that it should be used in all research situations, even when the observations are independent. Again, the beta error will be increased; since the observations are independent, there is no correlation with which to reduce the estimated standard error of difference.

In short, using the wrong statistical test increases the beta probability and therefore reduces the power of the test. A paired t is more powerful than is an independent t when the subjects are truly dependent on each other; the independent t is more powerful when the subjects are independently selected and assigned.

Some Cautions Regarding the Paired t

Although the paired t, as we have seen, is an appropriate test for analyzing results of matched-subjects and before-after designs, it is not the best test to apply to a before-after design with an independent control group. In this experimental design, as we know, two groups of subjects are randomly selected, and both groups are measured in the before and after conditions. The groups differ only in the extent to which they experience different levels of the independent variable. This design is becoming increasingly popular, and justifiably so.

As we learned previously, the use of the before-after design with separate control reduces the possibility of the occurrence of one of the major research errors, confounding the independent variable. We saw that with a one-group before-after design, factors other than the independent variable may inadvertently act on the subjects to change them. One example of such confounding is the Hawthorne effect, where the subjects change as a result not of the independent variable, but of the flattery and attention they receive from the researcher(s). With a separate control group, any possible confounding variables should influence both groups equally, and thus the pure effects of the independent variable can be more accurately factored out.

A Significant Change in the Control Group Before and After

With two groups, each measured twice, the paired t test may not always work. For example, although it is possible, despite the increase in alpha, to do t tests between the experimental group's pre- and post-measures and the control group's pre- and postmeasures, what happens when both t's are significant? That is, the control group may also change significantly from the before to the after condition, even though the change is much less than that observed for the experimental group. After all, the possibility that there might be a difference in the magnitude of the change scores between the two groups is the very reason for using a separate control group in the first place.

One of the most effective ways to analyze such data is to focus on the "change" scores (or the difference between the pre- and post-measures), since in this type of study the control group can also be expected to show a change. The question then is, Which group changed the most? Although some researchers have challenged the use of change scores because of the possibility of reliability problems, statistical techniques have been devised to counter such a possible threat (see Kerlinger, 1986, p. 311).

Researchers who argue in favor of the use of gain or change scores insist that regression toward the mean is no reason for not using the technique, unless the regression of the post scores on the pre-scores is not linear (Maris, 1998). That change scores can be reliable, in fact highly reliable, has been shown by Caruso and Cliff (2000) as well as Rogosa and Willet (1983). Those who argue against this technique suggest instead the use of an ANOVA mixed design (Sheskin, 1997). One must keep in mind, however, that if one of the groups scores at the very high end of the scale during the pre-condition, there may not be room for much more of an upward change for that group (review the section on the ceiling effect in Chapter 9).

For example, suppose that we wish to test a new method for teaching learning disabled children to read. Two groups are randomly selected and are given a reading test. One group (experimental) then undergoes six weeks of special training, while the other group (control) does not. At the end of the six-week period, both groups are given the reading test again. It is possible that the control group will improve because of maturation, outside influences, or whatever. However, the experimental group's improvement may be far greater—a clear indication that the training program works. The comparison, therefore, should be between each group's *change scores.* And since the two groups have been randomly and independently assigned, then the change scores themselves are independent of each other. We must remember that the paired t can only be used when scores can be paired off with each other. In this study, because the groups are independently selected, there is no way to match up the change scores. The appropriate technique in this case is to compare the change scores using the independent t, assessing the degrees of freedom on the basis of the number of separate change scores.

Example A researcher is interested in discovering whether a role-reversal procedure will influence men's attitudes toward women. Two samples of married men were *randomly* and *independently* selected. Both groups were given a test measuring their attitudes toward women's roles in our society (low scores indicating a traditional attitude and high scores a more enlightened attitude). The men in Group A were then instructed to reverse roles with their wives for the next two weekends (he doing her work and she doing his). The men in Group B were told nothing. Two weeks later both groups took the attitude test again. The data are as follows:

Group A			Group B		
Pre X_1	Post X_2	Change, $X_2 - X_1$	Pre X_1	Post X_2	Change, $X_2 - X_1$
140	152	12	160	161	1
150	156	6	149	151	2
163	165	2	139	149	10
140	150	10	148	152	4
146	148	2	133	136	3
130	135	5	143	144	1
150	155	5	139	146	7
143	163	20	145	160	15
140	155	15	158	160	2
160	162	2	139	152	13

Because it makes no difference to the *t* test whether the change scores are plus or minus, we can establish the change-score distribution with values $X_2 - X_1$, giving us positive values and making the calculation easier. Now we compare only the change-score distributions, using the independent *t*. We will call the change scores in the experimental group X_E and the change scores in the control group X_C.

Experimental Group		Control Group	
Change Score, X_E	Change Score², X_E^2	Change Score, X_C	Change Score², X_C^2
12	144	1	1
6	36	2	4
2	4	10	100
10	100	4	16
2	4	3	9
5	25	1	1
5	25	7	49
20	400	15	225
15	225	2	4
2	4	13	169
$\Sigma X_E = 79$	$\Sigma X_E^2 = 967$	$\Sigma X_C = 58$	$\Sigma X_C^2 = 578$

1. $M_E = 7.900$ $\quad M_C = 5.800$
2. $s_E = 6.173$ $\quad s_C = 5.181$
3. $SE_{M_E} = 1.952$ $\quad SE_{M_C} = 1.638$
4. $SE_D = 2.548$
5. $t = \dfrac{7.900 - 5.800}{2.548} = \dfrac{2.100}{2.548} = .824$

Was there a significant difference between the two groups regarding how much their attitudes changed? With 18 *df*, the tabled value of t at .05 is equal to 2.101, and since the calculated value of t did not meet or beat that value the null hypothesis would have to be accepted.

Conclusions: A two-sample independent t test was computed between two sets of change scores. The t ratio was not significant $t(18) = .824$, $p > .05$), ns.) The role reversal approach did not alter attitude scores.

Using the Paired t on a Repeated-Measures Design with a Matched Control Group

Another variation on the repeated-measures design theme is to create the separate control group by matching the subjects person by person with those in the experimental group. In this way it is assumed that the two groups are equivalent at the outset. The data from this type of design are handled in a fashion similar to that used for the R/M design with an independent control group, in that again the focus is on the change scores rather than the raw scores. However, since the change scores can be paired between the matched subjects in this variation, analysis of the data can be accomplished using the paired t. Also, notice that the value of the Pearson r in this next example gives us an indication of whether the matching process was successful.

Example Suppose that the two groups of men in the previous example had been matched on IQ (on the theory that a man's level of intelligence may be a factor in his willingness to change to a more enlightened attitude).

	Group A			Group B		
Matched Pair No.	Pre X_1	Post X_2	Change, $X_2 - X_1$	Pre X_1	Post X_2	Change, $X_2 - X_1$
1	163	165	2	160	161	1
2	150	156	6	149	151	2
3	140	152	12	139	149	10
4	150	155	5	148	152	4
5	130	135	5	133	136	3
6	146	148	2	143	144	1
7	140	150	10	139	146	7
8	143	163	20	145	160	15
9	160	162	2	158	160	2
10	140	155	15	139	152	13

The change scores are set up separately, as shown before.

Experimental Group		Control Group		
Change Score, X_E	Change Score², X_E^2	Change Score, X_C	Change Score², X_C^2	$X_E X_C$
2	4	1	1	2
6	36	2	4	12
12	144	10	100	120
5	25	4	16	20
5	25	3	9	15
2	4	1	1	2
10	100	7	49	70
20	400	15	225	300
2	4	2	4	4
15	225	13	169	195
$\Sigma X_E = 79$	$\Sigma X_E^2 = 967$	$\Sigma X_C = 58$	$\Sigma X_C^2 = 578$	$\Sigma X_E X_C = 740$

1. $M_E = 7.900$ $M_C = 5.800$
2. $s_E = 6.173$ $s_C = 5.181$

Notice that the means and standard deviations are exactly the same as were shown for the independent t analysis.

3. $r = .979$

4. $SE_{M_E} = 1.952$ $SE_{M_C} = 1.638$

5. $$SE_D = \sqrt{1.952^2 + 1.638^2 - 2(.979)(1.952)(1.638)}$$
$$= \sqrt{3.810 + 2.683 - 6.260} = \sqrt{.233}$$
$$= .483$$

6. $$t = \frac{7.900 - 5.800}{.483} = 4.348$$

For this analysis, the df are based on the number of *pairs* of change scores, minus one (10 – 1 = 9). The value for the paired t shows an increase over the value obtained from the independent t previously done on the same data. This is strong evidence that where there is a substantial correlation between pairs of scores, the paired t can be a very powerful test.

With 9 *df*, the tabled value of t at the .01 level is 3.250. Since our calculated t is clearly greater, 4.348, we reject the null hypothesis.

Conclusions: The paired t was computed on the change scores in this matched subjects study and was significant ($t(9) = 4.348, p < .01$). The role reversal procedure was effective in changing men's attitudes in a more positive direction after matching the men on the IQ variable.

WITHIN-SUBJECTS *F* RATIO

When a repeated-measures or matched-subjects design is being used and results in more than two sets of scores, the *t* test is no longer available. In this situation we return to ANOVA, the *F* ratio. Thus, when more than two distributions of scores are to be compared, the within-subjects *F* ratio becomes the statistical test of choice. Also, although a matched-subjects design can theoretically be set up involving virtually any number of groups, matching does become increasingly difficult as the number of groups increases. It can be difficult enough obtaining matched pairs of subjects, let alone matched trios or quartets. Incidentally, some statisticians demand proof that the matching is indeed relevant to the dependent variable, pointing out that unless the matching process is proven, ANOVA procedures for repeated-measures and matched-subjects designs are very different (Edwards, 1972).

Calculating the Within-Subjects F *Ratio*

Once it has been proven that the matching process has, in fact, produced equivalent groups of subjects, the basic computational method for the **within-subjects *F* ratio** is identical for data from both repeated-measures and matched-subjects designs. Therefore, the following example with a repeated-measures design serves also to illustrate the procedure if the design had utilized matched subjects. However, in this example, notice that each subject is serving in every treatment condition.

In a repeated-measures design, each subject therefore receives all levels of the independent variable, so there are always going to be more scores than there are subjects. This method is also referred to as the "treatments by subjects" design.

Example A researcher wishes to find out whether buying TV time for repeated showing of a political propaganda film is worth the cost. Do attitudes change more the more times a person sees the film, or does buying the repeated TV time simply give different people the opportunity to see it? An experiment is designed in which a group of registered voters is randomly selected and asked to complete an "Attitude Toward the Candidate" questionnaire (high scores indicate a pro attitude). Then they see the 20-minute film and fill out the questionnaire again. They see the film yet another time and again fill out the questionnaire. Thus, they fill out the questionnaire three times. The data are as follows:

	First Time		Second Time		Third Time		
Subject	X_1	X_1^2	X_2	X_2^2	X_3	X_3^2	ΣX_r
1	1	1	2	4	3	9	6
2	2	4	4	16	6	36	12
3	3	9	3	9	3	9	9
4	4	16	5	25	6	36	15
5	5	25	6	36	7	49	18
	Σ = 15	55	20	90	25	139	60

1. Calculate the correction factor, C.

$$C = \frac{(\Sigma X)^2}{N} = \frac{60^2}{15} = 240$$

2. Calculate the total sum of squares (as shown in Chapter 12).

$$SS_t = \Sigma X^2 - \frac{(\Sigma X)^2}{N} = \Sigma X^2 - C$$

$$= 55 + 90 + 139 - 240 = 284 - 240$$

$$= 44$$

3. Calculate the sum of squares between columns (shown in Chapter 12 as the SS_b).

$$SS_{bc} = \frac{(\Sigma X_1)^2}{n_1} + \frac{(\Sigma X_2)^2}{n_2} + \frac{(\Sigma X_3)^2}{n_3} - C$$

$$= \frac{15^2}{5} + \frac{20^2}{5} + \frac{25^2}{5} - 240 = \frac{225}{5} + \frac{400}{5} + \frac{625}{5} - 240$$

$$= 45 + 80 + 125 - 240 = 250 - 240 = 10$$

4. Calculate the sum of squares between rows. Add the scores across each row, so that $\Sigma X_{r1} = 6$, $\Sigma X_{r2} = 12$, and so on. Square each row total, divide by the number of scores in each row, and again subtract C.

$$SS_{br} = \frac{(\Sigma X_{r1})^2}{n_{r1}} + \frac{(\Sigma X_{r2})^2}{n_{r2}} + \frac{(\Sigma X_{r3})^2}{n_{r3}} + \frac{(\Sigma X_{r4})^2}{n_{r4}} + \frac{(\Sigma X_{r5})^2}{n_{r5}} - C$$

$$= \frac{6^2}{3} + \frac{12^2}{3} + \frac{9^2}{3} + \frac{15^2}{3} + \frac{18^2}{3} - 240$$

$$= \frac{36}{3} + \frac{144}{3} + \frac{81}{3} + \frac{225}{3} + \frac{324}{3} - 240$$

$$= 12 + 48 + 27 + 75 + 108 - 240$$

$$= 270 - 240 = 30$$

5. Calculate the residual ($r \times c$) sum of squares (this is the variability that is left over, after the column and row variabilities have been taken out).

$$SS_{r \times c} = SS_t - SS_{bc} - SS_{br}$$
$$= 44 - 10 - 30 = 4$$

6. Set up the ANOVA table, with df assigned as follows:

column df = number of columns minus one = 2

row df = number of rows minus one = 4

residual df = column df times row df = 8

We check the degrees of freedom by adding them together, $2 + 4 + 8 = 14$, to be sure that they equal $N - 1$, or $15 - 1 = 14$.

Source	*SS*	*df*	*MS*	F
Between columns	10	2	$\frac{SS}{df} = \frac{10}{2} = 5$	$\frac{MS_{bc}}{MS_{r \times c}} = \frac{5}{.50} = 10$
Residual ($r \times c$)	4	8	$\frac{SS}{df} = \frac{4}{8} = .50$	

Thus, the F ratio is equal to 10.00, calculated from the mean square between columns divided by the residual mean square.

Interpreting the Within-Subjects F Ratio

Since the numerator (mean square between columns) of the F ratio in the preceding example had 2 df and the denominator (residual mean square) had 8 df, we use these values to look up the critical value of F in Appendix Table G.

Across the top row of the table are the degrees of freedom associated with the numerator, which in this case equal 2, and down the far left column, the df associated with the denominator, which in this case equal 8. At the intersection of the two, we find, for an alpha level of .01, a critical F value of 8.65. Since our calculated F of 10.000 is greater than the tabled value, the null hypothesis should be rejected.

Conclusions: A one-way within-subjects ANOVA was computed, comparing political attitudes for subjects at three different times, after viewing a propaganda film once, then after seeing it again, and finally after viewing it for a third time. A significant F ratio was found ($F(2,8) = 10.000, p < .01$). As the film was repeatedly shown, the subjects' attitudes became increasingly positive.

WITHIN-SUBJECTS EFFECT SIZE

For the within-subjects F ratio, effect size is again (as in the other ANOVAs) calculated as an eta square. Eta square is equal to the sum of squares between columns divided by the sum of squares between columns PLUS the sum of squares for the residual. Remember again that the effect size should only be used when the F ratio is significant. (If the F ratio had not been significant, the effect size, of course, would have to be zero, since there was no effect.)

Thus, for the previous problem, the effect size is

$$\eta^2 = \frac{SS_{bc}}{SS_{bc} + SS_{r \times c}} = \frac{10}{10 + 4} = \frac{10}{14} = .714 \text{ (a strong effect).}$$

Within-Subjects F and Tukey's HSD

As was shown in Chapter 12, Tukey's HSD (honestly significant difference) test can be performed following the calculation of a significant F ratio. The HSD is used to determine, post hoc, exactly where the sample differences occurred. For the within-subjects F, Tukey's is performed by using k for the number of treatments and n for the number of subjects in one treatment condition and taking the df from the residual. For the preceding example, for which a significant F ratio of 10.00 was calculated, HSD at the .05 level was found as follows (using Appendix Table H):

$$\text{HSD} = \alpha_{.05}\sqrt{\frac{\text{MS}_{\text{residual}}}{n}}$$

$$= 4.04\sqrt{\frac{.50}{5}} = 4.04\sqrt{.10} = (4.04)(.316)$$

$$= 1.277$$

Looking at the three treatment means of 3.00, 4.00, and 5.00, we find that only the difference between the mean for the first treatment and the mean for the third treatment is significant (since a difference of at least 1.277 between means is required).

Importance of Correlation Within Subjects

Like the paired t, the within-subjects F is an extremely powerful statistical tool, *when there is, in fact, correlation within subjects across the rows.* If there is no correlation, the F ratio value may plummet into nonsignificance.

Example Using the data from the preceding example, we keep the column totals the same but rearrange the rows to eliminate the correlation.

X_1	X_1^2	X_2	X_2^2	X_3	X_3^2	ΣX_r
1	1	5	25	3	9	9
2	4	4	16	6	36	12
3	9	3	9	7	49	13
4	16	6	36	6	36	16
5	25	2	4	3	9	10
15	55	20	90	25	139	60

$$C = \frac{60^2}{15} = 240$$

$$SS_t = 284 - C = 284 - 240 = 44$$

$$SS_{bc} = \frac{15^2}{5} + \frac{20^2}{5} + \frac{25^2}{5} - 240 = 10$$

$$SS_{br} = \frac{9^2}{3} + \frac{12^2}{3} + \frac{13^2}{3} + \frac{16^2}{3} + \frac{10^2}{3} - 240 = 250 - 240 = 10$$

$$SS_{r\times c} = SS_t - SS_{bc} - SS_{br} = 44 - 10 - 10 = 24$$

Source	*SS*	*df*	*MS*	F
Between columns	10	2	5	$\frac{MS_{bc}}{MS_{r\times c}} = \frac{5}{3} = 1.667$
Residual ($r \times c$)	24	8	3	

$F_{.05(2.8)} = 4.46$

$F = 1.667$ Accept H_0; not significant.

Therefore, the same data with the rows rearranged to destroy the correlation result now in an acceptance of the null hypothesis.

Comparing the Within-Subjects F and the Factorial ANOVAs

The within-subjects F is in some ways similar to the factorial (two-way) ANOVA presented in Chapter 12. The factorial ANOVA uses the row variability to assess possible treatment effects, and the within-subjects F uses the same variability to get at possible differences among the subjects. Also, whereas the factorial ANOVA uses the mean square within groups as the error term, the F ratio's denominator, the within-subjects F, uses the residual or interaction mean square as its error term (which reflects the variability left over after any systematic variability has been removed). For other experimental designs, such as the repeated-measures with separate control group (either independent or matched), other ANOVA analyses are possible. These are presented in many advanced texts.

TESTING CORRELATED EXPERIMENTAL DATA

The experimental designs reviewed in this chapter have in common the fact that the data to be analyzed are, in some way, associated. The statistical tests used are all aimed at testing the hypothesis of difference (as they must be in experimental research), but by taking the score correlations into account, these tests become extremely sensitive. With the paired t or within-subjects F, very small differences may be found to be significant differences.

As we have seen, applying the t or F test for independent measures to data that are really correlated inevitably lowers the chance of rejecting the null hypothesis. Thus, when there is correlation, the researcher can take advantage of it, thereby reducing the beta probability and increasing the power of the test. The moral for the researcher is that to use the wrong test is to bet against yourself.

SUMMARY

Whenever subjects are paired, either with themselves in an R/M design or with equated partners in an M/S design, a correlation almost certainly results between the paired scores. For this reason, different statistical tests are used for analyses of these data than are used for independently selected samples. This chapter covers some of the statistical procedures for testing the hypothesis of difference when the data are in at least interval form and the samples are correlated.

When two distributions of interval scores are being compared and the data result from either an R/M or an M/S design, then the paired t must be used. For a given size difference between two sample means, the paired t will produce a higher numerical value than will the independent t. This is due to the fact that the estimated standard error of difference (the t ratio's denominator) is reduced in value by the subtraction of the correlation term. However, the use of the paired t also reduces the degrees of freedom by a factor of one-half. To some extent, then, the benefits from increasing the t value may be offset by the reduction in the degrees of freedom. Typically, however, the increased t value more than compensates for the loss in degrees of freedom. This means that when the data are correlated, the paired t is a more powerful test than is the independent t. (The more powerful the test, the less is the likelihood of accepting H_0 when it should have been rejected.) The independent t, however, is more powerful than the paired t when the samples are independent. This is because although independent samples should produce no correlation (and therefore no reduction in the estimated standard error of difference), the independent t does have more degrees of freedom. (The more degrees of freedom, the smaller is the t value needed to reject H_0.)

When the experiment involves an R/M design with a separate control group, then another approach to the analysis of the data must be used. Rather than comparing the individual raw scores, the "change" scores should be evaluated. If the separate control group is independent of the experimental group, the comparison of the change scores can be assessed via the independent t. If the subjects in the separate control group are matched to those in the experimental group, then the analysis of the data is accomplished by using the paired t ratio. When more than two sets of scores are involved, either from a repeated-measures or matched-subjects design, then the within-subjects F ratio (treatments by subjects) should be used, again assuming at least interval data. The within-subjects F ratio is a more powerful test than is the independent F when the experimental design is either repeated-measures or matched-subjects.

Key Terms

correlated samples
effect size
paired t ratio
within-subjects F ratio

PROBLEMS

1. A study was done to determine the effects of practice and fatigue on muscle strength. A random sample of 8 recruits on their first day at the police training academy was selected and asked to do as many push ups as possible (the pre-scores). After a five-minute break they were again asked to do as many push ups as they could (the post-scores).
   a. Test the hypothesis that there is a significant difference between the pre- and post-scores.
   b. If the difference is significant, which explanation would be more accurate as determined by these data, practice or fatigue?

Subject	*Pre*	*Post*
1	11	8
2	15	15
3	16	13
4	14	11

Continued

Note: Scores continued from previous page.

Subject	*Pre*	*Post*
5	10	11
6	15	10
7	6	6
8	11	8

2. Two groups of learning disabled children are randomly selected and matched person for person on the basis of WISC III IQ scores. Both groups are given a reading test and then assigned to either experimental or control conditions. The experimental group underwent six weeks of phonics training, whereas the control group did not. At the end of the six-week period, both groups were again tested for reading ability. Their pre- and post-scores were as follows:

	Experimental Group		*Control Group*	
Matched Pair No.	*Pre*	*Post*	*Pre*	*Post*
1	10	12	10	11
2	9	13	10	10
3	7	9	8	9
4	14	15	13	13
5	6	11	5	9
6	11	14	10	12
7	16	17	18	19
8	8	10	9	9
9	12	16	13	14
10	13	19	12	16

Test the hypothesis that special phonics training affects the reading ability of learning disabled children.

3. A study was done to investigate test anxiety as a function of number of college years completed. After having finished their freshman year, a random sample of 6 students was selected and given the CTAQ (College Test Anxiety Questionnaire). Their scores (high scores indicating greater levels of test anxiety) are reported in the A column. After completing their sophomore year, they were again tested (scores in the B column) and finally after their junior year they were tested for a third time (scores in the C column). Test the hypothesis that test anxiety differs as a function of college years completed.

Subject	*A*	*B*	*C*
1	15	12	11
2	17	15	13
3	18	16	15
4	17	17	14
5	20	18	16
6	15	13	15

4. A researcher wants to find out if background noise affects typing speed. Two groups of student typists were selected and matched on the basis of typing speed (error-free words per minute.). One group is then tested under high noise conditions and the other group in total silence. The data are as follows:

Pair	*Silence*	*High Noise*
1	50	42
2	65	60
3	72	65
4	90	85
5	48	50
6	62	60
7	75	60
8	50	51
9	68	59

Test the hypothesis.

5. A psychologist wishes to establish whether room color has any effect on frustration-induced anxiety among inmates at a minimum security prison. Three groups were selected and matched on the basis of their Taylor Manifest Anxiety scores. The subjects are brought into one of three rooms: one decorated in an off-blue (cool color), one in gray (neutral color), and one in bright red (warm color). Each subject is connected to a galvanometer (which measures the skin's electrical conductivity—presumably a measure of situational anxiety). A Correction Officer then reads the following to each group: Because of a rumored security problem you will all have to spend the next 24 hours in total lock-down. The GSR (Galvanic Skin Response) scores are then recorded (in decivolts), with high scores indicating greater anxiety.

Triad	*Blue Room*	*Gray Room*	*Red Room*
1	2	5	6
2	3	6	7
3	5	7	9
4	6	8	6
5	3	5	8
6	2	4	6
7	1	3	5

Test the hypothesis.

6. A researcher is interested in whether SI (Systematic Desensitization) can be used to reduce anger among violent offenders. A random sample of 10 inmates was selected from a prison population of inmates who had all been convicted of violent felonies. Each subject was tested on a standardized anger provocation scale and then given six weeks of SI training. They were then all tested again. The scores, with the subject's number included, were as follows:

Subject	Before	After
1	44	20
2	20	10
3	35	30
4	42	26
5	35	30
6	30	20
7	38	30
8	24	22
9	19	21
10	17	20
11	20	17
12	16	15
13	35	25
14	31	26
15	34	30
16	20	25

Test the hypothesis.

7. A researcher wished to discover whether a special course called "Attacking the SATs" will increase performance. A random sample of high school seniors was selected and given the verbal section of the SATs. They were then all enrolled in a six-month study course and again given the verbal SAT exam. The data are as follows:

Student No.	Pre SAT-V	Post SAT-V
1	650	640
2	710	750
3	430	510
4	650	590
5	430	500
6	560	570
7	330	480
8	500	490
9	430	500
10	650	740
11	390	410
12	650	600
13	630	690
14	490	510
15	670	550
16	380	330
17	420	490
18	580	600
19	520	530
20	480	510

a. Indicate the type of research and, if appropriate, the experimental design.
b. Indicate the IV and DV.
c. Were the results significant?
d. What confounding variables may have intruded on the pure effects of the IV?

8. For a given difference between the means and equal numbers of scores, which test, the paired t or the independent t, has more statistical power? (For this question, assume that for both tests all their assumptions are met.)

9. Of the three basic experimental designs, the between-subjects, the repeated-measures or the matched-subjects, on which type or types can the paired t be used?

10. When calculating the paired t ratio, what effect does a substantial correlation have on the size of the estimated standard error of difference?

11. State how the degrees of freedom compare between the independent and paired t ratios.

12. When calculating the paired t ratio, what effect does a substantial correlation have on the size of the resulting t ratio?

13. A within-subjects F ratio performed on data from a matched-subjects design results in a (higher or lower) F ratio than would be obtained by an independent, one-way ANOVA performed on the same data.

14. The fact that the paired t has fewer degrees of freedom than does its independent counterpart, *and that fact alone,* has what effect on the probability of achieving significance?

Fill in the blanks in problems 15 through 17.

15. In matched-subjects designs, the subjects should be equated on some variable(s) that is (are) related to the (dependent or independent) ________ variable.

16. When all its assumptions are met, the paired t is (more able or less able) ________ than the independent t to reject null when only a small difference exists between the sample means.

17. The paired t and the within-subjects F should be used only when the data are in the form of at least (interval, ordinal, or nominal) ________ measures.

True or False. Indicate either T or F for problems 18 through 30.

18. For equal numbers of scores, the paired t has more degrees of freedom than does the independent t.

19. The paired t has as its ultimate goal the detection of differences between two sets of interval measures when the data sets are correlated.

20. The paired t may test only the hypothesis of association, whereas the independent t may test the hypothesis of difference as well.

21. The more degrees of freedom a given t ratio has, the higher the likelihood of rejecting the null hypothesis.
22. For both t and F, whether from correlated or independent designs, the more subjects being tested, the greater is the number of degrees of freedom.
23. The higher a test's power, the less likely one is of committing the beta error.
24. The more powerful a test is, the greater the possibility of detecting differences when there really are differences in the population.
25. The paired t ratio may never be used for making population inferences.
26. Statistical tests that use interval data are inherently more powerful than are those that utilize ordinal or nominal data.
27. The Type 2 error occurs when the null hypothesis is rejected when it should have been accepted.
28. The Tukey test may only be applied to the within-subjects F ratio when the F has been shown to be significant.
29. Effect size may only be applied to the within-subjects F ratio when the F has been shown to be significant.
30. Effect size may only be applied to the paired t when the t ratio has been shown to be significant.

COMPUTER PROBLEMS

C1. A researcher wishes to test the hypothesis that the discovery teaching technique leads to higher levels of pupil learning than does the more traditional presentation. Two groups of sixth-grade students are selected and matched person for person on IQ. Both groups are presented with a geography lesson unit. Group A, the control group, is taught in a fairly traditional manner, whereby the teacher, in covering the subject area, asks a series of theoretical questions but quickly answers them herself. Group B, the experimental group, is exposed to the same material by the same teacher, but in this case she does not answer the rhetorical questions in the hope that this will stimulate pupil thinking and allow the pupils to discover for themselves certain basic relationships. Both groups are then tested on the geography content, and their retention scores (higher scores indicating greater retention) are as follows. Does the discovery teaching technique affect learning?

Subject #	Group A Control	Group B Experimental	Subject #	GroupA Control	GroupB Experimental
1.	100	98	3.	98	92
2.	82	80	4.	55	67

Continued

Note: Scores continued from previous page.

Subject #	Group A Control	Group B Experimental	Subject #	GroupA Control	GroupB Experimental
5.	75	80	19.	88	72
6.	66	77	20.	79	90
7.	82	98	21.	82	86
8.	65	75	22.	84	87
9.	42	67	23.	80	90
10.	76	80	84.	75	88
11.	82	95	25.	78	82
12.	64	74	26.	81	88
13.	82	80	27.	66	72
14.	83	87	28.	86	88
15.	78	75	29.	78	77
16.	84	88	30.	84	86
17.	81	83	31.	84	88
18.	67	77			

C2. Subjects were asked to take multiple-choice tests under three different conditions of crowding. In one condition (alone), the subject worked alone at a table. In another (three other subjects) each subject was at the same table with three other individuals, all working at the same task. In the third condition (seven other subjects) there were seven other people at the table, all working on the same task. The presentation order of the tasks was counterbalanced, and each subject participated in all three conditions. The dependent variable was the number of items answered correctly out of a total of 60 questions. Does crowding have an effect on the number of items answered correctly? The data follow.

Subject	Alone	Three Others	Seven Others
1	35	30	20
2	49	34	22
3	33	21	20
4	34	35	30
5	37	29	20
6	29	25	21
7	29	26	24
8	30	43	20
9	17	21	20
10	29	23	10
11	41	30	14
12	50	20	27
13	32	28	14
14	44	28	25
15	30	28	20

C3. A researcher suggests that a child's level of impulsivity can be reduced by training, thereby giving the child more of an internal locus of control. A large group of 12-year-old children was randomly selected and then split into two groups by equating them, person for person, on the Eysenck Extraversion subtest of the Eysenck Personality Inventory. Both groups were then given the Barrett Impulsivity Scale. Group A, the experimental group, was then given a week of Social Skills and Perception Training, while Group B, the control group, was not given this training. At the end of the week both groups were again tested on the Barrett Impulsivity Scale (higher scores mean higher levels of impulsivity). Does training reduce impulsivity? Their scores were as follows:

Matched Pair No.	Group A			Group B		
	Pre	Post	Change	Pre	Post	Change
1	55	53	−2	52	50	−2
2	25	20	−5	23	22	−1
3	50	49	−1	45	45	0
4	60	51	−9	52	50	−2
5	56	52	−4	55	54	−1
6	55	48	−7	48	46	−2
7	51	40	−11	42	38	−4
8	76	65	−11	76	66	−10
9	55	52	−3	58	55	−3
10	56	48	−8	60	50	−10
11	60	55	−5	64	60	−4
12	48	45	−3	50	44	−6
13	80	80	0	75	75	0
14	56	47	−9	54	45	−9
15	55	52	−3	51	50	−1
16	58	48	−10	55	49	−6
17	42	40	−2	45	43	−2
18	56	55	−1	60	57	−3
19	58	55	−3	52	50	−2
20	46	45	−1	48	46	−2

C4. A rat study was conducted to find out whether environmental inputs could affect brain growth. Sixteen laboratory baby rat twins were selected and randomly assigned to two groups. The rats were paired off on the basis of genetics, one rat from each twin pair in the experimental group and the other twin in the control group. The experimental group was then raised in a stimulating environment, in a cage equipped with ladders, running wheels, and other "rat toys." These animals were also let out of their cages for 30 minutes each day and allowed to explore new territory. They were also trained on a number of learning tasks and in general received a rich and varied array of stimulus inputs. The other group of rats was raised in a condition of extreme stimulus homogeneity. These rats lived alone in dimly lit cages, were rarely handled, and were never allowed to explore areas outside their cages. All animals received exactly the same diet. After 90 days, all the animals were sacrificed and their brains analyzed morphologically and chemically. For one comparison, the weight of each rat's cortex was recorded in milligrams. The data are as follows:

Pair No.	*Experimental Group*	*Control Group*
1	699	667
2	666	653
3	678	662
4	670	664
5	679	668
6	663	656
7	664	650
8	657	650
9	704	680
10	643	645
11	663	652
12	717	690
13	740	700
14	699	690
15	656	660
16	710	699

a. Test the hypothesis that the two groups differ significantly regarding cortical weight.
b. Was the matching process effective?

C5. A researcher is interested in the effects of marijuana on the pulse rate. A group of 20 college students was randomly selected and their pulse rates recorded. They were given (and smoked) one marijuana cigarette and 30 minutes later their pulse rates were again taken. The data follow:

Subject #	*Pre*	*Post*	*Subject #*	*Pre*	*Post*
1.	76	82	11.	73	78
2.	72	78	12.	60	70
3.	81	89	13.	70	76
4.	70	75	14.	76	74
5.	65	76	15.	71	79
6.	70	74	16.	72	80
7.	72	78	17.	72	78
8.	83	80	18.	74	80
9.	75	81	19.	76	86
10.	72	74	20.	70	74

Test the hypothesis.

As a sidebar to this research, it has been found that marijuana increases the risk of heart attacks for people over 40. Marijuana increases both pulse rate and blood pressure, and from one to two hours after ingestion, heart attack rates are increased by 50%, especially if ingestion is followed by any form of exercise—including sexual activity. (Geraci, 2000).

C6. A researcher conducts a pre-post study comparing the SAT scores of a random sample of 20 subjects before they took an SAT coaching course and then again after they took the course.

Subject #	*Pre*	*Post*	*Subject #*	*Pre*	*Post*
1.	500	510	11.	500	490
2.	510	500	12.	490	440
3.	510	510	13.	740	760
4.	550	570	14.	510	510
5.	370	370	15.	500	520
6.	600	620	16.	550	560
7.	270	250	17.	450	440
8.	700	710	18.	430	420
9.	510	520	19.	520	530
10.	490	500	20.	500	500

Test the hypothesis that the coaching course had a significant effect on their SAT scores.

C7. The researcher in the above problem suspects, on the basis of prior research, that perhaps only those who scored above 500 were aided by the coaching course. Thus, only those who scored 510 or better were selected from the original sample and their pre scores were compared with their post scores as follows:

Subject #	*Pre*	*Post*
2.	510	500
3.	510	510
4.	550	570
6.	600	620
8.	700	710
9.	510	520
13.	740	760
14.	510	510
16.	550	560
19.	520	530

Test the hypothesis that coaching is a more important factor for those who already have above-average scores.*

*This is a finding which is consistent with the literature, which shows that there is a larger coaching effect for individuals with higher precoaching scores than for those with average and below average scores, a fact that argues against the hope that coaching will necessarily narrow the gap between high- and low-scoring students (Sackett, et al., 2001).

Chapter

16

Nonparametrics Revisited: The Ordinal Case

As we learned in Chapter 9, the ordinal scale gives information regarding greater than or less than, but not how much greater or how much less. That is, a score with the rank of 1 is known to be greater than a score with the rank of 2, but whether 1 is inches or miles in front of 2 is not known. In short, then, for ordinal data, the distance between successive scale points is unknown.

In Chapter 11, we learned a technique for testing the hypothesis of association with ordinal data, the Spearman r_S. *In this chapter, tests of the hypothesis of difference for ordinal data are presented—one test for each of the basic research situations. As we know, all data gathered by any of the experimental methods must be analyzed by testing the hypothesis of difference. Data from the post-facto method, however, may be analyzed by testing either the hypothesis of association or the hypothesis of difference. When we test for association, we can use the* r_S, *but when we test for difference, one of the tests to follow may be used. Thus, each and every test in this chapter tests the hypothesis of difference. (For each of the research situations presented, a counterpart test suitable for interval data is mentioned. For instance, whereas the Spearman* r_S *is used to test for correlation with ordinal data, the Pearson* r *is used for the same purpose when the data are in interval form.)*

Ordinal Data Require Nonparametric Tests

All the statistical tests used on ordinal data are considered to be **nonparametric,** or distribution free. Whereas the major interval data tests, t, F, and r, all make careful assumptions regarding the characteristics of the population to which the results are to be generalized, the tests for ordinal (and as we saw in Chapter 13, nominal) data make no such assumptions. These tests do not make *any assumptions* regarding μ, the mean of the population, nor do they assume a normal distribution in the population. Therefore, if we obtain interval data from a population known not to be normal, these interval scores may be converted into ranks, and a test for ordinal data can then be performed. Although, as has been stated many times, the nonparametric tests are not as powerful as their parametric cousins, they are much safer when the population characteristics are at all suspect.

When to Use the Nonparametric Tests

As with the r_S, the nonparametric tests for ordinal data should be used whenever the distributions to be compared are such that

1. Both, or all, are originally presented in ordinal form.

2. One distribution is presented in ordinal form and the other(s) in interval form. In this situation, we convert the interval data to ordinal and perform a test for ordinal data. (We must never attempt, as was pointed out in Chapter 11, to convert ordinal to interval data.)

3. Both, or all, distributions are in at least interval form, but the populations from which the samples were selected are known to lack normality, or the samples show significant differences in variability.

In these situations, convert both (all) sets of interval scores into ordinal ranks, and use one of the nonparametric ordinal tests that are to be covered in this chapter. When converting interval scores into ordinal ranks, handle all tied scores as shown previously in Chapter 11.

MANN–WHITNEY *U* TEST FOR TWO ORDINAL DISTRIBUTIONS WITH INDEPENDENT SELECTION

With interval data, when the hypothesis of difference is to be tested between two independently selected samples, the obvious statistical test is the independent t. The ordinal answer to the independent t is the **Mann–Whitney *U* test.** Therefore, any time the research situation dictates the use of the independent t but the data are in ordinal form, the Mann–Whitney U can be used.

The Mann–Whitney U test assesses whether two sets of ranked scores are representative of the same population. If they are, the two distributions should be random and H_0 is accepted. If, however, the value of U detects a nonrandom pattern, then H_0 is rejected.

Calculating the Mann–Whitney U

As is true of the r_S, all the tests of ordinal data utilize a number of constants—values that do not change, regardless of the data.

Example A political analyst wishes to establish whether a difference in income exists between registered Republicans and registered Democrats. Random samples are selected of 10 Republicans and 11 Democrats, and the annual income for each subject is obtained. Because the income distribution in the population is known to be skewed, the interval scores are converted to ordinal ranks. In this process, the ranks are assigned to both sample distributions *combined* rather than ranking each distribution separately, as is done for the Spearman r_S. The reason we rank the combined distributions is to find out whether one set of ranks is significantly lower than the other.

The income scores and the resulting ranks are as follows:

Republicans		Democrats	
X_1	R_1	X_2	R_2
$40,000	8	$16,000	21
41,000	7	17,000	20
43,000	5	20,000	19
42,000	6	21,000	18
190,000	1	39,000	9
44,000	4	38,000	10
55,000	3	36,000	11
60,000	2	35,000	12
31,000	14	34,000	13
30,000	15	29,000	16
	$\Sigma R_1 = 65$	28,000	17
$n_1 = 10$		$n_2 = 11$	

To calculate the Mann–Whitney U, the only data values needed are those for R_1, n_1, and n_2. We carry out the following steps.

1. Add the ranks for the first distribution ($\Sigma R_1 = 65$). We use this value and the two sample sizes, n_1 and n_2, to solve for U.

$$U = n_1 n_2 + \frac{n_1(n_1 + 1)}{2} - \Sigma R_1$$

$$= (10)(11) + \frac{10(11)}{2} - 65$$

$$= 110 + 55 - 65 = 100$$

2. Using the value of U and the sample sizes again, solve for z_U.

$$z_U = \frac{U - n_1 n_2 / 2}{\sqrt{[n_1 n_2 (n_1 + n_2 + 1)] / 12}}$$

$$= \frac{100 - [(10)(11)] / 2}{\sqrt{[(10)(11)(10 + 11 + 1)] / 12}} = \frac{100 - 55}{\sqrt{2420 / 12}}$$

$$= \frac{45}{\sqrt{201.667}} = \frac{45}{14.201} = 3.169$$

3. Compare the obtained value of z_U with the z score value that excludes the extreme 1% of the distribution, $z_{.01} = \pm 2.58$. If z_U is equal to or greater than this value, reject the null hypothesis. (For an alpha error level of .05, compare our value with $z_{.05} = \pm 1.96$.)

$z_{.01} = \pm 2.58$

$z_U = 3.169$ Reject H_0; significant at $p < .01$.

Therefore the null hypothesis (H_0: $R_1 = R_2$) that the two sets of ranks represent a single population is rejected.

Conclusions: A Mann-Whitney test was computed to test for the difference in income levels between Democrats and Republicans. The Republicans earned more, ranked higher, with a mean rank of 6.500 compared to the Democrats lower mean rank of 15.091. The analysis showed a significant difference ($z = 3.169$, $p < .001$).

Sample Size and *U*. Note that in making the comparison between z_U and z, there are no degrees of freedom involved. This is because the distribution of U values, when n_1 and n_2 are each at least 9, is assumed to be close enough to normality to be directly compared to the z score distribution. Remember, though, that this is only true when there are *at least 9* ranked scores in each sample (for very small samples, that is, 8 or less, separate tables of critical U values are available) (Siegel & Castellan, 1988).

Interpreting the U Test Results

In the preceding example, because H_0 was rejected, we concluded that Republicans do earn more than Democrats. Great care, however, must be taken when interpreting the results. The example was a post-facto study, with the independent variable (party affiliation) an assigned subject variable rather than manipulated. Therefore, although it turned out that Republicans do earn more than Democrats, the causal basis of this difference was not established. Are persons who announce themselves as Republicans more apt to get promotions and raises? Or are persons who do get raises more apt to then become Republicans? Or could a third factor, perhaps parents' socioeconomic level, cause a person to be both a Republican and an earner of a higher income? (After all, if your father is president of General Motors, you may start out on a pay scale somewhat above minimum wage.) Finally, age was not controlled, so that it is possible that older persons, who earn more because of seniority, tend to be Republicans. Post-facto research can answer none of these questions with certainty.

Of course, since it tests the hypothesis of difference, the Mann–Whitney U can be used for the analysis of experimental data. However, as was the case with the independent t test, the experimental design must be completely randomized. The between-subjects design satisfies the independent samples restriction for the use of U.

KRUSKAL–WALLIS *H* TEST FOR THREE OR MORE ORDINAL DISTRIBUTIONS WITH INDEPENDENT SELECTION

With more than two sets of interval scores, and where the samples are independently selected, the hypothesis of difference is tested using the one-way ANOVA. When the data are ordinal, these same research situations are analyzed by the **Kruskal–Wallis *H* test.** The Kruskal–Wallis *H*, then, *is a one-way ANOVA for ordinal data.* Thus, whenever the independent variable has at least three levels and the data are ordinal, *H* is the proper statistic to use. Since *H* demands that the several sample groups be *independently* selected, then, of the various experimental designs, *H* can be applied only to studies in which the subjects have been independently assigned to the groups.

Also, as with the Mann–Whitney *U*, if the data represent skewed distributions, the interval scores from the combined distributions must be rank-ordered together. Whether the scores are ranked high to low or low to high does not affect the value of the Kruskal–Wallis *H*. The choice made here was to rank high to low, as was done previously with the Spearman r_S.

Calculating the Kruskal–Wallis H

Example Suppose that the Federal Aviation Administration is interested in discovering whether differences in flying ability are a function of pilot age. A dispute arises. One hypothesis states that younger persons, due to their better reflexes and general physical conditioning, are better pilots. Another hypothesis insists that older persons, due to their longer flying experience, are better pilots. A random sample of 24 licensed pilots is selected, with 6 pilots from each of four age categories. An FAA examiner takes each pilot on a test flight and rank-orders all 24 on their flight skills. Since the ranking is high to low in this study, the rank of 1 identifies the best pilot, and the rank of 24 designates the worst pilot.

Group 1 (21–30 Years Old), R_1	*Group 2* (31–40 Years Old), R_2	*Group 3* (41–50 Years Old), R_3	*Group 4* (51–60 Years Old), R_4
2	20	23	24
4	6	11	17
18	8	15	22
1	5	13	21
3	9	10	19
7	12	14	16
$\Sigma R_1 = 35$	$\Sigma R_2 = 60$	$\Sigma R_3 = 86$	$\Sigma R_4 = 119$
$n_1 = 6$	$n_2 = 6$	$n_3 = 6$	$n_4 = 6$

The only data values needed for this analysis are the sum of the ranks for each group and the group, or sample, sizes. We perform the *H* test in the following steps.

1. Add the ranks in each column (group) to obtain $\Sigma R_1 = 35$, $\Sigma R_2 = 60$, $\Sigma R_3 = 86$, and $\Sigma R_4 = 119$.
2. Substitute the values for ΣR, N, and n into the H equation and solve.

$$H = \frac{12}{N(N+1)}\left(\frac{(\Sigma R_1)^2}{n_1} + \frac{(\Sigma R_2)^2}{n_2} + \frac{(\Sigma R_3)^2}{n_3} + \frac{(\Sigma R_4)^2}{n_4}\right) - 3(N+1)$$

$$= \frac{12}{24(24+1)}\left(\frac{35^2}{6} + \frac{60^2}{6} + \frac{86^2}{6} + \frac{119^2}{6}\right) - 3(24+1)$$

$$= \frac{12}{24(25)}\left(\frac{1225}{6} + \frac{3600}{6} + \frac{7396}{6} + \frac{14,161}{6}\right) - 3(25)$$

$$= \frac{12}{600}(204.167 + 600 + 1232.667 + 2360.167) - 75$$

$$= .020(4397.001) - 75 = 87.940 - 75 = 12.940$$

3. Compare the calculated value of *H* with the critical values in the chi-square table (Appendix Table I). The *df* are equal to *k*, the number of columns (groups) minus one, that is $df = 4 - 1 = 3$. The table shows a value of 11.34 at an alpha level of .01 for 3 *df*. Since our calculated value is greater, 12.940, the null hypothesis is rejected.

Conclusions: A Kruskal-Wallis test was performed on pilot ratings based on age. The differences were significant, with younger pilots receiving significantly higher ratings than was the case for the older pilots ($H(3) = 12.940$, $p < .01$). It appears that younger pilots perform better than older pilots, but the key factor may not be age. We must be careful of our interpretation of these results, since this is again an example of post-facto research. The independent variable, age, was a subject variable, and, therefore, other factors may be involved. One may presume that the older (yet according to this study, worse) pilots have had more flying hours, but this is only conjecture. Perhaps some of the older pilots only recently earned their licenses and have few flying hours. More precise studies might be done where the pilots are matched according to age and then assigned to categories on the basis of flying time, or matched on flying time and then categorized on age. Studies of this nature, however, cannot be analyzed by *H*, since the use of *H* requires independent sample groups.

Sample Size and *H*. With at least three sample groups and a minimum of 6 subjects per sample, *H* may be assessed for significance with the chi square table. With smaller samples, a special table of *H* values is available (Siegel & Castellan, 1988), but attaining significance with these tiny samples becomes quite difficult. With three-group designs, the researcher should select a total sample of at least 18 subjects, 6 per group.

WILCOXON *T* TEST FOR TWO ORDINAL DISTRIBUTIONS WITH CORRELATED SELECTION

With interval data, when two correlated groups are to be compared for possible differences, the appropriate test is the paired t test. With ordinal data, the same situation can be handled by the **Wilcoxon T test.** This means that the analysis of ordinal data from either the repeated-measures or matched-subjects designs can be appropriately tested with the Wilcoxon T.

Procedure for the Wilcoxon T Test

The procedure for the Wilcoxon T is demonstrated in the following example.

Example Suppose that a golf pro creates a new method, including videotape replays, of teaching golf. A random sample of golfers at a certain country club is selected, and their average golf scores are ascertained. The subjects are then placed into matched groups on the basis of their average scores. That is, one golfer who averages 85 is placed in the experimental group, and another golfer who also averages 85 goes into the control group. The subjects in the experimental group are then given a week's instruction using the new teaching method; those in the control group are taught in the traditional way. At the end of the week's training, both groups play a round of golf and their scores are compared. In looking over the two sets of golf scores, it is discovered that the distributions are badly skewed, since in each group there were a few new members with extremely high scores. So despite the fact that the data are originally in interval form (golf scores), because of the skew, the ordinal Wilcoxon T test is chosen for the analysis.

The data for 10 matched pairs of subjects are as follows:

Pair	*Experimental Group (New Method),* X_1	*Control Group (Old Method),* X_2	*Difference,* $X_1 - X_2$	*Rank of Difference*	*Signed Rank*	*Ranks with Less Frequent Sign*
1	85	86	−1	1	−1	—
2	90	95	−5	6	−6	—
3	92	96	−4	5	−5	—
4	93	93	0	← (Dropped)		—
5	93	95	−2	2.5	−2.5	—
6	94	96	−2	2.5	−2.5	—
7	95	98	−3	4	−4	—
8	95	101	−6	7	−7	—
9	140	133	+7	8	+8	8
10	150	135	+15	9	+9	9
						$T = 17$

$T_{.05(9)} = 6$

$T = 17$ Accept H_0; not significant.

Important Note: With the Wilcoxon T, the null hypothesis is rejected *only* when the calculated T is equal to or *less than* the table value of T.

1. *Obtain the differences.* We set up the difference column, $X_1 - X_2$, being careful to retain the correct sign.

2. *Rank the differences.* We rank-order the absolute values of the differences. In this step, the sign of the difference is irrelevant. The difference value of –1 receives a rank of 1, or first place not because it was negative, but because it was the *smallest* difference. Note also that two of the differences (–2) are tied for second and third place. As for all conversions to ordinal ranks, we add the tied ranks (2 + 3), divide by the number of ranks tied in that position [(2 + 3)/2 = 2.5], and assign each the resulting average rank. Finally, whenever there is a zero difference between a pair of scores, as in the case of Pair 4 (where each subject scored a 93), the scores for these subjects are *dropped* from the analysis.

3. *Sign the ranks.* In this step, we simply affix the sign of the difference to the rank for that difference. Thus, the ranked differences appear in a separate column but now have whichever sign appears in the preceding difference column. Thus the difference of –1 (for Pair 1), ranked first, and the rank now gets a negative sign because the value in the difference column is negative. Similarly, the largest difference, which ranked ninth, is obtained from Pair 10 where the difference value is positive.

4. *Add the less frequent signed ranks.* Finally, we determine which sign, plus or minus, occurs less frequently among the ranks. The *plus* sign occurs less often (only twice, compared to seven minus signs). Then, to obtain the value of *T*, we merely add the ranks having the less frequent sign; $T = 17$.

5. *Check for significance.* We compare the calculated value of *T* with the critical table value of *T* with $N = 9$. (See Appendix Table J.)

Now, *unlike any other test* in this book, with the Wilcoxon *T* test the null hypothesis is rejected only when the calculated value of *T* is *equal to or less than the table value.* For *T*, smaller means more significant.

With an *N* of 9, we need a *T* value of 6 or lower to REJECT the null hypothesis. Since our calculated value was 17, the null hypothesis was accepted.

Conclusions: A Wilcoxon *T* test was computed on the golf scores of a matched groups of golfers, one group having had special training and the other with no training. No significant differences were found.

Sample Size and the Wilcoxon T The procedure just outlined for the Wilcoxon *T* is appropriate for use in many common research situations, where the sample sizes range from 6 to 25 pairs of scores. When more than 25 paired ranks are available, however, the distribution of Wilcoxon *T* values approaches normality. Then the following equation must be used:

$$z_T = \frac{T - [N(N+1)]/4}{\sqrt{[N(N+1)(2N+1)]/24}}$$

The resulting z_T value is compared with a critical z of ±1.96 for the .05 alpha error level or ±2.58 for the .01 alpha error level. The null hypothesis is rejected if the obtained value of z_T equals or exceeds one of these z values. This is exactly the same as the procedure described earlier for the Mann–Whitney U test with z_U.

FRIEDMAN ANOVA BY RANKS FOR THREE OR MORE ORDINAL DISTRIBUTIONS WITH CORRELATED SELECTION

When either matched-subjects or repeated-measures designs are used, and the hypothesis of difference is to be tested on three or more ordinal distributions, the appropriate test is the **Friedman ANOVA by ranks.** This is analogous to the within-subjects F when nonskewed distributions of interval data are involved.

In the example demonstrating the Mann–Whitney U test, the results informed us that Republicans do earn more money than Democrats. The interpretation of the results was unclear, however, because so many other variables were left uncontrolled, not the least of which was age. Now any age distribution is certainly composed of at least interval data, but like income distributions, it has to be badly skewed in the population. (There are far fewer 90-year-olds than there are 2-year-olds.) Therefore, whenever age or income is a dependent variable, we must consider converting to ordinal ranks before testing for significance.

Calculating the Friedman ANOVA by Ranks

Example A researcher is interested in whether there is an age difference among Independent, Democratic, and Republican voters. However, because older persons are apt to be earning more money, perhaps the key to the age–party affiliation relationship is economic. Perhaps richer persons are more apt to be Republicans, regardless of their age. To test this, random samples of 10 Independents, 10 Democrats, and 10 Republicans were selected and *matched on income.* That is, trios made up of 1 Independent, 1 Democrat, and 1 Republican, all earning roughly the same yearly income, are put together. They are then checked for age.

The data are as follows:

Triad	*Independent,* X_1	*Democrat,* X_2	*Republican,* X_3	*Independent,* R_1	*Democrat,* R_2	*Republican,* R_3
1	26	30	22	2	1	3
2	29	31	28	2	1	3
3	55	60	54	2	1	3
4	27	26	24	1	2	3
5	70	69	74	2	3	1
6	21	23	32	3	2	1

Continued

Note: Scores continued from previous page.

Triad	*Independent,* X_1	*Democrat,* X_2	*Republican,* X_3	*Independent,* R_1	*Democrat,* R_2	*Republican,* R_3
7	33	35	34	3	1	2
8	40	39	38	1	2	3
9	41	42	43	3	2	1
10	45	44	46	2	3	1
				$\Sigma R_1 = 21$	$\Sigma R_2 = 18$	$\Sigma R_3 = 21$

To solve the equation for the Friedman ANOVA, we need values for ΣR, N (the number of rows, that is, of matched groups of subjects), and k (the number of columns of ranked scores). We go through the following steps.

1. In each row of interval scores—that is, each triad—rank-order the scores from high to low. Thus, in the first row (Triad 1), the age score of 30 is ranked 1, the age score of 26 is ranked 2, and the age score of 22 is ranked 3. (There are only 3 scores per row here, so the ranks must run some combination of 1, 2, 3 in each and every row.)
2. Sum the columns of ranked scores to obtain

   $\Sigma R_1 = 21 \qquad \Sigma R_2 = 18 \qquad \Sigma R_3 = 21$
3. Plug in the values for ΣR_1, ΣR_2, and ΣR_3, along with the values for N (the number of rows), that is, 10, and for k (the number of rank columns), that is, 3, into the equation and solve for χ_r^2.

$$\chi_r^2 = \frac{12}{Nk(k+1)}[(\Sigma R_1)^2 + (\Sigma R_2)^2 + (\Sigma R_3)^2] - 3N(k+1)$$

$$= \frac{12}{(10)(3)(3+1)}[(21)^2 + (18)^2 + (21)^2] - 3(10)(3+1)$$

$$= \frac{12}{120}(441 + 324 + 441) - 120$$

$$= (.10)(1206) - 120 = 120.60 - 120 = .600$$

4. We check the calculated value of χ_r^2 against the table value (Appendix Table I) for df $= k - 1 = 2$. If the obtained value is equal to or greater than the critical value, we reject the null hypothesis.

In this case, we needed a value of 5.99 and calculated a value of only .600, meaning that the null hypothesis is accepted and a statistically significant difference could NOT be established.

Conclusions: The Friedman ANOVA by ranks was used to assess whether political affiliation differs according to age. The results were not significant (chi square(2) = .600, $p > .05$). The political affiliation of the subjects did not differ on the basis of age.

Here, the null hypothesis (H_0: $R_1 = R_2 = R_3$) is accepted. The three distributions of age scores represent a single population. Since the subjects were matched on income, this means that when income is not a factor, there are no age differences among persons of different political affiliations. This is another example of post-facto research, so even if the age differences had been found to be significant, only rather tentative conclusions could be drawn.

Sample Size and the Friedman ANOVA To use the chi square table for assessing the significance of χ_r^2, we must have a minimum of 10 scores per column when there are 3 columns of ranked scores. With 4 columns of ranked scores, only 5 scores per column are necessary. For smaller sample sizes, such as for an N of from 2 to 9 with a k of 3, or an N of from 2 to 4 with a k of 4, special tables are available (Siegel & Castellan, 1988).

ADVANTAGES AND DISADVANTAGES OF NONPARAMETRIC TESTS

None of the tests of ordinal data make any assumptions regarding the parameters of the population. For this reason, they are called nonparametric tests. Neither do these tests make any assumptions regarding the shape of the underlying population distribution. The population distributions can be skewed right, skewed left, or even bimodal, and these tests may still be used. For this reason, they are also called distribution free. Since they seem to be so safe (it is hard to violate assumptions if there are none), why not always use them instead of t and F? The answer is because they are less powerful.

Power is equal to $1 - \beta$ (one minus the beta error). Whenever beta error is increased, power is reduced. (Beta error is the probability of being wrong when accepting the null hypothesis, that is, accepting H_0 when it should have been rejected.) Nonparametric tests all tend to increase the beta error. That is, the nonparametrics are less sensitive to smaller differences, less able to detect that these differences might be significant. For equal sample sizes, a given difference between groups that proves to be significant with a t test might not be significant with a Mann–Whitney U test.

Thus, whenever the assumptions of the parametric tests can be reasonably met, we should use them. When the data are in ordinal form or when the distribution of interval scores is obviously skewed, we should use the nonparametric tests. As someone has said, good alternative hypotheses are sometimes hard to come by and should not be needlessly thrown away. The nonparametrics may be quick and easy to calculate, but the time saved does not equal the lost significance.

SUMMARY

Tests of significance for ordinal data are those that analyze measurements that are in rank-order form, that is, measures that provide information regarding greater than or less than status, but *not* how much greater or how much less. Since tests of ordinal data need not predict the population parameter, μ, they are collectively known as nonparametric tests. Although ordinal data may be tested for correlation (see Chapter 11), the focus of this chapter is on testing ordinal data to assess differences.

1. When two independently selected sets of ordinal scores are to be tested for differences, use the Mann–Whitney U test. The U test is the ordinal equivalent of the independent t test for interval data.

2. When three or more independently selected sets of ordinal scores are to be tested for differences, use the Kruskal–Wallis H test. The H test is the ordinal equivalent of the one-way ANOVA for interval data.

3. When two sets of ordinal scores are to be tested for differences and the samples are correlated, use the Wilcoxon T test. This T is the ordinal analog of the paired t test for interval data.

4. When three or more sets of ordinal scores are to be tested for differences and the samples are correlated, use the Friedman ANOVA by ranks. The Friedman ANOVA by ranks is the ordinal equivalent of the within-subjects F ratio for interval data.

These tests, all basically designed for ordinal data, can also be used on interval data when the underlying distributions are known to deviate significantly from normality, or when there are large differences in variability among the sample groups. To do this, the interval scores must first be converted to ordinal ranks.

Key Terms

Friedman ANOVA by ranks
Kruskal–Wallis *H* test
Mann–Whitney *U* test
nonparametric
Wilcoxon *T* test

PROBLEMS

1. Calculate a Mann–Whitney U for the following data obtained from two independent sets of ordinal measures.

R_1	R_2
1	2
3	5
4	8
6	11
7	14
9	16
10	18
12	17
13	15

2. Calculate a Kruskal–Wallis H for the following data obtained from three independent sets of ordinal measures.

R_1	R_2	R_3
1	2	4
3	6	8
5	10	11
7	13	17
9	15	21
12	16	20
14	18	19
22	23	24

3. Calculate a Wilcoxon T for the following data obtained from two correlated and skewed distributions of interval measures.

X_1	X_2
50	52
40	48
55	60
48	52
42	40
45	51
51	82
70	55

4. Calculate a Friedman ANOVA by ranks for the following data obtained from three correlated and skewed distributions of interval scores.

X_1	X_2	X_3
15	17	21
14	16	25
16	18	15
15	17	19
15	18	20
16	18	19
12	14	20
14	21	20
16	35	21
31	11	57

5. A researcher is interested in establishing whether type of background music affects how ice-skating performances are judged. A random sample of 24 skaters is selected from among the students of a large skating club. The skaters are randomly assigned to one of three conditions: A, jazz music; B, classical music; C, no music. Each skater

performs for 10 minutes. All are rank-ordered by judges as to their performance. (Each skater receives the median rank assigned by three judges.)

Condition A, R_1	*Condition B,* R_2	*Condition C,* R_3
3	1	8
4	2	15
6	5	18
9	7	21
11	10	20
13	12	19
16	14	17
22	23	24

Test the hypothesis.

6. Hypothesis: There is a significant difference in the Breathalyzer readings from suspected drunken drivers as a function of the time between apprehension and the onset of the testing procedure. A random sample of 10 suspected drunken drivers is tested on the Breathalyzer under three conditions: first, 20 minutes after their arrival at the police station (suspects must be observed for 20 minutes to verify that they do not place anything, solid or liquid, in their mouths before testing begins); second, after 1 hour and 20 minutes; third, after 2 hours and 20 minutes. (Scores on the Breathalyzer range from 0 to .40 and indicate the percentage of alcohol in the bloodstream. Thus, these scores are in interval form. Since a score of .35 indicates a comatose condition, few subjects score that high. Thus, the distribution is typically skewed to the right.)

Subject	*First Condition*	*Second Condition*	*Third Condition*
A	.12	.11	.10
B	.17	.15	.14
C	.12	.11	.10
D	.11	.10	.09
E	.15	.14	.13
F	.12	.11	.10
G	.13	.12	.11
H	.20	.19	.16
I	.30	.28	.25
J	.15	.14	.13

Test the hypothesis.

7. A researcher is interested in discovering whether competency test scores for high school teachers can be increased by the introduction of a three-day in-service workshop in the teacher's own specialty area. A random sample of 10 high school teachers is selected. They are given the competency test, sent to the workshop, and then given the competency test again. A nonparametric statistical analysis of the data is undertaken, as it became obvious that the distribution of test scores is severely skewed to the left.

Teacher	*Before Scores*	*After Scores*
1	84	87
2	88	89
3	85	83
4	92	96
5	90	92
6	88	90
7	78	75
8	41	40
9	87	90
10	31	29

Test the hypothesis.

8. Hypothesis: Students are more likely to show verbal aggressiveness (assertively challenging the professor, etc.) in small classes than they are in large classes. Random samples of 10 large classes (50 students or more) and 10 small classes (less than 25 students) are selected. A three-judge panel visits each classroom on five separate occasions and then rank-orders all classes as to the extent of student assertiveness.

Small Class	*Large Class*
2	6
7	13
11	5
3	9
10	18
1	17
4	20
8	19
12	16
14	15

Test the hypothesis.

For the research situations described in problems 9 through 12, indicate which statistical test would be most appropriate.

9. Data: ordinal
   Design: two groups, between-subjects
10. Data: skewed, interval
    Design: one group, repeated-measures
11. Data: skewed interval
    Design: four groups of matched subjects
12. Data: ordinal
    Design: three groups, between subjects

True or False. Indicate either T or F for problems 13 through 17.

13. The Mann–Whitney U test is less powerful than the independent t test.
14. The ordinal tests of significance can be used on skewed interval data.
15. With equal sample sizes, nonparametric tests are just as powerful as parametric tests.
16. The larger the numerical value of the Wilcoxon T, the higher is the likelihood of achieving significance.
17. Nonparametric tests can never be used on interval data, no matter what the shape of the underlying distribution.

Fill in the blanks in problems 18 through 25.

18. The ordinal equivalent of the paired t test is the ________.
19. A less powerful statistical test is one in which it is less likely that the ________ hypothesis will be rejected.
20. The ordinal equivalent of the within-subjects ANOVA is the ________.
21. The ordinal equivalent of the Pearson r is the ________.
22. The ordinal equivalent of the one-way ANOVA is the ________.
23. The ordinal equivalent of the two-sample independent t test is the ________.
24. If the researcher had interval data from a repeated-measures design, with three sets of scores, and discovered that the scores from each measure were badly skewed, then the appropriate statistical test would be the ________.
25. If the researcher had interval data from a between-subjects design, with three sets of scores, and discovered that the scores from each group were badly skewed, then the appropriate statistical test would be the ________.

Chapter

17

Tests and Measurements

By this time in your college career you have probably been tested literally thousands of times. From the day you were born, and for many of you even before that, you have been measured and evaluated on myriad variables, from perhaps, in utero, whether you were a boy or a girl to, more recently, that bewildering College Adjustment Inventory you took during freshman orientation week. In this chapter we home in on the concepts that underlie the tests and measures that are used to orchestrate the underlying theme song that runs throughout all psychology and education—individual differences. Now that the basic statistical techniques have been covered, you should be putting them together to better understand the meaning and reality of psychological and educational testing. What started back in Chapter 1 as a softly muted theme now becomes a loud and practical crescendo. In other words, this chapter will be directed toward telling you what to do now that you know the score.

In January of 2002 President Bush and key members of congress unveiled a new educational initiative which, among other things, requires that all children in grades three to eight be tested each year in the basic skills deemed necessary for full participation in a democratic society. The federal role in local education has become far more insistent since the passage of this new bill, although the states are allowed leeway to set their own local standards rather than following a one-shoe-fits-all approach.

Testing is already a big business in the United States, and the new law will only make it bigger. Adding to this testing surge are at least three other developments: (1) A new test, developed as a successor to the NTE (National Teacher's Exam) by ETS (Educational Testing Service, the folks who brought you the SAT) for individuals attempting to enter the teaching profession. (2) The continued testing, as proposed by the National Board for Professional Teaching Standards, of working teachers who must maintain their licenses. Such tests must be constructed with great care and sensitivity and must be able to face the legal challenges both from those teachers who don't pass the tests and from advocates for students who have been disadvantaged by teachers who are clearly not competent (Dwyer, 1991).

(3) Federal legislation regarding disabilities, The Individuals with Disabilities Education Act (IDEA), Public Law 101–476, was enacted in the 1990s and puts further pressure on the testing enterprise. Millions of children, from age three on, must now be assessed for the presence of a disability, and the children so identified become eligible for special education and related services. This process involves more than just the administration of a standardized educational and psychological test and must include observations, interviews, and medical assessments. This psychological and educational testing, however, must be done on an individual *basis, and the test instruments must be proven to be both reliable and valid. This means that a test may not be used to rate a student in an area for which the test has not been designed. Furthermore, the tests must not be racially or culturally discriminatory (Waterman, 1994).*

Whether one agrees or disagrees with this proliferation of testing, the practice of measuring individual performance and/or ability has become an inevitable fact of life, especially in the field of psychology and education. Since the dawn of the twenty-first century, over 70 million primary and high school students are taking more than 100 million standardized tests each year, and even this number doesn't factor in the countless teacher-made tests to which America's schoolchildren are constantly exposed. A byproduct of this extensive use of tests is a presumed confidence in their accuracy. Our goal in this chapter is to point out where and under what circumstances that confidence is most probably warranted and where it is clearly not. Throughout this chapter, keep in mind that an "understanding of how measures in behavioral research are constructed is essential to evaluating the worth of that research" (Gillespie, 1994, p. 5).

NORM AND CRITERION REFERENCING: RELATIVE VERSUS ABSOLUTE PERFORMANCE MEASURES

The tests used in psychology and education may use either norm-referenced or criterion-referenced scoring systems. The difference is fundamental. In a norm-referenced test an individual's performance is compared to the average performance of the entire test-taking population. For example, on an IQ test an individual's score is not reported in terms of the absolute number of correct answers given but, instead, on the basis of a comparison between that person's performance and the *average* performance of all the individuals of the same age who have previously taken the same test. This procedure, in fact, was precisely the same as that used way back at the turn of the century by that giant in the field of intelligence testing, Alfred Binet. Binet used the term mental age to describe his scoring method. To establish a student's mental age, Binet would give his test to large numbers of children of various ages. Then the average performance for a given age became the benchmark for evaluating a given student's performance. In effect, Binet would find out how many items the average eight-year-old (having tested thousands of eight-year-olds) could answer correctly, and then a child of any age who answered the same number of items was assigned a mental age of eight. The mental-age technique, as with all norm-referenced scoring systems, thus provides information regarding a person's *relative* standing. Most of the nationally standardized, mass-produced tests used in psychology and education today are of this type. Intelligence tests, achievement tests, aptitude tests, personality tests, and so on are accompanied by a set of national norms with which each individual's performance can be compared.

The criterion-referenced test, however, is *not* based on relative performance but on *absolute* performance. For this reason it is also called a mastery test by educational psychologists, since the focus is on what absolute fraction of the material presented on the test the test taker has conquered. For example, the Federal Aviation Administration gives a test to each prospective pilot before he or she is allowed to solo. The applicant must answer at least 70% of the questions correctly before taking to the air without the instructor. Thus, the FAA demands that the student pilot know the large majority of the material on the test, regardless of how others have performed. The FAA reasons that it's not enough to have a high relative standing on the test, since it's possible that the majority of those taking the test were so ignorant of flying techniques and procedures that they might kill themselves on takeoff. It's like the conductor of a fine orchestra demanding that the musicians pass an absolute yardstick of talent before being assigned to a chair. It's not enough to say that this is the best flute player out of a tone-deaf group.

Teachers tend not to prefer standardized, norm-referenced tests, usually saying that their particular instructional mission should not be narrowed to those areas sampled by the test. This reluctance seems to be based on the fact that the content of the course cannot be under teacher control when standardized tests are being used as judge and jury of both the teacher and the students. Nor are teachers enamored of the tests being used for teacher certification purposes, which seems to put teachers in the ironic position of arguing that tests are not a good way of judging people.

Critics of standardized testing allege that these tests create an artificial classroom, since it forces teacher to "teach to the test," especially when the children's test scores are used not just to evaluate the children but also the quality of the teacher's work. In fact there have been charges that teachers have gone one step further and, instead of just teaching to the test, have taught the test itself. When the teaching is too narrowly focused on the direct answers to test question, trivialization of the curriculum surely follows. But if the general content of the domain under question is being taught, the test evaluation could still be fair and also remain both reliable and valid. In any case it is true that teachers over the years admit that they have consistently taught to their own tests, and of course they should. It may really be an issue of whose test is being taught to whom. To test students on material they haven't yet been exposed to would surely be an exercise in folly.

Some tests, called high-stakes tests are those whose scores have a profound impact on the lives of the test-takers since they can be used to determine who will or will not be selected for jobs, colleges, law schools, medical schools, licenses, certification and a variety of important credentials (Sackett et al., 2001). Such tests as the SAT, which helps determine who will be accepted in college and even who might receive financial aid, or the National Teacher Test, which determines who will be allowed into the teaching profession, or state-mandated educational achievement tests which decide who is going to receive a high-school diploma are examples of high-stakes tests (Gutloff, 1999). As of now, there are 47 states that. have established standards for what students at various grade levels should know. And in about half of these the test results determine such crucial things as promotion and graduation (Smith, 2001). Because of the importance attached to the results of these tests, it is critical that any elements of racial/minority bias be removed.

THE PROBLEM OF BIAS

Ethnic bias may at times creep into standardized tests. Several years ago the word *regatta* was used in one of the College Board tests. This resulted in a storm of protest, since that is not a word that is often bandied about in minority neighborhoods. Since that time, the Educational Testing Service has been careful to a review each word for any such signs of cultural bias. ETS insists, "test results cannot be judged in isolation from the unequal outcomes produced by our educational, economic and social systems" (Educational Testing Service, 1991, p. 3). In other words a fair and accurate test will in fact reflect the quality of academic preparation, and if it doesn't, it will not be a true indicator of educational accomplishments for any test-taker, no matter what the student's background might be. The concerns of minorities have also been addressed through the use of what is called race norming, in which test scores within minority groups are compared only with those from the same minority population rather than across the board (Helms, 1992). Some have felt that race norming is especially important when dealing with any of the cognitive ability tests, as opposed to aptitude or personality tests. Except for the cognitive tests, sub-group differences are minimized since measures of interpersonal skills and personality show only small differences between racial and ethnic groups. The problem is that the interpersonal skills tests do not seem to predict academic achievement as well as the cognitive tests do. Race norming may not, however, be the wave of the future since recent court decisions and voter initiatives show a growing trend away from preference-based forms of affirmative action (Sackett et al., 2001).

Multiple-Choice Testing

What about multiple-choice tests, or, as many poorly prepared students like to call them, "multiple-guess tests"? One of the criticisms of the MC test is that it rewards rote memorization rather than true understanding. This can certainly happen if the test is poorly designed, but when thoroughly researched and carefully prepared, the MC test can assess a person's ability to apply concepts to problem-solving situations. Rather than breaking up the unity of knowledge and isolating the pieces, as the critics typically charge, a *well-designed* multiple-choice test, such as the SAT, demands that the student be able to understand concepts and bring facts together. This is especially true of the SAT's reading comprehension section. The research evidence clearly shows that the SAT verbal score shares much in common with IQ, the correlation between them being an extremely high +.80 (Flynn, 1987). What about essay questions? There is the fear that standardized tests based only on essay questions and writing samples may have an adverse effect on minority groups (Science Agenda, 1990). Verbally adept but uninformed students may bluff their way through an essay exam (Aiken, 1988), but this is a tactic that is used less often by minorities. Although essay exams can often illuminate the student's thought process in more detailed form, the teacher with a large class of widely varying abilities, interests, and needs may have to rely on the multiple-choice test. It not only helps to produce reliable scores, but, more importantly, it gives the teacher more free time to work with individual students who may have been identified by the tests as needing extra help. It is often all too easy to blame standardized MC testing for deteriorating academic skills and overlook the more immediate culprits, overcrowded classrooms and underpaid teachers.

TEST RELIABILITY, VALIDITY, AND MEASUREMENT THEORY

As mentioned back in Chapter 9, measurement is based on the assigning of numbers to observations (of persons, events, or things), and the way these numbers are assigned determines the scale of measurement being used (whether it be nominal, ordinal, interval, or ratio). For these measurements to be useful and provide meaning to the researcher, they must also satisfy two fundamental criteria, reliability and validity.

It is known that all measurements contain some degree of error. For example, if you are measuring a piece of wood before cutting, your measure would be contaminated by such factors as how straight the end of the ruler is, the thickness of the lead in the pencil that you use to make your mark, the angle that you're viewing from, and even the acuity of your vision. In fact, if you continued measuring that same piece of wood, you are very likely to get different results; your pencil wears down, your head moves and you change your point of observation, and so forth. Each individual score, therefore, is said to be composed of both true score and error score.

$$X_{\text{observed}} = X_{\text{true}} + X_{\text{error}}$$

The true score is the theoretically genuine measure and is technically defined as the mean value of an infinite number of measures. The error score, however, is assumed to be random; that is, the deviation should be just as likely to create an overestimate as an underestimate of the true score. Because error is assumed to be random, the mean deviation will be zero since the error itself will be positive just as often as negative. The less the error, the more the observed score is composed of true score and the more *consistent* the measures will turn out to be. **Reliability,** then, is the proportion of true score to the observed score, and the higher the proportion is, the more consistent will be the resulting measures. Reliability, thus, *results* in consistent measures. It must always be kept firmly in mind, however, that reliability refers to a specific measuring instrument applied to a specific population under specific conditions.

$$\text{reliability} = \frac{X_{\text{true}}}{X_{\text{observed}}}$$

Reliability and Interval Data

To assess the reliability of any test based on at least interval data, we can use the Pearson r, and the resulting correlation will be our estimate of the reliability coefficient. The correlation must be significant in order to be used to establish reliability, and the strength of the correlation provides information regarding the dependability of the measures involved.

Test-Retest Reliability

The most intuitively obvious of the reliability techniques is called the **test-retest** method: We simply give the test, wait for a given amount of time to elapse, and then give the same test again. To establish whether the measures are consistent, a correlation is computed between the results obtained on the first administration of the test with those from the second testing situation. When a correlation is used to assess reliability, it is symbolized as r_{tt} (the correlation of a *t*est with a *t*est). Although popular, this method is not above criticism, since there are at least two uncontrolled variables that may contaminate the resulting correlation: practice and/or fatigue. Some subjects may do significantly better the second time, simply because of the practice effect. Perhaps some of

Never mind the standard error of measurement or the concurrent validity. Did I pass the test?

them spent the intervening time thinking over several of their previous answers or were even curious enough to ask someone how to do a given problem that they had omitted or done wrong the first time. This can lead to differences that are not truly random, since perhaps not all the subjects were motivated enough to think about the test and take advantage of the practice opportunity. Also, if the time interval is too short, some of the subjects may be simply too tired or bored to concentrate as much during the second administration of the test. Here, again, this could lead to systematic differences between those who fatigue easily and those who retain their powers of concentration despite their fatigue. Even worse, some subjects might even be ill during the second test date and not perform up to their capabilities or not show up at all for the retest. The test-retest method has also been seen as providing an estimate of the test's stability, especially when the assumption is met that the test-takers have not changed between test administrations in terms of the quality the test measures (McIntire & Miller, 2000)

Example To obtain the test-retest reliability for the Figure-Ground Test, a test of field dependence or one's ability to search out embedded figures from a background design, use the following data. The test was given to 10 normal adults, and the mean on the first administration was 15.000 with an s of 3.621. On the second administration (six weeks later), the mean was 16.000 with an s of 3.859. The correlation is established using the Pearson r equation shown in Chapter 11.

	Test		Retest		
Subject	X	X^2	Y	Y^2	XY
1	15	225.000	16	256.000	240.000
2	12	144.000	15	225.000	180.000
3	11	121.000	12	144.000	132.000
4	20	400.000	22	484.000	440.000
5	14	196.000	18	324.000	252.000
6	16	256.000	15	225.000	240.000
7	17	289.000	15	225.000	255.000
8	16	256.000	17	289.000	272.000
9	9	81.000	9	81.000	81.000
10	20	400.000	21	441.000	420.000

Mean $M_x = 15.000$; Mean $M_y = 16.000$.
$s_x = 3.621$; $s_y = 3.859$.
$r_{tt} = 0.891$.

With 8 df, the Pearson r is significant at an alpha of .01 (a table value of .765 was needed). Thus, the reliability is established at .891, certainly an acceptable value for this type of test. While it is true that as a group the scores did increase, the individuals largely retained their relative positions.

Alternate- or Parallel-Form Reliability

In this method, a separate but equal form of the original test is used for the comparison. That is, another version of the test is created that contains the same kinds of items, with the same levels of difficulty and selected from the same population of items as was used for the first test. The results of this test are then correlated with those of the original test. The practice effect is not as noticeable with this technique, since the subject is not seeing the same items over again. To prevent fatigue, however, it is wise not to give the alternate form on the same day as was used for the original test.

Example The Sound Recognition Test is a test for a condition known as auditory agnosia, or a person's ability to recognize familiar environmental sounds, such as a bell, a whistle, chimes, or crowd sounds. There are two forms of the test, A and B, with a total of 13 items per test. Scoring is based on allowing up to 3 points per item, making 39 the highest possible score. A group of normal five-year-old children was selected and given form A. Then the next day they were given form B. The data were as follows:

	Form A		*Form B*		
Subject	X	X^2	Y	Y^2	XY
1	30.000	900.000	29.000	841.000	870.000
2	25.000	625.000	22.000	484.000	550.000
3	35.000	1225.000	31.000	961.000	1085.000
4	22.000	484.000	25.000	625.000	550.000
5	38.000	1444.000	36.000	1296.000	1368.000
6	31.000	961.000	31.000	961.000	961.000
7	29.000	841.000	30.000	900.000	870.000
8	28.000	784.000	30.000	900.000	840.000
9	32.000	1024.000	34.000	1156.000	1088.000
10	30.000	900.000	32.000	1024.000	960.000

Mean $M_x = 30.000$; Mean $M_y = 30.000$.
$s_x = 4.570$; $s_y = 4.055$.
$r_{tt} = 0.851$.

Again, with 8 df, the Pearson r is significant at an alpha of .01 (a table value of .765 was needed). Thus, the alternate-form reliability is established at .851, which again is certainly an acceptable value for a test of this type.

Internal-Consistency Reliability and the Split-Half Method

The methods to be discussed next are called *internal-consistency* techniques because what is being assessed in this case is the consistency that occurs *within* the test, rather than between two separate administrations of the same test or separate administrations of alternate forms. One way to get at this type of reliability is to split the whole test into two halves (**the split-half**) so that only one administration of the test is required. To even out the practice-fatigue problem, the test is typically split by comparing the results of the odd-numbered items with those of the even-numbered items. The theory here is that a subject will be just as practiced (or fatigued) on item 1 as on item 2 or item 97 as on item 98. Also, this alternating format aids in controlling for increasing item difficulty. (If the test were simply split by comparing, say, the first 50 items with the last 50 items, the practice-fatigue variables would still be present and still be uncontrolled, especially if all the easy items were in the first half.) When the reliability coefficient has been calculated on a split-half, odd-even comparison, the reliability of the test is *underestimated* (since the number of items has been effectively cut in half).

Spearman-Brown

To find out what the reliability would have been if the whole test were used (which is certainly the final test goal), we apply the **Spearman–Brown prophecy formula** for estimating the reliability of a test that has been doubled in size. For example, if a 100-item test is assessed for reliability by comparing the 50 odd items with the 50 even items, the correlation has been obtained on only 50 pairs of scores. After all, when the test is finally produced and standardized, it won't just be restricted to the odd-numbered items.

$$r_{SB} = \frac{(n)(r)}{1 + (n - 1)(r)}$$

where r_{SB} equals the Spearman–Brown prophesied reliability coefficient, n equals the number of times the test is to be increased, and r equals the obtained self-correlation between the two halves of the test. Thus, if the self-correlation between the two halves of the test were .80, we could determine the effect on reliability of doubling the test (back into its whole form).

$$r_{SB} = \frac{(2)(.80)}{1 + (1)(.80)} = \frac{1.60}{1.80}$$

$$r = .889$$

Or if the self-correlation were only .70, we could determine the effect on reliability if the test were tripled in length.

$$r_{SB} = \frac{(3)(.70)}{1 + (2)(.70)} = \frac{2.10}{2.40}$$

$$r = 0.875$$

This equation can also be used to estimate how long a test must be to attain a certain reliability. For example, if the self-correlation were .80, we could determine how long the test must be to get an r_{SB} of .95.

$$n = \frac{r_{SB}(1 - r)}{r(1 - r_{SB})} = \frac{.95(1 - .80)}{.80(1 - .95)} = \frac{(.95)(.20)}{(.80)(.05)} = \frac{.190}{.040}$$

$$= 4.750$$

Thus, the test would have to be increased by 4.75 times to achieve a predicted reliability of .95.

Example The KeyMath test is an example of a split-half with a Spearman–Brown correction used to assess strengths and weaknesses in a number of math and math-related areas. Scores range from a low of 40 to a high of 136. A sample of 10 fourth-grade children, all aged nine years, was selected and given the test. The scaled scores for the odd and even items follow:

	Odd		Even		
Subject	X	X^2	Y	Y^2	XY
1	125	15,625.000	120	14,400.000	15,000.000
2	100	10,000.000	100	10,000.000	10,000.000
3	75	5,625.000	80	6,400.000	6,000.000
4	90	8,100.000	95	9,025.000	8,550.000
5	110	12,100.000	105	11,025.000	11,550.000
6	95	9,025.000	90	8,100.000	8,550.000
7	105	11,025.000	110	12,100.000	11,550.000
8	97	9,409.000	98	9,604.000	9,506.000
9	103	10,609.000	102	10,404.000	10,506.000
10	100	10,000.000	100	10,000.000	10,000.000
Sums	1000.000		1000.000		

Mean $M_x = 100.000$; Mean $M_y = 100.000$.
$s_x = 12.987$; $s_y = 10.842$.
Pearson $r = 0.956$.

With 8 df, the Pearson r is significant at an alpha of .01 (a table value of .765 was needed). Since the r is significant, we now apply the Spearman–Brown as follows:

$$r_{SB} = \frac{(2)(.956)}{1 + .956} = \frac{1.912}{1.956}$$
$$= .978$$

Thus, the reliability is now established at .978, which is certainly a lofty value and signifies a high degree of reliability for this test.

There are times, however, when adding an item to a test actually decreases the test's reliability, as when the new item lacks its own reliability and doesn't relate to the construct being measured. Thus, it is important to note that the Spearman–Brown equation suggests the following: that reliability increases both with test length *and* with the reliability of each separate item (Li, Rosenthal, & Rubin, 1996).

Internal-Consistency Reliability and the Kuder–Richardson Reliability Formula (K–R 21)

The split-half reliability technique is one way of establishing a measure of internal consistency. Other internal-consistency reliability techniques include those developed by Kuder and Richardson and are called, not surprisingly, the **Kuder–Richardson reliability** formulas. The one to be discussed here is their **formula 21.** It is quick and easy to calculate, and when used appropriately, it can be extremely useful in determining a test's overall reliability. Unlike the split-half method, which splits the test just once, the K–R 21 estimates the reliability of a test that has been split into *all possible halves,* and it automatically corrects for the splits without any need for a Spearman–Brown type of adjustment. In fact, K–R 21 produces the mean of all the possible split-half correlation coefficients. To calculate this value, all you need to know about the test is the number of items, the standard deviation of the entire test, and the mean of the entire test. The formula is as follows:

$$\text{K–R 21} = \frac{[(SD^2)(k)] - [(M)(k - M)]}{SD^2(k - 1)}$$

where M equals the mean on the entire test, k equals the number of items on the test, and SD is the standard deviation of the entire test.

Let's take an example. The mean on a given test is 80 with an *SD* of 10. The test is composed of a total of 100 items.

$$\begin{aligned}\text{K–R 21} &= \frac{[(10^2)(100)] - [(80)(100 - 80)]}{10^2(100 - 1)} \\ &= \frac{[(100)(100)] - [(80)(20)]}{100(99)} \\ &= \frac{10,000 - 1600}{9900} \\ &= \frac{8400}{9900} \\ &= .848\end{aligned}$$

which, depending on the type of test, can be a strong reliability value.

To use the K–R 21 the following criteria should be kept in mind.

1. The entire test should be aimed at tapping a single domain. If the test is not clearly focused on a single underlying concept, the reliability value will be underestimated.
2. The test is scored on the basis of each item being either right or wrong.
3. All items have about the same degree of difficulty. The formula works best (produces its highest reliability estimate) when the difficulty index is

approximately .50 for each item. More will be said later regarding the important issue of item difficulty.

Internal-Consistency Reliability and ANOVA

If you have completed the material on ANOVA in Chapters 12 and 15, especially the material on the within-subjects ANOVA in Chapter 15, you may find that the F ratio can become your best friend when assessing internal-consistency test reliability. In this section we will present two ANOVA techniques for establishing the r_{tt}, one for the situation in which the items are dichotomously scored, either right or wrong, agree-disagree (nominal data), and also one in which the items are scored on a continuous scale (interval data). In both cases the eventual r_{tt} is the reliability reported. Since the ANOVA procedures compare every item with the total score, there is no further need for any Spearman–Brown correction.

Cyril Hoyt Method

When the items are scored on the basis of being either correct or incorrect, the following ANOVA variation (first suggested by that great Minnesota statistician Cyril Hoyt) can be used. For this example, assume that there are five subjects and four items and that we are using 1 for correct and zero for incorrect (obviously these are fictional data, but if you can do small-data problems, you can always use the same techniques for more realistic large-data problems).

	Items				
Subjects	***a***	***b***	***c***	***d***	**ΣX rows**
1	1	1	1	1	4
2	1	1	1	1	4
3	1	1	1	0	3
4	1	1	0	0	2
5	1	0	0	0	1
ΣX cols.	5	4	3	2	14

1. Get the correction factor, where N = number of scores.

$$C = \frac{(\Sigma X)^2}{N} = \frac{14^2}{20} = 9.800$$

2. Find the SS total by simply counting the number of right and wrong answers.

$$SS_t = \frac{(\text{right})(\text{wrong})}{\text{right} + \text{wrong}} = \frac{(14)(6)}{14 + 6} = 4.200$$

3. Get the SS between columns, which will now represent the SS for items.

$$SS_{\text{items}} = \frac{(\Sigma X_1)^2}{n_1} + \frac{(\Sigma X_2)^2}{n_2} + \frac{(\Sigma X_3)^2}{n_3} + \frac{(\Sigma X_4)^2}{n_4} - C$$

$$= \frac{5^2}{5} + \frac{4^2}{5} + \frac{3^2}{5} + \frac{2^2}{5} - C = 10.8 - 9.8 = 1.000$$

4. Find the SS between rows, which will now represent the SS for subjects.

$$SS_{\text{subjects}} = \frac{(\Sigma X_{r1})^2}{n_{r1}} + \frac{(\Sigma X_{r2})^2}{n_{r2}} + \frac{(\Sigma X_{r3})^2}{n_{r3}} + \frac{(\Sigma X_{r4})^2}{n_{r4}} + \frac{(\Sigma X_{r5})^2}{n_{r5}} - C$$

$$= \frac{4^2}{4} + \frac{4^2}{4} + \frac{3^2}{4} + \frac{2^2}{4} + \frac{1^2}{4} - C = 11.500 - 9.800 = 1.700$$

5. Get the $r \times c$ (residual) by subtraction.

$$SS_{r\times c} = SS_t - SS_{\text{items}} - SS_{\text{subjects}} = 4.200 - 1.000 - 1.700 = 1.500$$

6. Set up the summary table, where column df = number of columns – 1 = 3; row df = number of rows – 1 = 4; and residual df = column df × row df = 12.

Source	SS	df	MS	F
Between columns (items)	1.000	3	.333	$\frac{MS_{\text{subjects}}}{MS_{\text{residual}}} = 3.40$
Between rows (subjects)	1.700	4	.425	
Residual ($r \times c$)	1.500	12	.125	

Note: This F ratio is calculated on the basis of the MS for subjects (rows) in the numerator (as opposed to placing the columns in the numerator, as we did for the straight within-subjects F ratio back in Chapter 15). However, we still use the residual mean square for the denominator.

$$F = \frac{MS_{\text{subjects}}}{MS_{\text{residual}}}$$

$F = .425/.125 = 3.40$ ($F_{.05}$ with 4 and 12 df = 3.26). Reject H_0; the F ratio is significant and we can proceed to calculate the reliability.

$$r_{tt} = 1 - \frac{MS_{\text{residual}}}{MS_{\text{subjects}}}$$

$$= 1 - \frac{.125}{.425} = 1 - .294 = .706$$

This test is showing fair to low reliability.

Had the F ratio not been significant, the reliability value should not be calculated since it cannot then be significant.

When the items are not simply scored as right versus wrong, then the following ANOVA can be used. Assume that we again have five subjects taking a four-item attitude test, but this time the test is scored on a Likert scale format; that is, each item is weighted from 1 to 5 (based on strongly agree, agree, neutral, disagree, and strongly disagree).

	Items								
Subjects	*a*	*a*²	*b*	*b*²	*c*	*c*²	*d*	*d*²	Σ**X rows**
1	6	36	6	36	5	25	4	16	21
2	4	16	6	36	5	25	3	9	18
3	4	16	4	16	4	16	2	4	14
4	3	9	1	1	4	16	2	4	10
5	1	1	2	4	1	1	1	1	5
Σ*X* cols	18	78	19	93	19	77	12	34	68

1. Get the correction factor, where N = number of scores.

$$C = \frac{(\Sigma X)^2}{N} = \frac{68^2}{20} = 231.200$$

2. Find the SS total by adding the squares and subtracting the correction factor.

$$SS_t = \Sigma X^2 - \frac{(\Sigma X)^2}{N} = \Sigma X^2 - C$$
$$= 288 - 231.200 = 56.800$$

3. Get the SS between columns, which will now represent the SS for items.

$$SS_{\text{items}} = \frac{(\Sigma X_1)^2}{n_1} + \frac{(\Sigma X_2)^2}{n_2} + \frac{(\Sigma X_3)^2}{n_3} + \frac{(\Sigma X_4)^2}{n_4} - C$$
$$= \frac{18^2}{5} + \frac{19^2}{5} + \frac{19^2}{5} + \frac{12^2}{5} - C = 238 - 231.200 = 6.800$$

4. Find the SS between rows, which will now represent the SS for subjects.

$$SS_{\text{subjects}} = \frac{(\Sigma X_{r1})^2}{n_{r1}} + \frac{(\Sigma X_{r2})^2}{n_{r2}} + \frac{(\Sigma X_{r3})^2}{n_{r3}} + \frac{(\Sigma X_{r4})^2}{n_{r4}} + \frac{(\Sigma X_{r5})^2}{n_{r5}} - C$$
$$= \frac{21^2}{4} + \frac{18^2}{4} + \frac{14^2}{4} + \frac{10^2}{4} + \frac{5^2}{4} - C = 271.500 - 231.200 = 40.300$$

5. Get the $r \times c$ (residual) by subtraction.

$$SS_{r \times c} = SS_t - SS_{\text{items}} - SS_{\text{subjects}} = 56.800 - 6.800 - 40.300 = 9.700$$

6. Set up the summary table, where column df = number of columns −1 = 3; row df = number of rows −1 = 4; and residual df = column df × row df = 12.

Source	*SS*	*df*	*MS*	F
Between columns (items)	6.800	3	2.267	$\frac{MS_{\text{subjects}}}{MS_{\text{residual}}} = 12.469$
Between rows (subjects)	40.300	4	10.075	
Residual ($r \times c$)	9.700	12	0.808	

$$F = \frac{MS_{\text{subjects}}}{MS_{\text{residual}}}$$

Note: The F ratio is again calculated on the basis of the MS for subjects (rows) in the numerator and the residual mean square in the denominator. $10.075/.808 = 12.469$ ($F_{.01}$ with 4 and 12 df = 5.41). Reject H_0; the F ratio is significant at $p < .01$ and we can proceed to calculate the reliability.

$$r_{tt} = 1 - \frac{MS_{\text{residual}}}{MS_{\text{subjects}}}$$

$$= 1 - \frac{.808}{10.075} = 1 - .08 = .92$$

This test is showing very high reliability.

Cronbach's Coefficient Alpha

For this situation, where the items are not scored dichotomously, it is also possible to use Cronbach's coefficient alpha (Cronbach, 1951), which is

$$\alpha = \left(\frac{N}{N-1}\right)\left(\frac{SD^z - \Sigma SD^z_{\text{items}}}{SD^z}\right)$$

where N = the number of items, SD^z = the overall variance on the entire test and ΣSD^z items = the sum of the individual item variances. Cronbach's alpha has the advantage of identifying which items are or are not contributing to the overall reliability, since each and every item has to be individually assessed for variability. Without a computer program it can be very labor intensive to calculate. However, with *SPSS* the procedure takes only seconds (once the data are typed in). Because of its flexibility, coefficient alpha is probably the most popular reliability procedure in use today (Osburn, 2000).

The Cyril Hoyt and Coefficient Alpha

Also, you can use the Cyril Hoyt reliability procedure shown above which will give you exactly the same value as will coefficient alpha. Thus, on the data shown above the coefficient alpha procedure also equals .92.

Increasing Reliability Estimates

The estimated reliability of a test can be increased by the following techniques (all of which are strongly related to each other).

1. Increase the length of the test by adding more items. The longer the test is, within reason, the higher the likely reliability. One should only add items that have been selected from the same population of items.
2. Increase the variance by making the test more discriminating. Add items that separate persons along a continuum, with some items that only the top scorers will get right, some only the worst scorers will get wrong, and so on up and down the line. Get rid of items that everybody gets right (as well as those that everyone gets wrong), since they are not providing any power to discriminate. A 2-item test obviously cannot yield as much variation as a 100-item test, nor will the 2-item test provide much information on the discriminating power of the items, other than at best perhaps sorting the above average from the below average.
3. Increase the size of the sample being tested. No matter how many items you have or how well your items discriminate, if there are not enough subjects of varying ranges of ability, the reliability of the test can be severely underestimated.

How High Must the Correlation Be?

First, before getting into the issue of how strong the correlation has to be to demonstrate test reliability, it is critical to remember once again that before the correlation can be generalized to the population, it must be proved significant. That is, the researcher has to really make two decisions. First, the statistician has to decide whether to reject the null hypothesis. Fortunately, this is rather cut-and-dried since, as you certainly know by now, tables are available for making that decision an automatic process. In short, the null hypothesis must be rejected and the correlation found to be significant before even considering whether to call the test reliable. Second, the researcher must determine how high the *significant correlation* has to be in order to feel comfortable with its level of reliability. Unfortunately, there is no absolute answer to this second point. The strength of the correlation denotes different things depending on the test itself, especially the type of test. For example, for an achievement test battery, reliabilities typically show an r_{tt} of over .90; some are as high as .98. Scholastic ability tests range around .90, aptitude tests just below .90, and attitude tests about .80. Some of the objective personality tests have even lower reliabilities, some being reported to be as low as .60. The main message here is that in comparing reliability coefficients for different tests, you try to compare tests of the same general type and purpose.

Too Much Consistency?

Sometimes test-retest correlation coefficients may be too high and indicate an inflexible test that is not truly measuring changes that may actually be taking place in the characteristics under study. Obviously, some traits do vary over time, and when this occurs, it is not unusual to find lowered reliability coefficients. In these cases, the lower reliability level is not the result of measurement error but is instead a true reflection of the fact that certain traits sometimes vary. In fact, there are times when important variables, such as health status, are *expected* to vary over time (Kaplan & Saccuzzo, 1993). In general, however, the higher the reliability is, the more accurate the test. A test with a reliability coefficient of near zero is only an instrument of sheer chance. If a test doesn't correlate with itself, don't expect it to correlate with anything else, because it won't.

Criterion Referencing and Reliability

On criterion-referenced tests, the usual methods for assessing **reliability** are not as useful because they typically produce very low reliability estimates. As we have seen, correlation coefficients will be low when the variability among the scores is small. With criterion-referenced tests, variability is usually restricted, since the researcher is interested in the number of individuals who reach criterion versus those who don't, rather than how much of a trait or characteristic the individuals possess. For criterion-referenced tests, then, a better approach is to compare those who reach mastery with those who don't, and one method for accomplishing this is through the use of a test-retest reliability formula suggested by Lindeman and Merenda and adapted here as

$$\text{CR} = \frac{(F_{1,2})(P_{1,2}) - (P_1)(P_2)}{(F_{1,2})(P_{1,2}) + v(F_{1,2} + P_{1,2} + v)}$$

$F_{1,2}$ = number who failed both administrations

$P_{1,2}$ = number who passed both administrations

P_1 = number who passed only the first administration

P_2 = number who passed only the second administration

v = smaller of the two values, P_1 or P_2

Suppose that we were assessing the reliability of a criterion-referenced test and we had a total of 150 subjects. Of those, 30 failed both tests ($F_{1,2} = 30$), 85 passed both tests ($P_{1,2} = 85$), 25 passed only the first test ($P_1 = 25$), and finally 10 passed only the second test ($P_2 = 10$); and since P_2 is smaller than P_1, then $v = 10$.

$$\text{CR} = \frac{(30)(85) - (25)(10)}{(30)(85) + 10(30 + 85 + 10)}$$
$$= .605$$

TEST VALIDITY

For a test to have any practical use, it must not only be reliable, but it also must be valid. Test **validity** attempts to answer the question, Does the test measure what it purports to measure? That is, if a test has been designed to measure musical aptitude, for example, a valid test measures just that and not some other extraneous variable(s). When you step on the scales, you want to know your weight, not your IQ or mechanical aptitude or some unknown quality. As we shall soon see, ferreting out validity is, in fact, more difficult than determining reliability. For example, the evidence for the validity of polygraph testing has been found (Abrams, 1989) and shows an accuracy of 90% with innocent suspects, which of course means that 10% of the innocents are blamed for lying when they are telling the truth. Would you like those odds?

Although there are a variety of validity types, our concentration will be largely on those methods that are primarily statistically oriented and can be analyzed with the bivariate correlations shown in this chapter. The discussion here involves *test* validity. Back in Chapter 9, we discussed *other* kinds of validity (internal and external) that apply to the experimental situation.

First, there are two types of nonstatistical validity that are so important that they do deserve special mention: face validity and content validity.

Face Validity

Face validity is determined not by statistical analysis but by simply looking the test over and judging whether it has the look and feel of what it intends to measure. This is not meant to minimize the importance of face validity. Since face validity is based on the general appearance of the test, a test will have more face validity if its look and feel are consistent with the construct under study. From a testing perspective, it really is important to have a test that looks on the surface like it can tap the characteristic under study. This surface appearance of the test may motivate the test takers, especially when the test produces an assumption of relevance. For example, a test for measuring a person's aptitude for banking might be more realistic and motivating if it asks questions about balance sheets and interest payments, rather than, say, the batting average of a certain major league hitter. Although both sets of questions may involve similar arithmetic skills, the fledgling banker may feel better about answering items of a financial nature, rather than those focused on sports. Also, with regard to the testing of teacher competence, some of the items require only minimal-level skills, but they shouldn't be written in a manner that makes them appear childish. It is felt that a test has face validity if the items are reasonably related to the perceived purpose of the test (Kaplan & Saccuzzo, 1993).

Content Validity

Content validity is based on whether the test items are a fair and representative sample of the general domain that the test was designed to evaluate. This is an especially important concern among those (teachers as well as professional test constructors) who create educational tests. If the test is supposed to be covering a student's understanding of a certain social studies unit, for example, the items should be largely restricted to that domain, and the test should not be scored on the basis of the student's penmanship, neatness, or writing ability. Like face validity, content validity is based on logic, intuition, and, of course, common sense, rather than on statistical tests of significance.

Statistical Validity

The correlation between the test and an independent criterion is called a validity coefficient, and for a test to have **statistical** validity, it must be first found to be reliable. In theory, *no test should correlate higher with another measure than it does with itself.*

The two methods chosen for this section are concurrent validity and predictive validity. In each case, when the Pearson r or Spearman r_s is used, the correct notation is r_{tc}; that is, the *t*est is being correlated against an independent *c*riterion.

Concurrent Validity. In this type of validity, the test scores are correlated with an already established and accepted measure of the construct under study. Thus, we examine the simultaneous relationship between the test and the criterion. **Concurrent validity** may be assessed when we use a test to evaluate how a person will do right away, such as using an employment screening test to predict performance on a job that might start later that same afternoon. For example, the Strong–Campbell Interest Inventory (SCII) bases its criterion on the patterns of interest among persons who are both satisfied and successful in their present careers (Kaplan & Saccuzzo, 1993). Concurrent validity could be established by creating a college test of, say, business aptitude and then giving the test to an outside population of businesspersons ranging from successful to unsuccessful. Using those scores as the criterion, you could then correlate them with the scores of college students who might be considering a business career. This was the method employed when the Minnesota Multiphasic Personality Inventory (MMPI) was validated. Since persons diagnosed as having specific psychiatric disorders typically made characteristic scores on certain groups of items, these same patterns were used as the criterion against which people now taking the test are to be judged.

The criterion may also be based on the scores on another test of the same construct that has already been established as valid. That is, we can correlate the new set of test scores with the scores from a test of known validity. This may seem somewhat redundant. Why, for example, develop a new test if it's simply going to do what the old test can already do? The answer is that the new test must in some way be better, that is, cheaper, shorter, more efficient, and perhaps even *more* predictive. As an example, suppose that we were trying to establish the validity of a new IQ test that is based on measuring the possible physiological correlates of intelligence. Can intelligence be measured in the absence of traditional IQ tests? The answer seems to be a tentative and cautious "maybe." Recent evidence has shown that there is a significant correlation between certain brain-wave measures (as well as measures of neural conduction speed and the rate at which glucose is metabolized within the brain) and traditional IQ scores (Matarazzo, 1992). Thus, the traditional IQ test measures are being used in these studies as the criterion variable against which the new measures are being correlated. If the new brain-wave tests continue to show validity promise, they could provide educators and psychologists with a breakthrough, since the new tests would not be so easily challenged on the issue of cultural fairness.

Predictive Validity. **Predictive validity** is based on the degree to which test scores can be used to predict future performance. Can we use, for example, the results of the Scholastic Assessment Test given in high school to predict how well a student will do later in college? Validity research of this type can be done as part of a longitudinal research effort in which the same people are tested at different points in their lives. We give the SAT to high school seniors, and then we monitor the group after their freshman year, sophomore year, and so on, continuing to run validity correlations, using college and even in some cases graduate school GPAs as the criteria. Validity coefficients for predicting academic performance drop dramatically as the groups being tested become more and more homogeneous. This is because both sets of scores (predictor and criterion) lose variability as these tested groups become more similar. Variability, as we have seen, is a crucial component of a high correlation. Thus, as the groups being tested become more homogeneous, the predictor variable will almost certainly lose variability. For example, as was suggested back in Chapter 11 during the discussion of the restricted range, the smaller and more selective the group being tested is, the lower will be the correlations of test scores with performance. Thus, the correlations between IQ and grades in school become lower and lower as we go up the academic ladder. Grades and IQ correlate in the .60s for elementary school children, the .50s for high school students, .40s for college students, and only in the .30s for graduate students. Such studies show that as the groups tested become smaller and more homogeneous, the correlations are lowered. Also, since the lower IQs are weeded out as the students progress toward more intellectually demanding experiences, these smaller and more homogeneous groups are produced by systematic, not random, exclusion. Thus, we would expect the GRE to have a lower predictive validity than does the SAT because typically only those undergrads with the highest GPAs even apply to graduate school. Also the criterion may also lose variability and begin to suffer from this same restricted-range phenomenon. For example, the criterion becomes severely restricted when graduate school grades are used. In the typical graduate school the only passing grades are As and Bs, since a C grade is usually treated like an F for undergraduates; that is, the course must be repeated. The correlation between GRE verbal scores and graduate school GPA is only .28, and between GRE analytic and graduate GPA it's .38 (*GRE Guide*, 1990).

As the criterion measure loses variability, it will also suffer a loss of reliability. Keep in mind that the criterion measure used to support a test's validity must itself be reliable. If the criterion measures are simply reflecting chance factors, they will not correlate with the test scores or with anything else. In fact, there is a quick little equation that can be used for assessing the theoretical limit for the correlation between any two tests. No correlation may be greater than $\sqrt{(r_{tt_1})(r_{tt_2})}$. The maximum correlation is, thus, the square root of the product of the two reliability values. If, for example, the reliability of a certain test were .90 but the reliability of the criterion variable was only .10, then the highest possible validity coefficient would be only $\sqrt{(.90)(.10)} = \sqrt{.09} = .30$. Thus, a test that itself is extremely reliable loses statistical validity when it is being compared to an unreliable criterion, or, to put it another way, it becomes a test in search of a criterion. We know that a highly reliable test is measuring something, by golly, but it's not always obvious what that something is. The following story, although obviously apocryphal, nevertheless makes the point.

A test of musical aptitude was introduced, and the test-retest correlations taken on elementary school children produced reliability coefficients in the mid .90s. It also possessed a goodly supply of face validity, being duly shrink-wrapped in packets that contained pictures of Beethoven, Bach, and a large G clef. It also seemed to possess content validity, since part of the test demanded that the subject discriminate among various recorded tones. Unfortunately, it didn't seem to correlate at all with musical ability, the highest r_{tc} correlation being in the range of −.03 to +.05. When children who had taken the test were given music lessons, the test failed to predict which children would profit from the training. Then one day the teacher noticed that one of the children who had scored highest on the test had recently built and was showing off in school an absolutely beautiful wren house. The teacher then selected one of the students who had scored lowest on the test and asked him to build a birdhouse. The result, brought to school two days later, was one of the worst wren houses to ever grace the planet, a contraption so ugly that no self-respecting wren would ever venture within miles of it. All the children were then asked to construct wren houses, and their musical-aptitude test scores correlated .85 with the judge's ratings of their carpentry efforts. The test was then repackaged; off came the G clef and the bust of Beethoven, and on went pictures of beautifully plumaged members of the feathered kingdom. The test was now not only reliable, but statistically valid as well. Unfortunately, there's no money in a BBAT ("birdhouse-building ability test"), and the test and its authors were never heard from again.

The User's Responsibility

When using a test for a particular psychological or educational objective, it is up to the test publisher to supply information to the user regarding the way the test was standardized and its reliability and validity. However, it's the responsibility of the user of the test to become familiar with this information and to make the final determination regarding whether the test is appropriate for the job at hand (Geisinger, 1992).

Example A researcher wishes to establish the reliability and validity for a new 30-item personality test that attempts to measure a person's aptitude for leadership. A random sample of 10 college students was selected and given the entire test. The researcher then separated the scores obtained on the odd items from those achieved on the even items. Finally, each student was then evaluated by a panel of judges in a "leaderless-group" simulation and rank-ordered on the basis of the amount of leadership potential displayed.

a. What was the reliability for the whole test?
b. Was the test statistically valid?

Subject	*Odd,* X	X^2	*Even,* Y	Y^2	XY	*Score Total,* X + Y	*Score Ranks,* R_1	*Judges' Ranks,* R_2	d	d^2
A	12	144	13	169	156	25	2	3	1	1
B	5	25	3	9	15	8	10	9	1	1
C	10	100	8	64	80	18	8	10	2	4
D	9	81	10	100	90	19	6.5	5	1.5	2.25
E	11	121	10	100	110	21	3.5	1	2.5	6.25
F	14	196	13	169	182	27	1	2	1	1
G	9	81	10	100	90	19	6.5	8	1.5	2.25
H	5	25	4	16	20	9	9	7	2	4
I	10	100	11	121	110	21	3.5	6	2.5	6.25
J	9	81	11	121	99	20	5	4	1	1
	94	954	93	969	952					29

$M_x = 94/10 = 9.400$; $M_y = 93/10 = 9.300$
$s_x = 2.797$; $s_y = 3.401$

a. $r = 0.909$; $r_{.01(8)} = .765$ Reject H_0 at $p < .01$.

The split-half reliability is therefore equal to 0.909. Correcting for only using the self-correlation, the Spearman–Brown now yields

$$r_{SB} = \frac{(2)(.909)}{1 + (1)(.909)} = .952$$

b. For the validity, we use the Spearman r_S (since the criterion measure was in ordinal form).

$$r_S = 1 - \frac{6\Sigma d^2}{N(N^2 - 1)} = \frac{(6)(29)}{(10)(99)} = \frac{174}{990}$$

$$= 1 - .176$$

$$= 0.824$$

$$r_{S.01(10)} = .794 \qquad \text{Reject } H_0 \text{ at } p < .01.$$

We have now established a statistical validity of .824 and may conclude that the test is indeed valid. The technique employed here was called predictive validity, since the test scores were correlated with the sample's later performance in the leaderless-group situation.

Validity coefficients should be examined in the context of how and on whom the testing was done. For example, suppose that a certain test has been shown to have a high degree of predictive validity when used to estimate how persons might perform in business leadership positions. Perhaps, however, the test was standardized at a time when only white males were taking the test, and on this population the test may still be valid. However, it may not be valid for testing other groups, such as women or minorities, partly because it was never designed to sample items that might include supervising these groups and also because the test was never standardized on samples from these populations.

Construct Validity

Finally, there is a type of validity that has become increasingly popular called construct validity. A construct, sometimes called a hypothetical or theoretical variable, is a broad and abstract concept that refers to an attribute or even a cluster of attributes that an individual might be assumed to possess. Sometimes the construct is even thought to have only a theoretical rather than an empirical existence. Examples of constructs include anxiety, intelligence, motivation, and honesty. Another example of construct validity comes from the field of cognitive psychology where researchers have been using a system approach to integrate cognitive theory and abstract reasoning into test development (Embertson, 1998). Although other validity techniques, as has been shown, can be accomplished by running a single correlational study, **construct validity** requires a long series of studies, each having a hypothesis based on the nature of the construct. Thus, **construct validity** consists of a great deal of evidence gathered from the testing of a number of different hypotheses that have all been aimed at the same underlying trait (Sprinthall, Schmutte, & Sirois, 1990).

As Fraenkel and Wallen (2000) point out, establishing construct validity involves three important steps:

1. The variable being measures must be clearly defined.
2. There must be a statement as to how people will differ regarding the theory underlying the variable.
3. The hypothesis (or theory) must rest on logically and empirically tested principles.

This is sometime done through a procedure called *factor analysis* (see Chapter 14 and/or the glossary entry). Factor analysis is also used in test development, and in general there have been two approaches. Exploratory Factor Analysis (EFA) is used in the first stages of test development and Confirmatory Factor Analysis (CFA) is used later in the process, often as one of the possible validity approaches (McIntire & Miller, 2000).

The importance of both the reliability and validity of test measures should now be rather obvious. One of the problems inherent in the use of the polygraph test has been its questionable reliability and validity. To be of any real use in the courtroom, the polygraph must be consistently accurate in determining whether or not the subject is lying. It must continue producing dependable results (reliability), and it must detect when the subject lies and when the subject is telling the truth (validity).

ITEM ANALYSIS

Up to this point, the discussion of test effectiveness has been focused on overall performance on the entire test. However, some tests may have both overall reliability and overall validity and yet still differ in effectiveness due to the characteristics of the individual items. IRT, or item response theory, attempts to assess two related test parameters: item difficulty and item discrimination. That is, the strength of a test can be evaluated by **item analysis,** isolating and assessing each individual item on the basis of both item difficulty and item discrimination. These procedures view each item as a subtest in a larger composite test. Item response theory has been described not only for its statistical basis but also for its practical applications as they relate to test standardization and adaptive testing (Aiken, 1988). Using IRT, test construction becomes a dynamic process where new items are entered into the domain while items with less predictive power are dropped. This allows for continual test improvement (Embertson & Reise, 2000).

Item Difficulty

To measure **item difficulty,** the test can be given to a representative sample of the population in question, and each item is evaluated on the basis of the *proportion* of persons who correctly answered that particular item. If a certain item were correctly answered by 15% of the sample, that item would be recorded as having a difficulty index of .15. On the other hand, if an item is correctly answered by 95% of the sample, it would be assigned a difficulty index of .95. Notice that these proportions are actually inversely related to the concept of difficulty, since the lower the numerical value of the index, the more difficult the test item. In fact, some researchers have even asked that the term be renamed "item easiness." The difficulty index should be kept at approximately .50 for the majority of items. Although this sounds as though it would set up a situation that would not differentiate the more able students from the less able, the fact is that the 50% getting the item right may be composed of somewhat different persons on each item. Also, having too many items of varying difficulty produces lower reliability estimates. Despite the overall 50% rule, some items should tend toward the easy side, especially near the beginning of the test, if for no other reason than to keep the less able student motivated enough to continue. Also, some more difficult items should be included in order to help distinguish between the stars and the superstars. In any event, it is wise not to use any item whose difficulty index is greater than .80 or below .20. A good ability or achievement test, for example, should provide for a range of difficulty, from a few items that only the highest performers can correctly answer down to a few items that most test takers can handle.

Item Discrimination

Another important aspect of an item's usefulness is based on its ability to discriminate between those who do well on the entire test and those who do poorly. **Item discrimination** is determined by comparing a person's performance on each of the individual items with that same person's performance on the entire test. That is, do persons who correctly answer, say, item 5 achieve a high total score, whereas those who fail that same item wind up with a low overall score? The procedure used for this evaluation is to compare *for each item* the performances of those whose overall scores put them among the highest third of the group with those who scored among the lowest third. The two proportions are compared, and the resulting difference is called the DI, or discrimination index. For example,

Item No.	Proportion Correct for Top Third	Proportion Correct for Bottom Third	Discrimination Index
1	.90	.60	.30
2	.85	.50	.35
3	.50	.20	.30
4	.40	.70	–.30

Notice in this example that for item number 4 the lower third of the group has a higher proportion of correct responses than does the top third. When this occurs, the item is said to be a *negative discriminator* and is usually then either dropped from the test or rewritten. Negative discriminators have been found among certain poorly designed multiple-choice tests in which highly prepared students might spot an ambiguous distractor, a distractor that could easily be overlooked by students who are not as well prepared.

Point Biserial

One statistical technique for assessing the level of item discrimination is called the point-biserial correlation method. Using this formula, a researcher can determine the correlation between an interval scale measure, such as the result for the whole test, and a nominal measure, such as whether the individual item was answered correctly or incorrectly. That is, a measure that is continuous can be correlated with one that is genuinely dichotomized (divided into two categories), such as right versus wrong. Examples of other dichotomized variables are male versus female, college graduates versus noncollege graduates, and dorm students versus commuters.

The **point biserial** is a direct derivation of the Pearson *r*, which you already have learned to calculate. Whenever one of the variables, in this case the particular item, is scored as a dichotomous value, either zero or 1, then the Pearson *r* may become the point biserial. For example, on a given achievement test with a mean of 23.200 and a standard deviation of 2.209 (these values are for the whole test), 25 subjects were randomly selected and their scores recorded. For a given item, say item 12, we scored them as to whether they were correct or incorrect, a value of 1 for correct and a value of zero for incorrect. In this instance, the resulting Pearson *r* can be labeled as r_{pbis}, which is the reliability value for the particular item being analyzed. Then the Pearson *r* is as shown in Table 17.1. If we had chosen to arbitrarily use zero for correct and 1 for incorrect, the Pearson *r* point-biserial correlation would then have had a negative sign. The sign of the *r* depends on which of the dichotomized values (higher or lower) is linking with the higher scores on the continuous variable. Thus, if we had used the zero for correct in the above example, we would have then had to change the sign of the correlation and report it as positive.

Although this correlation is significant, its value as a reliability estimate is rather on the low side. In this case, the researcher should either drop or rewrite the item in question. One derivation of the point biserial, which is easy to use but keeps the background concepts offstage, is

$$r_{pb} = \left(\frac{M_1 - M_t}{SD}\right)\sqrt{\left(\frac{(p)}{1-p}\right)}$$

where M_1 = mean score on the test for those who answered the given item correctly

M_t = mean score on the *whole* test for everyone who took the test

SD = true standard deviation of the whole test for everyone who took the test

p = the proportion of all test takers who correctly answered the given item

Using the data from the above example, we saw that the mean for the whole test, M_t, was 23.200 and the *SD* was 2.209. The mean score for those test takers who got the correct answer (scored as 1) was 24.636, that is, 271 divided by the 11 subjects who an-

TABLE 17.1

	X	X^2	Y	Y^2	XY
	1.000	1.000	25.000	625.000	25.000
$N = 25$	1.000	1.000	27.000	729.000	27.000
	1.000	1.000	24.000	576.000	24.000
	1.000	1.000	23.000	529.000	23.000
	1.000	1.000	27.000	729.000	27.000
	1.000	1.000	21.000	441.000	21.000
	1.000	1.000	29.000	841.000	29.000
	1.000	1.000	24.000	576.000	24.000
	1.000	1.000	25.000	625.000	25.000
	1.000	1.000	22.000	484.000	22.000
	1.000	1.000	24.000	576.000	24.000
	0.000	0.000	21.000	441.000	0.000
	0.000	0.000	20.000	400.000	0.000
	0.000	0.000	21.000	441.000	0.000
	0.000	0.000	22.000	484.000	0.000
	0.000	0.000	23.000	529.000	0.000
	0.000	0.000	20.000	400.000	0.000
	0.000	0.000	24.000	576.000	0.000
	0.000	0.000	25.000	625.000	0.000
	0.000	0.000	22.000	484.000	0.000
	0.000	0.000	23.000	529.000	0.000
	0.000	0.000	21.000	441.000	0.000
	0.000	0.000	22.000	484.000	0.000
	0.000	0.000	22.000	484.000	0.000
	0.000	0.000	23.000	529.000	0.000
Sums	11.000	11.000	580.000	13,578.000	271.000

$\text{Mean}_1 = 0.440$, $SD_1 = 0.496$.
$\text{Mean}_2 = 23.200$; $SD_2 = 2.209$.
$r = 0.577$.
$r_{.05(23)} = .396$ Reject H_0.

swered the item correctly. Finally, since 11 out of 25 answered the item correctly, the p (proportion) value was 11/25 or .44.

$$
\begin{aligned}
r_{pb} &= \left(\frac{24.636 - 23.200}{2.209}\right)\sqrt{\frac{.44}{1 - .44}} \\
&= \left(\frac{1.436}{2.209}\right)\sqrt{\frac{.44}{.56}} \\
&= (.650)\sqrt{.786} \\
&= (.650)(.887) \\
&= .577
\end{aligned}
$$

Thus, the correlation between being correct on the *given item* and performance on the *entire test* is a significant but very modest .577. As you can see, when this procedure is applied to each and every item, the researcher gets the opportunity of determining which items have the highest reliability (and should be retained) and which items are low in reliability and should be either discarded or rewritten.

Standard Error of Measurement

As we have seen, once the reliability of the test has been found, it tells us the degree of consistency that we may expect among repeated test scores. But how can we use that information to understand the meaning of an individual score? For example, for a certain degree of reliability, how does that help us in interpreting, say, a specific IQ score or a specific SAT score? For this, we turn to a statistic called the standard error of measurement (or sometimes called the standard error of an obtained score). Like all standard errors, the **standard error of measurement** is an estimate of variability, and in this case it is used to predict the standard deviation of all the possible scores a given individual might obtain around that same person's true score. The true score is the score that the researcher would obtain by taking the average score a given individual would receive on an infinite number of trials. Thus, although the true score is theoretical, we may determine by how much the obtained score might deviate from the theoretically true value. The standard error of measurement, thus, combines information about the test's reliability as well as the variability occurring around the individual scores.

$$SEM = SD\sqrt{1 - r_{tt}}$$

To calculate the standard error of measurement, we use the reliability coefficient, r_{tt}, and the actual standard deviation of the distribution of obtained scores. Assume for a moment that on a given IQ test the reliability coefficient turned out to be .95, and the standard deviation of the IQ scores was 15. The standard error of measurement would then be

$$SEM = 15\sqrt{1 - .95} = 15\sqrt{.05} = (15)(.224) = 3.360$$

As a logic check here, the standard error of measurement *cannot* be greater than the standard deviation of the test.

Once we have that estimate, we can set up a confidence interval within which we feel a strong measure of assurance that the true score will be found.

$$\text{CI} = (\pm z)(SEM) + X$$

As usual, for any distribution that is assumed to be normal, the .95 confidence interval uses a z score of 1.96, and the .99 level uses 2.58. On the IQ test shown above, assume that an individual had an IQ of 106, and your job is to assess the degree to which that score may deviate around the true score. Setting up a .95 confidence interval, we have

$$\begin{aligned} \text{CI at .95} &= (\pm 1.96)(3.360) + 106 \\ &= +6.586 + 106 = 112.586 \\ &= -6.586 + 106 = 99.414 \end{aligned}$$

We may now feel confident that although the obtained score was 106, it probably reflects a true score that varies from about 99 on the low side to 113 on the high side. This confidence interval is also telling us by *how much an individual score might change on a retest.*

Example Suppose that a given test has a reliability of .90 and a standard deviation of 9; find the standard error of measurement.

$$SEM = 9\sqrt{1 - .90} = 9\sqrt{.10} = (9)(.316)$$
$$= 2.844$$

If a student were to get a score of 65 on the test, find the .95 confidence interval.

$$\begin{aligned} \text{CI at .95} &= (\pm 1.96)(2.844) + 65 \\ &= 5.574 + 65 = 70.574 \\ &= -5.574 + 65 = 59.426 \end{aligned}$$

Measurement error can also occur when there are systematic differences in the conditions under which the test is taken. For example, suppose some persons are allowed to take a test on the computer and thus have the advantage of a spell checker and calculator while other students aren't given this same opportunity (Camera, 2001). Such dramatic differences in conditions surely produce unfair advantages for some of the test-takers. Also there is what is called "transient error," or errors due to the test-taker's mood, emotional state, and general feeling on the day the test is taken. Those feelings might change and thus alter the score on subsequent test dates (Becker, 2000).

States are coming under increasing pressure to create tests that will prove whether a student has acquired enough basic knowledge to be awarded a high school diploma. Often there is not enough time devoted to insuring that these high-stakes tests are properly evaluated for both reliability and validity (Camara, 2001).

To ease your computational chores, Table 17.2 has been prepared to give the standard error of measurement readouts for selected standard deviation and reliability values. So for the test shown above whose reliability was .90 and whose standard deviation was 9, we look down the far left column until we get to the standard deviation of 9 and then across that row until we intersect .90 and read out the standard error of measurement as 2.846 (within rounding error), the same as the value shown above.

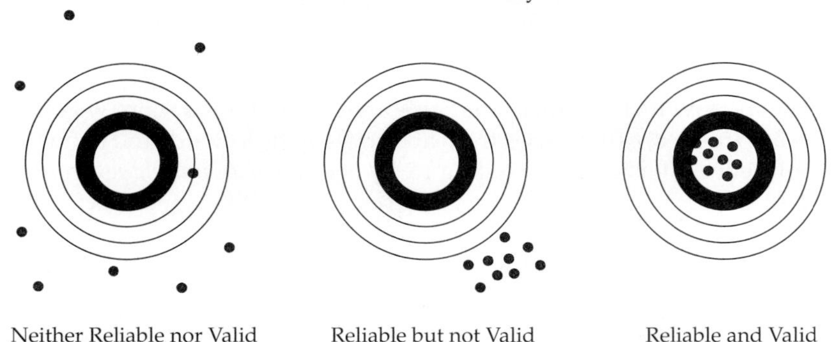

Notice that as the reliability values (top row) increase, the standard error of measurement decreases, and when the reliability reaches 1.00, the standard error of measurement becomes zero, meaning that in this highly unlikely scenario the obtained

TABLE 17.2 Table of the standard errors of measurement for combinations of standard deviations and reliability coefficients.

	r_{tt} =							
SD	*0.65*	*0.70*	*0.75*	*0.80*	*0.85*	*0.90*	*0.95*	*1.00*
3.00	1.775	1.643	1.500	1.342	1.162	0.949	0.671	0.000
4.00	2.366	2.191	2.000	1.789	1.549	1.265	0.894	0.000
5.00	2.958	2.739	2.500	2.236	1.936	1.581	1.118	0.000
6.00	3.550	3.286	3.000	2.683	2.324	1.897	1.342	0.000
7.00	4.141	3.834	3.500	3.130	2.711	2.214	1.565	0.000
8.00	4.733	4.382	4.000	3.578	3.098	2.530	1.789	0.000
9.00	5.324	4.930	4.500	4.025	3.486	2.846	2.012	0.000
10.00	5.916	5.477	5.000	4.472	3.873	3.162	2.236	0.000
11.00	6.508	6.025	5.500	4.919	4.260	3.479	2.460	0.000
12.00	7.099	6.573	6.000	5.367	4.648	3.795	2.683	0.000
13.00	7.691	7.120	6.500	5.814	5.035	4.111	2.907	0.000
14.00	8.283	7.668	7.000	6.261	5.422	4.427	3.130	0.000
15.00	8.874	8.216	7.500	6.708	5.809	4.743	3.354	0.000
16.00	9.466	8.764	8.000	7.155	6.197	5.060	3.578	0.000
17.00	10.057	9.311	8.500	7.603	6.584	5.376	3.801	0.000
18.00	10.649	9.859	9.000	8.050	6.971	5.692	4.025	0.000
19.00	11.241	10.407	9.500	8.497	7.359	6.008	4.249	0.000
20.00	11.832	10.954	10.000	8.944	7.746	6.325	4.472	0.000
21.00	12.424	11.502	10.500	9.391	8.133	6.641	4.696	0.000
22.00	13.015	12.050	11.000	9.839	8.521	6.957	4.919	0.000
23.00	13.607	12.598	11.500	10.286	8.908	7.273	5.143	0.000
24.00	14.199	13.145	12.000	10.733	9.295	7.589	5.367	0.000

score is exactly the same as the true score. Don't hold your breath waiting for reliabilities of 1.00, but the notion at least sheds some light on the standard error of measurement concept. Also, take note of the fact that as the standard deviations increase (far left column) for a given level of reliability, the standard error of measurement also increases. This must be the case, since the standard error of measurement is a variability value, just as is the standard deviation.

Example For a test with a reliability of .75 and a standard deviation of 6.00, find the standard error of measurement. See Table 17.2.

The standard error of measurement equals 3.000.

The standard error of measurement can also be used to estimate reliability through the use of a formula suggested many years ago by Harvard's P. J. Rulon (Rulon, 1939).

$$r_{tt} = 1 - \frac{SEM^2}{SD^2}$$

Using the data from the above problem, where we already know the reliability to be .75, we were given an *SD* of 6, and we found an *SEM* of 3.00. Using the Rulon formula to get back to the reliability, we find

$$\begin{aligned} r_{tt} &= 1 - \frac{3^2}{6^2} \\ &= 1 - \frac{9}{36} \\ &= 1 - .25 \\ &= .75 \end{aligned}$$

To shortcut the Rulon formula, you may also use the above table. Look down the far left column for an *SD* of 6 and across until you intersect with an *SEM* of 3. Then look directly above (to the top row) and read out the reliability value of .75.

Example Using the data from the problem on pages 472–473, where we already know the reliability to be at .95, let's now just use the standard error of measurement, which we found to be 3.36, and the standard deviation, given as 15.00, to get back to that reliability value.

$$
\begin{aligned}
r_{tt} &= 1 - \frac{SEM^2}{SD^2} \\
&= 1 - \frac{3.36^2}{15^2} \\
&= 1 - \frac{11.290}{225} \\
&= 1 - .050 \\
&= .950
\end{aligned}
$$

Or, by using the table, go down the far left column until you get to the 15 and across that row until you intersect with the closest value you can find to 3.36, which on the table is 3.354. Look directly above (to the top row) and find the r_{tt} value of .95.

Reliability and Validity

Reliability and validity together are used to establish the accuracy of measures and for determining how they should be used. It is important that the behavioral measures we use are as accurate as it's humanly possible to make them. Telling time by a sundial may give you a general idea as to whether it's morning or afternoon, but you probably wouldn't want to schedule an important appointment on that basis.

Test-Taking Accommodations

Because of disabilities of one kind or another, testing procedures now offer accommodations for persons suffering from assorted problems, ranging from learning disabilities, to asthma, anxiety, and/or other emotional problems. Although there are a number of accommodations now in use, (such as modifying print size, having a reader, using a tape recorder, offering frequent breaks etc.) the most common method is to extend the time to complete the test. This is done in order to "level the playing field to ensure assessments primarily measure the intended construct rather than sources resulting from the disability" (Camara, 2001). The most common procedure, which incidentally is used by the College Board tests, is to extend the testing time. The research on this is clear and should not be surprising—score gains increase as amount of time increases.

Looking Ahead

As statistical techniques continue to improve and as more is learned about testing in general, the accuracy of educational and psychological assessment will also improve. Even the best of our current measures involves error, even though our methods and techniques have improved vastly over the past century. The public, however, especially the affluent public, is enamored of test scores in the schools, and parents even house-hunt in areas where the schools have shown especially strong performances. Real estate agents have even used a school's test scores as a selling tool for that area, which, of course, affects the prices of the homes. Much, however, remains to be done. But, despite the lack of testing perfection, we should not throw out the baby with the bathwater. Anti-test and some civil rights groups have ironically fought against attempts to establish the reliability and validity of educational testing. In what one test expert describes as "shooting themselves in the foot," Fairtest (an advocacy group for fairness in testing) has led the charge to do away with reliability and validity when in fact the most basic step toward creating fair tests is to ensure their reliability and validity (Brown, 1994).

We must, however, keep firmly in mind that even the simplest test is not foolproof, and in the words of that testing giant of only a few years ago, Florence Goodenough (Goodenough, 1949, p. 540):

> Every testing device employed for mental measurement has its own set of implicit assumptions, its own limits of applicability and its own hazards of interpretations. Neglect of these factors through carelessness or ignorance on the part of the psychologist who makes use of tests for the guidance of human lives may have consequences nearly as serious as those resulting from similar neglect on the part of the physician who deals with their bodily ailments.

SUMMARY

Measurement is based on the assigning of numbers to observations (of persons, events, or things), and the way these numbers are assigned determines the scale of measurement being used (whether it be nominal, ordinal, interval, or ratio). For test purposes, measurements may be scored on either a norm-referenced or criterion-referenced system. Norm referencing relates an individual's score to the performance of the entire group to which that individual is to be compared, whereas criterion referencing bases its scoring on what absolute proportion of the material covered on the test has been mastered. Thus, norm referencing produces relative scores, whereas criterion referencing produces absolute scores. For these measurements to be useful and provide meaning to the researcher, they must also satisfy two fundamental criteria, reliability and validity, and correlation techniques can be used to assess both of these. Reliable measures are those that produce consistent results and afford the researcher with dependable measures. The higher the proportion of true score to observed or total score is, the higher the reliability. When the data are in interval form, the Pearson r can be used to assess reliability. Three methods mentioned for determining test reliability are (1) test-retest, (2) alternate forms, and (3) internal consistency. For the test-retest method, the correlation is computed by comparing the test scores on one administration of the test with the test scores on a second administration of the same test. For the alternate-form method, the correlation is used to compare scores on one form of the test with scores on a second version of the same test, called an alternate or parallel form. Internal consistency can be evaluated in several ways, one of the most popular being the split-half method. With the split-half technique, the test is deliberately divided into equal halves, usually on the basis of the odd- versus the even-numbered items. The correlation is then run by comparing scores on one-half of the test with scores on the second half. Since this correlation typically underestimates the true reliability of the whole test, a correctional equation, called the Spearman–Brown prophecy formula, is applied to more accurately determine the true reliability of the entire test. Other internal-consistency methods include the Kuder–Richardson formula 21. Unlike the split-half method, which splits the test just once, the K–R 21 estimates the reliability of a test that has been split into *all possible halves,* and it automatically corrects for the splits without any need for a Spearman–Brown type of adjustment.

Internal-consistency reliability can also be found on the basis of ANOVA. These techniques were created and modified by those great statisticians Sir Ronald Fisher in England, Cyril Hoyt at Minnesota, and P. J. Rulon at Harvard. Two ANOVA techniques for establishing the r_{tt} reliability were presented, one for the situation in which the items are dichotomously scored, either right or wrong, agree-disagree (nominal data), and one in which the items are scored on a continuous scale (interval data). In both cases, the reliability is eventually reported as an $r_{tt,}$ even though the significance of the r_{tt} is gauged on the basis of the F ratio. Since the ANOVA procedures compare every item with the total score, there is no further need for any Spearman–Brown correction.

Test reliability has been shown to be increased by (1) increasing the length of the test, (2) increasing the variability of the test measures (which makes the test more discriminating), and (3) increasing the size of the sample of test takers. Most nationally standardized tests report reliability coefficients of at least .80, but these vary according to what the test has been designed to measure. For example, achievement tests tend to have higher reliability coefficients than do attitude or personality tests.

Test validity is based on whether the test is in fact measuring what it has been designed to measure. Two types of statistical validity techniques were mentioned in the chapter: concurrent validity, in which scores on the test being validated are compared with the test scores achieved on a different, already established test (which had been designed to measure the same construct); and predictive validity, in which the test scores are correlated against some measure of future performance. In cases for which the measure of future performance is based on the rank orders assigned by judges, then the Spearman r_s should be used as the correlation of choice. Validity coefficients tend to be lower than those for reliability, especially when the criterion itself lacks a high degree of reliability or when the criterion group is so homogeneous as to produce a restricted range.

Test effectiveness can also be gauged on the basis of item analysis, such as item difficulty and item discrimination. To measure item difficulty, the test can be given to a representative sample of the population in question, and each item is evaluated on the basis of the *proportion* of persons who correctly answered that particular item; the higher the proportion of correct answers is, the easier the item. For a good test, the majority of items should be kept at a proportion level, called the difficulty index, of approximately .50. Item discrimination is used to evaluate the test's ability to discriminate among those who do well on the entire test versus those who do poorly. Item discrimination is determined by comparing a person's performance on each of the individual items with that same person's performance on the entire test. One statistical technique for assessing the level of item discrimination is called the point-biserial correlation method. Using this formula, a researcher can determine the correlation between an interval scale measure, such as the result for the whole test, and a nominal measure, such as whether the individual item was answered correctly or incorrectly.

The standard error of measurement (sometimes called the standard error of an obtained score) is an estimate of variability and is used to predict the standard deviation of all the possible scores a given individual might obtain around that same person's true score. It is linked to reliability, since the smaller the standard error of measurement, the higher the resulting reliability value. In fact, when the reliability reaches 1.00, the standard error of measurement becomes zero, meaning that in this case the obtained score is exactly the same as the true score.

Key Terms

Criterion-referenced reliability
Cronbach's alpha
Item analysis
- Item difficulty
- Item discrimination

Kuder–Richardson formula 21
Point biserial
Reliability
- Alternate-form
- Split-half
- Test–retest

Spearman–Brown prophecy formula
Standard error of measurement
Validity
- Construct
- Content
- Face
- Statistical: concurrent and predictive

PROBLEMS

1. The Children's Depression Inventory (CDI), developed from Beck's Depression Inventory, is a self-report instrument with 27 sets of statements. Higher scores indicate higher levels of depression, and the mean for grades 4 through 8 is 9.30, with an *SD* of 3.00. Scores above 13 are said to indicate a major depressive disorder (MDD), and very few false-positives have been reported. A random sample of 10 children was selected and given the CDI. A month later the children took the same test again. Was the test reliable?

Subject	*CDI Test Score*	*CDI Retest Score*	*Subject*	*CDI Test Score*	*CDI Retest Score*
1	9	8	6	7	5
2	11	12	7	9	9
3	7	8	8	4	6
4	13	11	9	15	13
5	8	9	10	10	10

2. A personality theorist was interested in the measurement of gregariousness. The NEO Personality Inventory has a G score as a measure of gregariousness. On the full

test the mean is 50, with an *SD* of 10. A random sample of 10 adolescents was selected and given the NEO, and their G scores were recorded. A week later they took the same test, and their G scores were recorded a second time. Estimate the reliability of the test.

Subject	*NEO G test*	*NEO G retest*
1	50	50
2	30	31
3	70	71
4	60	65
5	40	45
6	50	50
7	55	48
8	45	40
9	52	51
10	48	49

3. Another researcher, also interested in the NEO gregarious measure, used the same test shown in problem 2 but split the test into two halves, based on a comparison of the odd-numbered items with the even-numbered items. Again, a random sample of 10 adolescents was selected and their G scores were as follows:

Subject	*NEO G test (Odd Items)*	*NEO G Retest (Even Items)*
1	25	25
2	15	14
3	35	36
4	30	33
5	20	23
6	25	25
7	28	28
8	26	21
9	24	27
10	24	18

a. Find the Pearson r value between the two sets of scores.
b. Estimate the whole-test reliability for the NEO G test.

4. A test expert wishes to find the whole-test reliabilities for the following split-half correlations, all of which were calculated on the basis of a sample of 20 persons. For each of the following split-half correlations, find the whole-test reliability.
a. .83
b. .76
c. .92
d. .50
e. .40

5. A test constructor is interested in discovering how long a set of subtests must be to reach a reliability value of .90. For each subtest, the following significant correlations were found. Estimate how long each subtest should be to obtain the .90 reliability value.
   a. .82
   b. .60
   c. .75
   d. .72
   e. .69

6. A researcher wishes to establish the DI (discrimination index) for a series of five items contained in a new test being developed to assess math anxiety, called the Interest in Basic Math (IBM) test. Using the following data, calculate the DI for each item.

Item	*Proportion Correct for Top Third*	*Proportion Correct for Bottom Third*
a.	.80	.60
b.	.75	.50
c.	.60	.20
d.	.90	.70
e.	.30	.50

7. A test constructor wishes to determine the reliability of a certain item on a new test of math ability. The item was scored 1 for correct and zero for incorrect. Also, for each subject the total score on the test is reported.

Subject	*Item Score* X	*Total Score* Y
1	1	12
2	1	14
3	1	12
4	1	13
5	1	15
6	0	4
7	0	7
8	0	8
9	0	6
10	0	9

   Find the point-biserial reliability value for the selected item. (Use either method.)

8. A researcher wants to assess the validity of a new test of childhood aggressiveness (in hopes of identifying potential bullies). A random sample of 10 eight-year-old boys was selected and given the AIC (Aggressive Inventory for Children) test. Later, a group of three judges observed the children during recess play and rank-ordered each boy with regard to his perceived aggressiveness. The data follow:

Subject	AIC Score	Aggressive Rank
1	50	1
2	45	3
3	15	10
4	25	7
5	30	6
6	30	5
7	35	8
8	27	9
9	30	2
10	33	4

a. Is the test statistically valid?
b. What type of validity would this be?

9. For a given administration of the verbal SAT, the mean was 500 with an *SD* of 100. The test-retest reliability was .95. Find the standard error of measurement.
10. Using the standard error of measurement found above, find the confidence interval at the .95 level for the following SAT verbal scores.
    a. 300
    b. 500
    c. 550
    d. 650
11. Assume that for the SAT question shown in problem 9 the test-retest reliability had only been .80; find the new standard error of measurement.
12. Using the new standard error of measurement of problem 11, find the .95 confidence interval for the SAT verbal scores of
    a. 300
    b. 500
    c. 550
    d. 650
13. The standard error of measurement on a certain test has been found to be 15.03, with an *SD* of 50.00. Using the Rulon formula, find the r_{tt} (reliability value) for the test.
14. A certain test has been found to have an *SD* of 5.00 and a standard error of measurement of 3.00. Using the Rulon formula, find the r_{tt} (reliability value) for the test.
15. A new intelligence test is being standardized, and it has been found to have a mean of 100 and an *SD* of 15. The test is composed of 150 items. Find its K–R 21 internal-consistency reliability.
16. A new test of creativity is being standardized, and it has been found to have a mean of 35.37 and an *SD* of 5.09. The test is composed of 50 items. Find its K–R 21 internal-consistency reliability.
17. The reliability of a certain test has been found to be .90, whereas the reliability of the criterion is .70. What is the highest possible validity value for this test?

18. The reliability of a certain test has been found to be .85, whereas the reliability of the criterion is .60. What is the highest possible validity value for this test?

19. A testing specialist has created a four-item test of shyness and given it to a random sample of five subjects. The test is scored only on the basis of whether the item indicates shyness or whether it does not. It was scored as a 1 for shyness and a zero for no shyness. Using ANOVA, estimate the internal-consistency reliability of the test.

	Items			
Subjects	*a*	*b*	*c*	*d*
1	1	1	1	1
2	1	1	1	1
3	1	0	0	1
4	0	1	0	0
5	0	0	0	0

20. A new test, designed to test hyperactivity among K through fourth-grade males, was developed and you, as the statistical consultant, have been asked to assess its reliability. The test is composed of four items and was given to a random sample of five youngsters. The items on the test were scored from 1 to 10, with 10 showing the most hyperactivity. Using ANOVA, estimate the internal-consistency reliability of the test.

	Items			
Subjects	*a*	*b*	*c*	*d*
1	10	10	9	8
2	8	8	9	10
3	6	5	7	4
4	3	3	4	5
5	1	2	1	1

True or False. Indicate either T or F for problems 21 through 30.

21. The Spearman–Brown prophecy formula must always be used when assessing reliability via the test-retest technique.

22. The Kuder–Richardson technique (K–R 21) is used to determine internal-consistency reliability.

23. The Spearman r_s must be used instead of the Pearson r when trying to establish the validity of a test where the independent criterion is made up of ordinal data.

24. The standard error of measurement is related to reliability such that the higher the value of the standard error of measurement, the lower the resulting reliability value.

25. In general, increasing the number of test items typically increases the test's reliability.

26. The reliability of any test depends on the value of the correlation, regardless of whether the correlation is significant.

27. If a test is proved to be reliable, it must necessarily be valid.
28. In general, reliability values tend to be higher than validity values.
29. If the correlation between the two halves of a given test proves to be significant, then that value can be used in the Spearman–Brown for estimating total reliability.
30. The use of the Spearman–Brown following a split-half correlation will never lower the reliability estimate of the whole test.

Fill in the blanks with either *reliability* or *validity* for problems 31 to 38.

31. The correlation between scores on the test and scores on the retest is attempting to establish ________.
32. The correlation between scores on the test and scores on an alternate form of the same test is attempting to establish ________.
33. The correlation between scores on the test and scores on another, already standardized test is attempting to establish ________.
34. The correlation between the scores on one-half of the test with scores on the other half of the test is attempting to establish ________.
35. The correlation between the scores on the test with some independent measure of future performance is attempting to establish ________.
36. When the observed score is made up of a high percentage of true score and a low percentage of error score, we can say that the test is high in ________.
37. When the test is found in fact to be measuring what it was designed to measure, we can say that the test is high in ________.
38. When test scores are highly correlated with an independent criterion known to be measuring the same underlying construct, then the test can be said to be high in ________.

COMPUTER PROBLEMS

C1. A random group of 22 elementary school children was chosen and given the Eysenk Venturesomeness Test on a Friday morning at 8:30 A.M. That same afternoon at 3:00 P.M., the children were given the same test again. Their morning scores appear below under "Test" and their afternoon scores under "Retest."
  - *a.* Was the test reliable?
  - *b.* Explain your answer, indicating the type of reliability found and the kinds of problems associated with this technique. For example, were their afternoon scores in any way different and, if so, why do you suspect they changed?

During the next few weeks a panel of four judges, trained in child psychology, watched all the children while on the school premises, both in the classroom and outside during recess. They ranked all the children according to how adventurous and daring they were observed to be and assigned a median rank to each child. Compare the child's "Test" score with the judge's ranks.

  - *c.* Is the test statistically valid?

Subject	Test	Retest	Rank	Subject	Test	Retest	Rank
1	60	50	4	12	56	53	9
2	40	35	17	13	51	47	19
3	70	65	1	14	55	50	15
4	53	48	11	15	53	48	22
5	57	52	7	16	57	52	6
6	59	55	2	17	55	50	13
7	50	45	18	18	50	46	16
8	57	52	8	19	60	54	3
9	52	48	20	20	55	50	12
10	42	40	14	21	56	56	10
11	57	60	5	22	54	53	21

C2. A group of 15 college students was randomly selected from a population of students who had never flown an airplane and given the Pilot's Aptitude Test, a test designed by a manufacturer of aircraft for persons interested in general aviation. The test contained a total of 40 items. To control for practice and fatigue, the researcher decided to split the test into two halves, odd items versus even items. Calculate the split-half reliability of the test. What would the reliability have been if the whole test were used? Later, all the subjects were given flying lessons for a total of three months, and at the end of the training an independent evaluator (an instructor who had never seen any of the subjects previously and who did not know their scores on the aptitude test) took each student pilot for a short flight and ranked each student on his or her ability to fly the plane. Assess the validity of the total score on the aptitude test.

	Scores			Criterion
Subject No.	Odd X	Even Y	Total X + Y	Ranks R
1	19	18	37	4
2	14	16	30	1
3	4	5	9	14
4	12	13	25	2
5	11	9	20	12
6	13	10	23	6
7	3	2	5	13
8	16	15	31	3
9	10	7	17	10
10	8	11	19	15
11	9	12	21	7
12	13	8	21	11
13	11	13	24	8
14	12	14	26	5
15	12	11	23	9

C3. A researcher has constructed a new substance abuse test consisting of 10 items scored on a 5-point scale from strongly agree to strongly disagree. A random sample of 15 prison inmates is selected from a large house of correction. The data follow:

Subj#	Item 1	Item 2	Item 3	Item 4	Item 5	Item 6	Item 7	Item 8	Item 9	Item 10
1.	5	4	5	4	5	4	5	5	5	5
2.	1	2	2	3	2	3	2	3	3	2
3.	5	4	5	3	4	3	3	5	5	4
4.	3	3	3	4	4	3	2	4	3	3
5.	5	4	3	4	4	4	5	5	5	5
6.	2	2	1	3	2	2	2	1	1	2
7.	4	3	2	5	5	5	1	3	4	3
8.	4	3	3	3	3	3	2	4	3	3
9.	3	2	3	4	4	4	3	5	5	2
10.	3	1	2	2	2	3	2	3	3	1
11.	5	1	2	5	3	5	3	4	5	5
12.	4	3	3	3	5	5	1	3	4	3
13.	3	2	1	5	4	5	2	5	3	2
14.	4	2	2	4	4	3	3	4	4	2
15.	4	5	4	5	4	5	5	4	4	5

The researcher is interested in the internal consistency reliability. Compute Coefficient Alpha and indicate whether the results prove the test to be reliable.

Chapter

18

Computers and Statistical Analysis

Now, during the dawn of the twenty-first century, the computer revolution (at least as important as the Industrial Revolution) is fully upon us. Computers and their "magic chips" do everything from making out the payroll and scanning income tax returns to controlling VCRs, watches, and cameras. In fact, silicon chips run our lives. Chips are the brains and memories in TV sets, toasters, musical keyboards, radios, cars, telephones, and traffic lights. And even those things that contain no chips, such as tables and Rollerblades, were designed on computer-controlled machinery, shipped with bills generated by computers, stored in warehouses with computer-controlled inventory systems, and probably sold in a credit-card transaction that was verified by a computer. Are computers affecting our lives? Absolutely! Every time we make an airplane or hotel reservation, go to a bank . . . the list is endless, a computer is somewhere hiding in the wings, crunching out the data and KEEPING TRACK of you. For example, computes are in constant use by law enforcement, for obvious reasons, like checking licenses and auto registrations and spotting credit card fraud. But computers are also used for far more esoteric endeavors. Known as data mining, or knowledge management, specialized SPSS programs can cull huge volumes of data in search of abnormalities that might signal a terrorist attack (such as identifying persons taking flying lessons, especially those who want to learn to fly the plane, but not how to take off or land—as in the case of the twin-tower assassins).

Sometimes the computer can even become a stubborn opponent, which won't take no for an answer. In one celebrated case, a man kept getting a computerized bill for $0.00. Phone calls, visits to the medical center, letters, or whatever failed to stop this electronic harassment. Only when the man capitulated and sent in a check for $0.00 did the computer stop its cunning dunning. And just as there has been a tremendous increase in the use of computers, there has also been a dramatic drop in the cost of computers over the past few years. In fact, it has been estimated that had automobile technology kept pace with computer technology over the past 45 years, a

Rolls Royce would now cost less than a dollar, and it would get about 10 million miles per gallon. It's little wonder that the demands of society now virtually insist that every educated person become to some degree computer-literate, and perhaps nowhere is this message more compelling than in the field of statistical analysis. In order to organize and understand large groups of data, or even to run sophisticated statistical tests of significance on small-sample databases, the computer has fast become the statistician's best friend. The fact that the computer can handle such vast amounts of data has allowed statisticians to spend less time calculating and more time choosing suitable research designs and models (Cobb, 1991). These sophisticated techniques have provided the world with another arrow in the quiver of those wanting more analytical power.

If you are approaching the computer for the very first time, fear not. Even if you think a Web site is a spider's trap in your attic, don't worry. You don't have to be a rocket scientist to get the computer up and running, and if it is your first time, you'll probably be amazed at how little effort it will take to get the computer to give you an instant standard deviation or t *ratio. Think of it more as a calculator with a typewriter-like keyboard than as an imposing, science-fiction creation, laden with cryptic dials, blinking lights, and arcane operations. Years ago, computers were difficult to use and had to be hand-wired for even the simplest of calculations. Computer theorists and technicians often became so single-minded as to become, to put it kindly, detached from the real-life demands of ordinary existence. It is said that Norbert Wiener, the famous mathematician and computer pioneer, went home routinely after work one day despite being warned by his wife that on that day they were going to be in the process of moving. Finding his old house empty, he cautiously asked a little girl sitting out in the front yard: "Excuse me, but do you know where the famous computer expert Norbert Wiener lives?" "Yes I do, Daddy," she replied, "and Mommy sent me to take you there" (Jennings, 1990).*

In the late nineteenth century Sir Francis Galton argued plaintively that the field of statistics is one where scientists have to spend enormous amounts of time creating their own subject matter before beginning the arduous task of analysis. Said Galton, "the work of statisticians is that of the Israelites in Egypt—they must not only make the bricks but find the materials" (Kevles, 1984). One can only guess as to what analytic strides Galton and his friend, Karl Pearson, would have made had they been able to avail themselves of even the most inexpensive of today's microcomputers. As you will see, you don't have to be a computer expert to get results from a computer. You should, however, have at least a modicum of "computer literacy."

COMPUTER LITERACY

No doubt about it, "computer literacy" has become the catch phrase of the 21st century. Even first-grade children are being introduced to the computer and are receiving grades in computer science. To argue against computers today is viewed as akin to seeing the first automobiles roll down the road and saying, "I'd rather have a horse," or "The only accelerator I'll ever need is a buggy whip."

But what about this phrase "computer literacy"? How literate must one become to take advantage of today's technology? A few short years ago computer literacy was thought by many as being virtually synonymous with achieving an advanced degree in computer science. Students, it was thought, should all have courses in everything from machine language to FORTRAN, from computer architecture to advanced PASCAL, C++ and much, much more. Mercifully, that view is now held only by a minority, and computer literacy has been downsized to far more modest proportions. Programming skills are no longer seen as a necessary component. Computer literacy is now accepted as being more like automobile literacy . . . you need only to learn how to drive it, not all the electronics and physics involved in its operation. Also, today there are even programs, such as QUICKPRO, which have been designed to write programs for you. Instead of learning the programming language directly, you simply answer a series of easy questions that appear on your screen, and the program then writes the new program.

Taking a course or two in BASIC may be like taking a couple of French courses. Although it might be fun to go to a French restaurant and order your meal directly from the menu, if you really need a highly accurate translation of a technical journal article, you'd better hire an expert. Besides, in programming, a little learning could be a dangerous thing. If you start trying to break into a prepackaged program to change a line or two, you might just ruin part of the program. One enterprising young student broke into a "canned" statistical package in order to "make it better." Inadvertently, however, the student inserted an endless loop, and the program could no longer get past this spot. (A simple example of an endless loop is where the program at A says GOTO B, and at B the program says GOTO A.) Imagine how expensive this might be if you were using this program on a time-sharing mainframe. Even learning binary math is no longer seen as essential. Spending a lot of time polishing your ability to add, subtract, multiply, and divide in base two may become a hip dinnertime conversation piece but may not be all that much help in conquering a new word-processing, spreadsheet, statistical, or "game" program. The machine may have to "think" in on-off terms, but you don't. And speaking of game programs, they are not just for the chess players anymore. Computer programs have even created a world of high-tech titillation, where characters perform feats of on-screen electronic eroticism so hot as to threaten to melt your microprocessor (*Time*, 1991).

What, then, should you have to learn to become computer-literate as we cross into a new century? Minimal computer literacy should probably include the following:

1. Know the basic parts of the computer, from input to storage to output.

2. Know how to handle, format, and copy disks as well as how to protect them from damage.

3. Know at least one word-processing program, such as WordPerfect or Microsoft's Word (you'll wonder how you ever lived without it); a spreadsheet program, such as Lotus, and a math or statistical program, such as *SPSS*. Research strongly suggests that the use of these programs will enhance your math, logic, and writing skills.

4. Know how to handle the computer safely and intelligently, and learn as many of the basic damage-prevention techniques as possible. And speaking of damage, never leave your disk in a closed car on a hot summer day, and don't ever use a pen (unless it's felt-tipped) to write on the label after it has been placed on the disk. If you do have to use a ballpoint pen, write the label before sticking it on the disk. In one dramatic case, a computer-illiterate student is alleged to have run his floppy disk through a typewriter to label it and tried to make disk copies by using the photocopy machine.

5. Learn the basics of computer networking, and especially how to get yourself on the Internet. This is especially useful for high school and college students. The tools of research are no longer based just on library card catalogs. A computer network allows the student to electronically browse and retrieve articles and documents essential for the research process (Hoh, 1992).

THE STATISTICAL PROGRAMS

Who's in Charge?

The most powerful statistical programs, like SPSS, can handle virtually any job the professional statistician will ever be called on to perform. Hundreds of variables and many thousands of scores can be analyzed in what seem (from the manuals) to be almost endless ways. These programs also, however, can be used to generate the simplest of calculations, means, medians, and modes.

Despite the power of these statistical packages, you must constantly keep in mind, however, that you, not the computer, are the one who's in full charge of the analysis. The choice of which statistical test to perform is in your hands. You're the one who must remember that chi square demands nominal data, that the Friedman ANOVA asks for ordinal data, and that t, F, and r must be fed at least interval data. You're the one, also, who must remember how these values may be interpreted, and which statistical tests are appropriate to which research designs. In short, the computer will be your lightning-fast slave that will only do exactly what it is told to do. And if you tell it to do dumb things, it will surely provide you with quick, but incredibly dumb, solutions. In one example, a group of programmers prepared a new Pentagon computer to be used by a four-star general: "Will it be peace or war?" typed the general, and the computer answered, "Yes." "Yes what?" the general asked. "Yes, sir," the computer instantly responded.

You simply can't blindly trust any result the "genius computer" spits out. Here is a sample of the kinds of thorny situations created by the user, not the computer . . . for in each case the computer did exactly what it was told to do.

1. A student decided to do a correlational study using a large number of variables and a large sample size. Among the variables chosen was GPA, and among the subject descriptions was each subject's Social Security number. Later, the printout revealed a Pearson r –.016 between GPA and the social security number, a difficult value to interpret. The GPA may be an interval measure (or as SPSS calls it a scale value), but the social security number surely is not (and the computer program didn't care). The computer was simply following orders. Most programs are totally insensitive to errors in the scale of measurement. SPSS does ask, but most do not. They will quickly crunch the data and spew out uninterpretable values if you feed in uninterpretable data. It would be like asking for a measure of skewness on ordinal measures. The computer will oblige, but whatever does it mean?

2. A fledgling researcher decided to assign subjects to one of two groups according to gender. Each subject was coded on gender, a "1" representing females and a "2" representing males. The student then asked the computer to do a *t*-test, and, not surprisingly the two groups were found to differ significantly on gender—another embarrassingly difficult result on which to build a research paper.

3. In yet another example, a student familiar with a computer program for performing a *t*-test, but not with statistical logic, fed the computer two sets of scores for each of over 50 subjects. The first distribution listed their heights in inches, and the second, their weights in pounds. Since the *t* ratio was obviously significant, the only possible conclusion was that the subjects were heavier than they were tall. The program, of course, assumed that both sets of measures were of the same trait.

Finally, it is absolutely crucial that you read the manual carefully before using any of these popular statistical packages. Even for those familiar with statistical assumptions it is important to find out whether a given program's version of a particular analysis conforms to the user's expectations.

Computer Overkill

Sometimes the novice researcher, flushed with the excitement of having so many computer-generated techniques to choose from, simply enters the data and tries them all. In one case a cross-tabs was created, which included the following information on each subject: age, class, college major, sex, parental income, student's summer earnings, and much more. Now, Cross Tabs can certainly produce the table and, in this case, subtables within tables. But getting a table of age by class by sex by major by GPA, etc., would not only be enormous in size, but, even worse, literally impossible to understand.

Also, just because a given program can theoretically handle data from extremely complicated designs doesn't necessarily mean that the results are always readily open to interpretation. For example, most programs can easily analyze ANOVA designs with 10 or more dependent variables, but making sense out of the many interactions is at best harrowing to contemplate.

In designing experiments, it is well to be reminded of the acronym KISS—Keep It Simple, Stupid. The research design should always be simple enough to generate results that have the potential for answering the research question. The computer, alas, can perform far more calculations than are always logically possible to interpret. Therefore, try to avoid that dreaded disease called "conspicuous computing," which is an obsessive and irrational lust for overusing the computer in order to generate an aura of scientific sophistication.

ADA LOVELACE (NEE BYRON, 1815–1852)

Augusta Ada Lovelace, the daughter of the poet Lord Byron and Annabella Milbanke, was a woman way ahead of her time. Byron and his wife separated only a few weeks after Ada was born, and, in fact, Ada never again saw her illustrious father. That her father reminisced fondly of her, however, is shown in these few lines from Childe Harold

> Is thy face like thy mother's, my fair child
> Ada, sole daughter of my house and of my heart?
> When last I saw thy young blue eyes they smiled
> And then we parted, not as now we part,
> but with hope.

Her mother was a strong and highly-educated woman who had been well schooled in both science and mathematics. She took over the tutoring of young Ada, giving her extensive lessons in both math and music. It was soon apparent that Ada's mathematical gifts were so powerful that professional tutoring was needed. Augustus DeMorgan, famed at the time for his contributions to Boolean Algebra, was hired to teach Ada, and by age 8 she had flourished to the point where she was absorbed in higher-order mathematical proofs. When she was 17 she met and came under the influence of the important mathematician, Mary Somerville, whose own texts were used at Cambridge and who had been instrumental in translating the works of the French statistician Pierre Laplace. It was through Mrs. Somerville that Ada, at the age of only 17, met Charles Babbage, the eccentric inventor of the first digital computer. Babbage was far more comfortable with things and blueprints than he was with people, but he did take a liking to Ada and they became lifelong friends.

In 1835, Ada met and married William King who later inherited a noble title. The couple became known as the Earl and Countess of Lovelace. Ada had three children who were primarily raised by her husband and staff, since Ada was extremely busy writing notes and articles on Babbage's invention. Because it was considered to be undignified and unladylike to publish scientific papers, Ada signed her manuscripts modestly with the letters "A.A.L." It was almost 30 years before the scientific community recognized who A.A.L. really was. It was through Ada's genius that Babbage's machine became more fully appreciated and understood as the world's first programmable computer. She saw its potential more clearly than did Babbage, and wrote that his machine could be a general manipulator of symbols, and not just a mathematical calculator. Ada is now seen as being the first computer programmer in history, since she wrote software for calculating Bernoulli numbers on Babbage's machine. She also made some uncanny forecasts about the future of computing, even predicting that it would someday produce graphics and even compose music. In her honor, the U.S. Department of Defense in 1979 initiated a new version of the computer language Pascal, and named it Ada in honor of this remarkable woman. That she foresaw some of the exaggerations of artificial intelligence comes through loud and clear. She insisted that the computer can do whatever we order it to do, but it has no power of anticipating any analytical relations or truths.

On the negative side, Ada and Babbage, anxious to accumulate enough money to finance the construction of his computer, created a statistical system for what they thought could beat the odds at the race track. Although there were some early successes, the system failed in the long run, (you can beat a race, but not the races) and they both found themselves with massive gambling debts and a degree of social disgrace.

One of Ada's friends was the English author Charles Dickens, who, although a friend, distrusted Ada at times, especially her seeming powers of out-performing probability. In 1849, he wrote to her that strange things were happening to him, and he wondered if she had cast a spell and was haunting him, and if so "I hope you won't do so any longer."

Later that same year, at the age of only 36 (coincidentally the same age as her father had been at his death), Ada died of cervical cancer. One of the few non-family members to visit her deathbed was Charles Dickens (and one of the very last persons to see her alive). Her final wish was to be buried next to her father, even though she never knew him (other than as a neonate).

Ada's contributions to statistical computing have become legendary, and her accomplishments live on even today. The ARA, (Ada Resource Association) sponsors a programming contest each year for students who can write the "most readable, original, reusable and clear working Ada programs". Also, there is the Augusta Ada Lovelace Award presented to an individual(s) showing "extraordinary service to the computing community through their accomplishments and contributions on behalf of women in computing".

Bugs, Glitches, and Things That Go Bump in the Night

Despite their elegant sophistication, any of the statistical packages have the potential for producing bizarre and even scary results. In fact, under some conditions, you may get no results at all. Until your data have been saved, they simply reside in the volatile world of RAM (the computer's internal memory, called random access memory). If, for example, there were suddenly a power failure, all of the RAM data would disappear and be gone forever. This is an especially traumatic event if you've already spent hours on the data entry. A prudent rule to follow is to be disciplined enough to stop about every 15 minutes and save the files. Remember, when the power goes out, the lights will come back, but your data won't. Things, as you will quickly learn, can and do go wrong. Entry errors are probably the number one culprit, but other dangers are definitely lurking in the background. A speck of dust, especially on a hard drive, can even cause a total crash of the system and leave you looking around for a hand calculator or perhaps even an abacus. Particles of dust or smoke can also cause extremely subtle problems, either to the stored program or to your data files. Even the computer itself may have a hardware problem that affects only part of the program being run. These are the most insidious problems because most of the program will still run flawlessly, but one or two sections have been damaged and are suddenly producing wildly inaccurate results. The worst part of this tale is that since most of the program still yields results that do check out perfectly, you then, of course, assume that the entire program is still in working order. In listing a frightening series of output mistakes, all due to hardware problems, one statistician says:

> Two points stand out in all of this. First, none of these were bugs, that is, mistakes in the programming: they are all fundamental hardware problems. Whereas none of these were serious enough to cause a system crash, all were completely destructive of the user's aims. Second, virtually all of them were detected by users who were *closely and critically examining their output* and who challenged the computer professionals (Galway, 1993, p. 75).

Even with hardware in perfect condition, bugs can remain in programs for months or even years. A "bug" is a computer term for an error. A software bug is a programming error, whereas a hardware bug is a malfunction or design error in the computer itself or one of its components (Blissmer, 1988). The term bug has its origin in electrical engineering and was first used by Thomas Alva Edison in 1878 to refer to the "little faults and difficulties" in one of his designs. Then in 1899, a newspaper, the *Pall Mall Gazette*, reported that Edison was "implying that some imaginary insect had secreted itself inside his machine and was causing all the trouble" (Jennings, 1990). And trouble it will cause because hidden programming errors often don't announce themselves until you, the user, do something that the program doesn't expect. A fairly simple case would be when you enter a letter where the program has to have a number, or you mistakenly put a question mark before the value. Rather than aborting, or telling you that you've made an error, the program may go into convulsions and provide you with some rather weird answers. As Covvey and McAlister have said, "Most completed software, then, harbors error. In programs of any complexity, it is literally impossible to ever eliminate all bugs. An operating system, for example, will always contain several serious errors when first released. Efforts to correct bugs in systems programming may introduce new bugs. Only as users subject the system to an enormous variety of uses will some of the bugs emerge from the woodwork" (Covvey & McAlister, 1982).

In fact, one of the authors of the above quote created his own statistical package, a program that seemed to work so well that he passed it on to his colleagues in the department. About a year later, one of his colleagues detected a serious error in the independent t test, and they all then realized that for over a year they had been relying on a faulty piece of software. Some of the t ratios, alas, may have even been published. Journal editors, after all, don't have time to check calculations, even when they have the raw data. Therefore, the prudent user of any program should scan the results and do at least some spot-checking before rushing those results to some journal.

New Trends

The most popular trends in computing as we enter the twenty-first century are: laptops, wireless, mobile computing, networking, and E-books or digital texts (Boettcher, 2001). And, online college courses are becoming increasingly popular, but experts in the field insist that a textbook link should always be provided to supplement the computer material. Also, increasing numbers of students are bringing laptops to classes with them. When using the computer during college class time, the instructor is no longer just a lecturer, but instead becomes the mentor, helping students shape their understanding and insights—"the electronic equivalent of the scholar and student sitting on opposite ends of the proverbial log" (Johnstone, 2001). Finally, the use of computers for e-mail has literally grown exponentially, from fewer that 10 million e-mail accounts in 1990 to 900 million in 2001 (Hafner, 2001).

Memory

The memory capacity of computers has increased so much over the past few years that we have to learn or relearn our scientific prefixes to make sense of the huge sizes. We read about hard drives that are measured in gigabytes. To appreciate these sizes, note the following:

Prefix	*Power of 10*	*Units*	*Number*
kilo	3	thousands	1,000
mega	6	millions	1,000,000
giga	9	billions	1,000,000,000
tera	12	trillions	1,000,000,000,000

And if you think those are big, try expressing the

yotta 24 septillions 1 followed by 24 zeros.

Speed

Also the computing speeds have become so incredibly fast that we have to clock them with some extremely small numbers. Consider the following:

Prefix	*Power of 10*	*Units (in seconds)*	*Number*
milli	–3	one thousandth	.001
micro	–6	one millionth	.000001
nano	–9	one billionth	.000000001
pico	–12	one trillionth	.000000000001

And finally, for a speed that is almost incomprehensible, look at the

yocto –24 one septillionth .000000000000000000000001

Yocto! Just think, a yoctosecond . . . now that's fast, so fast in fact that at night you could turn off the lights in your room by flipping the switch and then within a yoctosecond be safely tucked in bed before it got dark.

Some Words of Caution. Because so much data can be run so speedily and cheaply by today's computer packages, one statistician has issued an urgent warning about the pitfalls that await the unwary user (Searle, 1989). The warning is especially compelling today because the modern computer provides the researcher with such a precise-looking, neatly printed output that its very appearance lends a strong aura of authority. Two important points bear repeating.

1. Don't run your data through a statistical package until after you have decided which analysis is appropriate and how the results might be interpreted. Don't simply try a certain analysis just because it's there. "Sidling up to a computer and ordering two runs of package XYZ and three from PQR is now as easy as (and can be equally as damaging as) bellying up to the bar in the local tavern and ordering two gins and three beers" (Searle, 1989, p. 189).

2. Don't be fooled into assuming that just because you can use a computer to run your data, you therefore must know the rules of statistical methods. Students (and even some faculty) must be made to realize that computing expertise and statistical expertise are not the same. Unfortunately, failure to heed this warning has resulted in a new genre of statistics courses, such as "Understanding Statistics Through Computer Analysis," taught by an instructor who is alleged to have said, "You don't need to understand the details of statistical methods in order to use them" (Searle, 1989). There are numerous ways to make unsubstantiated statistical claims, and a high-speed computer will just do it all the faster.

3. Don't assume that because your printout says that the P value for a certain significance test is equal to or less than zero that the alpha error has suddenly become a quaint relic of the past. Some computer programs indicate extremely low probability values in that fashion, rather than stringing together a long series of zeros after the decimal.

None of this is intended to make you overly cynical, just cautious. The major statistical programs have been thoroughly tested. Nevertheless, although you shouldn't be overly cynical, you should adopt a healthy skepticism and at least run your results through some logic checkpoints.

LOGIC CHECKPOINTS

When reading your printouts, you can implement a series of logic checks that should at least alert you to the possibility of some of the most flagrant types of errors. The following is a list of 15 computer printouts, all of which have been beset in some way with flawed results. In each case, this powerful, beautiful, technologically sophisticated machine has acted like a fast idiot, and your job as the smart leader is to ferret out its mistakes. At the end of this section is an analysis of each error and a chapter reference where you may find the appropriate logic check.

1.
```
The minimum value is            2.000
The maximum value is           10.000
The range is                    8.000
The mean is                     6.000
The true standard deviation
of this distribution is        12.828
```

2.
```
Please enter lower z score:     -1.45
Please enter upper z score:     -1.05
The area is 78.1913% or a probability of 0.7819
```

3.
```
Two-Sample Independent t Test
The standard deviation of the first set is      2.0548
and its standard error of the mean is           3.9189
The standard deviation of the second set is     2.4095
and its standard error of the mean is           1.0775
The value of the t statistic is                 -3.884
and the standard error of difference is         1.4162
Type return to continue
```

4.
```
Pearson Correlation
Enter two numbers per line
  1:   9  10
  2:   2   5
  3:  12  15
   .
   .
   .
 10:   4   6
The Pearson r correlation is 2.456
```

5. One-Way Analysis of Variance

Source of Variation	Degrees of Freedom	Sum of Squares	Mean Squares	F
Between (Treatment)	3	-147.000	49.000	42.000
Within (Error)	12	14.000	1.167	
Total	15	161.000		

6. Two-Way Analysis of Variance

Source of Variation	Degrees of Freedom	Sum of Squares	Mean Squares	F
Rows	1	344.450	344.450	120.860
Columns	1	54.450	54.450	19.105
Interaction	5	2.450	2.450	0.860
Within	16	45.600	2.850	

7. ```
Chi Square Tests
Enter the observed frequency first, then the expected frequency
 1 : 40 25
 2 : 30 25
 3 : 20 25
 4 : 10 25
The value of chi square is -20.000
```

8. ```
The minimum value is           200.000
The maximum value is           800.000
The range is                   600.000
The mean is                    500.000
The true standard deviation
of this distribution is        100.000
The distribution is severely leptokurtic
```

9. ```
The minimum value is   55.000
The maximum value is   145.000
The range is   90.000
The mean is   100.000
The median is   115.000
The true standard deviation
of this distribution is   15.000
The distribution is normal
```

10. ```
Two-Sample Independent t Test
The standard deviation of the first set is      14.970
and its standard error of the mean is            3.469
The standard deviation of the second set is    16.7245
and its standard error of the mean is            4.001
The value of the t statistic is                  2.967
and the standard error of difference is        -1.4162
Please type return to continue
```

11. ```
Calculating a Regression Line and the Pearson r
The regression equation is
             Y = 0.921 X + 1.421
The correlation is -0.861
```

12. ```
Within-Subjects F
Source of      Degrees of      Sum of       Mean        F
Variation      Freedom         Squares      Squares
Between
Columns        0               10.00        5.000       10.000
Residual       8                4.00         .500
```

13. Two-Way Analysis of Variance

Source of Variation	Degrees of Freedom	Sum of Squares	Mean Squares	*F*
Rows	2	0.620	0.310	0.139
Columns	2	229.672	114.836	25.706
Interaction	2	2.602	1.301	0.291
Within	54	241.235	4.467	

14. Chi Square Tests

```
-----entering the first row-----
Enter the two values  :   7   10
-----entering the second row-----
Enter the two values  :  15    8
The value of chi square is                   2.2827
The value of chi square modified with
the Yates correction factor is               4.0667
```

15. One-Sample *t* test

```
Mean and Standard Deviation Calculated
From Data
Please enter the hypothesized mean                 :  80.000
The sample mean is                                 :  80.000
The estimated population standard deviation is     :   8.489
The t statistic is                                 :  18.602
```

ANSWERS

1. The true standard deviation of 12.828 is too high. The true standard deviation of any set of sample scores may not be greater than half the range, or, in this case, 4.000. (*See Chapter 3.*)

2. The area of 78.1913% is too large. Since both *z* scores fall on the same side of the mean, the area between them cannot be greater than 50% and the probability may not be greater than .50. (*See Chapter 4.*)

3. The standard error of the mean for the first set of scores is too large. The value of the standard error of the mean cannot be greater than the standard deviation of the sample scores. (*See Chapter 7.*)

4. The Pearson *r* value of 2.456 exceeds its limits. The Pearson *r* may never exceed 1.00. (*See Chapter 11.*)

5. The one-way analysis of variance is producing a negative sum of squares of –147.000. Since the sum of squares is a measure of variability, it may never be negative. (*See Chapter 12.*)

6. The two-way analysis of variance is indicating an error among the degrees of freedom. If the degrees of freedom for the rows and columns are correct, then the interaction must equal 1, not 5. (*See Chapter 12.*)

7. Without running a calculation check on the value of chi square, you still can tell that the answer of –20.00 cannot be correct. No chi square value may ever be negative. (*See Chapter 13.*)

8. If the standard deviation of this distribution is correct, then the distribution cannot be severely leptokurtic. If, however, the distribution is leptokurtic, then the computer-generated standard deviation is too high. (*See Chapter 3.*)

9. If the distribution is normal, then the computer values for the mean and/or median are incorrect. If, however, the mean and median values are accurate, then the distribution is not normal. (*See Chapter 4.*)

10. The standard error of difference, since it estimates variability, cannot ever be negative. The computer solution of –1.4162 must be incorrect. (*See Chapter 10.*)

11. Since the regression equation is indicating a positive slope of 0.921, then the correlation of –0.861 cannot be correct. When the slope is positive, the correlation must also be positive. If the correlation in this printout is correct, then the slope value must be wrong. (*See Chapter 14.*)

12. The within-subjects F is showing zero degrees of freedom between columns. This must be an incorrect value, since an ANOVA of this type is based on repeated measures, and the measures should be set up in the columns. (*See Chapter 15.*)

13. For this two-way analysis of variance, the degrees of freedom are being allocated incorrectly. The interaction df of 2 must be in error if the row and column df's are correct. Or perhaps the df's are incorrect for either the rows, the columns, or both. (*See Chapter 12.*)

14. For any 2×2 chi square, the value for the Yates correction may never be greater than the value of the chi square itself. Thus, either the chi square value is too low or the Yates correction is too high. (*See Chapter 13.*)

15. If the sample and hypothesized means have been correctly entered, then the t ratio is too high. Since there is no difference between the means, the t ratio itself must be equal to zero. Or if the t ratio is correct, one or both of the means must be in error. (*See Chapter 10.*)

RECOMMENDED READING

If your university has an SPSS license and the program is loaded on the server, we recommend that you read:

George, D., & Mallery, B. (1999). *SPSS for windows step by step: A simple guide and reference.* Needham Heights, MA: Allyn-Bacon.

Mertler, C. A. & Vannata, R. A. (2001). *Advanced and multivariate statistical methods*. Los Angeles, CA: Pyrczak Publishing

Chapter

19

Research Simulations: Choosing the Correct Statistical Test

*In this final chapter, all the various statistical tests, both parametric and nonparametric, for all the measurement scales—nominal, ordinal, interval—will be matched with a series of simulated research situations. Since all the interval data statistical tests—*t, F, r, *and so on—can also handle ratio data, we will not make the interval-ratio distinction throughout this chapter. For the purist, the phrase "at least interval data" may be inserted wherever interval data are mentioned. The ability to calculate the tests is important, but the more meaningful competency consists of knowing when to calculate which one. This decision making can be greatly simplified by using the checklist and charts included in this chapter. With these aids, the decision as to which test to use becomes totally automatic. When confronted with any problem, the only way we can get the right answer is to ask the right question. The specific checklist questions, when asked and answered in sequence, serve to narrow the range of appropriate statistical tests until there is one—one best test for the particular research situation. A flowchart is presented for each form of data, and as you answer the questions and follow the various routes offered on the chart, the eventual solution should, it is hoped, become clear.*

METHODOLOGY: RESEARCH'S BOTTOM LINE

Before discussing the three specific questions, we must consider the one overriding factor that literally transcends all the others. That one is the bottom-line question as to what type of research is involved—experimental or post facto. Although the answer to this question does not affect which statistical test to apply, it does dictate the type of conclusions that can validly be drawn from the results of the study.

We should not be like the researcher who, so the story goes, trained groups of grasshoppers to respond on command. When the researcher yelled "Jump!" they all obediently jumped. Then, each day the researcher pulled out one leg from each of the grasshoppers, again cried "Jump!" and measured how high they leaped. After six sequential trials, when the grasshoppers had all been totally delegged, the researcher found that, despite the strident command, the grasshoppers would no longer jump. A series of fancy, repeated-measures F ratios was calculated, and all were found to be highly significant. The researcher's conclusion? When one pulls out the legs of a grasshopper, it becomes deaf. Selecting the appropriate statistical test and then drawing erroneous conclusions is like winning the daily double at the races and then having your pocket picked on the way home—an empty victory, indeed.

Drawing logically derived conclusions does not depend on which statistical test has been used, but rather on the type of research methodology—experimental or post facto. And this categorization depends solely on the independent variable. When the independent variable is manipulated, the research is experimental. When the independent variable is an assigned-subject characteristic, the research is post facto.

There are two types of post-facto research. One type tests the hypothesis of difference—groups already assumed to differ on some important variable(s) are compared in order to discover whether they also differ systematically on other variables. Another type of post-facto research tests the hypothesis of association by attempting to find out whether a correlation exists between certain variables. As we have learned, correlation does not prove causation, but *neither does it rule it out.* A discovered correlation may later be shown, through an experimental method, to be involved in causation. For example, there is obviously a high, positive correlation between ambient temperature and the amount a person sweats. However, other important factors are also involved, such as the amount of humidity, the person's body size, and how active the person is. An experiment can be set up in which these other factors (humidity, amount of activity) are held constant (not allowed to be variables), and heat and *only heat* is manipulated as the independent variable. If the subjects now exhibit significantly different sweat levels as a function of changes in the heat level (*manipulated* independent variable), then it is probable that a cause-and-effect relationship has been uncovered. Thus, post-facto research, although designed to lead to better-than-chance predictions, may provide important clues that might later be used to hone in on causation.

CHECKLIST QUESTIONS

Question 1: What Scale of Measurement Has Been Used? The choice of the correct statistical test depends heavily on which scale of measurement the data represent. When the data are in nominal form—that is, presented as frequencies of occurrence within mutually exclusive categories—then chi square or one of its many variations should be used. Chi square, as is true with all tests for nominal and ordinal data, is a nonparametric test. As such, it is able to sidestep any questions concerning the shape, manner, or form of the population distribution. (All the nonparametric tests are also called distribution-free tests.)

When the scores are in ordinal form—that is, the distances between successive scale points are not known or are known not to be equal—then the ordinal tests, U, H, T, χ_r^2, and r_S, must be used. These tests of ordinal data are called nonparametrics because no assumptions regarding the population characteristics are made. As a result, these tests are also appropriate for interval data when the underlying distribution of interval scores is not normal. In this situation, the interval scores are first rank-ordered and then the test analysis is performed.

Finally, when the scores are in interval form—that is, the distances between successive scale points are assumed to be equal—then, in most instances, t, F, and r should be used. We limit this rule to most instances because there is one dramatic exception, and that is when the interval scores do not distribute normally in the population. Fortunately, most measures of human attributes *do*, in fact, distribute normally, but the exceptions, such as age and income, must be constantly watched for. The tests of interval data are called parametric tests because they assume some knowledge of the population, for example, that its underlying distribution is normal. Also, they are used to estimate certain population parameters, such as the mean and the standard deviation.

Question 2: Which Hypothesis Has Been Tested? There are only two hypotheses that can be statistically tested—the hypothesis of difference and the hypothesis of association. Whenever the research is experimental, then the hypothesis of difference is the one that must be tested. This hypothesis states that the populations from which the sample groups have been selected are in some way different from each other. If, however, the research is post facto, then the hypothesis under scrutiny might be one of either difference or association. The hypothesis of association states that a correlation exists in the population from which the sample has been selected. This correlation may exist between different measures taken on the same group of subjects (for example, a single group of subjects being measured on both height and weight) or between the same measure taken on different subjects (for example, obtaining IQ scores from pairs of identical twins). Testing the hypothesis of association requires different statistical tests than does testing the hypothesis of difference.

Question 3: If the Hypothesis of Difference Has Been Tested, Are the Samples Independent or Correlated?* Whenever the hypothesis of difference is tested, whether in experimental or post-facto research, it must be clearly determined whether the sample groups are independent or correlated. If the selection of one sample is in no way influenced by the selection of another, then the samples are independent. This occurs when each sample is randomly selected. If, on the other hand, the subjects to be measured are in any way paired off, either by using the same subject more than once or by equating subjects on the basis of some relevant variable, then the groups are correlated. Attempting to isolate differences between correlated sample measures requires

*Pertains *only* when the hypothesis of difference is being tested.

different statistical tests than when analyzing differences between independent sample measures.

Question 4: How Many Sets of Measures Are Involved? When only two measures are being compared, one set of statistical tests is available, but when more than two sets of measures have been taken, other choices can be made. For example, when two groups of interval scores are being tested for difference, the *t* test is available, whereas when three or more groups of interval scores are being compared, the *F* ratio must be used. In experimental research the number of measurement sets always refers to the measures taken on the dependent variable.

CRITICAL-DECISION POINTS

Once the checklist questions are answered, we turn to the three flowcharts provided here—one for each of the three scales of measurement. Each flowchart presents a series of critical-decision points going from top to bottom and based on the checklist questions just outlined. At each critical-decision point, the answer to the next question dictates which route should be followed to the next point. The routes, branching at each decision point, lead inevitably to an appropriate statistical test.

RESEARCH SIMULATIONS: FROM A TO Z

This section presents a series of research simulations, a total of 26, each designed to be used with the checklist questions and the accompanying flowcharts (Figures 19.1 to 19.3) to aid you in choosing an appropriate statistical test for each situation.

Simulation A

It has been traditional for the man rather than the woman to receive the check when a couple dines out. A researcher wondered whether this would still be true if the woman was clearly in charge, asking for the wine list, and so on. A large random sample of restaurants was selected. One couple was used in all restaurants, but in half the man assumed the traditional in-charge role, and in the other half the woman was in charge. The data were in the form of the number of times the check was presented to each member of the couple.

Analyzing the Methodology. This is experimental research; the independent variable (man or woman in charge) was clearly manipulated. The dependent variable was whether the man or the woman received the check. Thus, if differences in the dependent variable occur here, they can be attributed to the action of the independent variable. Of the experimental designs, this was a between-subjects, after-only design. There was no pre-test given, they were only measured once, and no matching took place.

FIGURE 19.1

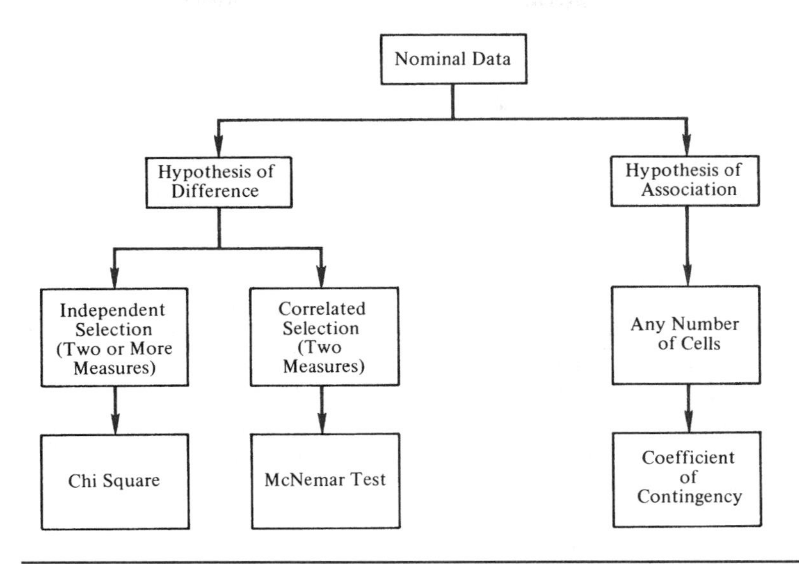

FIGURE 19.2

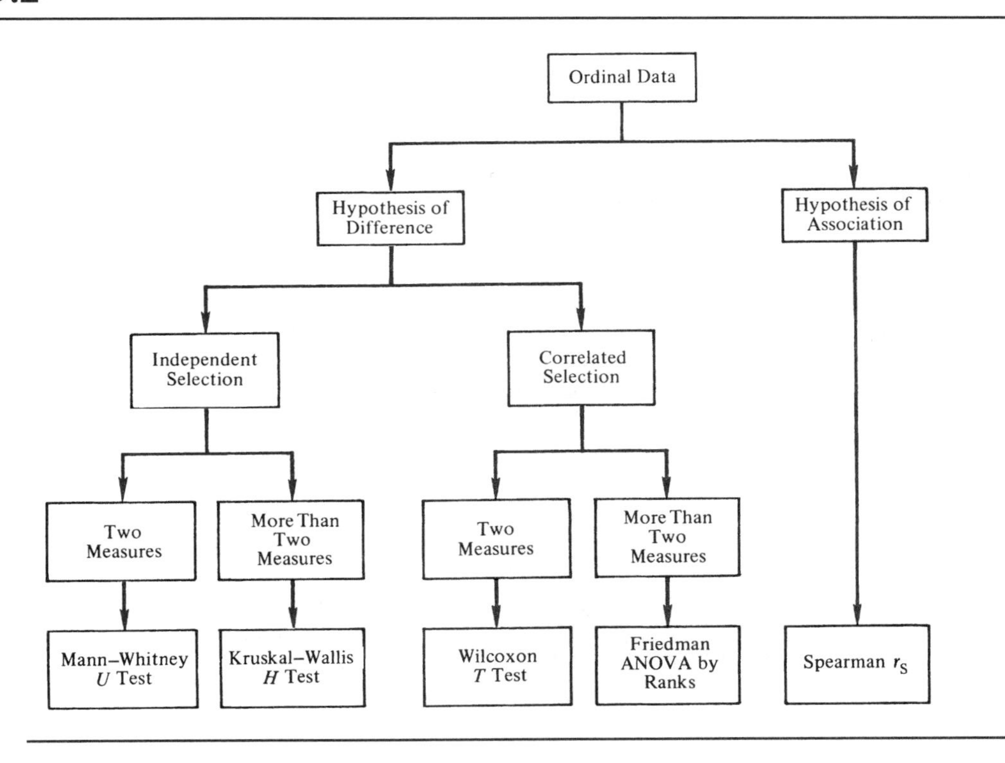

FIGURE 19.3

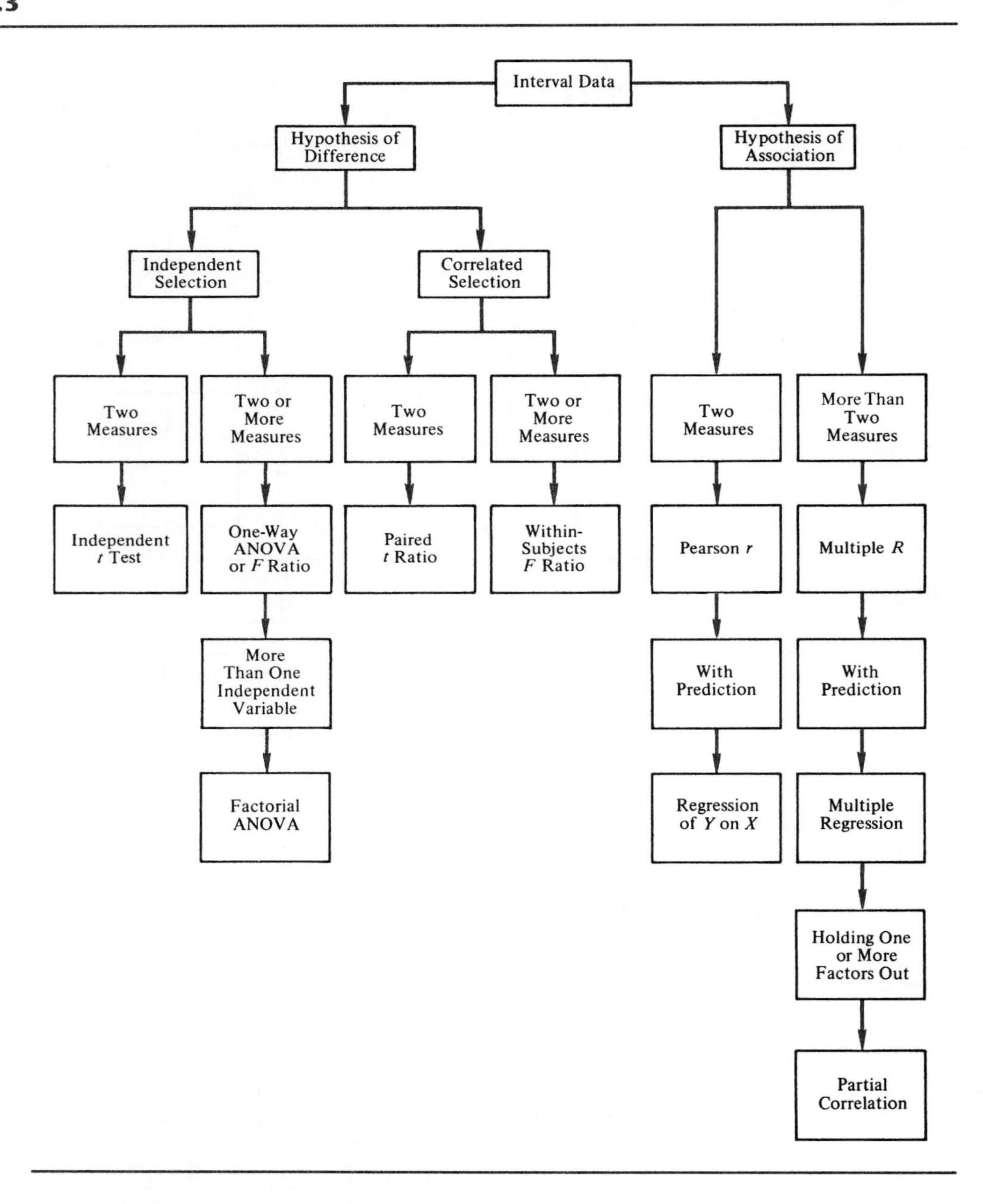
Interval Data
Hypothesis of Difference
Hypothesis of Association
Independent Selection
Correlated Selection
Two Measures
Two or More Measures
Two Measures
Two or More Measures
Two Measures
More Than Two Measures
Independent t Test
One-Way ANOVA or F Ratio
Paired t Ratio
Within-Subjects F Ratio
Pearson r
Multiple R
More Than One Independent Variable
With Prediction
With Prediction
Factorial ANOVA
Regression of Y on X
Multiple Regression
Holding One or More Factors Out
Partial Correlation

Answering the Checklist Questions: The Critical Decisions

1. *Scale of measurement?* The measurements in this study are in the form of nominal data—the responses being categorized on the basis of which member of the couple received the check. Use Figure 19.1.

2. *Hypothesis?* The hypothesis being tested is that of difference—whether or not a difference as to who received the check would occur as a function of who appeared to be in charge. (Since this was the experimental method, it had to test the hypothesis of difference.)

3. *If the hypothesis of difference is tested, are the groups independent or correlated samples?* The subjects (waiters or waitresses) were independently selected, since the restaurants were randomly chosen.

4. *How many sets of measurements?* There are two sets of nominal measurements based on whether the man or the woman gets the check.

Solution. Use chi square, in this instance a 2 × 2 chi square, with the independent variable (who is in charge) in the rows and the dependent variable (who gets the check) in the columns.

	Woman gets check	***Man gets check***	
Man in charge	a	b	
Woman in charge	c	d	
			N

Simulation B

A researcher, noting the positive correlation between socioeconomic status and amount of education, assumes there will be a difference in the amount of TV viewing by the principal wage earners in high and low socioeconomically placed families. A random sample of families is selected and categorized as having either high or low socioeconomic status on the basis of a number of measures (income of principal wage earner, occupation, and location of residence). The principal wage earners of all families were then contacted and asked for their hours per week of TV viewing. The hours per week for each socioeconomic class were compared.

Analyzing the Methodology. This is an example of post-facto research; the independent variable (socioeconomic status) is a subject variable, not manipulated. Regardless of whether the statistical test is significant, no definitive cause-and-effect statement can be made. Even if the upper-class group is shown to watch significantly less, the reason for this difference can only be suspected. (Perhaps upper-class persons read more, or work more hours, or have expensive time-consuming hobbies, or . . . the list of possibilities is endless.)

Answering the Checklist Questions: The Critical Decisions

1. *Scale of measurement?* The measurements here are in at least interval form. If one person watches TV for 25 hours and another for 24, we know not only that one person spends more time watching TV but also how much more time. With no evidence to the contrary, the distribution of number of hours spent watching TV is assumed to be normal. Use Figure 19.3.

2. *Hypothesis?* The hypothesis tested is that of difference—that the two groups would *differ* as to the extent that they watched TV.

3. *If the hypothesis of difference is tested, are the groups independent or correlated samples?* The two groups in this case are independent of each other, each being randomly selected and then assigned to the high or low socioeconomic status.

4. *How many sets of measurements?* There are two sets of measurements, one for upper status and one for lower status.

Solution. Use the independent *t* test. Had it been predicted that one group would watch more TV than the other, then test as a one-tail *t*.

Simulation C

A researcher wished to test the hypothesis that taller men are more likely than shorter men to be judged as leaders. A random sample of 30-year-old men was selected and measured for height. The men were then brought before a panel of personnel managers and rank-ordered on the basis of perceived leadership qualities. Each subject was assigned the median rank of the panel's decisions.

Analyzing the Methodology. This is post-facto research. The independent variable (height) was not manipulated (although it could have been by using elevator shoes, or even hidden stilts). Therefore, the issue of causation is not relevant here, although the probability of making accurate leadership predictions is very much at issue.

Answering the Checklist Questions: The Critical Decisions

1. *Scale of measurement?* Despite using the panel's consensus on rankings, the data are still in ordinal form. Although we may know that Subject 1 is perceived to be imbued with more of the leadership image than is Subject 2, we do not know how much more. Also, although height is clearly an interval measurement, it must be converted to an ordinal rank before the statistical analysis can be completed (the men are simply rank-ordered as to height). Use Figure 19.2.

2. *Hypothesis?* The hypothesis being tested here is that of association. It is gratuitous to try and establish a difference between such already disparate measures as height and perceived leadership qualities.

3. *If the hypothesis of difference is tested, are the groups independent or correlated samples?* Not applicable; the hypothesis of association is being tested.

4. *How many sets of measurements?* There are two sets of measurements, height and leadership.

Solution. Use the Spearman r_S. Note that in this problem both age and sex were ruled out as variables, since only men of the same age were selected.

Simulation D

A market researcher, working for a manufacturer of hair coloring, wished to establish whether blondes do indeed have more fun. A large random sample of female college sophomores was selected and categorized as to hair color—blonde, brunette, or red. Each subject was then asked to answer yes or no to the question, "On balance, would you say you've been having fun this semester?"

Analyzing the Methodology. This is post-facto research. The independent variable (hair color) was a subject variable, not manipulated. (Here again, this could have been designed as an experiment, by perhaps randomly dividing the brunettes into two groups and then giving all members of one group a blonde tint and allowing the other group to remain as they were.) As a post-facto study, however, no cause-and-effect statement can be made. The groups may differ on a variety of other variables related to happiness such as number of dates or grade-point average. Also, it may be that some of the women are already using the blonde tint, and this itself may be a function of the woman's personality. Perhaps socially oriented women are more apt to use a tint, and to be more optimistic.

Answering the Checklist Questions: The Critical Decisions

1. *Scale of measurement?* Since the women are measured on the basis of answering the question either yes or no, and then the frequencies of these answers are tallied, the data are nominal. Use Figure 19.1.

2. *Hypothesis?* This is one of those studies in which the hypothesis could conceivably be classified either way—difference or association. It is likely that the researcher meant this as a difference study since the hypothesis was stated as, "Blondes have more fun." (Had it been the hypothesis of association, it probably should have been stated as, "There is a correlation between hair color and having fun.")

3. *If the hypothesis of difference is tested, are the groups independent or correlated samples?* The samples (blondes, brunettes, and redheads) are independent of one another. The fact that one woman was placed in the blonde group neither caused another woman to be placed in nor precluded her placement in a different group.

4. *How many sets of measurements?* There are two sets of scores based on whether the subjects report having or not having fun.

Solution. Use chi square, in this case a 3 × 2 chi square.

	Having fun	*Not having fun*
Blonde	a	b
Brunette	c	d
Redhead	e	f

N

Since this is a 3 × 2 chi square, the special 2 × 2 computational method cannot be used. This means that all of the values for f_e (frequencies expected) must be separately calculated. Had this been a test of association, the chi square could have been followed up with the coefficient of contingency.

Simulation E

A researcher wished to test the hypothesis that older men sleep less than younger men do. Random samples of 30-year-old men, 50-year-old men, and 70-year-old men were selected. Each subject was brought to a sleep laboratory and measured as to how many hours of sleep per night occurred.

Analyzing the Methodology. This is a post-facto study; the independent variable (age) was a subject variable, not manipulated. (Age can, of course, never be a manipulated independent variable.) Even if the results prove to be significant, great care must be taken in interpreting them. If it is found that the older men sleep less, it may be they did so as young men too. As youngsters, these men may have been more apt to rise early as a result of the differing cultural patterns that typified their younger days. (This is actually cross-sectional research, the hazards of which have been pointed out.)

Answering the Checklist Questions: The Critical Decisions

1. *Scale of measurement?* The dependent variable (hours of sleep) provides at least interval data, with an underlying distribution that is probably normal. Use Figure 19.3.

2. *Hypothesis?* The hypothesis being tested is one of difference—that different age groups have *different* sleep habits.

3. *If the hypothesis of difference is tested, are the groups independent or correlated samples?* These sample groups are independent of one another. The fact that a given man is selected for the 30-year-old group has no bearing on who is being selected in the 50- or 70-year-old group.

4. *How many sets of measurements?* There are three sets, one for each age group.

Solution. Use the one-way ANOVA, the *F* ratio. If *F* proves to be significant, proceed with Tukey's HSD.

Simulation F

A researcher wanted to test the hypothesis that racial prejudice is a function of personal authoritarianism. A random sample of college students was selected and measured on the F scale, an index of personal rigidity and authoritarianism. All subjects were then given the A-S (for Anti-Semitic) scale, a measure of prejudice toward Jews.

Analyzing the Methodology. This is a post-facto study; the independent variable (authoritarianism) is a subject variable, not manipulated. Even if the hypothesis is validated, there will be no way to tell whether authoritarianism affects prejudice or prejudice affects authoritarianism. It is even possible that a third variable, such as a family's child-rearing practices, produces both authoritarianism and prejudice.

Answering the Checklist Questions: The Critical Decisions

1. *Scale of measurement?* Both the F and A-S scales are considered to be interval measures distributed normally in the population. Use Figure 19.3.

2. *Hypothesis?* The hypothesis in this case is one of association. (One group is being measured on two different response dimensions.) We can never test for differences between completely unrelated measures.

3. *If the hypothesis of difference is tested, are the groups independent or correlated samples?* Not applicable; the hypothesis of association is being tested.

4. *How many sets of measurements?* There are two sets, one for F scale and one for A-S scores.

Solution. Use the Pearson *r*. If found to be significant, the *r* could be followed by a regression equation, with which specific A-S scores could be predicted from given F scores.

Simulation G

A researcher working for a large corporation wished to test the hypothesis that the company's toothpaste, containing fluoride, reduces dental caries. A random sample of 18-year-olds was selected, and all subjects were checked for caries. A dentist then filled the cavities for all subjects having them. For the next three years, all subjects received free monthly supplies of the fluoride toothpaste. Finally, at age 21, the subjects were again checked for dental caries. The researcher then compared the number of persons with caries found in the first dental checkup with the number of persons with caries found at age 21.

Analyzing the Methodology. This is experimental research, repeated-measures design (in this case, before-after). The independent variable (toothpaste) is manipulated (rather than being assigned on the basis of whether the subjects, on their own, were using it). This is, however, a shaky design because maturation is a variable and may confound the independent variable. Perhaps 18-year-olds, as a group, are more apt to have caries (the cavity-prone years) than are 21-year-olds. Perhaps, therefore, there might have been a significant result even without the introduction of the independent variable. It would have been better to have had a separate control group, checked for caries at both ages but given, instead of the fluoride brand, toothpaste that looked the same but did not contain fluoride.

Answering the Checklist Questions: The Critical Decisions

1. *Scale of measurement?* These are nominal data. The subjects were categorized, as a group, at each age as to whether caries were detected. We do not know whether any subject had more or fewer caries than another, only whether any caries were present. Frequencies of occurrence for each age were compared. Had the subjects been rank-ordered in terms of amount and/or severity of caries, the data would have been ordinal. Perhaps even interval measures could have been designed, where the subjects received scaled scores based on the number and severity of caries found. Use Figure 19.1.

2. *Hypothesis?* This is the hypothesis of difference—the assumption being that after using the fluoride toothpaste, the frequency of caries within the group will diminish.

3. *If the hypothesis of difference is tested, are the groups independent or correlated samples?* The groups are as correlated as possible, since the same group was used as its own control. The samples are always correlated in a before-after design.

4. *How many sets of measurements?* There are two sets, one taken at age 18 and the other at age 21.

Solution. Use the McNemar test, which is a chi square based on the change scores.

Simulation H

A researcher wished to find out whether the perception of a person's height depends on that person's perceived status. A random sample of army inductees was selected and equally divided into four groups. An actor gave a short address to each group separately, extolling the joys of army life. For the first group, the actor was dressed as a private; for the second, as a sergeant; for the third, as a captain; and, finally, for the fourth group, as a colonel. The inductees were asked to fill out a questionnaire evaluating the speech. Among the questions was one asking for an estimate of the lecturer's height.

Analyzing the Methodology. This is experimental research, between-subjects (after-only) design. The independent variable (perceived status) was manipulated by having the same actor wear different uniforms. Using the same actor was a good idea, since otherwise differences in personal characteristics might have confounded the independent variable.

Answering the Checklist Questions: The Critical Decisions

1. *Scale of measurement?* The dependent variable (estimated height) provided at least interval data in this study. Use Figure 19.3.

2. *Hypothesis?* The researcher is looking for *differences* in the height estimates made by the four groups.

3. *If the hypothesis of difference is tested, are the groups independent or correlated samples?* The four groups are independent of one another. The selection of one soldier had no bearing on whether another was or was not selected. This is an after-only experimental design, in which sample groups must be independent of one another.

4. *How many sets of measurements?* There are four sets, one for each treatment condition.

Solution. Use the one-way ANOVA, the F ratio. If F is significant, follow it up with Tukey's HSD.

Simulation I

A researcher suspects that both meaningfulness *and* length of presentation affect word retention, as scored on a standardized test. Four groups of fifth-grade students were randomly selected and then assigned to different treatment conditions. Group A was given low-meaningful words and a 2-second presentation time. Group B was given a list of high-meaningful words and also a 2-second presentation time. Group C was given low-meaningful words and a 5-second presentation time, whereas Group D received a high-meaningful list and a 5-second presentation time. The results of their retention scores were analyzed.

Analyzing the Methodology. This is experimental methodology with *two* independent variables, both having been manipulated by the researcher.

Answering the Checklist Questions: The Critical Decisions

1. *Scale of measurement?* Standardized retention scores come in at least interval form.

2. *Hypothesis?* The researcher is attempting to discover if there is a difference in the retention scores.

3. *If the hypothesis of difference is tested, are the groups independent or correlated samples?* Since each of the four groups was randomly selected and assigned separately to the various conditions, the samples are independent.

4. *How many sets of measurements?* There are four sets of measures, one set for each group.

Solution. Use the factorial ANOVA and be especially careful to look for the possibility of a significant interaction.

Simulation J

A researcher for an electronics corporation wished to establish whether, other things being equal, the tonal quality of a hi-fi set is judged to be better as the size of the speaker enclosure is increased. A random sample of subjects was selected and asked to listen to the same CD played on "different sound systems." Actually, the amplifier, the size and quality of the speaker and baffle, and so on remained the same. Only the size of the speaker enclosure was allowed to vary. Three enclosure sizes were used—small, medium, and large. The subjects were asked to rank-order their preferences, from 1 (best) to 3 (worst). The order in which the subjects were presented with the various speaker sizes was counterbalanced, so that some subjects had the large enclosure first, others the small enclosure first, and so on.

Analyzing the Methodology. This is experimental research, repeated-measures design (before-after-after). The independent variable (enclosure size) was manipulated by the experimenter. If significant results are obtained, cause-and-effect inferences can be made.

Answering the Checklist Questions: The Critical Decisions

1. *Scale of measurement?* The dependent variable (judgment of tonal quality) is in ordinal form, that is, the subjects' rank-ordering of the three listening conditions. Use Figure 19.2.

2. *Hypothesis?* This study tests the hypothesis of difference—that different judgments of sound quality will occur as enclosure size is changed.

3. *If the hypothesis of difference is tested, are the groups independent or correlated samples?* The groups in this study are definitely correlated, as the same group is used in all three treatment conditions. Groups are always correlated in repeated-measures designs.

4. *How many sets of measurements?* There are three sets, one for each treatment condition.

Solution. Use the Friedman ANOVA by ranks. Compare the ranks (1, 2, and 3) of each subject under the three listening conditions.

Simulation K

A researcher was interested in establishing whether attendance in a preschool program affects the social maturity level of children. A random sample of 30 kindergarten children was selected and watched closely by trained observers for one full week. The children were then rank-ordered on the basis of perceived social maturity. The children were then divided into two groups on the basis of whether or not they had previously attended a day-care center.

Analyzing the Methodology. This is post-facto research; each child's parents, not the researcher, decided whether the child would attend a day-care center. The independent variable (whether the child attended the day-care center) was thus a subject variable, not manipulated. Perhaps parents are more apt to send a less mature (or, more mature, who knows?) child to a day-care center in the first place.

Answering the Checklist Questions: The Critical Decisions

1. *Scale of measurement?* This is an example of ordinal data, since the dependent variable is the ranking of the child's social maturity level. Use Figure 19.2.

2. *Hypothesis?* This is the hypothesis of difference—that maturity levels should differ as a function of day-care attendance.

3. *If the hypothesis of difference is tested, are the groups independent or correlated samples?* The groups are independent of each other. The fact that one child was placed in a given category had no influence on where another child was placed.

4. *How many sets of measurements?* There are two sets, those of children who had attended the day-care center versus those who had not.

Solution. Use the Mann–Whitney U test, which compares the ranks of two independent groups.

Simulation L

A researcher is interested in whether coaching can have any effect on math SAT scores. A group of 100 high school seniors was randomly selected from a large metropolitan school district. The group was then randomly divided into two subgroups. One group was given three months of daily coaching in those math skills deemed important to the SAT, while the other group spent the same amount of time each day watching reruns of the TV show "Happy Days." At the end of the three-month period, all students took the SAT and their math scores were compared.

Analyzing the Methodology. This is experimental methodology, the independent variable being based on whether the students received the coaching.

Answering the Checklist Questions: The Critical Decisions

1. *Scale of measurement?* The data from the SAT come in at least interval form.

2. *Hypothesis?* The researcher is testing the hypothesis of difference.

3. *If the hypothesis of difference is tested, are the groups independent or correlated samples?* The two groups are independent, since the selection of one student being random had no influence on the selection of other students.

4. *How many sets of measurements?* There are two sets of measures, the SAT scores from each of the two groups.

Solution. Use the independent t test.

Simulation M

A study was designed to test whether presenting one side or both sides of an argument is more effective in changing attitudes. Perhaps presenting just the pro side would be more effective because an audience might not be fully aware of the anti side. Or perhaps to appear impartial and to avoid having members of the audience go over to the anti side and therefore tune out the pro message, it would be more effective to at least present some of the anti arguments.

A large random sample was selected, and the subjects were assigned to one of two conditions. Group A heard only the pro side of the issue, whereas Group B heard the entire pro side plus a few minutes of anti arguments. Both presentations were made by the same person. A questionnaire, tapping attitudes toward the issue, was then filled out by each subject.

Analyzing the Methodology. This is experimental research. The independent variable (one-sided versus two-sided presentations) was manipulated by the researcher. As no matching occurred and no attitude testing was done prior to the presentation, this was a between-subjects (after-only) design. If the results prove significant, a causal inference could be drawn.

Answering the Checklist Questions: The Critical Decisions

1. *Scale of measurement?* The questionnaire was scored as interval data, and the assumption of a normal distribution was made. Use Figure 19.3.

2. *Hypothesis?* As in all experimental research, the hypothesis of difference was tested. Presumably, *differences* in attitudes between the two groups can be attributed to the independent variable.

3. *If the hypothesis of difference is tested, are the groups independent or correlated samples?* As is true of all after-only experimental designs, the groups were independently selected.

4. *How many sets of measurements?* There are two sets of scores to be compared.

Solution. Use the independent t. As no prediction regarding direction was even suggested, check the t as a two-tail test.

Simulation N

A researcher wished to test the hypothesis that male business majors earn more in later life than do either male liberal arts or education majors. A random sample of alumni was selected from the university files from each of the three subject major categories. To attempt to control for length of experience on the job, all subjects were selected from the same graduating class—the class that graduated 10 years ago. All the selected alumni were contacted and asked to indicate their yearly incomes. The men were promised that the information would be held in strict confidence and would not be given to the chairman of the upcoming alumni fund drive. Because a few of the subjects reported enormously high incomes, the resulting distribution became so skewed that it was decided to rank-order the incomes.

Analyzing the Methodology. This is post-facto research. The independent variable, college major, was a subject variable, not manipulated.

Answering the Checklist Questions: The Critical Decisions

1. *Scale of measurement?* Although income is an interval measurement, the skewed distribution forced a rank-ordering of the data, thus creating a series of ordinal measures. Use Figure 19.2.

2. *Hypothesis?* The researcher was testing for differences among the income ranks of the three groups.

3. *If the hypothesis of difference is tested, are the groups independent or correlated samples?* The groups are independent. The assignment of alumni into subject major categories is strictly independent. The selection of one person from the "education" category did not demand or preclude another subject being selected from the "liberal arts" category.

4. *How many sets of measurements?* There are three sets, one for each of the subject major categories.

Solution. Use the Kruskal–Wallis *H* test for three or more independent groups and ordinal data.

Simulation O

A researcher wished to increase the predictability of student pilot scores on the FAA's written general aviation exam. Dependable relationships were found to exist between number of hours of ground school and FAA exam scores and also between IQ and exam scores. Finally, a small, but significant, relationship was found to exist between IQ and number of hours of ground school. (Note that for the private pilot's license, the number of hours of ground school is not fixed by the FAA. A few student pilots put in many hours, and a few study on their own and never attend at all.)

Analyzing the Methodology. This is post-facto research. Although the pilots did experience different conditions (attending ground school or not), this was their choice, not the choice of the experimenter. Also, IQ can never be a manipulated variable. Thus, the two independent variables (ground school and IQ) were subject variables, not manipulated.

Answering the Checklist Questions: The Critical Decisions

1. *Scale of measurement?* The three measures (hours of ground school, IQ, and scores on the FAA exam) all yielded at least interval scores. The distributions all appear to be normal. Use Figure 19.3.

2. *Hypothesis?* The researcher has attempted to test for associations among the three measures.

3. *If the hypothesis of difference is tested, are the groups independent or correlated samples?* Not applicable; the hypothesis of association is being tested.

4. *How many sets of measurements?* There are three sets—hours of ground school, FAA examination scores, and IQ.

Solution. Use the multiple R. The three separate values of the Pearson r (between ground school and the exam scores, between IQ and the exam scores, and between IQ and ground school) should all be used together to determine whether their combinations increase the predictive efficiency. If the value of the multiple R is larger than the separate correlations with the exam scores, solve the multiple regression equation.

Simulation P

A firearms manufacturer hired a researcher to establish whether a new handgun increases accuracy. A group of law enforcement agents was randomly selected and brought to the firing range. First, all subjects used the same traditional service revolver, and their error scores (in inches from the bull's-eye) were determined. Then, they all fired again, using the new weapon, and their error scores were again determined.

Analyzing the Methodology. This is experimental research; the independent variable (type of weapon) was manipulated by the experimenter. Since the same subjects are used in both treatment conditions, the design is repeated-measures (before-after). This is not the best design for this study because it is possible that scores might improve the second time as a result of practice. This could act to confound the independent variable. It would have been better to set up a separate group that used the old weapon twice, another group that used the new weapon first and then the old weapon, and another group that used the new weapon twice.

Answering the Checklist Questions: The Critical Decisions

1. *Scale of measurement?* The dependent variable (error measured on the basis of inches from the bull's-eye) provides at least interval data. The researcher claimed a normal distribution for these error scores. Use Figure 19.3. Had the distribution not been normal, a different statistical test should have been used. (See Solution.)

2. *Hypothesis?* As in all experimental research, the hypothesis of difference (between error scores) was tested.

3. *If the hypothesis of difference is tested, are the groups independent or correlated samples?* The groups are correlated; the same group is used as its own control.

4. *How many sets of measurements?* There are two sets, one for each treatment condition.

Solution. Use the paired t ratio, probably as a one-tail test since the manufacturer undoubtedly has some reason for believing in the superiority of the new weapon. (If the results had gone the other way, they possibly would have been filed away in the back of a drawer.) Had the distribution of error scores been skewed (many officers hitting the bull's-eye, but a few missing the target altogether), then the scores should have been rank-ordered and the Wilcoxon T test performed.

Simulation Q

An investigator wished to establish whether a dependable relationship exists between height at age 3 and height at age 21. A random sample of 3-year-olds was selected, and height measures were taken on each. The researcher then patiently waited 18 years and measured the subjects again.

Analyzing the Methodology. This is post-facto research; the independent variable (height at age 3) is a subject variable, not manipulated. This is also called longitudinal research, since the same subjects are followed through the years and are used again. (A less patient researcher could have obtained adult heights and then checked personal records for the infant heights.)

Answering the Checklist Questions: The Critical Decisions

1. *Scale of measurement?* The data are in at least interval form, and the distributions for each age level are probably normal. Use Figure 19.3.

2. *Hypothesis?* This is strictly the hypothesis of association. (Testing the hypothesis of difference in this situation—that is, that 21-year-olds are significantly taller than 3-year-olds—would hardly add much to the book of knowledge.)

3. *If the hypothesis of difference is tested, are the groups independent or correlated samples?* Not applicable; the hypothesis of association is being tested.

4. *How many sets of measurements?* There are two sets of measurements, one taken at age 3 and the other at age 21.

Solution. Use the Pearson r. If it is significant, set up the regression equation of Y on X. Thus predictions of adult height can be made from height at age 3.

Simulation R

Some researchers have suspected that because of academic and other frustrations, adolescents labeled as LD (learning disabled) would have more symptoms of depression and even possibly higher levels of suicidal ideation than would non-LD adolescents. Two groups of 16-year-old students, one labeled LD and the other non-LD (50 male adolescents in each group), were selected on the basis of a certain school district's records. All students were then given the SIQ-JR (Reynolds Suicide Ideational Questionnaire), and the results were as follows:

LD Range (0 to 90)		Non-LD Range (0 to 90)	
M	= 15.21	*M*	= 12.33
SD	= 17.32	*SD*	= 16.28
Mdn	= 7.92	Mdn	= 6.34
Sk	= +2.40	Sk	= +2.45

Analyzing the Methodology. This is post-facto research, as the IV in this study, LD versus non-LD, is clearly a subject variable. If the results prove to be significant, predictions, but not direct cause and effect, become more viable.

Answering the Checklist Questions: The Critical Decisions

1. *Scale of measurement?* The measurements are in interval form, but since the mean is so much higher than the median and the standard deviation is large, relative to the mean, the distributions are significantly skewed to the right. Notice that with means of only 15 and 12 and *no negative* scores (the ranges were 0 to 90), the distribution must be skewed to the high side. If these distributions were to approach normality, the range of scores would have to have been from about –37 to +61 for the non-LD and –37 to +67 for the LD (see Chapter 3). With severe skews of this sort, reported as +2.40 and +2.45, respectively, the interval data should be converted to ordinal.

2. *Hypothesis?* The researcher is testing for differences in suicidal ideation between the two groups.

3. *If the hypothesis of difference is tested, are the groups independent or correlated samples?* The groups are independent.

4. *How many sets of measurements?* The researcher was comparing two sets of measurements.

Solution. Use the Mann–Whitney U test for detecting differences between two sets of ordinal scores.

Simulation S

A cultural anthropologist became interested in discovering whether differences in the age of menarche (the age when young women have their first menstrual cycle) are a function of climate. Two groups of young women were selected—one from a northern climate (Norway) and one from a southern climate (Italy). The subjects were matched according to both height and weight, and their ages at menarche were compared. The age distribution was found to be skewed.

Analyzing the Methodology. This is post-facto research; the independent variable (climate) was a subject variable, not manipulated. Thus, even if significance is established, no positive causal statement can be made. The subjects obviously differ on a host of variables (diet, genetic background, medical care, etc.) other than climate.

Answering the Checklist Questions: The Critical Decisions

1. *Scale of measurement?* Although age is at least an interval measure, the lack of normality in the underlying distribution forces a conversion of the age scores into ordinal data. Use Figure 19.2.

2. *Hypothesis?* The researcher is testing the hypothesis of difference—that age at menarche *differs* as a function of climate.

3. *If the hypothesis of difference is tested, are the groups independent or correlated samples?* The groups are correlated, having been matched on both height and weight.

4. *How many sets of measurements?* There are two sets of measurements, one taken in Norway and the other in Italy.

Solution. Use the Wilcoxon T test.

Simulation T

A researcher wanted to find out whether IQ is a function of family size. The speculation was that among families with fewer children, each child receives more parental attention and intellectual stimulation and should therefore have a higher IQ than would a child reared in a larger family. A large random sample of two-child families was selected as well as a similar sample of six-child families. The IQs of all children were measured, and the two sample groups were compared.

Analyzing the Methodology. This is post-facto research. The independent variable (family size) was a subject variable, not manipulated. (Natural forces or their own decision, not the decision of the experimenter, determined which families had small or large numbers of children.) Thus, even if significance is established, the causal factor remains in the realm of speculation. Could it be, instead, that lower-IQ parents have more children?

Answering the Checklist Questions: The Critical Decisions

1. *Scale of measurement?* IQ scores are considered to be interval measures, and the underlying distribution to be fairly normal. Use Figure 19.3.

2. *Hypothesis?* The researcher is looking for IQ *differences* among children from small and large families.

3. *If the hypothesis of difference is tested, are the groups independent or correlated samples?* The sample groups are independent. The selection of a given family depended on its size, not on whether or not some other family had been selected.

4. *How many sets of measurements?* There are two sets of IQ scores, one taken from large families and one taken from small families.

Solution. Use the independent t. If the score for each child is to be used separately, use the equation for unequal values of N (there are three times as many IQ scores in the six-child families). If the children's IQ scores are to be averaged within each family, then equal values of N can be maintained.

Simulation U

A rat study was conducted to find out whether environmental inputs could affect brain growth. A group of 16 laboratory rat twins was selected at birth and randomly assigned to the two groups. The rats were thus paired off on the basis of genetic inputs, one rat from each twin-pair in the experimental group and the other twin in the control group. The experimental group was then raised in a stimulating environment, in a cage equipped with ladders, running wheels, and other "rat toys." These animals were also let out of their cages for 30 minutes each day and allowed to explore new territory. They were also trained on a number of learning tasks and in general received a rich and varied array of stimulus inputs. The other group of rats was raised in a condition of extreme stimulus homogeneity. These rats lived alone in dimly lit cages, were rarely handled, and were never allowed to explore areas outside their cages. All animals received exactly the same diet. After 90 days all the animals were sacrificed and their brains analyzed morphologically and chemically. The weight of each rat's cortex was then recorded in milligrams, and the two groups were compared on the basis of cortical weight.

Analyzing the Methodology. This is experimental research, with the independent variable, environmental inputs, manipulated by the researcher. If the results prove to be significant, cause-and-effect statements become possible.

Answering the Checklist Questions: The Critical Decisions

1. *Scale of measurement?* The data are in at least interval form.

2. *Hypothesis?* The researcher is testing the hypothesis of difference.

3. *If the hypothesis of difference is tested, are the groups independent or correlated samples?* The groups were deliberately correlated by the researcher, since all the rats were paired off on the basis of twinship.

4. *How many sets of measurements?* There were two sets of measurements, one set from each group.

Solution. Use the paired t.

Simulation V

A political analyst attempted to find out whether the political slant of a newspaper affects the voting preference of its readers. In a large eastern city, a random sample of homes was selected where Newspaper L (Liberal) was delivered. Also a random sample of homes receiving Newspaper C (Conservative) was selected. The voting preference of each head of household was obtained and categorized as Republican, Democrat, or Other.

Analyzing the Methodology. This is post-facto research. The subjects themselves chose which newspaper (independent variable) to have delivered. If significant results are obtained, does it mean the newspaper affected voting preference, or was voting preference the key to which newspaper was ordered?

Answering the Checklist Questions: The Critical Decisions

1. *Scale of measurement?* This is the nominal case. The measures are in the form of *how many* persons subscribe to which newspaper and *how many* persons vote in which category. Frequency of occurrence within mutually exclusive categories defines the nominal case. Use Figure 19.1.

2. *Hypothesis?* The researcher is interested in differences in voting preference.

3. *If the hypothesis of difference is tested, are the groups independent or correlated samples?* The groups are independent of one another. The choice of which group a subject was placed in was determined by the newspaper being delivered, not by the group another subject was placed in.

4. *How many sets of measurements?* There are three sets based on voting preference.

Solution. Use chi square set up as a 2×3.

		Voting Preference		
		Republican	Democrat	Other
Newspaper	Liberal	a	b	c
	Conservative	d	e	f

N

Simulation W

An investigator tried to shed light on the hypothesis that perception shapes attitudes. A large random sample was selected, and each subject was then randomly assigned to one of three groups. Each group then heard an identical speech, given by the same speaker, on the topic of land reform in Cuba. To Group A, the speaker was introduced as a prominent political scientist; to Group B, as a member of the U.S. State Department; and to Group C, as a member of the Cuban delegation to the UN. After the speech, the subjects all took an "Attitude Toward Cuba" test. The distribution of test scores was normal.

Analyzing the Methodology. This is experimental research. The independent variable (perception of the speaker) was manipulated by the experimenter. Other factors, such as content of speech personality of speaker, and were not allowed to vary. If significance is obtained, a causal factor may be isolated.

Answering the Checklist Questions: The Critical Decisions

1. *Scale of measurement?* Attitude test scores provide interval data. As has been stated, the scores distribute normally. Use Figure 19.3.

2. *Hypothesis?* The researcher is testing the hypothesis of difference—that different attitudes will result from different perceptions.

3. *If the hypothesis of difference is tested, are the groups independent or correlated samples?* The three sample groups are independent of one another.

4. *How many sets of measurements?* There are three sets, one from each treatment condition.

Solution. Use the one-way ANOVA, the F ratio. If F is significant, follow it up with Tukey's HSD.

Simulation X

A researcher for the Registry of Motor Vehicles became interested in whether recidivism among convicted drunken drivers is affected by the judicial outcome. A random sample of convicted drunken drivers was selected. The drivers were then randomly divided into two groups. The members of Group A received heavy fines and temporarily lost their licenses. The members of Group B were not fined but were placed in a six-month rehabilitation program and temporarily lost their licenses. Two years later, all subjects were checked for repeat convictions. The groups were compared on the basis of the number of subjects in each group found to have and not have repeat convictions.

Analyzing the Methodology. This is experimental research, between-subjects (after-only). The independent variable (whether subjects go into the rehabilitation program) is manipulated by the researcher. The potential for uncovering a causal relationship is thus available.

Answering the Checklist Questions: The Critical Decisions

1. *Scale of measurement?* The dependent variable (repeat convictions versus no repeat convictions) is measured in nominal form. Use Figure 19.1.

2. *Hypothesis?* The researcher is testing the hypothesis of difference—that new conviction ratios differ as a function of which group the subjects were in.

3. *If the hypothesis of difference is tested, are the groups independent or correlated samples?* The groups are independent, the subjects having been randomly assigned to the two groups.

4. *How many sets of measurements?* There are two sets based on whether the subjects participated in rehabilitation programs.

Solution. Use chi square. This is a 2×2 chi square.

Simulation Y

A sports physiologist was interested in whether extended periods of jogging reduce the resting pulse rate. A random sample of accounting majors (none of whom had ever been involved in jogging) was selected from the senior class at a large university. Resting pulse rates were taken on every subject at the following intervals: first, before starting the jogging program; second, after 4 weeks of the program; third, after 8 weeks; and, finally, after 12 weeks.

Analyzing the Methodology. This is experimental research; the independent variable (amount of time spent jogging) was manipulated by the experimenter. Significant results of this study would pave the way for a cause-and-effect conclusion.

Answering the Checklist Questions: The Critical Decisions

1. *Scale of measurement?* The dependent variable (resting pulse rate) provides normally distributed data, which are at least interval. Use Figure 19.3.

2. *Hypothesis?* The researcher is testing the hypothesis of difference, as is always the case in experimental research.

3. *If the hypothesis of difference is tested, are the groups independent or correlated samples?* The groups are correlated; the same subjects were used in each treatment condition.

4. *How many sets of measurements?* There are four sets of measurements of pulse rate.

Solution. Use the within-subjects F ratio for repeated measures.

Simulation Z

An investigator wished to test the hypothesis that reading speed is a function of how extensively a student reads. A random sample of high school seniors was selected in September, and the subjects were asked how many books they had read during the summer. The subjects were then categorized in the following groups: group 1, no books read; group 2, one to three books; group 3, four to six books; and group 4, more than six books. Reading speed tests were then administered, the scores being in the form of words per minute.

Analyzing the Methodology. This is post-facto research; the independent variable (number of books read) was not manipulated. The subjects were *assigned* to groups on the basis of how many books they themselves had chosen to read. A unidirectional interpretation of this study will therefore be impossible. Does extensive reading increase reading speed, or do fast readers prefer to read more? Or is it a combination of the two? One can never know from this study.

Answering the Checklist Questions: The Critical Decisions

1. *Scale of measurement?* The dependent variable (reading speed in words per minute) provides at least interval data. The distribution is close enough to normality to use interval tests. Use Figure 19.3.

2. *Hypothesis?* The researcher is looking for *differences* in reading speed among the subjects.

3. *If the hypothesis of difference is tested, are the groups independent or correlated samples?* The groups are independent of one another. The placement of a subject in a given group depends on the number of books read, not on the placement of some other subject.

4. *How many sets of measurements?* There are four sets of measures of reading speed.

Solution. Use the one-way ANOVA, the F ratio. If F is significant, proceed with Tukey's HSD.

THE RESEARCH ENTERPRISE

Now that we have met the research simulations from A to Z, choosing the correct test for analysis of the data from 26 studies, we need to take time for a little reflection on the whole idea of statistical research. While confronting each research situation, we were involved in the case study learning approach; that is, these specific examples should provide the fodder for some more global generalizations. There are three especially important general characteristics that must be highlighted here.

Statistical Research Is Empirical. In each research problem, measurements were taken, and the data from these measures were analyzed for possible significance. Measurements always imply *observation*, and observation is the key to the empirical approach. Whenever knowledge comes to us through the direct observation of the world around us (as opposed to knowledge obtained only through the powers of reason—without direct, sensory experience), we are using empirical methods. Statistical research must therefore always be empirical. The research enterprise demands good, solid, specific, empirical observations. No statistical analysis, no matter how elegant and sophisticated, can compensate for bad data. Remember, GIGO (garbage in, garbage out).

Statistical Research Is Inductive. The logical technique of *induction*—that is, arguing from the specific to the general—is absolutely crucial to statistical research. The empirical measures mentioned must be *specific*. Each separate measure must be individually observed. The statistical analysis then determines whether these specific sample measures can be generalized to the population. Philosophers call this the *inductive* *leap,* and although, as we have seen, statistical generalizations may be erroneous, at least their probability of error can be assessed. This is really the main thrust of statistical analysis—determining the probability of error of the induced generalizations.

Statistical Research Should Be Interpretable. Once the data are gathered and the appropriate statistical test used for the data analysis, the results should be open to clear, unequivocal interpretation. It is of little use to do an elaborate study involving enormous amounts of empirical data, only to find that the resulting generalizations are laden with ambiguities. This is like saying "the operation was a success, but the doctor died." Ambiguities are almost certain to arise unless great care is taken in structuring the logic and design of the research. No clear statistical analysis is possible, for example, if the measures taken on some of the subjects are correlated and the measures taken on other subjects are independent. Neither can unequivocal interpretations of experimental data be obtained if the independent variable has been confounded.

A FINAL THOUGHT: THE BURDEN OF PROOF

In the world of science, certain ground rules have been long established. One of the most important of these is that the burden of proof is on the innovator. Whenever a new hypothesis is introduced, it is up to the person making the introduction to give accompanying proof. It is not left to the scientific community to disprove it. If an "innovator" were to come along and loudly proclaim that there are indeed unicorns living in some remote region of Nepal, it would be the duty of that individual to produce the evidence. It is not enough for the proponent of the theory simply to offer the challenge: "Prove me wrong." Other scientists are not expected to drop their laptops and test tubes and hurry off to Nepal to disprove a new unicorn theory. Even if this were done, and many years later the now wizened scientists returned empty-handed, the innovator could smugly reply, "You didn't look behind the right rock." Just as in a court of law the burden of proof is on the prosecutor, so, too, in the "court" of scientific inquiry, the person making the charge must come up with the proof.

It has been a long road from means and medians to factorial ANOVAs and multiple regressions. However, the road has been well lit and logically straight. You have come a long way—you can read research studies, and you can do some research. Of course, there are more advanced topics to be covered before you can consider doing a senior research project, but, just as certainly, these topics will not be as forbidding as they once may have seemed. You may not yet have all the answers, but you are now in a position to know what many of the questions are.

The Binomial Case

This special unit will cover the basic elements of the binomial distribution and how it may be used in the various research situations. The focus will still be on the practical applications of the binomial case and will not suddenly involve you in a heavy course in higher mathematical theory. The math will be kept as absolutely simple as possible. You and your trusty calculator will find that solving binomial problems will be no more painful than doing z *scores or a Pearson* r.

This unit is divided into three sections, with problems presented at the end of each of these sections. (Answers to all the problems, not just the odd items, are given.) In Section 1 we cover the ABCs of binomial probability and the relationship between the binomial distribution and the standard normal curve. Section 2 involves hypothesis testing, using the z *test to evaluate differences between sample proportions and known population proportions. In Section 3 we again look at the testing of hypotheses, but this time the spotlight will be on the difference between one sample proportion when it is being pitted against the proportion occurring in a second sample. The unit concludes with a brief look at the relationship between the binomial test and the chi square test.*

The unit may be undertaken in its entirety if you have already read Chapter 6 on probability, Chapter 7 on the z *test, and Chapter 13 on chi square. If not, the sections may be treated separately and read as each of the appropriate text chapters has been completed: that is, read Section 1 after Chapter 6, Section 2 after Chapter 7, and Section 3 after Chapter 13.*

In any case, rest assured that some understanding of the binomial case will definitely increase your overall appreciation of probability theory, hypothesis testing, and perhaps even, if you're so inclined, general skill in the "games of chance."

SECTION ONE

Binomial Probability

When an event has only two possible outcomes on each of a number of independent occasions, it is called *binomial* (literally, "two names" or "two terms"). The coin-flipping problems discussed in Chapter 6 are examples of binomial problems. You may find that a more detailed look at these kinds of problems, however, is not only worthwhile in its own right but may also make the normal probability curve more understandable.

When tossing a coin, there are obviously only two possible outcomes, a head (H) or a tail (T), and, as we have seen, each of these outcomes has a probability of .50. Also, as was shown, if we want the probability of tossing two coins and getting both to turn up heads, we use the MULT-AND rule and multiply the separate probabilities.

With two coins, $p = .50 \times .50 = .50^2 = .25$

With three coins, $.50 \times .50 \times .50 = .50^3 = .125$

Now, although this procedure works perfectly well when asking for the probability of *all heads* or *all tails,* it won't directly handle the situation of finding the probability of getting, say, 2 heads out of 5 tosses. In this type of problem we can't simply multiply the separate probabilities because now we have *more* tosses than hits. (A "hit" is a success, or when you call the outcome, say a head, and then actually get that outcome, that is, the head turns up.) For problems like these we use the following formula, where n is equal to the number of coins, r is equal to the number of hits, and ! stands for "factorial." Factorial means successively multiplying a given number by one less than the preceding number until you get to 1. Hence, $5! = 5 \times 4 \times 3 \times 2 \times 1 = 120$. Also, by convention, $0! = 1$.

$$p = \frac{n!}{(r!)(n-r)!} p^r q^{n-r}$$

where p is the probability of the event occurring or, as we have called it, the probability of a "hit," and q is the probability of the event not occurring, or the probability of a "miss." Therefore, q is always equal to $p - 1$.

In the coin-tossing example, p must always equal .50, and q, which is $1 - p$, must also be equal to .50. Thus, $p = q = .50$, and the equation can be written as

$$p = \frac{n!}{(r!)(n-r)!} (.50)^r (.50)^{n-r}$$

This formula works just as well when assessing the probability of *all* heads out of n tosses as it does in the situation where we have more coins than hits. For example, to get the probability of *all* heads with 3 coins, which as we saw was .125, this formula gives us

$$p = \frac{3!}{(3!)(0!)}(.50)^3(.50)^0$$

$$= \frac{6}{(6)(1)}(.125)(1) = .125$$

$\uparrow$–(any value taken to the 0 power = 1)

Or, to get the probability of getting exactly 2 heads out of 3 coins,

$$p = \frac{3!}{(2!)(1!)}(.50)^2(.50)^1$$

$$= \frac{6}{(2)(1)}(.25)(.50) = .375$$

The same analysis can be made on data obtained from a true-false exam, in which the number of "true" items is equal to the number of "false" items. If we assume there are 8 questions, then, on the basis of chance, the probability of selecting 5 "true" items would be

$$p = \frac{8!}{(5!)(3!)}(.50)^5(.50)^3$$

$$= \frac{40{,}320}{(120)(6)}(.03125)(.1250) = .2188$$

To appreciate how helpful this formula is, we will now apply it to the seemingly simple situation of including only 4 items, rather than 8. Even here, however, the preceding formula is a blessing, since working out all those separate possibilities can get rather complicated. Assume that you are taking a four-item true-false test, where the possible outcomes total 2^4, or 16. The following shows these possible outcomes:

All possible outcomes (total number of events) for a 4-item true-false test:

	T	T	T	T	F	T	T	T	F	F	F	T	F	F	F	F
	T	F	T	T	T	T	F	F	T	T	F	F	F	F	T	F
	T	T	F	T	T	F	T	F	T	F	T	F	F	T	F	F
	T	T	T	F	T	F	F	T	F	T	T	F	T	F	F	F
Number of true "hits"	4	3	3	3	3	2	2	2	2	2	2	1	1	1	1	0

Number of "Hits"	Specific Outcomes	P-Specific Outcomes/Total Possible Outcomes
4	1	1/16 = .0625
3	4	4/16 = .2500
2	6	6/16 = .3750
1	4	4/16 = .2500
0	1	1/16 = .0625

Or, to find the probability of getting 2 hits, which the table shows as .3750, the use of the preceding formula simplifies the problem to

$$\begin{aligned} p &= \frac{n!}{(r!)(n-r)!}p^r q^{n-r} \\ &= \frac{4!}{(2!)(2!)}(.50)^2(.50)^2 = \frac{24}{4}(.25)(.25) \\ &= 6(.25)(.25) = .3750 \end{aligned}$$

Discrete and Continuous Distributions

The binomial distribution is a discrete distribution. That is, the steps between adjacent values are totally separated. If you flip 3 coins, you can get 0 heads, 1 head, 2 heads, or 3 heads, but you can't get 2½ heads or 1.395 heads. A continuous distribution, on the other hand, yields values that may fall at any point along an unseparated scale of points. Therefore, when fractional values are permissible and have meaning, we are dealing with a continuous measure. Height, for example, is measured on a continuous scale so that a person's height can take on a value of 65 inches, or 65.25 inches, or 65.2514 inches (or you can go out as many decimal places as the accuracy of the measurement technique will allow). Since so many measures are continuous, we tend to round them at some convenient level, like the nearest inch, or nearest degree (for temperature), or nearest dollar (for income). Because the number of decimal places chosen for any continuous measure is, in the final analysis, arbitrary, mathematicians describe each continuous value as falling within an interval of values. If length is measured to the nearest inch, for example, a measure of 72 inches is assumed to be somewhere in an interval ranging from 71.5 inches to 72.5 inches. If, however, length were being measured only to the nearest foot, a value of 6 feet is assumed to lie between 5.5 feet and 6.5 feet. The z distribution is on a continuous scale, whereas the binomial, as shown, is discrete.

Binomial Distribution

Let's look at the resulting distribution when 8 coins are tossed and, as is always the case with fair coins, where $p = q = .50$. The probability of getting a head (in this case a "hit") is

Exact Number of Hits								
0	1	2	3	4	5	6	7	8
p = .0039	.0313	.1094	.2188	.2734	.2188	.1094	.0313	.0039

Notice that the probabilities at the end points, 0 heads and 8 heads, are extremely small, whereas the probability in the middle, 4 heads, is much higher. Thus, the probabilities get increasingly less as you go away from the middle of the distribution in either direction. Using our previous formula, let's prove the probability value for 4 heads (which the table lists as .2734).

$$p = \frac{n!}{(r!)(n-r)!}(.50)^r(.50)^{n-r}$$

$$= \frac{8}{(4!)(4!)}(.50)^4(.50)^4 = \frac{40320}{(24)(24)}(.0625)(.0625)$$

$$= .70(.0625)(.0625) = .2734$$

If we were to display this distribution as a histogram, it would look like this (with the normal distribution superimposed).

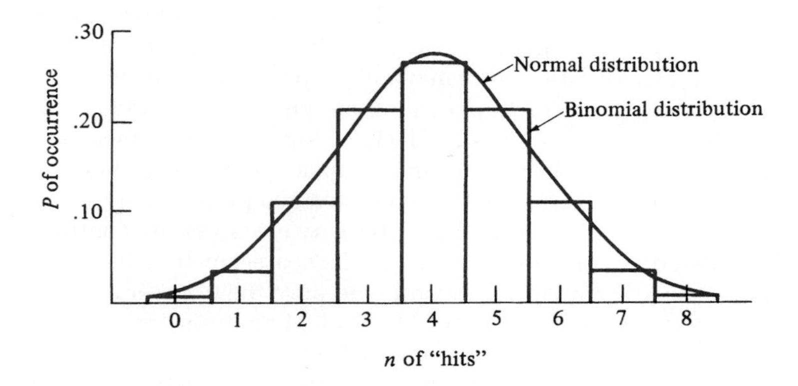

Binomial and Normal Distributions: A Comparison

As you can see, this distribution has the same general shape as the normal distribution. In fact, as n in the binomial increases and the steps between the adjacent bars in the histogram become ever smaller, the binomial distribution, which is discrete, tends to resemble ever more closely the z distribution (which, as you know, is continuous). In fact, the approximation of the binomial to the z is fairly close, even with an n as small as 8, *as long as the p value is .50* (or close to it). As p departs from .50, the approximation is still possible, but it then must take high n values.

The mean and standard deviation of a binomial distribution for a sample of n trials can be calculated on the basis of the following:

$M = np$ (n equals the number of events or "trials" and p the probability of a hit)

Thus, with 8 coins and a p of .50, $M = 4$:

$$M = 8(.50) = 4$$

$SD = \sqrt{npq}$ (q equals $1 - p$, or the probability of a "miss")

With the 8 coins, then, we get a standard deviation of 1.414:

$$SD = \sqrt{(8)(.50)(.50)} = \sqrt{2} = 1.414$$

With the mean and standard deviation known, we can now calculate z scores for the various numbers of "hits," and from these z scores we can use the normal distribution to approximate the exact probabilities given under the binomial distribution. It must again be pointed out, however, that the z distribution is continuous. A continuous value, such as the mean value of 4, is really some value that lies between 3.5 and 4.5, which are its literal limits. Using a value of 4, then, is in effect like saying that 4 is the midpoint of a continuous interval whose real limits are 3.50 and 4.50.

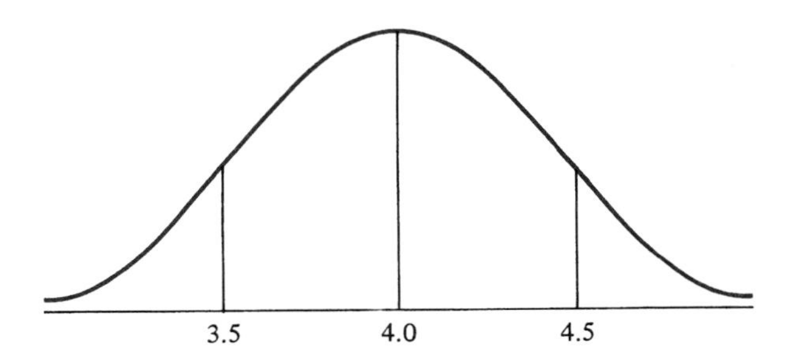

To find the normal distribution's probability of an event falling between the scores 3.50 and 4.50, we first calculate the two z scores.

$$z = \frac{X - M}{SD}$$

$$= \frac{4.50 - 4.00}{1.414} = .35$$

$$= \frac{3.50 - 4.00}{1.414} = -.35$$

Under the normal curve, as shown in Appendix Table A, the percentage values are 13.68% for each of these z scores, and the area between the z scores is 13.68% + 13.68% = 27.36%. Converting to a probability value, we get .2736. As can be seen, this is an extremely close approximation of the probability value of .2734 for the 4 hits shown under the binomial distribution on page 559.

Beware the Hustler: Dice Anyone?

As already stated, when the P value departs from .50, the normal approximation is possible only with a large number of trials. Let's now roll a fair die 180 times. Since each die has six sides, we should expect any specific number—say, a 3—to occur one-sixth of the time (p) and not to occur five-sixths of the time (q). Thus, the expected number of hits is 180(1/6) = 30, and the expected number of misses is 180(5/6) = 150. The standard deviation approximation will then be

$$SD = \sqrt{npq} = \sqrt{180\left(\frac{1}{6}\right)\left(\frac{5}{6}\right)} = 5$$

Let's now examine the distribution formed by *two* dice, and the array formed for each possible result. Since each die has six sides, the total number of combinations with two dice is 6 times 6, or 36. Under the number 2, for example, the 1 indicates that a 2 can be rolled in only *one* way. Notice that a 7 can be rolled in *six* ways. Let's look the table over carefully, and then watch a "hustler" (a cheater) who has spotted a "mark" (someone who is seen as an easy victim).

Number Rolled										
2	3	4	5	6	7	8	9	10	11	12
Ways to Roll										
1	2	3	4	5	6	5	4	3	2	1
Probability Values										
.027	.055	.083	.111	.138	.167	.138	.111	.083	.055	.027

The hustler can then say to the mark, "Even money you won't roll a 6 before a 7, *or* an 8 before a 7. *Which do you want?*" (The term *mark* also refers to a sucker, about whom P. T. Barnum is alleged to have said, "There's one born every minute.") The mark should decline because although the choice appears to offer two positive opportunities (6 or 8) to only one negative opportunity (7), by asking the mark "which do you want?" the choice is cut in half. For the mark, this is a bad bet *in the long run.*

Let's examine the distribution shown in the figure. Notice that there are six ways to roll a 7 and only five ways to roll a 6 or an 8. Thus, the probability of the 7 appearing is 6/36 or .167, which is greater than the probability of either the 6 or the 8, each at 5/36, or .138. The hustler who plays these probabilities long enough has to be a winner. Suppose, however, that the mark declines the offer. The hustler can then say,

"Okay, I'll switch sides and choose either a 6 or an 8 before I roll a 7." Notice that the bet hasn't really been reversed but instead has been *changed* so that *either* a 6 or an 8 becomes a winner, and the hustler isn't even declaring which one has to come first. There are 10 ways to roll either a 6 or an 8, $P = .277$, and only six ways to roll a 7, $P = .167$. Guess how this is going to turn out.

This general scenario has been used by the professional gambler and TV hustler, Harry Anderson (Anderson & Pipkin, 1989). This is the same Harry Anderson who has made a small fortune by spotting a mark in a bar, covering a drink with his hat, and then betting he can down the drink without touching the hat. As soon as the mark takes the bet, Harry immediately squats down under the table and makes a series of ridiculous gurgling sounds. Reappearing, and wiping his mouth in a satisfied manner, he announces to the mark that the drink is gone. As soon as the mark lifts the hat to check the validity of Harry's statement, Harry quickly grabs the drink and quaffs it down. True to his bet, Harry Anderson never touched the hat.

SECTION ONE PROBLEMS

1. Determine the probability of tossing all heads with 5 coins.

   Answer: .03125

2. Find the probability of getting 3 heads with 5 coins.

   Answer: .3125

3. Using z scores, find the normal curve's approximation of the probability of obtaining 7 hits using a total of 8 coins. Compare this answer with the exact probability shown under the binomial distribution.

   Answer: .03 (a very close approximation)

4. On a 10-item true-false test, where the number of true items equals the number of false items, what is the exact probability of getting
   a. 4 true items?
   b. 5 true items?

   (For this problem, use the binomial formula shown on page 560.)

   Answers:
   a. .2047
   b. .2460

5. Using the data from problem 4, use z approximations.
   a. Find the mean of the distribution.
   b. Find the standard deviation of the distribution.
   c. Find the probability of getting 4 true items.
   d. Find the probability of getting 5 true items.
   e. Compare these normal curve approximations with the exact probabilities found in problem 4.

Answers:
a. Mean equals 5.
b. Standard deviation equals 1.581.
c. .2035
d. .2510
e. Approximations are extremely close.

SECTION TWO

z Test and Binomial Proportions

In some studies the researcher is confronted with the problem of comparing the observed proportions that occur in a sample with those true proportions known to occur in the population. For example, it is known that in an infinite population of tosses using an unbiased coin, the true proportion of heads is 1/2, or .50. Suppose, however, that a friend slipped in a biased coin (loaded to produce more than its share of heads), and your job as the statistician is to show the likelihood of such a coin conforming to a chance explanation.

As shown on page 560, the standard deviation of the binomial distribution for n events can be calculated as

$$SD = \sqrt{npq}$$

where n is the number of events and p is the probability of a "hit,"

$$q = p - 1, \qquad \text{or the probability of not getting a "hit"}$$

However, in the case of proportions, where the total population of events can be infinite, the *population* standard deviation is defined as

$$\sigma = \sqrt{pq}$$

We can now use this population standard deviation to calculate the standard error of the entire sampling distribution of proportions, just as we did when calculating the standard error of the mean, by dividing through by the square root of n. This standard error of the proportion will be symbolized as σ_p.

$$\sigma_p = \frac{\sigma}{\sqrt{n}} = \frac{\sqrt{pq}}{\sqrt{n}} = \sqrt{\frac{pq}{n}}$$

Getting back to our friend with the loaded coin, assume that on a run of 20 tosses this coin has produced heads 16 times, for an observed proportion of 16/20 = .80. Although it is known that an unbiased coin produces a population proportion of .50 heads, your friend still insists that this was just a streak of luck and that the coin is perfectly fair.

To test this allegation, first calculate the standard error of the proportion for the true population parameter.

$$\sigma_p = \sqrt{\frac{pq}{n}} = \sqrt{\frac{(.50)(.50)}{n}} = \sqrt{\frac{.25}{20}} = \sqrt{.0125} = .1118$$

Second, find the z score, where p_s is the proportion occurring in the sample, and P_t is the true proportion known to occur in the population.

$$z = \frac{p_s - p_t}{\sigma_p} = \frac{.80 - .50}{.1118} = \frac{.30}{.1118} = 2.6832$$

1. Null hypothesis: H_0: $p_s = p_t$ (where p_s is the hypothesized proportion found in the sample and P_t is the known population proportion)
2. Alternative hypothesis H_a: $p_s - p_t$
3. Critical value of z at .01 = ±2.58
4. Calculated value of z test:

   $z = 2.6832$ Reject H_0.

We thus reject the hypothesis that our friend's coin is representative of the population of unbiased coins, and we conclude instead that the coin is indeed probably "loaded."

Remember, as we pointed out earlier, the binomial approximation of the z distribution is valid only when the number of events is sufficiently large (at least 8) and the p value is close to .50.

SECTION TWO PROBLEMS

1. Assume that the proportion of women and men on the faculty of a large university is equal, $p = .50$ and $q = .50$. However, among the 30 departmental chairpersons, there are only 9 women. Test the hypothesis that bias exists in the selection of chairpersons.

   Answer: $z = -2.1905$, z at .05 = ±1.96. Reject the null hypothesis, and conclude that bias probably does exist.

2. Assume that the proportion of true answers is equal to the proportion of false answers on a 50-item true-false exam in botany. One student, who claims that he hasn't studied for the test for even one minute and in general says he knows nothing about botany, answers 30 of the items correctly. Test the hypothesis that he probably did sneak a look at the text and does know more about botany than he's willing to admit.

   Answer: $z = 1.43$, z at .05 = ±1.96. Accept H_0; the student is apparently telling the truth. He didn't study, and he sure doesn't know anything about botany.

3. Assume that the proportion of boys and girls in a large high school is equal, $p = .50$ and $q = .50$. Among a sample of 16 students elected to serve as class officers, however, there are only 4 girls. Test the hypothesis that bias exists in the election of class officers.

   Answer: $z = -2.00$, z at $.05 = \pm 1.96$. Reject H_0; significant at $p < .05$. It appears that bias does exist.

SECTION THREE

Testing the Difference Between Two Sample Proportions

Just as the binomial z test can assess whether differences exist between those proportions that have been observed in the sample and those that are known to exist in the population, z can also be used to test for differences between two independent *sample* proportions. To use this form of z, it is assumed that the sample sizes must be fairly large and the parameter proportions of p and q not too different from .50. One rule that can be used for testing this assumption is that if both samples have an N of at least 100, then the binomial z is appropriate. If, however, either of the samples has an N of less than 100 and the p value is either greater than .90 or less than .10, then this test should not be used.

For example, a researcher wishes to find out whether there is a difference between Republicans and Democrats regarding their opinions on the statement, "More tax money should be used to support welfare programs." A random sample of 100 Democrats was selected ($N = 100$) and a second random sample of 80 Republicans was also selected ($N = 80$). Of the Democrats, 60 agreed with the statement and 40 disagreed. Of the Republicans, 30 agreed and 50 disagreed.

First, we calculate the two proportions, p_1 and p_2. To obtain these values, simply calculate the proportion in each group who "agree," or $60/100 = .60$ for p_1 (the Democrats) and $30/80 = .375$ for p_2 (the Republicans).

Next, calculate the standard error for each proportion, using the equation

$$\sigma_p = \sqrt{\frac{pq}{n}}$$

For sample 1, the Democrats, this would be

$$\sigma_{p1} = \sqrt{\frac{pq}{n}} = \sqrt{\frac{(.60)(.40)}{100}} = \sqrt{\frac{.24}{100}} = \sqrt{.0024} = .0490$$

And for sample 2, the Republicans, it would be

$$\sigma_{p2} = \sqrt{\frac{pq}{n}} = \sqrt{\frac{(.375)(.625)}{80}} = \sqrt{\frac{.2344}{80}} = \sqrt{.0029} = .0539$$

Both these values are then used to obtain the standard error of the difference between two proportions, or σ_{DP}

$$\sigma_{DP} = \sqrt{\sigma_{p_1}{}^2 + \sigma_{p_2}{}^2} = \sqrt{.0490^2 + .0539^2} = \sqrt{.0024 + .0029}$$

$$= \sqrt{.0053} = .0748$$

Finally, use p_1, the proportion of Democrats who "agreed," .60, and p_2, the proportion of Republicans who "agreed," .375, and calculate z as follows:

$$z = \frac{p_1 - p_2}{\sigma_{DP}} = \frac{.60 - .375}{.0748} = 3.008$$

$$= 3.008$$

1. Null hypothesis H_0: $p_1 = p_2$ (the sample proportions are equal)
2. Alternative hypothesis H_a: $p_1 \neq p_2$ (the sample proportions are not equal)
3. Critical value of z at .01 $= \pm 2.58$
4. Calculated value of z test $= 3.008$ Reject H_0, $p < .01$, and conclude that the samples are significantly different.

Chi Square and Proportions

For those of you who have already completed Chapter 13, you may prefer a chi square analysis of these types of proportion problems. For example, since with one degree of freedom, chi square is equal to z^2, the proportion problems just shown could be done via the chi square.

First, to test whether a sample proportion is consistent with a known population proportion, the problem shown on page 565 produced a significant z of 2.68. Setting these data in a chi square table would result in

	Heads		Tails	
f_o	16		4	
f_e	10		10	← (They should split evenly in the population.)
$f_o - f_e$	6		−6	
$(f_o - f_e)^2$	36		36	
$(f_o - f_e)^2/f_e$	3.60	+	3.60	$= x^2 = 7.20$
				$x^2_{.01(1)} = 6.64$ Reject H_0; significant at $p < .01$.

The z value of 2.68, shown on page 534, when squared will equal chi square ($2.6832^2 = 7.20$).

Or, to test the difference between two independent samples, the problem shown above would be solved by chi square as follows:

	Agree	Disagree	
Democrats	a 60	b 40	100
Republicans	c 30	d 50	80
	90	90	180

	a	*b*	*c*	*d*		
f_o	60	40	30	50		
f_e	50	50	40	40		
$f_o - f_e$	10	–10	–10	10		
$(f_o - f_e)^2$	100	100	100	100		
$(f_o - f_e)^2$						
$(f_o - f_e)^2/f_e$	2.00 +	2.00 +	2.50 +	2.50	= $x^2 = 9.00$	
					$x^2_{.01(1)} = 6.64$	Reject H_0; significant at $p < .01$.

The z value for this problem, shown on page 536, was 3.008, which when squared equals, within rounding errors, the chi square value of 9.00.

SECTION THREE PROBLEMS

1. A political scientist wishes to assess the possible difference between white- and blue-collar workers regarding their preferences for a certain political candidate. A sample of 50 blue-collar workers was randomly selected, and 30 of those said they would vote for the candidate. A random sample of 50 white-collar workers was selected, and 20 said they would vote for the candidate. Do the voting proportions in the two sample groups differ significantly?

   Answer: (For all the answers in this section, the z^2 values will equal chi square *within rounding errors.*) The difference is significant: $z = 2.00$, $z_{.05} = \pm 1.96$. Reject H_0; $p < .05$. Or, as a chi square, $x^2 = 4.00$, $x^2_{.05(1)} = 3.84$. Reject H_0; significant at $p < .05$.

2. A sociologist is interested in discovering whether a difference exists between the sample proportions of men and women autoworkers regarding attitudes toward labor unions. Random samples of 100 men and 100 women were selected from the assembly-line workers at a large auto plant. The workers were asked whether they agreed with the statement, "Only through collective bargaining can a worker expect a fair wage." Of the men, 70 agreed, but of the women, only 60 agreed. Test the hypothesis that men and women differ regarding attitudes toward unions.

   Answer: The difference is not significant: $z = 1.49$, $z_{.05} = \pm 1.96$. Accept H_0. Or, as a chi square, $x^2 = 2.18$, $x^2_{.05(1)} = 3.84$. Accept H_0.

3. Two groups of rats were randomly selected. Group A, consisting of 100 rats, was fed a diet containing glutamic acid. Group B, also consisting of 100 rats, was fed an identical diet but without containing the glutamic acid supplement. Both groups were then given 25 reinforced trials on a certain maze. The results of this experiment are as follows:

	Error Free	*Not Error Free*
Group A	55	45
Group B	40	60

Test the hypothesis that glutamic acid affected maze learning among these rats.

Answer: The difference is significant: $z = 2.14$, $z_{.05} = \pm 1.96$. Reject H_0. Or, as a chi square, $x^2 = 4.51$, $x^2_{.05(1)} = 3.84$. Reject H_0.

4. A market researcher sent a free sample-size box of a certain dishwashing detergent to 200 subjects chosen at random. A separate random sample of 200 subjects was also chosen and received a discount coupon toward the purchase of the product. Six weeks later, all subjects were contacted and data were gathered on the number of subjects from each group who had actually gone out and purchased the product.

	Bought Product	*Did Not Buy Product*
Sampled group	40	160
Coupon group	50	150

Determine whether a significant difference exists between the two groups regarding their purchase of the dishwashing product.

Answer: The difference is not significant: $z = -1.1995$, $z_{.05} = \pm 1.96$. Accept H_0. Or, as a chi square, $x^2 = 1.43$, $x^2_{.05(1)} = 3.84$. Accept H_0.

5. A psychologist is interested in whether there is a gender difference regarding the manner in which children play in the schoolyard. To investigate this question, data were gathered for the previous year on the number of girls versus boys who had broken at least one bone while playing during recess. By checking the records of a small school district, information was provided on a total of 150 girls and 200 boys.

	Broken Bone	*No Broken Bone*
Girls	30	120
Boys	60	140

Answer: The difference is significant: $z = -2.1186$, $z_{.05} = \pm 1.96$. Reject H_0. Or, as a chi square, $x^2 = 4.48$, $x^2_{.05(1)} = 3.84$. Reject H_0.

Appendix A

PART A: MOMENTS OF THE CURVE

1. In any unimodal distribution, the first moment defines the mean, where the average deviations equal zero. A deviation score equals the raw score minus the mean, $x = X - M$.

$$m_1 = \frac{\Sigma x}{N} = 0$$

2. The second moment defines the variance, so that

$$m_2 = \frac{\Sigma x^2}{N}$$

The square root of the variance equals the standard deviation.

$$SD = \sqrt{\frac{\Sigma x^2}{N}}$$

3. The third moment defines skewness, so that

$$m_3 = \frac{\Sigma x^3}{N}$$

As a standard value,

$$\text{Sk} = \frac{m_3}{SD^3}$$

Any standard value greater than ±1.00 indicates marked skewness. (Note: Cubing the deviations allows for Sk to take on negative values.)

4. The fourth moment defines kurtosis, such that

$$m_4 = \frac{\Sigma x^4}{N}$$

As a standard value, which will assess the amount of deviation from mesokurtosis,

$$\text{Ku} = \frac{m_4}{SD^4} - 3.00$$

(Note: This is because $m_4/SD^4 = 3.00$ defines mesokurtosis.)

Leptokurtic distributions will yield a positive standard value, whereas platykurtic distributions produce a negative standard value. Extreme leptokurtic curves produce a positive value greater than +2.00, such as +2.50, and extreme platykurtic curves a negative value greater than –2.00, such as –2.50.

PART B: WRITING UP THE STATISTICAL RESULTS

Whenever you have to prepare and type your results for a term paper, research report, or (it's possible) journal article, the following rules should be observed. Although statistical symbols are written in italics when being typeset for print, the symbols are underlined when typed for a paper. (In fact, for a paper, do not even use the italics function if using a word processor.) The most common symbols are N for total number of subjects, M for sample mean, df for degrees of freedom, SS for sums of squares, SE for standard error, t for t ratio, F for F ratio, a for alpha error, r for Pearson correlation, p for probability (also for the success probability of a binomial variable), z for a standard score, R for the multiple correlation, and SEM for the standard error of measurement.

The t Test

For both the independent and paired t tests, the means, standard deviations and sample sizes for each group should be reported. Use N for the total sample size and n for the number in a limited portion of the sample, such as the number of subjects in the various subgroups.

TABLE 1 Means and Standard Deviations

	M	SD	n
Group 1	50	7.50	50
Group 2	44	8.60	50

Then the results of the t are indicated as follows (with the t followed by the degrees of freedom and then the probability value): The difference between the means was statistically significant: t (98) = 3.68, p < .01.

Suppose, however, that the means had been closer together, such as

TABLE 2 Means and Standard Deviations

	M	SD	n
Group 1	50	7.50	50
Group 2	47	8.60	50

The difference between the means was not statistically significant (ns). $\underline{t}(98) = 1.84$, ns.

In reporting $\underline{t}$ tests, you should also indicate in words whether you used a paired or independent $\underline{t}$ test and either in words or by symbol whether it is a one- or two-tailed $\underline{t}$. The APA publication manual suggests using symbols, such as asterisks, for the two-tailed $\underline{p}$ values and an alternative, such as a dagger, for the one-tailed values.

Pearson $\underline{r}$

Presentations of Pearson $\underline{r}$ values are

$\underline{r}$ (102) = −.96, $\underline{p} < .01$

$\underline{r}$ (10) = .09, ns

Use the same format for the multiple $\underline{R}$ or for the partial correlation.

$\underline{R}$ (93) = .89, $\underline{p} < .01$

Chi Square

Place the data in the contingency table, with the independent variable in the rows and the dependent variable in the columns.

TABLE 3

	Passed	Failed
Group A (was trained)	50	30
Group B (was not trained)	20	40

Then

$\chi^2(1, \underline{N} = 140) = 11.67, \underline{p} < .01$

Or, if the frequencies had been closer,

TABLE 4

	Passed	Failed
Group A (was trained)	50	40
Group B (was not trained)	30	30

Then

$\chi^2(1, N = 150) = .45$, ns

ANOVA (One-Way)

For the one-way ANOVA, the means for each group should be reported and the summary table then includes the following:

TABLE 5

Source of Variation	df	Sums of Squares	Mean Square	F
Between groups	3	54.187	18.062	6.827
Within groups	12	31.750	2.646	
Total	15	85.938		

$F(3,12) = 6.83$, $p < .01$
$M_1 = 6.50$, $M_2 = 7.75$, $M_3 = 8.50$, $M_4 = 11.50$

Or, if the means were closer together, then

TABLE 6

Source of Variation	df	Sums of Squares	Mean Square	F
Between groups	3	10.687	3.563	1.346
Within groups	12	31.750	2.646	
Total	15	42.437		

$F\ (3,12) = 1.35$, ns
$M_1 = 6.50$, $M_2 = 7.75$, $M_3 = 8.50$, $M_4 = 8.50$

Factorial ANOVA

For a two-way ANOVA, indicate the results as follows:

TABLE 7

	Factor A		
Factor B	High Temperature	Low Temperature	Row Means
Exercise	M = 8.33	M = 5.33	M = 6.83
No exercise	M = 3.67	M = 1.83	M = 2.75
Columns means	M = 6.00	M = 3.58	

TABLE 8

Source of Variation	df	Sums of Squares	Mean Square	F
Between rows	1	100.042	100.042	118.861
Between columns	1	35.042	35.042	41.634
Interaction ($r \times c$)	1	2.042	2.042	2.426
Within	20	16.833	0.842	

$F_r(1,20) = 118.86$, $p < .01$
$F_c(1,20) = 41.63$, $p < .01$
$F_{r \times c}(1,20) = 2.43$, ns

PART C: TABLES

TABLE A Percent of area under the normal curve between the mean and *z*.

z	*.00*	*.01*	*.02*	*.03*	*.04*	*.05*	*.06*	*.07*	*.08*	*.09*
0.0	00.00	00.40	00.80	01.20	01.60	01.99	02.39	02.79	03.19	03.59
0.1	03.98	04.38	04.78	05.17	05.57	05.96	06.36	06.75	07.14	07.53
0.2	07.93	08.32	08.71	09.10	09.48	09.87	10.26	10.64	11.03	11.41
0.3	11.79	12.17	12.55	12.93	13.31	13.68	14.06	14.43	14.80	15.17
0.4	15.54	15.91	16.28	16.64	17.00	17.36	17.72	18.08	18.44	18.79
0.5	19.15	19.50	19.85	20.19	20.54	20.88	21.23	21.57	21.90	22.24
0.6	22.57	22.91	23.24	23.57	23.89	24.22	24.54	24.86	25.17	25.49
0.7	25.80	26.11	26.42	26.73	27.04	27.34	27.64	27.94	28.23	28.52
0.8	28.81	29.10	29.39	29.67	29.95	30.23	30.51	30.78	31.06	31.33
0.9	31.59	31.86	32.12	32.38	32.64	32.90	33.15	33.40	33.65	33.89
1.0	34.13	34.38	34.61	34.85	35.08	35.31	35.54	35.77	35.99	36.21
1.1	36.43	36.65	36.86	37.08	37.29	37.49	37.70	37.90	38.10	38.30
1.2	38.49	38.69	38.88	39.07	39.25	39.44	39.62	39.80	39.97	40.15
1.3	40.32	40.49	40.66	40.82	40.99	41.15	41.31	41.47	41.62	41.77
1.4	41.92	42.07	42.22	42.36	42.51	42.65	42.79	42.92	43.06	43.19
1.5	43.32	43.45	43.57	43.70	43.83	43.94	44.06	44.18	44.29	44.41
1.6	44.52	44.63	44.74	44.84	44.95	45.05	45.15	45.25	45.35	45.45
1.7	45.54	45.64	45.73	45.82	45.91	45.99	46.08	46.16	46.25	46.33
1.8	46.41	46.49	46.56	46.64	46.71	46.78	46.86	46.93	46.99	47.06
1.9	47.13	47.19	47.26	47.32	47.38	47.44	47.50	47.56	47.61	47.67
2.0	47.72	47.78	47.83	47.88	47.93	47.98	48.03	48.08	48.12	48.17
2.1	48.21	48.26	48.30	48.34	48.38	48.42	48.46	48.50	48.54	48.57
2.2	48.61	48.64	48.68	48.71	48.75	48.78	48.81	48.84	48.87	48.90
2.3	48.93	48.96	48.98	49.01	49.04	49.06	49.09	49.11	49.13	49.16
2.4	49.18	49.20	49.22	49.25	49.27	49.29	49.31	49.32	49.34	49.36
2.5	49.38	49.40	49.41	49.43	49.45	49.46	49.48	49.49	49.51	49.52
2.6	49.53	49.55	49.56	49.57	49.59	49.60	49.61	49.62	49.63	49.64
2.7	49.65	49.66	49.67	49.68	49.69	49.70	49.71	49.72	49.73	49.74
2.8	49.74	49.75	49.76	49.77	49.77	49.78	49.79	49.79	49.80	49.81
2.9	49.81	49.82	49.82	49.83	49.84	49.84	49.85	49.85	49.86	49.86
3.0	49.87									
4.0	49.997									

Source: Karl Pearson, *Tables for Statisticians and Biometricians,* Cambridge University Press, London, pp. 98–101, by permission of the Biometrika Trustees.

TABLE B Conversion table—percentiles to *z* scores.

Percentile	*z Score*	*Percentile*	*z Score*	*Percentile*	*z Score*	*Percentile*	*z Score*
1st	–2.41	26th	–0.64	51st	0.03	76th	0.71
2nd	–2.05	27th	–0.61	52nd	0.05	77th	0.74
3rd	–1.88	28th	–0.58	53rd	0.08	78th	0.77
4th	–1.75	29th	–0.55	54th	0.10	79th	0.81
5th	–1.65	30th	–0.52	55th	0.13	80th	0.84
6th	–1.56	31st	–0.50	56th	0.15	81st	0.88
7th	–1.48	32nd	–0.47	57th	0.18	82nd	0.92
8th	–1.41	33rd	–0.44	58th	0.20	83rd	0.95
9th	–1.34	34th	–0.41	59th	0.23	84th	1.00
10th	–1.28	35th	–0.39	60th	0.25	85th	1.04
11th	–1.23	36th	–0.36	61st	0.28	86th	1.08
12th	–1.18	37th	–0.33	62nd	0.31	87th	1.13
13th	–1.13	38th	–0.31	63rd	0.33	88th	1.18
14th	–1.08	39th	–0.28	64th	0.36	89th	1.23
15th	–1.04	40th	–0.25	65th	0.39	90th	1.28
16th	–1.00	41st	–0.23	66th	0.41	91st	1.34
17th	–0.95	42nd	–0.20	67th	0.44	92nd	1.41
18th	–0.92	43rd	–0.18	68th	0.47	93rd	1.48
19th	–0.88	44th	–0.15	69th	0.50	94th	1.56
20th	–0.84	45th	–0.13	70th	0.52	95th	1.65
21st	–0.81	46th	–0.10	71st	0.55	96th	1.75
22nd	–0.77	47th	–0.08	72nd	0.58	97th	1.88
23rd	–0.74	48th	–0.05	73rd	0.61	98th	2.05
24th	–0.71	49th	–0.03	74th	0.64	99th	2.41
25th	–0.67	50th	0.00	75th	0.67	100th	∞

TABLE C Critical values of *t*.

	Level of Significance for Two-Tail Test				
df	*.05*	*.01*	df	*.05*	*.01*
1	12.706	63.657	28	2.048	2.763
2	4.303	9.925	29	2.045	2.756
3	3.182	5.841	30	2.042	2.750
4	2.776	4.604	31	2.040	2.744
5	2.571	4.032	32	2.037	2.738
6	2.447	3.707	33	2.034	2.733
7	2.365	3.499	34	2.032	2.728
8	2.306	3.355	35	2.030	2.724
9	2.262	3.250	36	2.028	2.720
10	2.228	3.169	37	2.026	2.715
11	2.201	3.106	38	2.024	2.712
12	2.179	3.055	39	2.023	2.708
13	2.160	3.012	40	2.021	2.704
14	2.145	2.977	45	2.014	2.690
15	2.131	2.947	50	2.009	2.678
16	2.120	2.921	55	2.004	2.668
17	2.110	2.898	60	2.000	2.660
18	2.101	2.878	70	1.994	2.648
19	2.093	2.861	80	1.990	2.639
20	2.086	2.845	90	1.987	2.632
21	2.080	2.831	100	1.984	2.626
22	2.074	2.819	120	1.980	2.617
23	2.069	2.807	200	1.972	2.601
24	2.064	2.797	500	1.965	2.586
25	2.060	2.787	1000	1.962	2.581
26	2.056	2.779	∞	1.960	2.576
27	2.052	2.771			

TABLE D Critical values of *t*.

	Level of Significance for One-Tail Test				
df	***.05***	***.01***	**df**	***.05***	***.01***
1	6.314	31.821	18	1.734	2.552
2	2.920	6.965	19	1.729	2.539
3	2.353	4.541	20	1.725	2.528
4	2.132	3.747	21	1.721	2.518
5	2.015	3.365	22	1.717	2.508
6	1.943	3.143	23	1.714	2.500
7	1.895	2.998	24	1.711	2.492
8	1.860	2.896	25	1.708	2.485
9	1.833	2.821	26	1.706	2.479
10	1.812	2.764	27	1.703	2.473
11	1.796	2.718	28	1.701	2.467
12	1.782	2.681	29	1.699	2.462
13	1.771	2.650	30	1.697	2.457
14	1.761	2.624	40	1.684	2.423
15	1.753	2.602	60	1.671	2.390
16	1.746	2.583	120	1.658	2.358
17	1.740	2.567	∞	1.645	2.326

Source: Abridged from Table III of Fisher and Yates, *Statistical Tables for Biological, Agricultural, and Medical Research,* Longman Group Ltd., London (previously published by Oliver and Boyd Ltd., Edinburgh), and by permission of the authors and publishers.

TABLE E Critical values of r for the Pearson correlation coefficient (degrees of freedom = number of pairs of scores – 2).

df	$\alpha = .05$	$\alpha = .01$	*df*	$\alpha = .05$	$\alpha = .01$
1	0.997	0.999	32	0.339	0.436
2	0.950	0.990	34	0.329	0.424
3	0.878	0.959	35	0.325	0.418
4	0.811	0.917	36	0.320	0.413
5	0.754	0.874	38	0.312	0.403
6	0.707	0.834	40	0.304	0.393
7	0.666	0.798	42	0.297	0.384
8	0.632	0.765	44	0.291	0.376
9	0.602	0.735	45	0.288	0.372
10	0.576	0.708	46	0.284	0.368
11	0.553	0.684	48	0.279	0.361
12	0.532	0.661	50	0.273	0.354
13	0.514	0.641	55	0.261	0.338
14	0.497	0.623	60	0.250	0.325
15	0.482	0.606	65	0.241	0.313
16	0.468	0.590	70	0.232	0.302
17	0.456	0.575	75	0.224	0.292
18	0.444	0.561	80	0.217	0.283
19	0.433	0.549	85	0.211	0.275
20	0.423	0.537	90	0.205	0.267
21	0.413	0.526	95	0.200	0.260
22	0.404	0.515	100	0.195	0.254
23	0.396	0.505	125	0.174	0.228
24	0.388	0.496	150	0.159	0.208
25	0.381	0.487	175	0.148	0.193
26	0.374	0.479	200	0.138	0.181
27	0.367	0.471	300	0.113	0.148
28	0.361	0.463	400	0.098	0.128
29	0.355	0.456	500	0.088	0.115
30	0.349	0.449	1000	0.062	0.081

If your calculated r is equal to or greater than table r, reject H_0. If your value of degrees of freedom is not listed, use table r for the next smaller value of degrees of freedom. (See Chapter 11.)

Source: Table VI of Fisher and Yates, *Statistical Tables for Biological, Agricultural, and Medical Research,* 6th ed., 1974, Longman Group Ltd., London (previously published by Oliver and Boyd Ltd., Edinburgh), and by permission of the authors and publishers.

TABLE F Critical values for the Spearman rank-order correlation coefficient (N = number of pairs of scores).

N	*.05*	*.01*	N	*.05*	*.01*
5	1.000		18	.474	.600
6	.886	1.000	19	.460	.585
7	.786	.929	20	.447	.570
8	.715	.881	21	.437	.556
9	.700	.834	22	.426	.544
10	.649	.794	23	.417	.532
11	.619	.764	24	.407	.521
12	.588	.735	25	.399	.511
13	.561	.704	26	.391	.501
14	.539	.680	27	.383	.493
15	.522	.658	28	.376	.484
16	.503	.636	29	.369	.475
17	.488	.618	30	.363	.467

Source: Glasser, G. J., and R. F. Winter, "Critical Values of the Coefficient of Rank Correlation for Testing the Hypothesis of Independence," *Biometrika,* 48, 444 (1961).

TABLE G Critical values of F for the analysis of variance.

Degrees of Freedom for Denominator	*Degrees of Freedom for Numerator*											
	1	*2*	*3*	*4*	*5*	*6*	*7*	*8*	*9*	*10*	*11*	*12*
1	161	200	216	225	230	234	237	239	241	242	243	244
	4052	**4999**	**5403**	**5625**	**5764**	**5859**	**5928**	**5981**	**6022**	**6056**	**6082**	**6106**
2	18.51	19.00	19.16	19.25	19.30	19.33	19.36	19.37	19.38	19.39	19.40	19.41
	98.49	**99.01**	**99.17**	**99.25**	**99.30**	**99.33**	**99.34**	**99.36**	**99.38**	**99.40**	**99.41**	**99.42**
3	10.13	9.55	9.28	9.12	9.01	8.94	8.88	8.84	8.81	8.78	8.76	8.74
	34.12	**30.81**	**29.46**	**28.71**	**28.24**	**27.91**	**27.67**	**27.49**	**27.34**	**27.23**	**27.13**	**27.05**
4	7.71	6.94	6.59	6.39	6.26	6.16	6.09	6.04	6.00	5.96	5.93	5.91
	21.20	**18.00**	**16.69**	**15.98**	**15.52**	**15.21**	**14.98**	**14.80**	**14.66**	**14.54**	**14.45**	**14.37**
5	6.61	5.79	5.41	5.19	5.05	4.95	4.88	4.82	4.78	4.74	4.70	4.68
	16.26	**13.27**	**12.06**	**11.39**	**10.97**	**10.67**	**10.45**	**10.27**	**10.15**	**10.05**	**9.96**	**9.89**
6	5.99	5.14	4.76	4.53	4.39	4.28	4.21	4.15	4.10	4.06	4.03	4.00
	13.74	**10.92**	**9.78**	**9.15**	**8.75**	**8.47**	**8.26**	**8.10**	**7.98**	**7.87**	**7.79**	**7.72**
7	5.59	4.74	4.35	4.12	3.97	3.87	3.79	3.73	3.68	3.63	3.60	3.57
	12.25	**9.55**	**8.45**	**7.85**	**7.46**	**7.19**	**7.00**	**6.84**	**6.71**	**6.62**	**6.54**	**6.47**
8	5.32	4.46	4.07	3.84	3.69	3.58	3.50	3.44	3.39	3.34	3.31	3.28
	11.26	**8.65**	**7.59**	**7.01**	**6.63**	**6.37**	**6.19**	**6.03**	**5.91**	**5.82**	**5.74**	**5.67**
9	5.12	4.26	3.86	3.63	3.48	3.37	3.29	3.23	3.18	3.13	3.10	3.07
	10.56	**8.02**	**6.99**	**6.42**	**6.06**	**5.80**	**5.62**	**5.47**	**5.35**	**5.26**	**5.18**	**5.11**
10	4.96	4.10	3.71	3.48	3.33	3.22	3.14	3.07	3.02	2.97	2.94	2.91
	10.04	**7.56**	**6.55**	**5.99**	**5.64**	**5.39**	**5.21**	**5.06**	**4.95**	**4.85**	**4.78**	**4.71**
11	4.84	3.98	3.59	3.36	3.20	3.09	3.01	2.95	2.90	2.86	2.82	2.79
	9.65	**7.20**	**6.22**	**5.67**	**5.32**	**5.07**	**4.88**	**4.74**	**4.63**	**4.54**	**4.46**	**4.40**
12	4.75	3.88	3.49	3.26	3.11	3.00	2.92	2.85	2.80	2.76	2.72	2.69
	9.33	**6.93**	**5.95**	**5.41**	**5.06**	**4.82**	**4.65**	**4.50**	**4.39**	**4.30**	**4.22**	**4.16**
13	4.67	3.80	3.41	3.18	3.02	2.92	2.84	2.77	2.72	2.67	2.63	2.60
	9.07	**6.70**	**5.74**	**5.20**	**4.86**	**4.62**	**4.44**	**4.30**	**4.19**	**4.10**	**4.02**	**3.96**
14	4.60	3.74	3.34	3.11	2.96	2.85	2.77	2.70	2.65	2.60	2.56	2.53
	8.86	**6.51**	**5.56**	**5.03**	**4.69**	**4.46**	**4.28**	**4.14**	**4.03**	**3.94**	**3.86**	**3.80**
15	4.54	3.68	3.29	3.06	2.90	2.79	2.70	2.64	2.59	2.55	2.51	2.48
	8.68	**6.36**	**5.42**	**4.89**	**4.56**	**4.32**	**4.14**	**4.00**	**3.89**	**3.80**	**3.73**	**3.67**
16	4.49	3.63	3.24	3.01	2.85	2.74	2.66	2.59	2.54	2.49	2.45	2.42
	8.53	**6.23**	**5.29**	**4.77**	**4.44**	**4.20**	**4.03**	**3.89**	**3.78**	**3.69**	**3.61**	**3.55**
17	4.45	3.59	3.20	2.96	2.81	2.70	2.62	2.55	2.50	2.45	2.41	2.38
	8.40	**6.11**	**5.18**	**4.67**	**4.34**	**4.10**	**3.93**	**3.79**	**3.68**	**3.59**	**3.52**	**3.45**
18	4.41	3.55	3.16	2.93	2.77	2.66	2.58	2.51	2.46	2.41	2.37	2.34
	8.28	**6.01**	**5.09**	**4.58**	**4.25**	**4.01**	**3.85**	**3.71**	**3.60**	**3.51**	**3.44**	**3.37**
19	4.38	3.52	3.13	2.90	2.74	2.63	2.55	2.48	2.43	2.38	2.34	2.31
	8.18	**5.93**	**5.01**	**4.50**	**4.17**	**3.94**	**3.77**	**3.63**	**3.52**	**3.43**	**3.36**	**3.30**
20	4.35	3.49	3.10	2.87	2.71	2.60	2.52	2.45	2.40	2.35	2.31	2.28
	8.10	**5.85**	**4.94**	**4.43**	**4.10**	**3.87**	**3.71**	**3.56**	**3.45**	**3.37**	**3.30**	**3.23**
21	4.32	3.47	3.07	2.84	2.68	2.57	2.49	2.42	2.37	2.32	2.28	2.25
	8.02	**5.78**	**4.87**	**4.37**	**4.04**	**3.81**	**3.65**	**3.51**	**3.40**	**3.31**	**3.24**	**3.17**

(continued)

TABLE G *(continued)*

Degrees of Freedom for Denominator	Degrees of Freedom for Numerator											
	1	2	3	4	5	6	7	8	9	10	11	12
22	4.30	3.44	3.05	2.82	2.66	2.55	2.47	2.40	2.35	2.30	2.26	2.23
	7.94	**5.72**	**4.82**	**4.31**	**3.99**	**3.76**	**3.59**	**3.45**	**3.35**	**3.26**	**3.18**	**3.12**
23	4.28	3.42	3.03	2.80	2.64	2.53	2.45	2.38	2.32	2.28	2.24	2.20
	7.88	**5.66**	**4.76**	**4.26**	**3.94**	**3.71**	**3.54**	**3.41**	**3.30**	**3.21**	**3.14**	**3.07**
24	4.26	3.40	3.01	2.78	2.62	2.51	2.43	2.36	2.30	2.26	2.22	2.18
	7.82	**5.61**	**4.72**	**4.22**	**3.90**	**3.67**	**3.50**	**3.36**	**3.25**	**3.17**	**3.09**	**3.03**
25	4.24	3.38	2.99	2.76	2.60	2.49	2.41	2.34	2.28	2.24	2.20	2.16
	7.77	**5.57**	**4.68**	**4.18**	**3.86**	**3.63**	**3.46**	**3.32**	**3.21**	**3.13**	**3.05**	**2.99**
26	4.22	3.37	2.89	2.74	2.59	2.47	2.39	2.32	2.27	2.22	2.18	2.15
	7.72	**5.53**	**4.64**	**4.14**	**3.82**	**3.59**	**3.42**	**3.29**	**3.17**	**3.09**	**3.02**	**2.96**
27	4.21	3.35	2.96	2.73	2.57	2.46	2.37	2.30	2.25	2.20	2.16	2.13
	7.68	**5.49**	**4.60**	**4.11**	**3.79**	**3.56**	**3.39**	**3.26**	**3.14**	**3.06**	**2.98**	**2.93**
28	4.20	3.34	2.95	2.71	2.56	2.44	2.36	2.29	2.24	2.19	2.15	2.12
	7.64	**5.45**	**4.57**	**4.07**	**3.76**	**3.53**	**3.36**	**3.23**	**3.11**	**3.03**	**2.95**	**2.90**
29	4.18	3.33	2.93	2.70	2.54	2.43	2.35	2.28	2.22	2.18	2.14	2.10
	7.60	**5.52**	**4.54**	**4.04**	**3.73**	**3.50**	**3.32**	**3.20**	**3.08**	**3.00**	**2.92**	**2.87**
30	4.17	3.32	2.92	2.69	2.53	2.42	2.34	2.27	2.21	2.16	2.12	2.09
	7.56	**5.39**	**4.51**	**4.02**	**3.70**	**3.47**	**3.30**	**3.17**	**3.06**	**2.98**	**2.90**	**2.84**
32	4.15	3.30	2.90	2.67	2.51	2.40	2.32	2.25	2.19	2.14	2.10	2.07
	7.50	**5.34**	**4.46**	**3.97**	**3.66**	**3.42**	**3.25**	**3.12**	**3.01**	**2.94**	**2.86**	**2.80**
34	4.13	3.28	2.88	2.65	2.49	2.38	2.30	2.23	2.17	2.12	2.08	2.05
	7.44	**5.29**	**4.42**	**3.93**	**3.61**	**3.38**	**3.21**	**3.08**	**2.97**	**2.89**	**2.82**	**2.76**
36	4.11	3.26	2.86	2.63	2.48	2.36	2.28	2.21	2.15	2.10	2.06	2.03
	7.39	**5.25**	**4.38**	**3.89**	**3.58**	**3.35**	**3.18**	**3.04**	**2.94**	**2.86**	**2.78**	**2.72**
38	4.10	3.25	2.85	2.62	2.46	2.35	2.26	2.19	2.14	2.09	2.05	2.02
	7.35	**5.21**	**4.34**	**3.86**	**3.54**	**3.32**	**3.15**	**3.02**	**2.91**	**2.82**	**2.75**	**2.69**
40	4.08	3.23	2.84	2.61	2.45	2.34	2.25	2.18	2.12	2.07	2.04	2.00
	7.31	**5.18**	**4.31**	**3.83**	**3.51**	**3.29**	**3.12**	**2.99**	**2.88**	**2.80**	**2.73**	**2.66**
42	4.07	3.22	2.83	2.59	2.44	2.32	2.24	2.17	2.11	2.06	2.02	1.90
	7.27	**5.15**	**4.29**	**3.80**	**3.49**	**3.26**	**3.10**	**2.96**	**2.86**	**2.77**	**2.70**	**2.64**
44	4.06	3.21	2.82	2.58	2.43	2.31	2.23	2.16	2.10	2.05	2.01	1.98
	7.24	**5.12**	**4.26**	**3.78**	**3.46**	**3.24**	**3.07**	**2.94**	**2.84**	**2.75**	**2.68**	**2.62**
46	4.05	3.20	2.81	2.57	2.42	2.30	2.22	2.14	2.09	2.04	2.00	1.97
	7.21	**5.10**	**4.24**	**3.76**	**3.44**	**3.22**	**3.05**	**2.92**	**2.82**	**2.73**	**2.66**	**2.60**
48	4.04	3.19	2.80	2.56	2.41	2.30	2.21	2.14	2.08	2.03	1.99	1.96
	7.19	**5.08**	**4.22**	**3.74**	**3.42**	**3.20**	**3.04**	**2.90**	**2.80**	**2.71**	**2.64**	**2.58**
50	4.03	3.18	2.79	2.56	2.40	2.29	2.20	2.13	2.07	2.02	1.98	1.95
	7.17	**5.06**	**4.20**	**3.72**	**3.41**	**3.18**	**3.02**	**2.88**	**2.78**	**2.70**	**2.62**	**2.56**
55	4.02	3.17	2.78	2.54	2.38	2.27	2.18	2.11	2.05	2.00	1.97	1.93
	7.12	**5.01**	**4.16**	**3.68**	**3.37**	**3.15**	**2.98**	**2.85**	**2.75**	**2.66**	**2.59**	**2.53**
60	4.00	3.15	2.76	2.52	2.37	2.25	2.17	2.10	2.04	1.99	1.95	1.92
	7.08	**4.98**	**4.13**	**3.65**	**3.34**	**3.12**	**2.95**	**2.82**	**2.72**	**2.63**	**2.56**	**2.50**

TABLE G *(continued)*

Degrees of Freedom for Denominator	Degrees of Freedom for Numerator											
	1	*2*	*3*	*4*	*5*	*6*	*7*	*8*	*9*	*10*	*11*	*12*
65	3.99	3.14	2.75	2.51	2.36	2.24	2.15	2.08	2.02	1.98	1.94	1.90
	7.04	**4.95**	**4.10**	**3.62**	**3.31**	**3.09**	**2.93**	**2.79**	**2.70**	**2.61**	**2.54**	**2.47**
70	3.98	3.13	2.74	2.50	2.35	2.22	2.14	2.07	2.01	1.97	1.93	1.89
	7.01	**4.92**	**4.08**	**3.60**	**3.29**	**3.07**	**2.91**	**2.77**	**2.67**	**2.59**	**2.51**	**2.45**
80	3.96	3.11	2.72	2.48	2.33	2.21	2.12	2.05	1.99	1.95	1.91	1.88
	6.96	**4.88**	**4.04**	**3.56**	**3.25**	**3.04**	**2.87**	**2.74**	**2.64**	**2.55**	**2.48**	**2.41**
100	3.94	3.09	2.70	2.46	2.30	2.19	2.10	2.03	1.97	1.92	1.88	1.85
	6.90	**4.82**	**3.98**	**3.51**	**3.20**	**2.99**	**2.82**	**2.69**	**2.59**	**2.51**	**2.43**	**2.36**
125	3.92	3.07	2.68	2.44	2.29	2.17	2.08	2.01	1.95	1.90	1.86	1.83
	6.84	**4.78**	**3.94**	**3.47**	**3.17**	**2.95**	**2.79**	**2.65**	**2.56**	**2.47**	**2.40**	**2.33**
150	3.91	3.06	2.67	2.43	2.27	2.16	2.07	2.00	1.94	1.89	1.85	1.82
	6.81	**4.75**	**3.91**	**3.44**	**3.13**	**2.92**	**2.76**	**2.62**	**2.53**	**2.44**	**2.37**	**2.30**
200	3.89	3.04	2.65	2.41	2.26	2.14	2.05	1.98	1.92	1.87	1.83	1.80
	6.76	**4.71**	**3.88**	**3.41**	**3.11**	**2.90**	**2.73**	**2.60**	**2.50**	**2.41**	**2.34**	**2.28**
400	3.86	3.02	2.62	2.39	2.23	2.12	2.03	1.96	1.90	1.85	1.81	1.78
	6.70	**4.66**	**3.83**	**3.36**	**3.06**	**2.85**	**2.69**	**2.55**	**2.46**	**2.37**	**2.29**	**2.23**
1000	3.85	3.00	2.61	2.38	2.22	2.10	2.02	1.95	1.89	1.84	1.80	1.76
	6.66	**4.62**	**3.80**	**3.34**	**3.04**	**2.82**	**2.66**	**2.53**	**2.43**	**2.34**	**2.26**	**2.20**
∞	3.84	2.99	2.60	2.37	2.21	2.09	2.01	1.94	1.88	1.83	1.79	1.75
	6.64	**4.60**	**3.78**	**3.32**	**3.02**	**2.80**	**2.64**	**2.51**	**2.41**	**2.32**	**2.24**	**2.18**

Values of F for $\alpha = .05$ are given in lightface type, and values for F for $\alpha = .01$ are given in boldface type.
Source: George W. Snedecor and William G. Cochran, *Statistical Methods,* 1980, Iowa State University Press, Ames, Iowa 50010. Reprinted by permission.

TABLE H Percentage points of the studentized range (critical values for Tukey's HSD).

MS_w df	α	k = Number of Means 2	3	4	5	6	7	8	9	10	11
5	.05	3.64	4.60	5.22	5.67	6.03	6.33	6.58	6.80	6.99	7.17
	.01	5.70	6.98	7.80	8.42	8.91	9.32	9.67	9.97	10.24	10.48
6	.05	3.46	4.34	4.90	5.30	5.63	5.90	6.12	6.32	6.49	6.65
	.01	5.24	6.33	7.03	7.56	7.97	8.32	8.61	8.87	9.10	9.30
7	.05	3.34	4.16	4.68	5.06	5.36	5.61	5.82	6.00	6.16	6.30
	.01	4.95	5.92	6.54	7.01	7.37	7.68	7.94	8.17	8.37	8.55
8	.05	3.26	4.04	4.53	4.89	5.17	5.40	5.60	5.77	5.92	6.05
	.01	4.75	5.64	6.20	6.62	6.96	7.24	7.47	7.68	7.86	8.03
9	.05	3.20	3.95	4.41	4.76	5.02	5.24	5.43	5.59	5.74	5.87
	.01	4.60	5.43	5.96	6.35	6.66	6.91	7.13	7.33	7.49	7.65
10	.05	3.15	3.88	4.33	4.65	4.91	5.12	5.30	5.46	5.60	5.72
	.01	4.48	5.27	5.77	6.14	6.43	6.67	6.87	7.05	7.21	7.36
11	.05	3.11	3.82	4.26	4.57	4.82	5.03	5.20	5.35	5.49	5.61
	.01	4.39	5.15	5.62	5.97	6.25	6.48	6.67	6.84	6.99	7.13
12	.05	3.08	3.77	4.20	4.51	4.75	4.95	5.12	5.27	5.39	5.51
	.01	4.32	5.05	5.50	5.84	6.10	6.32	6.51	6.67	6.81	6.94
13	.05	3.06	3.73	4.15	4.45	4.69	4.88	5.05	5.19	5.32	5.43
	.01	4.26	4.96	5.40	5.73	5.98	6.19	6.37	6.53	6.67	6.79
14	.05	3.03	3.70	4.11	4.41	4.64	4.83	4.99	5.13	5.25	5.36
	.01	4.21	4.89	5.32	5.63	5.88	6.08	6.26	6.41	6.54	6.66
15	.05	3.01	3.67	4.08	4.37	4.59	4.78	4.94	5.08	5.20	5.31
	.01	4.17	4.84	5.25	5.56	5.80	5.99	6.16	6.31	6.44	6.55
16	.05	3.00	3.65	4.05	4.33	4.56	4.74	4.90	5.03	5.15	5.26
	.01	4.13	4.79	5.19	5.49	5.72	5.92	6.08	6.22	6.35	6.46
17	.05	2.98	3.63	4.02	4.30	4.52	4.70	4.86	4.99	5.11	5.21
	.01	4.10	4.74	5.14	5.43	5.66	5.85	6.01	6.15	6.27	6.38
18	.05	2.97	3.61	4.00	4.28	4.49	4.67	4.82	4.96	5.07	5.17
	.01	4.07	4.70	5.09	5.38	5.60	5.79	5.94	6.08	6.20	6.31
19	.05	2.96	3.59	3.98	4.25	4.47	4.65	4.79	4.92	5.04	5.14
	.01	4.05	4.67	5.05	5.33	5.55	5.73	5.89	6.02	6.14	6.25
20	.05	2.95	3.58	3.96	4.23	4.45	4.62	4.77	4.90	5.01	5.11
	.01	4.02	4.64	5.02	5.29	5.51	5.69	5.84	5.97	6.09	6.19
24	.05	2.92	3.53	3.90	4.17	4.37	4.54	4.68	4.81	4.92	5.01
	.01	3.96	4.55	4.91	5.17	5.37	5.54	5.69	5.81	5.92	6.02
30	.05	2.89	3.49	3.85	4.10	4.30	4.46	4.60	4.72	4.82	4.92
	.01	3.89	4.45	4.80	5.05	5.24	5.40	5.54	5.65	5.76	5.85
40	.05	2.86	3.44	3.79	4.04	4.23	4.39	4.52	4.63	4.73	4.82
	.01	3.82	4.37	4.70	4.93	5.11	5.26	5.39	5.50	5.60	5.69
60	.05	2.83	3.40	3.74	3.98	4.16	4.31	4.44	4.55	4.65	4.73
	.01	3.76	4.28	4.59	4.82	4.99	5.13	5.25	5.36	5.45	5.53
120	.05	2.80	3.36	3.68	3.92	4.10	4.24	4.36	4.47	4.56	4.64
	.01	3.70	4.20	4.50	4.71	4.87	5.01	5.12	5.21	5.30	5.37
∞	.05	2.77	3.31	3.63	3.86	4.03	4.17	4.29	4.39	4.47	4.55
	.01	3.64	4.12	4.40	4.60	4.76	4.88	4.99	5.08	5.16	5.23

Source: E. S. Pearson and H. O. Hartley, *Biometrika Tables for Statisticians,* Vol. 1, 3rd ed., 1966, Cambridge Press, New York, by permission of the Biometrika Trustees.

TABLE I Critical values of chi square.

df	*.05*	*.01*	*df*	*.05*	*.01*
1	3.84	6.64	16	26.30	32.00
2	5.99	9.21	17	27.59	33.41
3	7.82	11.34	18	28.87	34.80
4	9.49	13.28	19	30.14	36.19
5	11.07	15.09	20	31.41	37.57
6	12.59	16.81	21	32.67	38.93
7	14.07	18.48	22	33.92	40.29
8	15.51	20.09	23	35.17	41.64
9	16.92	21.67	24	36.42	42.98
10	18.31	23.21	25	37.65	44.31
11	19.68	24.72	26	38.88	45.64
12	21.03	26.22	27	40.11	46.96
13	22.36	27.69	28	41.34	48.28
14	23.68	29.14	29	42.56	49.59
15	25.00	30.58	30	43.77	50.89

Source: Abridged from Table IV of Fisher and Yates, *Statistical Tables for Biological, Agricultural, and Medical Research,* Longman Group Ltd., London (previously published by Oliver and Boyd Ltd., Edinburgh), and by permission of the authors and publishers.

TABLE J Values of *T* at the .05 and .01 levels of significance in the Wilcoxon signed-ranks test.

N	*.05*	*.01*	N	*.05*	*.01*
6	0	–	16	30	20
7	2	–	17	35	23
8	4	0	18	40	28
9	6	2	19	46	32
10	8	3	20	52	38
11	11	5	21	59	43
12	14	7	22	66	49
13	17	10	23	73	55
14	21	13	24	81	58
15	25	16	25	89	61

Source: Abridged from Table I of G. Wilcoxon, *Some Rapid Approximate Statistical Procedures,* 1949, American Cyanamid Co., New York, with the permission of the publisher.

SPSS: The A B Cs

Appendix B: Program Instructions for SPSS

Student Version 11

BOPPING AROUND SPSS

To run through the examples in this manual, you must use your copy of Sprinthall's Basic Statistical Analysis, *7th edition*, published by Allyn-Bacon. All the data sets will come from this source. However, before doing any statistical analysis, it is wise to spend a brief time just getting used to the SPSS format. SPSS for Windows uses a regular spreadsheet arrangement, just like Lotus, Quattro, and Excel. That is, there are columns (which are headed by variable names), and rows (which refer to the cases or the subject numbers). Also, keep in mind that the intersection of a column and a row is called a cell. The opening window in SPSS is called the Data Editor and produces the spreadsheet that will be used for all your data entries. When SPSS is started, the first box to appear will ask you to check whether you want to "run the tutorial," "type in data," "run an existing query," "create a new query," or "open an existing file." Since you don't yet have an existing file, check "type in data" and click OK. You will now see the spreadsheet for inputting your data. At the top of this screen it will say

UNTITLED Data Editor

Before Entering Data

In order to enter data, you must *first* define the variable. Start by double-clicking the top of the first column (where it says var), and you will now see another spreadsheet screen, called the "variable view." It is headed by:

Name Type Width Decimals Label Values Missing Columns Align Measure

This is where you will name your variable and specify all of its characteristics. For example, in the first row under *name* type in subj# and leave the rest of that row in its default settings. Then right below, in the second row still in the first column under *name*, type in the word *scores*. Next, click on the cell in the next column to the right (under

Type) and click on the little gray box. A pop-up box will then appear that allows you to specify a variety of choices. Leave the check mark (the default setting) next to numeric. This will be your choice 99% of the time, because you will want the vast majority of your entries to be treated as numbers. The only time you would want to check "string" is when you are using letters only or combining letters and numbers, such as entering a subject's name alone or the name and then perhaps an identification number. For width, the default setting is 8, which will work most of the time. However, when typing a long string, such as a long last name or identification number, you will need to increase the width of the column. For decimal places, the default setting is 2, but your instructor may want you to increase it to 3 or perhaps even more. The next column, "Label," allows you to add text to any variable whose name is not immediately obvious. For instance, you have defined your variable as iq, but with Label you can indicate precisely that the score is a WAIS-III test score and was obtained on such and such a date. The Label option lets you use up to 256 characters, and its width will change automatically as you keep entering more characters. For the problems to follow you can leave this option blank. When you click the next column, "Values," another pop-up screen appears and you can identify your variables by number. For example, suppose you are using 0 for male and 1 for female, this is where you enter the codes, and this is where the program will REMEMBER for you which was which. The next column, "Missing," is used when a subject has refused to answer one of your questions, perhaps ethnic background or even gender. You may want to code these with a number, such as 10 or 11. The next column, "Columns," allows you to set the width of each of your columns. As already mentioned, the default setting is 8, but for many of your variables you may need only 3 or 4 spaces. The fewer spaces you use, the more of your spreadsheet will be visible on a single screen. Finally, under "Measure" you can choose which measurement scale your data are portraying. "Scale" is used to cover both ratio and interval data, and ordinal and nominal will become obvious as we continue through our lesson. When you finish, click the data view tab at the bottom of the screen and your data editor screen will reappear.

Entering Data

Your spreadsheet now has two columns labeled and available for data entry. Go to the top of the first column, which, if you followed the directions shown above, will now be under Subj#, and type in 1, then scroll down and type in 2, until you have numbered 5 cases. Now put the cursor under the second column, which you previously set up as scores, and enter the data on page 163.

Subj#	*Scores*
1.	10
2.	8
3.	6
4.	4
5.	2

Once the data are entered, check to see if you have any data entry errors. If you find an error, put the cursor over that score, and type in the correct value and enter it.

SAVING THE FILE

Much wasted time and many long lamentations have resulted from forgetting to save the file. Don't let it happen to you. Once the data are in, click on file in the top left corner of the screen and then click Save as. The screen then allows you to select where you want to save the file. Click the pull-down menu beside the word Desktop. Your screen will offer you the following choices: Desktop, My Documents, My Computer, 3.5 Floppy (A), (C), (D), and Network Neighborhood. Be sure to have a disk in your A drive and save it to your floppy by clicking the 3.5 drive A. Then on a line a few spaces below, next to file name, type in the file name you've created. Since it's on page 163, let's call it page163, so you will know which data set it contains. Now click save, and you are ready to roll.

RETRIEVING THE FILE

For practice, click file and exit. Now load the SPSS program again, click Open an existing data source and OK. Now click the 3.5 drive A, and you will find your file name, page163. Click on the file name and Open. The data will quickly appear. Your screen will look like this:

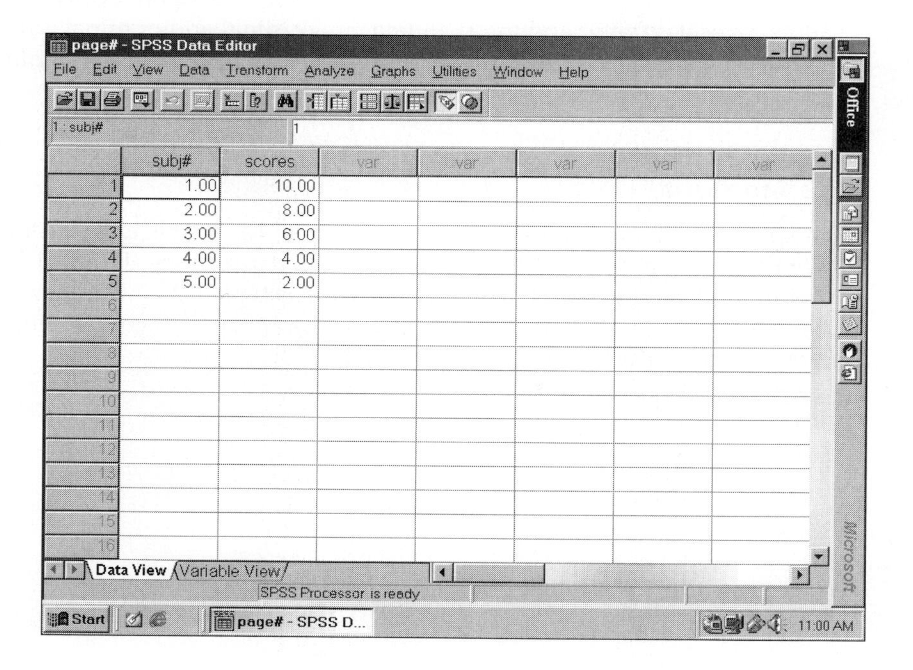

Without getting too far ahead, it might be interesting to see a little of what SPSS can do for you. Click Analyze, then Descriptive Statistics and your screen will look like this:

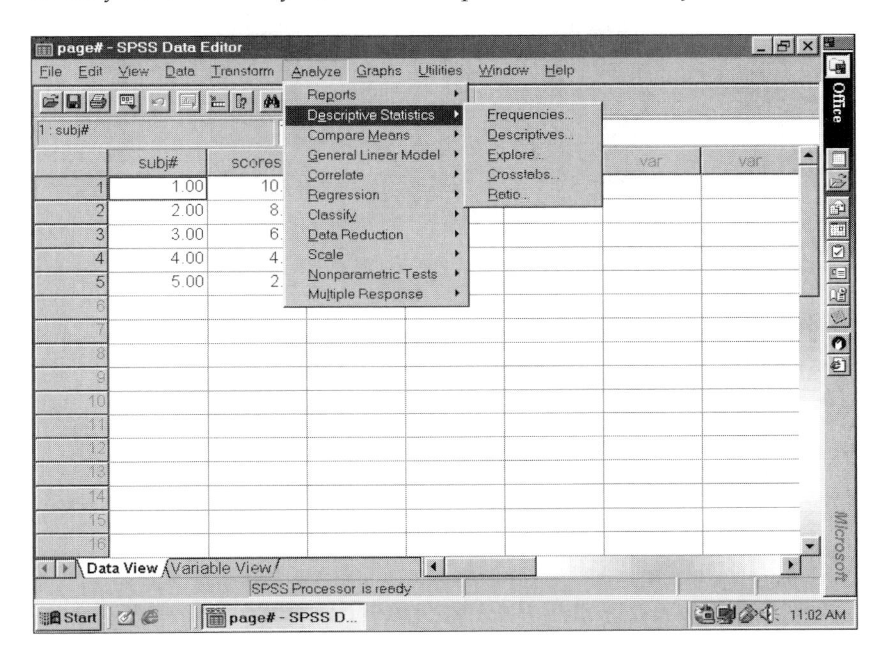

Then click Descriptives. Next, click on scores and then click the small arrowhead key in the middle of the screen. Your screen will look like this.

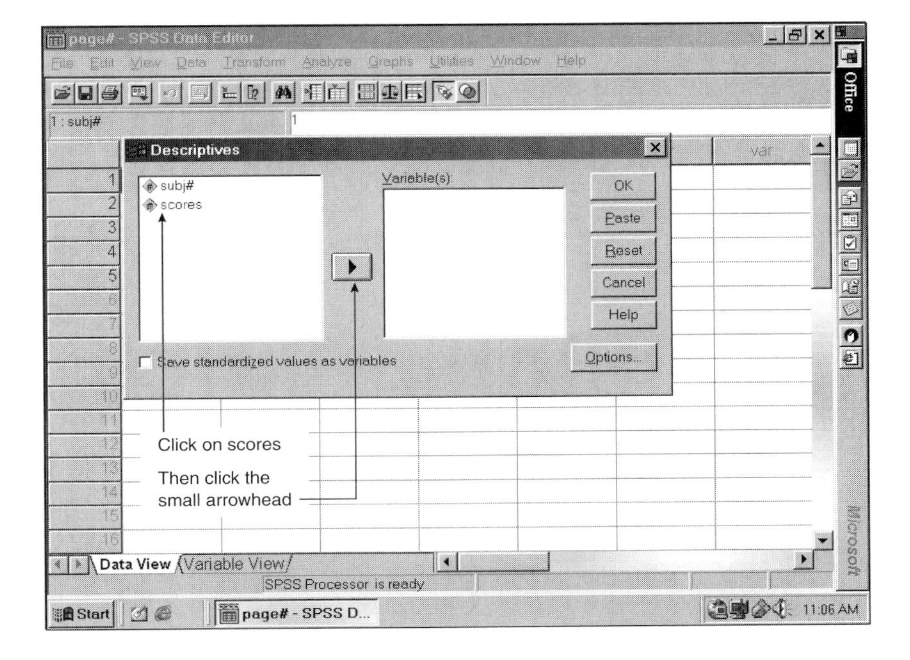

This will move the scores into the box on the right, which is shown in the next screen.

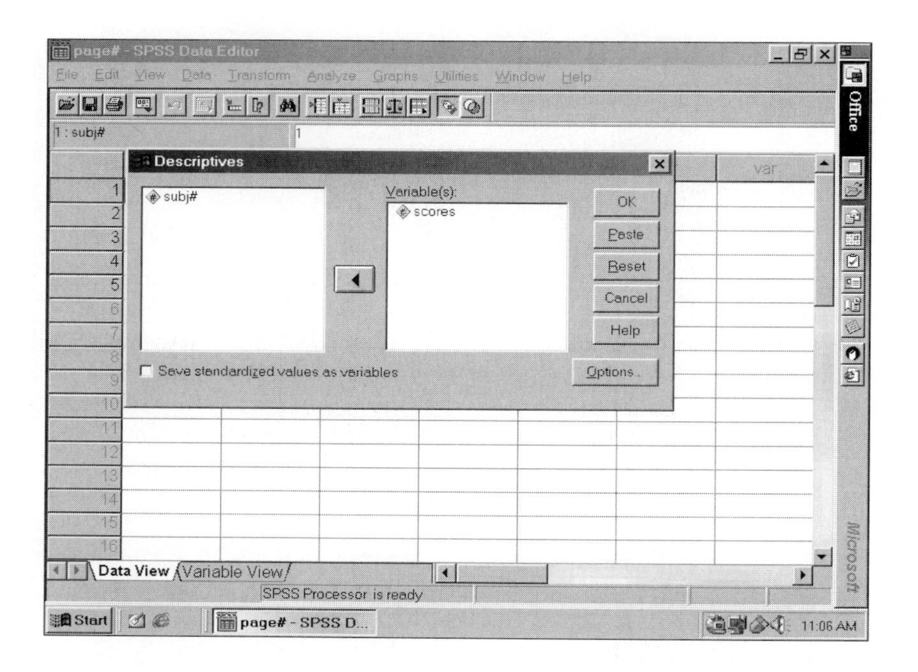

Now click on Options and check the Mean, the Std, the Maximum, and the Minimum. Click continue, and then OK. This will bring up the second important SPSS window, called the Output Viewer. This is where you will see the results of your various statistical calculations, and it's also here that you can click on the print icon on the SPSS toolbar to get a hard copy of your analyses. Or, if you prefer, you can click on file and then click print.

OUTPUT

For the above data, the output navigator will look like this:

Descriptives

Descriptive Statistics

	N	Minimum	Maximum	Mean	Std. Deviation
SCORES Valid N (listwise)	5 5	2.00	10.00	6.0000	3.1623

Notice that the standard deviation being provided is the estimated standard deviation for the population. Read the discussion on page 162 to understand the difference between the estimated population standard deviation and the true standard deviation in the sample. If you need the true *SD* of the sample, simply take the estimated *SD* (in this case 3.1623) and multiply it by the square root of (N – 1/N). Thus, 3.1623 times the square root of 4/5 = 2.828. Before leaving this screen, SPSS will ask if you want to save the content of the output viewer. Click OK only if you want to save these results to your disk in a separate file. Otherwise click NO, especially if you already have the results printed out.

Selected Output

In the sections to follow, you will be reading about SELECTED OUTPUT. Thus, you will be given only the most basic results for your interpretation. SPSS, remember, is used not only by students new to the field, but also by world-class statisticians interpreting very sophisticated research strategies. So that you won't be drowning in the large number of output pages SPSS provides, we will focus only on those results appropriate to the level of sophistication needed by the average student.

Changing the Column Width

When you bring up the Define Variable box, you will see at the lower right-side of the screen a button called Column Format. If you click this, you are given the opportunity to change the width of the column from its default setting of 8 all the way up to 256. Also, it is here that you can change the alignment from the default setting of right to either left or center. Be sure to use your mouse to highlight the column you want to change BEFORE you attempt the changes.

Inserting New Rows or Columns

There may be times when you decide to add a new row or column to your spreadsheet. For example, let's say you have a column for age and next to it a column for IQ. You decide you would like to put a column for gender in-between these two. Move the cursor to the column which is directly to the RIGHT of where you want the new column to be. Click Data and then Insert Variable. The new variable is going to automatically be called VAR00001, so go back to the variable view screen and type in gender in the appropriate row. To insert a new row the procedure is virtually the same except that after clicking Data you must check Insert Case. Keep your cursor on the cell just above the place where you want the new row. This could be useful if you notice that you forgot to put in the score for Subject # 5, but you have already inadvertently used that row for Subject # 6.

Deleting a Column

To delete a column, click on the top of the column you want to get rid of. Click Edit and then click Clear.

Deleting a Row

To delete a row, click on the far left side of the row in question, and use the mouse to drag across the row. Click Edit and click Clear.

Moving and Copying Data

Move. This is the procedure to use if you want to move a section of the spreadsheet from one place to another. Use the mouse to drag over the area to be moved (it must be a rectangular area), and then click Edit and Cut. Move the cursor to the cell in the UPPER LEFT HAND CORNER of where you want the data to go. Click Edit, and click Paste.

Copy. If you want to copy a section of the spreadsheet to another location, so that you will have it in both locations, then you must again begin by dragging over the area you want to copy. Then click Edit and Copy. Again move the cursor to the upper left-hand corner of where you want to put the data and click Edit and Paste.

Statistical Analyses

Descriptive Statistics. Please use the data from problem #C1, found on page 65.

1. Double-click the top of the first column where it says VAR and when the variable view screen appears, type in Subj# in the first row and IQ in the second. Now click the tab at the bottom called "data view" to get back to the data screen.
2. Type in the data as follows:

Subj#	*iq*
1.	100
2.	100
3.	86
.	
23.	108
24.	119
25.	79

Now before going any further click file. Then *save the data* by clicking on *save as* and then click on the arrow to the right of the pull-down menu that says desktop. Select the 3.5 Floppy drive A. Then toward the bottom, next to the line that reads the file name, type in the file name we will now create. Let's call it page65, so you will know which data set it contains. You are now ready to click Save, and at that point your data become safely stored.

3. Click Analyze, then Descriptive Statistics.
4. Click Descriptives, and click on IQ.

5. Move IQ into the Variables box by clicking on the little arrowhead in the middle of the screen.
6. Click Options and check Mean, Sum. Std. deviation, Variance, Range, Minimum, Maximum, SE mean, Kurtosis, and Skewness.
7. Click Continue and OK.

Selected Output:

The first output to appear will be the following:

Descriptive Statistics

	N *Statistic*	*Range* *Statistic*	*Minimum* *Statistic*	*Maximum* *Statistic*	*Sum* *Statistic*
IQ	25	90.00	55.00	145.00	2500.00
Valid *N* (listwise)	25				

If you don't see all that follows, keep scrolling to the right until you do.

Descriptive Statistics

	Mean *Statistic*	*Std. Error*	*Std. Dev.* *Statistic*	*Variance* *Statistic*
IQ	100.0000	3.0621	15.3107	234.417
Valid N (listwise)				

Descriptive Statistics

	Skewness		*Kurtosis*	
	Statistic	*Std.Error*	*Statistic*	*Std.Error*
IQ	–.030	.464	4.993	.902

The first output screen indicates that the range is 90.00, with a minimum score of 55.00 and a maximum score of 145.00. Also, the sum of all the IQ scores is 2500.00. Next, the screen shows that the mean is 100.00, and its standard error (the standard error of the mean is 3.0621). The estimated population standard deviation is shown as 15.3107 and the variance is 234.417. The third screen produces results for both skewness and kurtosis. The skewness value of –.030 indicates that this is a symmetrical distribution, with virtually no skewness, since the value is so close to zero. The kurtosis of 4.993 (which is a positive value) shows that the distribution is leptokurtic. See pages 570 and 571. Now click the print icon and your results will be printed to hard copy. Before leaving this output, the screen will ask if you want to Save Contents of Output 1. Since you now have a printed copy, you should probably answer NO.

SORTING THE DATA

There are numerous times when it is necessary to sort the data into some kind of meaningful order. Let's say you've spent an hour entering hundreds of IQ scores and you want to put them in order (either high-to-low or low-to-high). Click Data and Sort Cases. When the Sort Cases box appears, a list of the variables on your spread sheet is shown. Click on IQ (which is the file you just created) and move it, via the > button over into the Sort By box. Now check whether it's ascending or descending. For the IQ scores let's put the highest score at the top, so check descending and click OK. You will then have sorted the entire spread-sheet and you will NOT have lost your subject-data linkages.

Finding Standard Scores

To find the *z* score for each raw score, please use the same data set as shown above. To bring up the data, click on retrieve, A drive and page 65.

1. Click Analyze, Descriptive statistics, and Descriptives. When the Descriptives Box appears, check "Save standardized values as variables" (which is in the lower-left corner).
3. Click on Windows and check OK. Then check page65 Data Editor to get back to the spreadsheet. You will now find that right beside your original IQ scores is a new column called ziq. For each raw score the corresponding *z* score is now shown.

Subj#	iq	ziq
1	100	.00000
2	100	.00000
3	86	–.91439
.		
.		
23	108	.52251
24	119	1.24096
25	79	–1.37159

Graphing the Data Set

1. Click Analyze, Descriptives, and Frequencies. When the Frequencies box comes up, move IQ into the Variable(s) box, and
2. Click Charts, which will then bring up the Charts box. There are three types of charts available—Bar charts, Pie charts, and Histograms.
3. Check the Chart Type you wish to examine, and then at the bottom select whether you want the data presented in frequency or percentage form. Click Continue and OK.

Try creating charts using frequency, and then go back and use percentage. Unless you're dealing with very large data sets, frequency is the common option for Bar

Charts and Histograms, whereas percentages are more clearly portrayed in the pie chart. Under Histogram, you will notice an option asking if you want it shown With Normal Curve. If you check that option, the histogram will be shown with a normal curve overlay, and you will be able to get a general picture of how closely your data set approximates normality. In this case it is clear that the distribution is leptokurtic (as was indicated above).

INFERENTIAL STATISTICS

Single Sample t Test

For this example, please use the data on pages 175–176.

Subj#	*Scores*
1	98
2	85
3	82
.	
.	
.	
8	79
9	78
10	64

Save the file as p175. Can this sample represent a population whose mean is 80?

1. Double-click the top of the first column and when the variable view screen appears type in Subj# in the first row and Scores in the second. Now click the tab at the bottom called "data view" to get back to the data screen.
2. Enter data shown above.
3. Click Analyze. Compare means, One Sample *t*.
4. Put scores in as Test Variable.
5. Enter Test Value, which in this case is 80 (the parameter mean).
6. OK

SELECTED OUTPUT

ONE SAMPLE TEST

TEST VALUE = 80

T *TEST*	t	df	*sig*	*Mean Diff*
SCORES	.268	9	.794	.7000

Conclusion: A single sample *t* was computed comparing the sample mean with a known population mean of 80. The difference was not significant. Because no significant difference was found ($t(9) = .268, p > .05$), the null hypothesis was accepted. This sample could be representative of the known population.

Two-Sample t

Please use problem #3 on page 262. Is there a difference in performance as a result of varying light conditions? Two independent random samples were selected and timed as to how long it took (in hours) to solder 100 transistors. Group A worked in low light conditions, while Group B worked under high light conditions.

Double-click the top of the first column and when the Variable View screen comes up type subj# in the first row, group in the second row, and scores in the third row. Now click the data view tab to get back to the data screen. Data are

LoLight Group A	*HiLight Group B*
6.5	5.0
5.0	4.0
.	.
.	.
4.5	3.7
6.2	4.2
5.3	3.7

Rearrange the data such that Group A is labelled as 1 and Group B as 2, and enter the following scores:

Subj#	*Group*	*Scores*
1	1	6.5
2	1	5.0
.		
.		
5	1	4.5
6	1	6.2
7	1	5.3
8	2	5.0
9	2	4.0
.		
.		
12	2	3.7
13	2	4.2
14	2	3.7

Save as p262.

1. Click Analyze, Compare Means, and Independent Samples *T* test.
2. Put Scores in Test Variable Box.
3. Put Group in as Grouping Variable.
4. Click Define Groups.
5. In the Group 1 box type 1.
6. In the Group 2 box type 2.

Click Continue.
OK

SELECTED OUTPUT

T-Test

Group Statistics

	GROUP	N	Mean	Std Deviation	Std Error Mean
SCORES	1.00	7	5.0857	.9856	.3725
	2.00	7	3.8714	.6264	.2368

Independent Samples Test
Levene's Test for
Equality of Variances

		F	Sig
SCORES	Equal variances assumed	1.876	.196
	Equal variances not assumed		

t-test for equality of means

		t	df	Sig (2 tailed)	Mean Difference
SCORES	Equal variances assumed	2.751	12	.018	1.2143
	Equal variances not assumed	2.751	10.167	.020	1.2143

t-test for equality of means

		Std. Error Difference	95% Confidence Interval of the Difference	
			Lower	Upper
SCORES	Equal variances assumed	.4414	.2526	2.1760
	Equal variances not assumed	.4414	.2330	2.1956

Conclusion. A two-sample independent t test was computed and found to be significant ($t(12) = 2.751, p < .05$). Levene's test indicated equal variances ($F = 1.876, p > .05$). The null hypothesis was rejected and it was concluded that the high-light group took significantly less time to complete the job and therefore worked at a quicker pace.

Levene's test ($F = 1.876$, $p > .05$) indicates that the assumption of homogeneous variances was justified.

*Note that the output also produced the Interval Estimate for the difference. See text page 245.

Pearson r

Please use problem #1 on page 301. Is there a dependable relationship between party loyalty and number of hours worked per week for the party's candidate? That is, can hours worked be predicted from a Party Loyalty test? For this procedure the scores are listed side by side.

1. Double-click the top of the first column and when the variable view screen appears type in Subj# in the first row, Loyalty in the second and Hours in the third. Now click the "data view" tab at the bottom to get back to the data screen.

Enter the scores as follows:

Subject#	*Loyalty*	*Hours*
1	18	30
2	3	5
3	7	6
.		
.		
7	12	10
8	8	6
9	16	25
10	5	6

Save as p301.

2. Click Analyze.
3. Correlation
4. Bivariate
5. Click on both variables, loyalty and hours, and put them in the Variables Box.
6. Click Pearson.
7. Click two-tailed and make sure the flag is checked for all significant correlations.
8. Click OK.

SELECTED OUTPUT

	LOYALTY	*HOURS*
Pearson Correlation	1.000	.907
Significance (two-tailed)		.000
N	10	10

Conclusion: A Pearson correlation was computed between scores on a Party Loyalty test and the number of hours per week worked for the candidate. A strong, positive correlation was found ($r(8) = .907$, $p < .01$). Since a significant linear correlation was found, it can be concluded that the more loyal a person is the more hours that person did work for the candidate.

At this point, it is instructive to create a scatter plot.

1. Click graphs and select Scatter.
2. When a scatter plot box appears, check Simple and click Define. Now you will see the main scatter plot box, and move hours into the Y axis box, since that is the dependent variable and move loyalty into the X axis box, since that is the independent variable.
3. Click Titles and in line 1, type Hours as a function of Loyalty.
4. Click OK and there it is. In order to see the whole scatter plot, you may have to scroll to the right.

Notice as you examine this scatter plot, the array of data points slopes from lower left to upper right (which is the case with a positive correlation). Had the correlation been negative, the data points would have sloped from upper left to lower right.

Spearman Correlation

With ordinal data or skewed interval data, the Spearman should be chosen as the appropriate correlation tool. Please use the example on the top of page 298. Is there a significant relationship among members of the alumni association between the amount donated to the college's fund drive and the age of the alumnus? Now enter the data on the top of the page, which shows both Donation and Age converted to ordinal form.

Subject#	*Donation*	*Age*
1.	12	9
2.	9	6
3.	1	2
.		
.		
10.	8	8
11.	5	4
12.	10	10

Save as p298.

1. Double-click the top of the first column and when the variable view screen appears type in Subj# in the first row, Donation in the second, and Age in the third. Now click the "data view" tab at the bottom to get back to the data screen.
2. Click Analyze.

3. Click Correlate.
4. Click Bivariate.
5. Move the two variables, Donation and Age over into the Variables box.
6. Check Spearman and toggle off the check at Pearson.
7. Click two-tailed and check the flag for significant correlations.
8. Click OK.

SELECTED OUTPUT

Look at the section that is headed by Nonparametric Correlations

		DONATION	*AGE*
Spearman's rho			
DONATION	Correlation Coefficient	1.000	.790
	Sig (two-tailed)		.002
	N	12	12
AGE	Correlation Coefficient	.790	1.000
	Sig (two-tailed)	.000	
	N	12	12

Conclusion: A Spearman correlation was computed between the amount of donation and the age of a group of college alumni. A positive correlation was found ($r(12) = .790$, $p < .01$). The correlation was significant and indicated that donation amounts are positively correlated with age. Older persons donated more to their college than did younger persons.

One-Way ANOVA

Please use problem #4, on page 345, where shock was manipulated as the IV and retention levels were recorded as the DV. There were three levels of the IV.

1. Double-click the top of the first column and when the variable view screen appears, type in Group in the first row and Scores in the second. Now, click the "data view" tab at the bottom to get back to the data screen.
2. Define variables, Scores as DV and Group as IV and rearrange the data from this:

None	*Medium*	*High*
24	20	16
.	.	.
24	20	15
20	17	16

To this:

Group	*Scores*
1	24
1	.
1	.
1	24
1	20
2	20
2	.
2	.
2	20
2	17
3	16
3	.
3	.
3	15
3	16

Save as p345#4.

1. Click Analyze, then General Linear model, then Univariate.
2. Then place scores as DV and groups as fixed factor.
3. Click Options, Descriptive Statistics, and Estimates of Effect Size.
4. Continue.
5. Click Post Hoc and move Group into the box for Post Hoc Tests.
6. Click Tukey and Continue.
7. Click OK.

SELECTED OUTPUT
Univariate Analysis of Variance
Between Subjects Factors

		N
GROUP	1.00	5
	2.00	5
	3.00	5

(Which gives you the size of each sample).

Descriptive Statistics

Dependent Variable: SCORES

GROUP	*MEAN*	*Std Deviation*	*N*
1.00	24.6000	3.5777	5
2.00	19.4000	1.9494	5
3.00	14.6000	1.6733	5
Total	19.5333	4.8383	15

(which gives you all three Means and SDs as well as the total Mean and SD.)

Tests of Between Subjects

DV Scores Source	*TypeIIISS*	*df*	*MS*	*F*	*Sig*	*Partial Eta Squared*
Corrected Model	250.133	2	125.067	19.340	.000	.763
Intercept	5723.267	1	5723.267	885.041	.000	.987
GROUP	250.133	2	125.067	19.340	.000	.763
Error	77.600	12	6.467			
Total	6051.000	15				
Corrected Total	327.733	14				

Here we see the GROUP results, and the F ratio of 19.340, which is clearly significant.

Multiple Comparisons

Dependent Variable: SCORES

Tukey HSD

					95% Confidence Interval	
GROUP	*GROUP*	*Mean Difference*	*Std Error*	*Sig*	*Lower Bound*	*Upper Bound*
1.00	2.00	5.200	1.6083	.018	.9092	9.4908
	3.00	10.000	1.6083	.000	5.7092	14.2908
2.00	1.00	–5.200	1.6083	.018	–9.4908	–.9092
	3.00	4.800	1.6083	.029	.5092	9.0908
3.00	1.00	–10.000	1.6083	.000	–14.2908	–5.7092
	2.00	–4.800	1.6083	.029	–9.0908	–.5092

In the above table we find every possible combination of group differences. Each group is being compared with every other group. Notice that in this case, all mean differences are significant, at least at the .05 level. Finally we get

Homogeneous Subsets

		SCORES		
		Subset		
GROUP	N	1	2	3
3.00	5	14.800		
2.00	5		19.400	
3.00	5			24.600

Only the means for homogeneous subsets are being displayed, meaning that all three samples pass the homogeneity test.

Conclusions: A one-way ANOVA was computed comparing the retention scores of subjects who were tested under varying shock levels from low shock to medium shock to high shock. A significant difference was found between the groups ($F(2,12) = 19.340$, $p < .01$). A post-hoc analysis was conducted using Tukey's HSD, and all three groups were found to differ significantly from each other, Groups 1 ($m = 24.600$) was significantly higher than Group 2 ($m = 19.400, p < .05$), which, in turn, was significantly higher than Group 3 ($m = 14.600$, $p < .05$). Finally the difference between groups 1 and 3 was significant ($p < .01$). Further tests indicated homogeneous variances within the groups.

Factorial ANOVA

The factorial ANOVA is still a univariate analysis, since there is only one dependent variable. Unlike the one-way ANOVA, however, this procedure allows for more than just a single IV. Thus, we can select out the main effects for each IV, as well as for any interactions. Please use the data from problem #7 on page 347. For this exercise, please label temperature Hi as 1, and temperature Lo as 2, and label exercise as 1, and no exercise as 2

The GSR (DV) Scores are as follows:

	Hi Temp	Lo Temp
	9	7
	.	.
Exercise	.	.
	10	5
	7	4
	8	5
	5	3
	.	.
No Exer	.	.
	4	1
	3	2
	4	1

1. Double-click the top of the first column and when the variable view screen appears, type in Subj# in the first row, Temperature in the second, Exercise in the third, and gsr in the fourth row. Now click the "data view" tab at the bottom to get back to the data screen.
2. Then rearrange the data as shown below:

Subject #	*Temp*	*Exercise*	*GSR*
1	1	1	9
2	1	1	.
3	1	1	.
4	1	1	10
5	1	1	7
6	1	1	8
7	1	2	5
8	1	2	.
9	1	2	.
10	1	2	4
11	1	2	3
12	1	2	4
13	2	1	7
14	2	1	.
15	2	1	.
16	2	1	5
17	2	1	4
18	2	1	5
19	2	2	3
20	2	2	.
21	2	2	.
22	2	2	1
23	2	2	2
24	2	2	1

Save as p347#7

1. Click Analyze, General Linear, and Univariate.
2. Enter GSR as the DV.
3. Put temperature and exercise in as FIXED FACTORS (the two IVs).
4. Click Options, and then move exercise, temperature, and exercise x temperature into the display means box. Also check Descriptive Statistics and Estimates of Effect Size.

Important Note

If either of the IVs had been set at more than two levels then Click Post-Hoc, and put whichever IVs are at three or more levels into the "Post-Hoc Tests for" box and check Tukey. In this problem we don't need Tukey, because with only two levels of the IVs, the F ratio alone will tell you if there is a significant difference.

5. Check Continue and click OK.

SELECTED OUTPUT FOR FACTORIAL ANOVA

Univariate Analysis of Variance

Between-Subjects Factors

		N
TEMP	1.00	12
	2.00	12
Exercise	1.00	12
	2.00	12

Descriptive Statistics

Dependent Variable GSR

TEMP	*EXERCISE*	*MEAN*	*Std Deviation*	*N*
1.00	1.00	8.3333	1.0328	6
	2.00	3.6667	.8165	6
	Total	6.0000	2.5937	12
2.00	1.00	5.3333	1.0328	6
	2.00	1.8333	.7528	6
	Total	3.5833	2.0207	12
Total	1.00	6.8333	1.8505	12
	2.00	2.7500	1.2154	12
	Total	4.7917	2.5872	24

This screen gives us all the means, separately and in combination. We get the mean for each treatment condition, as well as the row mean, column mean, and overall mean. We are also given all the standard deviation combinations. This table becomes a valuable source for us when graphing the results.

Tests for Between-Subjects

Dependent Variable GSR

Source	*TypeIIISS*	*df*	*MS*	*F*	*Sig*	*Partial Eta Squared*
Corrected Model	137.125	3	45.708	54.307	.000	.891
Intercept	551.042	1	551.042	654.703	.000	.970
TEMP	35.042	1	35.042	41.634	.000	.676
EXERCISE	100.042	1	100.042	118.861	.000	.856
TEMP*EXERCISE	2.042	1	2.042	2.426	.135	.108
Error	16.833	20	.842			
Total	705.000	24				
Corrected Total	153.958	23				

This is the table that provides the heart of the analysis. We interpret the F ratios for the two main effects, TEMP ($F = 41.634$) and EXERCISE ($F = 118.861$), as well as for the interaction, TEMP*EXERCISE ($F = 2.426$)

Conclusions: A 2 x 2 between-subjects factorial ANOVA was conducted in an attempt to discover possible differences in GSR scores as a function of the two independent variables, temperature and exercise. A significant main effect for temperature was found ($F(1,20) = 41.634$, $p < .01$), as well as a significant main effect for exercise ($F(1,20) = 118.861$, $p < .01$). Subjects had higher GSR scores the higher the temperature ($m = 6.000$ for high temperature and $m = 3.5833$ for low temperature). They also had higher GSR scores if they participated in the exercise program, $m = 6.833$ versus $m = 2.7500$ if they did not participate. There was no significant interaction effect ($F(1,20) = 2.426$, $p > .05$). Neither of the main effects was significantly influenced by the other.

Before graphing the results, it is helpful to manually set up a matrix, using the data from the means table shown previously. If you remember that Hi Temp is 1 and Low Temp is 2, and that Exercise is 1 and No-Exercise is 2, then the table looks like this:

	TEMPERATURE	
	1. High	*2. Low*
Exercise 1.	8.3333	5.3333
No Exercise 2.	3.6667	1.8333

Using SPSS for Graphing Your Results

Follow the procedure shown previously, but this time when you get to step 6 Click Continue and then INSTEAD OF OK,

7. Click Plots, and move temperature into the horizontal box and exercise into the separate lines box.
8. Click Add.
9. Click Continue and OK.

Chi-Square

The 1 by k Chi Square

To illustrate the 1 × k Chi Square, turn to page 356 and use the data from the table of radio station preferences. There are four stations, A,B,C, and D, with observed frequencies of 40, 30, 20 and 10.

Enter the data as follows:

1. Double-click the top of the first column and when the variable view screen appears, type station in the first row and number in the second row. Now click the tab at the bottom called "data view" to get back to the data screen.
2. Enter the data as follows:

Station	*Number*
1	40
2	30
3	20
4	10

Now file-save by saving as p356.

3. Click data and weight cases.
4. The next screen asks whether to weight cases, and you should check the second option, "weight cases by."
5. Click number and move it into the Frequency Variable box.
6. Click Analyze and Nonparametric Tests.
7. Click Chi-Square, and put station in Test Variable List and check All Categories equal.
8. Click OK.

Selected Output

NPar Tests

Chi-Square Test

Frequencies

	Station *Observed* N (f_o)	*Expected* N (f_e)	*Residual* ($f_o - f_e$)
1.00	40	25	15.0
2.00	30	25	5.0
3.00	20	25	–5.0
4.00	10	25	–15.0
Total	100		

Test Statistic

	STATION
Chi-Square	20.000
df	3
Asymp. Sig.	.000

0 cells (0%) have expected frequencies less than 5.

Conclusions: A goodness-of-fit Chi-Square was computed comparing the frequency of occurrence of a sample of TV viewers according to four radio station categories. A significant difference was found between the observed and expected values (chi-square(3) = 20.000, $p < .01$.).

The r × k Chi Square

For the $r \times k$ chi square illustration, please use the problem on page 374 (since it involves both the Chi Square itself as well as the Coefficient of Contingency, which is a nominal measure of both correlation and effect size). The 2 × 2 contingency table is as follows:

	Incarcerated Again		
	Yes	*No*	*Total*
Treatment Program	30	70	100
No Treatment	62	38	100

Enter the data by using the following steps: Double-click the top of the first column and when the variable view screen appears, type in treatment in the first row, incarceration in the second, and number in the third row. Now click the "data view" tab at the bottom and get back to the data screen. Notice that the number 30 is in treat row 1 and incarc column 1. The number 70 is in treatment row 1, but is in incarc column 2 and so on. Now enter the data as follows, and save as p374:

Treat	*Incarc*	*Number*
1	1	30
1	2	70
2	1	62
2	2	38

1. Click Data and then select Weight Cases (at the bottom of the menu).
2. Click weight cases by and then check the "weight cases by" option.
3. Move number into the Frequency Variable box and click OK.
4. Click Analyze, Descriptives, and Crosstabs.
5. Put treat in rows and incarc in columns.
6. Click statistics and then check both Chi-Square and C, the coefficient of contingency.
7. Click Continue and OK.

Selected Output

Crosstabs

Case Processing Summary

	Cases					
	Valid		*Missing*		*Total*	
	N	*Percent*	N	*Percent*	N	*Percent*
TREAT*INCARC	200	100.0%	0	.0%	200	100.0%

TREAT*INCARC Crosstabulation

Count

		INCARC		
		1.00	2.00	Total
TREAT	1.00	30	70	100
	2.00	62	38	100
Total		92	108	200

Chi-Square Tests

	Value	df	Asymp. Sig. (Two-sided)	Exact Sig. (Two-sided)	Exact Sig. (One-sided)
Pearson Chi-Square	20.612	1	.000		
Continuity Correct*	19.344	1	.000		
Likelihood Ratio	20.992	1			
Fisher's Exact				.000	.000
Linear by Linear Association	20.509	1	.000		
N of Valid Cases	200				

*Computed only for a 2 × 2 Table

0 Cells have expected count less than 5.

Symmetric Measures

	Value	Approx Sig.
Nominal by Nominal Contingency Coeff.	.306	.000
N of Valid Cases	200	

Conclusions: A 2 × 2 chi-square was computed comparing the frequency of recidivism between inmates who had received treatment and those who had not. A significant difference was found between the two groups (chi-square(1) = 20.612, $p < .01$). (Had the Chi-Square been used to test for correlation, the C (Contingency Coefficient) of .306 would then be included).

Multiple Regression

Please use problem #18 on page 422.

1. Double-click the top of the first column and when the variable view screen appears, type in Student# in the first row, *ES* in the second row, S in the third row and V in the fourth row. Now click the "data view" tab at the bottom to get back to the data screen.

These refer to V (Venturesomeness), ES (Experience Seeking), and S (Sociability from Eysenck Personality Test).

For this example, we are going to predict *V* as the dependent variable.

Student#	*ES*	*S*	*V*
1	16	5	8
2	19	6	8
3	15	7	7
4	15	6	8
5	14	5	6
.			
.			
.			
.			
15	14	2	4
16	15	7	12
17	17	5	4
18	26	14	17
19	16	5	8
20	14	6	7

Save as p422.

1. Click Analyze, Regression, Linear.
2. Click *V* (since that is the variable being predicted) and put in as the DV.
3. Click *ES* and put in as the first IV.
4. Click *S* and put in as the second IV.
5. Click "statistics"—bottom middle of screen and click descriptives (keep estimates and model fit).
6. Continue.
7. Click OK.

SELECTED OUTPUT

Descriptive Statistics

	Mean	*SD*	*N*
V	7.6500	3.0483	20
ES	15.0000	3.9068	20
S	6.1000	2.2455	20

Correlations

		V	*ES*	*S*
Pearson Correlation	*V*	1.000	.716	.774
	ES	.716	1.000	.516
	S	.774	.516	1.000

Model Summary

Model	*R*	*R square*	*Adj R square*	*SE estimate*
1	.858	.736	.705	1.6558

ANOVA

	SS	*df*	*MS*	*F*	*Sig*
Regression	129.943	2	64.971	23.698	.000
Residual	46.607	17	2.742		
Total	176.550	19			

Coefficients

	Unstandardized Coefficients		*Standardized Coefficient*		
Model	*B*	*Std Error*	*Beta*	*t*	*Sig*
a (Constant)	−1.967	1.540		−1.277	.219
b1 *ES*	.336	.114	.431	2.964	.009
b2 *S*	.749	.197	.552	3.793	.001

Dependent Variable *V* score

Notice that the output gives us four important tables. First, we get the matrix of internal correlations, showing the Pearson r relationships among the three variables, *V*, *ES* and *S*. Then, under Model Summary, we get both the multiple R (.858) and R square (.736). Then, under ANOVA, we find the significance of the multiple R. Finally under coefficients we get the regression equation. The constant of −1.967 is the a value or intercept, and the b values are the slopes, or the weights associated with the regression equation. Thus, the regression equation is read as follows.

$$Y_{\text{mult pred}} = .336 + .749 - 1.967$$

This is the equation for predicting *V* values on the basis of *ES* and *S* scores. The Beta values are the *standardized weights* (based on a mean of 0 and an *SD* of 1.00) and indicate the relative importance of the independent variables. Notice in this case that S was a stronger predictor, .552, than *ES*, .431. The *t* values give the significance levels of the weights.

Conclusions: A multiple linear correlation was computed to predict *V* (Venturesomeness) from both *ES* (Experience Seeking) and *S* (Sociability). The *R* was found to be .858, and was significant ($F(2,17) = 23.698, p < .01$). The linear combination of *ES* and *S* was a significant predictor of *V*.

Partial Correlation

The problem to be addressed here, shown on page 421, involves evaluating the correlation between reading speed and reading comprehension, with the influence of IQ being partialled out or controlled. Enter the data as follows:

1. Double-click the top of the first column and when the variable view screen appears, type in rs in the first row, comp in the second, and iq in the third row. Now click the "data view" tab at the bottom to get back to the data screen.

Then enter the following:

rs	*Comp*	*iq*
55	50	100
35	45	95
75	60	135
.	.	.
.	.	.
.	.	.
42	48	100

Save as partial OR PAGE421.

1. Click Analyze.
2. Click Correlate.
3. Click Partial and put rs and comp in the Variables box and iq in the controlling for box.
4. Check Two-tailed and click OK.

Selected Output

Partial Corr

. . . . Partial Correlation Coefficient . . . .

Controlling for.. IQ

	RS	*COMP*
RS	1.0000 (0)	.1246 (7) P = .749
COMP	.1246 (7) *p* = .749	1.0000 (0)

(Coefficient / (D.F.) / Two-tailed Significance

Conclusions: A partial correlation coefficient was computed between Reading Speed and Reading Comprehension with the IQ variable being partialled out. The resulting correlation ($r(7) = .1246$, $p > .05$) was not significant. The two variables, Reading Speed and Reading Comprehension, are independent of each other when the IQ variable is statistically controlled.

*The p value of .749 (on page 612) shows the significance level to be FAR above .05, so high as to indicate the chance of alpha error being almost 75%.

Finally, it would be instructive if you were to run a Pearson correlation (the procedure for which was shown previously) between reading speed (*rs*) and comprehension (comp). You will find that the correlation is .785, which is significant at .008. Thus, when IQ was controlled for, that strong and highly significant correlation virtually disappeared.

Paired t

Please use problem #1 on page 448. Notice that as was the case with the Pearson r (previously shown), scores are listed side by side.

1. Double-click the top of the first column and when the variable view screen appears, type in Subj# in the first row, pre in the second row, and post in the third row. Now click the "data view" tab at the bottom and get back to the data screen.
2. Then enter the measures

Subject#	*Pre*	*Post*
1	11	8
2	15	15
.		
.		
.		
.		
7	6	6
8	11	8

Save as p448.

3. Click Analyze.
4. Compare Means.
5. Paired t
6. Current Selection, click pre and post and then click them both into paired variables.
7. Click OK.

SELECTED OUTPUT

PAIRED SAMPLES CORRELATION

Pair 1 PRE&POST	N	Corr	Sig
	8	.792	.019

PRE-POST	Mean	SD	SE_{mean}	95% Confidence Lower	95% Confidence Upper	t	df	Sig
	2.000	2.0702	.7319	.2693	3.7307	2.733	7	.029

Conclusion: The paired t was computed on this pre-post study. The Pearson r correlation was found to be significant (r,(6) = .792, $p < .05$). The paired t was significant ($t(7) = 2.733$, $p < .05$). The scores were significantly lower in the post test.

Within-Subjects Repeated Measure ANOVA

Please use the example on page 442.

1. Double-click the top of the first column and when the variable view screen appears, type subject# in the first row, first in the second row, second in the third row, and third in the fourth row. Now click the "data view" tab at the bottom to get back to the data screen.
2. Enter data

Subject#	First	Second	Third
1	1	2	3
2	2	4	6
3	3	3	3
4	4	5	6
5	5	6	7

Save as p442.

3. Click Analyze.
4. General Linear
5. GLM repeated measures
6. Type time in Factor Name and 3 in number of levels.
7. Click ADD, and Define. Then click on first, second, third, and move to within Subjects box. Click options and then descriptive statistics.
8. Continue and click OK.

Important Note: If your version of SPSS does not have "GLM Repeated Measures," go to your computer lab and inquire as to whether the full version of SPSS is available to you. If so, you will be able to complete the preceding problem in the lab.

SELECTED OUTPUT

Descriptive Statistics

	Mean	*Std Deviation*	*N*
FIRST	3.0000	1.5811	5
SECOND	4.0000	1.5811	5
THIRD	5.0000	1.8708	5

Tests of Within-Subjects Effects

Source		*SS*	*df*	*MS*	*F*	*Sig*
TIME	sphericity assumed	10.000	2	5.000	10.000	.034
Error (TIME)	sphericity assumed	4.000	8	.500		

Note: Sphericity is a term used in within-subjects ANOVAs to indicate that the variances of the *differences* for the paired measures are fairly equal. Unless this assumption is met the within-subjects F ratio may lead to ambiguous results. SPSS uses the Mauchly test to determine sphericity.

Conclusions: A one-way within-subjects ANOVA was computed, comparing political attitudes for subjects at three different times, after viewing a propaganda film once, then after seeing it again, and finally after viewing it for a third time. A significant F ratio was found ($F(2,8) = 10.000, p < .01$). As the film was repeatedly shown, the subjects' attitudes became increasingly positive.

Since SPSS will not do a post-hoc on this design, refer to page 444. You will discover that the only significant difference occurred between the first and third time periods. For the next four tests, the Ordinal tests, you may have to use your college's lab computers, assuming they have the full version of SPSS.

The Ordinal Tests

The Mann-Whitney Test

Turn to page 459 and use the data for income. Call Republicans Group1 and Democrats Group2. Double-click the top of the first column and when the variable view screen appears, type income in the first row and group in the second row. Now click the "data view" tab at the bottom to get back to the data screen.

Then enter the following:

Line the data up as follows:

Income	*Group*
40000	1
41000	1
43000	1

Data continued on next page.

Income	*Group*
42000	1
190000	1
.	
.	
30000	1
16000	2
17000	2
20000	2
21000	2
.	
.	
.	
28000	2

Save as p459.

1. Click Analyze.
2. Nonparametrics
3. Click 2 Independent Samples.
4. Put income in Test Variable and group in grouping variable.
5. Click Mann-Whitney.
6. Define Variables, typing a 1 for Group1 and a 2 for Group2.
7. Continue and click OK.

Selected Output

NPar Tests

Mann-Whitney Test

	Ranks			
	GROUP	*N*	*Mean Rank*	*Sum of Ranks*
INCOME	1.00	10	15.50	155.00
	2.00	11	6.91	76.00
	Total	21		

Test Statistics

	INCOME
Mann-Whitney	10.000
Wilcoxon W	76.000
Z	–3.169
Asympt. Sig (two-tailed)	.002
Exact Sig [2* (one-tailed Sig.)]	.001

Don't be alarmed because of the Mann-Whitney value of 10 (versus the book value of 100). The equation works out the same (except for the sign) and both the program and the book produce the exact same Z value of 3.169. Since this is a two-tailed test, the sign of the Z is irrelevant and we get a significance level of .002.

Conclusions: A Mann-Whitney test was computed to test for the difference in income levels between Democrats and Republicans. The Republicans earned more, ranked higher, with a mean rank of 6.91 compared to the Democrats' lower mean rank of 15.50. The analysis showed a significant difference ($z = -3.169$, $p < .001$).

Kruskal-Wallis H *Test*

Using the data on page 461, double-click the top of the first column and when the variable view screen appears type Group 1 in the first row, Group 2 in the the second row, Group 3 in the third row, and Group 4 in the fourth row. Now click the "data view" tab at the bottom to get back to the data screen. We have set the data up in groups so that the 21–30 ages are in Group 1, the 31–40s in Group 2, the 41–50s in Group 3 and the 51–60s in Group 4. The set-up follows:

Ranks	*Group*
2	1
4	1
18	1
.	1
.	1
.	1
20	2
6	2
8	2
.	2
.	2
.	2
23	3
11	3
15	3
.	3
.	3
.	3
24	4
17	4
22	4
.	4
.	4
16	4

Save as p461.

1. Click Analyze.
2. Click Nonparametrics.
3. Click K Independent Samples.
4. Put Rank in Test Variable List and Group in Grouping Variable.
5. Define Range and type 1 for Minimum and 4 for Maximum.
6. Continue and click OK.

Selected Output

NPar Tests

Kruskal-Wallis Test

Ranks

	Group	*N*	*Mean Rank*
RANKS	1.00	6	5.83
	2.00	6	10.00
	3.00	6	14.33
	4.00	6	19.83
	Total	24	

Test Statistic

	RANKS
Chi-Square	12.940
df	3
Asymp. Sig.	.005

Conclusions: A Kruskal-Wallis test was performed on pilot ratings based on age. The differences were significant, with the younger pilots receiving significantly higher ratings than was the case for the older pilots ($H(3) = 12.940$, $p < .01$).

Wilcoxon T

Use the data on page 463 and double-click the top of the first column and when the variable view screen appears type pair in the first row, Group1 in the second row and Group2 in the third row. Now click the "data view" tab at the bottom to get back to the data screen.

Enter the data as follows:

Pair	*Group1*	*Group2*
1.	85	86
2.	90	95
3.	92	96

Continued

Pair	*Group1*	*Group2*
.		
.		
.		
9.	140	133
10.	150	135

Save as p463.

1. Click Analyze.
2. Click Nonparametrics, 2 Related Sample.
3. Put BOTH group1 and group2 into the Test Pairs List.
4. Click OK.

Selected Output

NPar Tests

Wilcoxon Signed Ranks Test

Ranks

		N	*Mean Rank*	*Sum of Ranks*
GROUP2 – GROUP1	Negative Ranks	2	8.50	17.00 = T
	Positive Ranks	7	4.00	28.00
	Ties	1		
	Total	10		

Test Statistic

	GROUP2 GROUP1
Z	–.652
Asympt. Sig. (two-tailed)	.514

Conclusions: A Wilcoxon T test was computed on the golf scores of a matched groups of golfers, one group having had special training and the other with no training. No significant differences were found ($Z = -.652$, $p > .05$).

Note that if the T value of 17.00 had been used (shown above as the sum of the less frequent ranks) then the conclusions (based on a p value > .05) would have been exactly the same.

Friedman ANOVA

Use the data on page 465. Double-click the top of the first column and when the variable view screen appears type trial in the first row, group1 (for independent) in the second row, group2 (for Democrat) in the third row, and group4 (for Republicans) in the fourth row. Now click the "data view" tab at the bottom to get back to the data screen.

Type the data as follows:

Trial	*Group1*	*Group2*	*Group3*
1.	26	30	22
2.	29	31	28
3.	55	60	54
.			
.			
.			
.			
8.	40	39	38
9.	41	42	43
10.	45	44	46

Save as p465.

1. Click Analyze.
2. Click Nonparametrics.
3. Click K Related Samples.
4. Put all three groups into Test Variables.
5. Click Friedman.
6. Click OK.

Selected Output.

NPar Tests

Friedman Test

Ranks

	Mean Rank
GROUP 1	1.90
GROUP 2	2.20
GROUP 3	1.90

	Test Statistics
N	10
Chi-Square	.600
df	2
Asymp. Sig.	.741

Conclusions: The Friedman ANOVA by ranks was used to assess whether political affiliation differs according to age. The results were not significant (chi-square(2) = .600, $p > .05$). The political affiliation of the subjects did not differ on the basis of age.

Reliability and Validity

For the first reliability problem, turn to the data on page 479 in which the reliability of the figure-ground test is to be assessed. Since this is a test-retest analysis, use the procedure for the Pearson *r*.

1. Double-click the top of the first column and when the variable view screen appears, type in Subj# in the first row, test in the second, and retest in the third. Now click the "data view" tab at the bottom to get back to the data screen.

Enter the scores as follows:

Subject#	*Test*	*Retest*
1	15	16
2	12	15
3	11	12
.		
8	16	17
9	9	9
10	21	21

Save as p479.

2. Click Analyze.
3. Correlation
4. Bi Variate
5. Click on both variables, test and retest, and put them in the Variables Box.
6. Click Pearson.
7. Click OK.

SELECTED OUTPUT

Correlations

	TEST	*RETEST*
TEST Pearson Correlation	1.000	.891
Sig. (two-tailed)		.001
N	10	10

Correlation is significant at the 0.01 level (two-tailed).

Split Half Reliability

For this analysis, use the data on page 482, where a split-half reliability is being calculated. With SPSS, you don't have to compute the original Pearson *r* and then correct it. Instead you can go directly to the Corrected Spearman-Brown value.

1. Double-click the top of the first column and when the variable view screen appears type in Subj# in the first row, odd in the second, and even in the third. Now click the "data view" tab at the bottom to get back to the data screen.

Enter the scores as follows:

Subject#	Odd	Even
1	125	120
2	100	100
3	75	80
.		
8	97	98
9	103	102
10	100	100

Save as p482.

2. Click Analyze.
3. Click Scale.
4. Reliability
5. Click on both variables, odd and even, and put them in the Variables Box.
6. Go to the pull-down menu next to "model" and move down to split-half.
7. Click Split Half.
8. Click OK.

SELECTED OUTPUT

RELIABILITY ANALYSIS—SCALE (SPLIT)

Reliability Coefficients

N of Cases = 10.0
Correlation between forms = .9564
Guttman Split-half = .9696

N of Items = 2
Equal length Spearman-Brown = .9777
Unequal-length Spearman-Brown = .9777

Item Analysis

Coefficient Alpha

Cronbach's alpha will give precisely the same value as will the Cyril Hoyt. For this example, then, use the Cyril Hoyt data found on page 486. Double-click the top of the first column and then on the variable view screen type in Subject# in the first row, itema (for item a) in the second, and so on down to itemd. Then click the tab back to the data view screen and enter the scores as shown below.

Subj#	Item a	Item b	Item c	Item d
1.	6	6	5	4
2.	4	6	5	3
.	.	.	.	.
.	.	.	.	.
5.	1	2	1	1

Save as p486.

1. Click Analyze then Scale.
2. Click Reliability Analysis
3. Move all four items into the items box.
3. Click Alpha.
4. Click OK.

Selected Output

Reliability

Reliability Analysis- Scale (alpha)

Reliability Coefficient

N of cases = 5.0 *N* of items = 4.0

Alpha = .9198

Validity

For this problem, use the data found on page 494.

1. Double-click the top of the first column and when the variable view screen appears type in Subject in the first row, ranks1 (for the score ranks) in the second and ranks2 (for the judges ranks) in the third. Now click the "data view" tab at the bottom to get back to the data screen.

Enter the scores as follows:

Subject	Ranks1	Ranks2
1	2	3
2	10	9
3	8	10
.		
8	9	7
9	3.5	6
10	5	4

Save as p494.

2. Click Analyze.
3. Correlation

4. Bi Variate
5. Click on both variables, ranks1 and ranks2, and put them in the Variables Box.
6. Blank out the Pearson box and click on the Spearman box.
7. Click OK.

SELECTED OUTPUT
Nonparametric Correlations

		RANKS1	*RANKS2*
Spearman's rho	RANKS1 Correlation Coefficient	1.000	.823
	Sig (2-tailed)		.003
	N	10	10

Glossary

Abscissa The horizontal or X axis of the coordinate system. On a frequency distribution, the abscissa typically measures the variable in question (performance measures), whereas the Y axis (ordinate) represents the frequency of occurrence.

Alpha Level The probability of committing a Type 1 error. The level is usually set beforehand and should not be higher than .05.

Alternative Hypothesis The opposite of the null hypothesis. The alternative hypothesis states that chance has been ruled out—that there are population differences (when testing the hypothesis of difference) or that a correlation does exist in the population (when testing the hypothesis of association).

Analysis of Covariance (ANCOVA) A statistical procedure designed to control the effects of any variable(s) that is known to be correlated with the dependent variable. ANCOVA is typically used as an after-the-fact adjustment for controlling any possible differences that may have already existed between comparison groups.

Analysis of Variance Statistical test of significance developed by Sir Ronald Fisher. It is also called the F ratio, or ANOVA, for ANalysis Of VAriance. The test is designed to establish whether a significant (nonchance) difference exists among several sample means. Statistically, it is the ratio of the variance occurring between the sample means to the variance occurring within the sample groups. A large F ratio—that is, when the variance between is larger than the variance within—usually indicates a nonchance or significant difference.

Asymptote The actual limit of the continuous graph that the curve approaches but never reaches (except at infinity). On a normal curve, the graph approaches but never reaches the abscissa. On a learning curve, the curve approaches but never reaches the flat plateau. Such curves are asymptotic.

Beta Coefficient (*b*) (or **Slope**) In a scatter plot, the slope of the regression line of Y on X indicates how much change on the Y variable accompanies a one-unit change in the X variable. When the slope is positive (lower left to upper right), Y will show an increase as X increases, whereas a negative slope (upper left to lower right) indicates a decrease in Y is accompanying an increase in X. In the regression equation, $Y = bX + a$, the slope is symbolized by the b term.

$$b = \frac{rSD_y}{SD_x} \text{ or } \frac{rs_y}{s_x}$$

Beta Level The probability of committing a Type 2 error.

Bias Systematic or nonrandom sampling error. Occurs when the difference between M and μ is consistently in *one direction.* Bias results when samples *are not* representative of the population.

Bonferroni Test A test designed to control type 1 error in an ANOVA. It tests across all pairwise comparisons by dividing the set alpha level by the number of comparisons. If there were three groups and three comparisons, .05 divided by 3 = .017. To get significance, each t ratio must now reach .017 to be significant. With a two-group t ratio, there is no need to conduct the Bonferroni, since there is only one comparison, M1 versus M2, and dividing .05 by 1 equals .05.

Canonical Correlation A correlation technique where a set of IVs (predictor variables) are used to predict a set of DVs (criterion variables).

Ceiling Effect Occurs when scores on the DV are so high to begin with that there is virtually no room left for them to go up any further. On the other side is what might be called the "cellar effect," where the DV scores are so low to begin with that there is little likelihood of any further decline. This problem is especially apparent in before-after research designs, even when there is a separate control group.

Central Limit Theorem The theoretical statement that when sample means are selected randomly from a single population, the means will distribute as an approximation of the normal distribution, even if the population distribution deviates from normality. The theorem assumes that sample sizes are relatively large (at least 30) and that they are all randomly selected from *one* population.

Central Tendency (Measures of) A statistical term used for describing the typical, middle, or central scores in a distribution of scores. Measures of central tendency are used when the researcher wants to describe a group as a whole with a view toward characterizing that group on the basis of its most common measurement. The researcher wishes to know what score best represents a group of differing scores. The three measures of central tendency are the mean (or arithmetic average), the median (or the midpoint of the distribution), and the mode (the most frequently occurring score in the distribution).

Chi Square (χ^2) A statistical test of significance used to determine whether or not frequency differences have occurred on the basis of chance. Chi square requires that the data be in nominal form, or the actual number of cases (frequency of occurrence) that fall into two or more discrete categories. It is considered to be a nonparametric test (no population assumptions are required for its use). The basic equation is as follows:

$$\chi^2 = \Sigma \frac{(f_o - f_e)^2}{f_e}$$

where f_o denotes the frequencies actually observed and f_e the frequencies expected on the basis of chance.

Coefficient of Contingency (*C*) A test of correlation on nominal data sorted into any number of independent cells.

$$C = \sqrt{\frac{\chi^2}{N + \chi^2}}$$

Coefficient of Determination (r^2) A method for determining what proportion of the information about *Y* is contained in *X*; found by squaring the Pearson *r*.

Colinear In correlational research, colinear variables are those that correlate so highly that it can be assumed they are probably measuring the same underlying construct (since they carry redundant information).

Combination Research (also called **Treatment by Levels**) A factorial design in which at least one of the IVs is manipulated and at least one other IV is assigned as a subject variable.

Confidence Interval The range of predicted values within which one presumes with a stated degree of confidence that the true parameter will fall. Usually, confidence intervals are determined on the basis of a probability value of .95 (95% certainty) or .99 (99% certainty). Since sample values differ, even when drawn from the same population, one can never be certain that any particular interval contains the parameter.

Construct Validity A type of test validity for items that have been designed to measure the degree to which an individual possesses differing amounts of an underlying construct (such as shyness or dogmatism). Test scores are analyzed, usually in a long series of studies, on the basis of how well they conform to the underlying construct under study.

Content Validity A type of test validity based on whether the test items are a fair and representative sample of the general area that the test was designed to evaluate. If the test is supposed to be covering a student's understanding of mathematics, for example, the items should be largely restricted to that domain, and the test should not be scored on the basis of the student's penmanship or neatness.

Control Group In experimental research, the control group is the comparison group, or the group that ideally receives zero magnitude of the independent variable. The use of a control group is critical in evaluating the pure effects of the independent variable on the measured responses of the subjects.

Correlated (or **Dependent**) **Samples** In experimental research, two or more samples that are *not selected independently.* The selection of one sample determines who will be selected for the other sample(s), as in a matched-subjects design.

Correlation Coefficient A quantitative formulation of the relationship existing among two or more variables. A correlation is said to be positive when high scores on one variable associate with high scores on another variable, and low scores on the first variable associate with low scores on the second. A correlation is said to be negative when high scores on the first variable associate with low scores on the second, and vice versa. Correlation coefficients range in value from +1.00 to –1.00. Correlation coefficients falling near the zero point indicate no consistent relationship among the measured variables. In social research, the correlation coefficient is usually based on taking several response measures of *one group of subjects.*

Correlation Matrix Correlations among several variables can be shown in a correlation matrix. It is a square, symmetrical array where each row and each column represent a different variable. The point where the row and column intersect is called a cell, and this is where the correlation between the variables is shown.

Counterbalancing A technique used in repeated-measures designs to help prevent confounding of the independent variable by evenly distributing sequencing effects across all treatment conditions.

Cross-Sectional Research Type of nonexperimental research, sometimes used to obtain data on possible growth trends in a population. The researcher selects a sample (cross-section) at one age level—say, 20-year-olds—and compares these measurements with those taken on a sample of older subjects—say, 65-year-olds. Comparisons of this type are often misleading (today's 20-year-olds may have very different environmental backgrounds—educational experience, for example—than the 65-year-old subjects have).

Cyril Hoyt Reliability Method An internal consistency technique for assessing the reliability of a test through the use of ANOVA. The Cyril Hoyt can be used with either categorical or interval data sets.

Deciles Divisions of a distribution representing tenths, the first decile representing the 10th percentile, and so on. The 5th decile, therefore, equals the 50th percentile, the 2nd quartile, and the median.

Degrees of Freedom (df) With interval (or ratio) data, degrees of freedom refer to the number of scores free to vary after certain restrictions have been placed on the data. With six scores and a restriction that the sum of these scores must equal a specified value, then five of these scores are free to take on any value whereas the sixth score is fixed (not free to vary). In inferential statistics, the larger the size of the sample, the larger is the number of degrees of freedom. With nominal data, degrees of freedom depend *not on the size of the sample,* but on the number of categories in which the observations are allocated. Degrees of freedom are here based on the number of frequency values free to vary after the sum of the frequencies from all the cells has been fixed.

Dependent Variable In any antecedent-consequent relationship, the consequent variable is called the dependent variable. The dependent variable is a measure of the output side of the input-output relationship. In experimental research, the dependent variable is the possible effect half of the cause-and-effect relationship, whereas in correlational research, the dependent variable is the measure being predicted and is called the *criterion variable.* In the social sciences, the dependent variable is usually a response measure.

Descriptive Statistics Techniques for describing and summarizing data in abbreviated, symbolic form; shorthand symbols for describing large amounts of data.

Deviation Score (x) The difference between a single score and the mean of the distribution. It is found by subtracting the mean, M, from the score X. The deviation score $(X - M)$ is symbolized as x. Thus, $x = X - M$.

Discriminant Analysis The opposite of an ANOVA or MANOVA (which attempt to discover if group differences produce DV differences). With discriminant analysis the DV scores are instead used to predict which group the subjects came from. The goal is to accurately predict a subject's group membership on the basis of that subject's DV score(s).

Distribution The arrangement of measured scores in order of magnitude. Listing scores in distribution form allows the researcher to notice general trends more readily than with an unordered set of raw scores. A *frequency distribution* is a listing of each score achieved, together with the number of individuals receiving that score. When graphing frequency distributions, one usually indicates the scores on the horizontal axis (abscissa) and the frequency of occurrence on the vertical axis (ordinate).

Double-Blind Study A method used by researchers to reduce one form of experimental error. In a double-blind study neither the person conducting the study nor the subjects are aware of which group is the experimental group and which the control. This helps to prevent any unconscious bias on the part of the experimenter or any contaminating motivational sets on the part of the subjects.

Effect Size The difference between two population means in units of the *population* standard deviations. Thus, if we were to raise population scores on the verbal section of the SAT by, say, 30 points (from 500 to 530), and since we know that the population standard deviation for this test is 100, then the effect size would be $(530 - 500)/100 = .30$. The larger the effect size, the higher is the likelihood of detecting the population differences through the use of inferential statistical techniques. A t test or an ANOVA may show a significant difference even though the difference in the population is so weak as to be trivial. The two major statistical tests for computing effect size are d for the t test and eta square for ANOVA.

Experimental Design Techniques used in experimental research for creating equivalent groups of subjects. There are three basic experimental designs: (1) Between-Subjects, where subjects are randomly assigned to control and experimental groups and groups are completely independent. In this design the dependent variable is typically measured after the introduction of the independent variable; (2) Repeated-Measures, where a group of subjects is used as its own control and the dependent variable is measured both before and after the introduction of the independent variable; and (3) Matched-Subjects, where subjects are equated person for person on some variable deemed relevant to the dependent variable.

Experimental Research Research conducted using the experimental method, where an independent variable is manipulated (stimulus) in order to bring about a change in the dependent variable (response). Using this method, the experimenter is allowed the opportunity for making cause-and-effect inferences. Experimental research requires careful controls in order to establish the pure effects of the independent variable. Equivalent groups of subjects are formed, then exposed to different stimulus conditions, and then measured to see if differences can be observed.

External Validity The extent to which an experimental finding can be projected to the population at large. An experiment is high in external validity when the sample is representative of the population and when it simulates real-life conditions.

Factor Analysis A correlational attempt to reduce the number of separate variables by placing those with high intercorrelations into clusters or *factors*, or sometimes called components. A correlation matrix can be set up, and the researcher looks for the highest intercorrelations in an attempt to find out if there are fewer, more basic variables that might better explain the larger number of variables in the matrix. The goal is to isolate the factors so that they could then be used to represent fewer underlying constructs. The variables that correlate highly are called colinear, whereas those that remain independent are called orthogonal.

Factorial ANOVA As opposed to a one-way ANOVA, the factorial ANOVA allows for the analysis of data when more than one independent variable is involved. Results can be analyzed on the basis of the main effects of each independent variable or on the basis of the possible interaction among the independent variables. Data to be analyzed should be in at least interval form.

Fisher, Sir Ronald (1890–1962) English mathematician and statistician who developed the analysis of variance technique, or F (for Fisher) ratio.

Frequency Distribution Curve Graphing procedure where measures, such as raw scores or z scores, are plotted on the X axis (abscissa) and frequency of occurrence on the Y axis (ordinate).

Frequency Polygon A graphic display of data where single points are plotted above the measures of performance. The height where the point is placed indicates the frequency of occurrence. The points are connected by a series of straight lines.

Friedman ANOVA by Ranks (χ^2_r) A test of the hypothesis of difference on ordinal data when either the sample groups have been matched or a single sample has been repeatedly measured. The Friedman ANOVA is the ordinal counterpart of the within-subjects F.

$$\chi_r^2 = \frac{12}{Nk(k+1)}\left(\Sigma R_1^2 + \Sigma R_2^2 + \Sigma R_3^2 + \cdots\right) - 3N(k+1)$$

Galton, Sir Francis (1822–1911) The "father of intelligence testing" and the creator of the concept of individual differences. Galton also introduced the concepts of regression and correlation (although it was left to his friend and colleague Karl Pearson to work out the mathematical equations).

Gambler's Fallacy An erroneous assumption that independent events are somehow related. If a coin is flipped 10 times and comes up heads each of those times, the gambler's fallacy predicts that it is virtually certain for the coin to come up tails on the next flip. Since each coin flip is independent of the preceding one, the probability remains the same (.50) for each and every coin flip, regardless of what has happened in the past. The gambler remembers the past, but the coin does not.

Gauss, Carl Friedrich (1777–1855) German mathematician credited with having originated the normal curve. For this reason the normal curve is often called the Gaussian curve.

Gossett, William Sealy (1876–1937) Using the pen name "Student," Gossett, while working for the Guinness Brewing Company in Ireland, developed the technique of using sample data to predict population parameters, which led to the development of the t test.

Halo Effect A research error arising from the fact that people who are viewed positively on one trait are often also thought to have many other positive traits. Advertisers depend on this effect when they use famous personalities to endorse various products—anyone who can throw touchdown passes *must be* an expert in evaluating razor blades. Researchers must guard against the halo effect, as it will contaminate the independent variable.

Hawthorne Effect A major research error due to response differences resulting not from the action of the independent variable, but from the flattery or attention paid to the subjects by the experimenter. Typically, the potential for this error is inherent in any study using the before-after experimental design without an adequate control group. Any research, for example, where persons are measured, then subjected to some form of training, then measured again should be viewed with suspicion unless an appropriate control group is used—that is, an equivalent group that is measured, *then not subjected to the training*, and then measured again. Only then can the researcher be reasonably confident that the response differences are due to the pure effects of the independent variable.

Histogram A graphic representation of data in which a series of rectangles (bars) is drawn above the measure of performance. The height of each bar indicates the frequency of occurrence.

Homogeneity of Variance An assumption of both the t and F ratios, which demands that the variability within each of the sample groups being compared should be fairly similar.

Homoscedasticity The fact that the standard deviations of the Y scores along the regression line should be fairly equal. Otherwise, the standard error of estimate is not a valid index of accuracy.

Inclusion Area The middlemost area of the normal curve, included between two z scores equidistant from the mean. Because of the symmetry of the normal curve, the middlemost area includes, in equal proportions, the area immediately to the left of the mean and the area immediately to the right of the mean.

Independent Variable In any antecedent-consequent relationship, the antecedent variable is called the independent variable. Independent variables may be manipulated or assigned. A manipulated independent variable occurs when the researcher deliberately alters the environmental conditions to which subjects are being exposed. An assigned independent variable occurs when the researcher categorizes subjects on the basis of some preexisting trait.

Whether the independent variable is manipulated or assigned determines whether the research is experimental (manipulated independent variable) or post facto (assigned independent variable). In experimental research, the independent variable may be the causal half of the cause-and-effect relationship. In correlational research, the independent variable is the measure from which the prediction will be made and is thus called the *predictor* variable.

Inductive Fallacy An error in logic resulting from overgeneralizing on the basis of too few observations. The inductive fallacy occurs when one assumes that all members of a class have a certain characteristic because one member of that class has it. It would be fallacious to assume that all Mongolians are liars on the basis of having met one Mongolian who was a liar.

Inferential (Predictive) Statistics Techniques for using the measurements taken on samples to predict the characteristics of the population—the use of descriptive statistics for inferring parameters.

Interaction Effect When two or more independent variables are involved (as in an ANOVA), the interaction effect is produced by the combined influence of these independent variables working in concert.

Interclass Correlations Correlations between measures sharing the same measurement class, such as correlating IQ scores with IQ scores between twins, or correlating height measures with height measures for individuals at different ages, or test scores with the same test's scores taken at different times. The same metric and variance are found in both measures. That is, interclass correlations are used to compare apples with apples.

Interdecile Range Those scores that include the middlemost 80% of a distribution, or the difference between the first and ninth deciles.

Internal Validity The extent to which the results of an experiment can unambiguously identify the cause-and-effect relationship. A high degree of internal validity indicates that the experiment is relatively free of the contaminating effects of confounding variables—thus allowing for a clear interpretation of the pure effects of the independent variable(s).

Interquartile Range Those scores that include the middlemost 50% of a distribution, or the difference between the first and third quartiles.

Interval Data Data (measurements) in which values are assigned such that both the order of the numbers and the *intervals* between numbers are known. Thus, interval data provide not only information regarding greater than or less than status but also information as to how much greater or less than.

Intraclass Correlations Correlations between variables that share neither the same metric nor the same variance, as when measuring the relationship between IQ points (a class of measures representing intelligence) and grade-point average (a class of measures representing actual achievement). That is, intraclass correlations are used to compare apples with oranges.

Item Analysis The attempt to assess two related and important test parameters: item difficulty and item discrimination.

Item Difficulty Test items are evaluated on the basis of the *proportion* of persons who correctly answered a particular item. For example, an item correctly answered by only 15% of the sample would be recorded as having a difficulty index of .15. The lower the value is, the more difficult the item.

Item Discrimination Test items are evaluated on the basis of their ability to discriminate among those who do well on the entire test versus those who do poorly. Item discrimination is determined by comparing a person's performance on each of the individual items with that same person's performance on the entire test.

Kruskal–Wallis *H* Test A test of the hypothesis of difference on ordinal data among at least three independently selected random samples. The *H* test is the ordinal counterpart of the one-way ANOVA.

$$H = \frac{12}{N(N+1)}\left(\frac{\Sigma R_1^2}{n_1} + \frac{\Sigma R_2^2}{n_2} + \frac{\Sigma R_3^2}{n_3} + \cdots\right) - 3(N+1)$$

Kuder–Richardson Reliability (K–R 21) An internal consistency technique for assessing the reliability of a test that has been split into *all possible halves* and automatically corrects for the splits without any need for further adjustment.

Kurtosis (Ku) The state or degree of the curvature of a unimodal frequency distribution. Kurtosis refers to the peakedness or flatness of the curve.

Leptokurtic Distribution A unimodal frequency distribution in which the curve is relatively peaked—most of the scores occur in the middle of the distribution—with very few scores occurring in the tails. A leptokurtic distribution yields a relatively small standard deviation.

Levene's Test This is a test of homogeneity of variance among all DV measures. If the significance level is greater than .05, homogeneity can be assumed. If the significance level is .05 or less and the samples are of equal sizes, the *t* or *F* ratio may give erroneous impressions. SPSS gives significance levels for *t* or *F* for both assumptions, equal variances assumed or not assumed.

Longitudinal Research A type of post-facto research in which subjects are measured repeatedly throughout their lives in order to obtain data on possible trends in growth and development. Terman's* massive study of growth trends among intellectually gifted children is an example of this research technique. The study, begun in the early 1920s, is still in progress today and is still providing science with new data. Longitudinal research requires great patience on the part of the investigator, but the obtained data are considered to be more valid than those obtained using the cross-sectional approach.

Main Effects When two or more independent variables are involved (as in an ANOVA), the main effects are produced by the action of each independent variable working separately.

Mann–Whitney *U* Test A test on ordinal data of the hypothesis of difference between two independently selected random samples. The *U* test is the ordinal counterpart of the independent *t* test.

$$z_U = \frac{U - [(n_1)(n_2)]/2}{\sqrt{[n_1 n_2 (n_1 + n_2 + 1)]/12}}$$

MANOVA A type of analysis of variance that tests whether two or more IVs can significantly impact two or more DVs, as opposed to a straight factorial ANOVA where several IVs were used to establish a possible difference on a *single* DV.

*L. M. Terman, *Genetic Studies of Genius* (Stanford, Calif.: Stanford University Press, 1925, 1926, 1930, 1947, 1959).

McNemar Test Technique, developed by the statistician Quinn McNemar, that uses chi square for the analysis of nominal data from correlated samples.

$$\chi^2 = \frac{|a - d|^2}{a + d}$$

Mean (*M*) A measure of central tendency specifying the arithmetic average. Scores are added and then divided by the number of cases.

$$M = \frac{\Sigma X}{N}$$

The mean is best used when the distribution of scores is balanced and unimodal. In a normal distribution, the mean coincides with the median and the mode. When the entire population of scores is used, the mean is designated by the Greek letter μ (mu).

Measurement A method of quantifying observations by assigning numbers to them on the basis of specific rules. The rules chosen determine which scale of measurement is being used: nominal, ordinal, interval, or ratio.

Median (Mdn) A measure of central tendency that specifies the middlemost point in an ordered set of scores. The median always represents the 50th percentile. It is the most valid measure of central tendency whenever the distribution is skewed.

Mesokurtic A unimodal frequency distribution whose curve is normal. (*See* Normal Curve.)

Mixed Designs (or **Split Plot**) Research design where one of the measures is independent (between-subjects) and the other is correlated (within-subjects). For example, one of the IVs might be training method and the other might be number of trials.

Mode (Mo) A measure of central tendency that specifies the most frequently occurring score in a distribution of scores. When there are two most common points, the distribution is said to be bimodal.

Monotonic Function Where as one variable increases, the other either increases or decreases without reversal, such as the linear relationship between height and weight (a monotonic function may approach an asymptote but never becomes curvilinear).

Multiple *R* A single numerical value that quantifies the correlation among three or more variables. The equation for a three-variable multiple R is as follows:

$$R_{y \cdot 1,2} = \sqrt{\frac{r_{y,1}^2 + r_{y,2}^2 - 2r_{y,1}r_{y,2}r_{1,2}}{1 - r_{1,2}^2}}$$

Multiple Regression Technique using the multiple R for making predictions of one variable given measures on two or more others. It requires the calculation of the intercept (a) and also at least two slopes (b_1 and b_2). For the three-variable situation, the multiple regression equation is as follows:

$$Y_{\text{Mpred}} = b_1X_1 + b_2X_2 + a$$

Nominal (or **Categorical) Data** Data (measurements) in which numbers are used to label discrete, mutually exclusive categories; nose-counting data, which focus on the frequency of occurrence within independent categories.

Nonparametrics Statistical tests that neither predict the population parameter, μ, nor make any assumptions regarding the normality of the underlying population distribution. These tests may be run on ordinal or nominal data and typically have less power than do the parametric tests.

Normal Curve A frequency distribution curve resulting when scores are plotted on the horizontal axis (X) and frequency of occurrence is plotted on the vertical axis (Y). The normal curve is a theoretical curve shaped like a bell and fulfills the following conditions: (1) most of the scores cluster around the center, and as we move away from the center in either direction, there are fewer and fewer scores; (2) the scores fall into a symmetrical shape—each half of the curve is a mirror image of the other; (3) the mean, median, and mode all fall at precisely the same point, the center; (4) there are constant area characteristics regarding the standard deviation; and (5) the curve is asymptotic to the abscissa.

Null (or **"Chance") Hypothesis** The assumption that the results are simply due to chance. When testing the hypothesis of difference, the null hypothesis states that no real differences exist in the population from which the samples were drawn. When testing the hypothesis of association, the null hypothesis states that the correlation in the population is equal to zero (does not exist).

Odds The chances *against* a specific event occurring. When the odds are 5 to 1, for example, it means that the event will *not* occur five times for each single time that it will occur.

One-Tail (Directional) Test The use of only one tail of the distribution for testing the null hypothesis. For example, with the t test, if the direction of the difference has already been specified in the alternative hypothesis, then the critical value of the t statistic, for a given number of degrees of freedom, is taken from only one side of its distribution.

Ordinal Data Rank-ordered data, that is, derived only from the order of the numbers, not the differences between them. Ordinal measures provide information regarding greater than or less than status, but *not* how much greater or less.

Ordinate The vertical or Y axis in the coordinate system. On a frequency distribution, the ordinate indicates the frequency of occurrence.

Orthogonal In correlational research, orthogonal measures are independent, uncorrelated, since one variable carries no information about nor can be predicted from the other variable (as opposed to colinear variables). Also, in experimental research orthogonal comparisons can be made with a factorial ANOVA when each level of each independent variable is combined with each level of all other independent variables.

Paired *t* Ratio Statistical test of the hypothesis of difference between *two sample means* where the sample selection is not independent. The paired t (also called correlated t) requires interval data and is typically used when the design has been before-after or matched-subjects.

$$t = \frac{M_1 - M_2}{\sqrt{SE_{M_1}^2 + SE_{M_2}^2 - 2r_{1\cdot 2}SE_{M_1}SE_{M_2}}}$$

Parameter Any measure obtained by having measured the entire population. Parameters are therefore usually inferred rather than directly measured.

Partial Correlation Correlation technique that allows for the ruling out of the possible effects of one or more variables on the relationship among the remaining variables. In the three-variable situation, the partial correlation rules out the influence of the third variable on the correlation between the remaining two variables. The equation for partialing out the influence of the third variable is as follows:

$$r_{y,1\cdot 2} = \frac{r_{y,1} - r_{y,2}r_{1,2}}{\sqrt{(1 - r_{y,2}^2)(1 - r_{1,2}^2)}}$$

Pascal, Blaise (1623–1662) French mathematician and philosopher who introduced the concepts of probability and random events.

Path Analysis Sophisticated correlational techniques have been used in a causal modeling method called path analysis. Although path analysis is a correlational procedure, it is used to test a set of hypothesized cause-and-effect relationships among a series of variables that are logically ordered on the basis of time. Since, as logic suggests, a causal variable must precede (in time) a variable it is supposed to influence, the correlational analysis is done on a set of variables, each presumed to show a causal ordering. The attempt is made to find out whether a given variable is being influenced by the variables that precede it and then, in turn, is influencing the variables that follow it. A "path" diagram is drawn that portrays the assumed direction of the various relationships. Although not as definitive a proof of causation as when the independent variable is experimentally manipulated, path analysis takes us a long step forward from the naïve extrapolations of causation that at one time were taken from simple bivariate correlations.

Pearson, Karl (1857–1936) English mathematician and colleague of Sir Francis Galton. It was Pearson who translated Galton's ideas on correlation into precise mathematical terms, creating the equation for the product moment correlation coefficient, or the Pearson *r*.

Pearson *r* Statistical technique introduced by Karl Pearson for showing the degree of linear relationship between two variables. Also called the product moment correlation coefficient, it is used to test the hypothesis of association, that is, whether there is a relationship between two sets of measurements. The Pearson *r* can be calculated as follows:

$$r = \frac{(\Sigma XY) / N - (M_x)(M_y)}{SD_x SD_y},$$

or

$$\frac{\dfrac{\Sigma XY - (M_x)(M_y)(N)}{N - 1}}{s_x s_y}$$

Computed correlation values range from +1.00 (perfect positive correlation) through zero to −1.00 (perfect negative correlation). The farther the Pearson *r* is from zero, whether in a positive or negative direction, the stronger is the relationship between the two variables. The Pearson *r* can be used for making better-than-chance predictions but should not be used alone for isolating causal factors.

Percentiles (or **Centiles**) The percentage of cases falling below a given score. Thus, if an individual scores at the 95th percentile, that individual has exceeded 95% of all persons taking that particular test. If test scores are normally distributed, and if the standard deviation of the distribution is known, percentile scores can easily be converted to the resulting *z* scores.

Percentile Rank The value that indicates a given percentile. A point at the 75th percentile is said to have a percentile rank of 75.

Platykurtic Distribution A unimodal frequency distribution in which the curve is relatively flat. Large numbers of scores appear in both tails of the distribution. A platykurtic distribution of scores yields a relatively large standard deviation.

Point Biserial A statistical technique for assessing the level of item discrimination. Using this method, a researcher can determine the correlation between an interval scale measure, such as the result for the whole test, and a nominal measure, such as whether an individual item was answered correctly or incorrectly.

Point Estimate The use of a sample value for predicting a *single* population value. For example, the use of the sample mean for estimating μ is a point estimate.

Point of Inflection In a continuous graph, the point of inflection is where the curve changes from positive acceleration to negative acceleration, or from concave upward to concave downward. On the normal curve, the points of inflection on either side of the mean are exactly one *SD* unit from the mean.

Point of Intercept (*a*) In a scatter plot, the point of intercept is the location where the regression line of *Y* on X crosses the ordinate, or the value of *Y* when *X* is equal to zero (or its minimum value). In the regression equation, $Y = bX + a$, the intercept is symbolized by the *a* term.

Population The entire number of persons, things, or events (observations) having at *least* one trait in common. Populations may be limited (finite) or unlimited (infinite).

Post-Facto Research A type of research that, while not allowing for direct cause-and-effect conclusions, does allow the researcher to make better-than-chance predictions. In such research, subjects are measured on one response dimension, and these measurements are compared with other trait or response measures.

Power (1 – β) A measure of the sensitivity of a statistical test. The more powerful a test is, the less is the likelihood of committing the beta error (accepting the null hypothesis when it should have been rejected). The higher a test's power, the higher is the probability of a small difference or a small correlation being found to be significant.

Primary Variance Variability of the dependent variable that is assumed to have been produced by the direct action of the independent variable.

Probability (*P*) The statement as to the number of times a specific event, *s*, can occur out of the total possible number of events, *t*.

$$P = \frac{s}{t}$$

Probability should be expressed in decimal form. Thus, a probability of 1/20 is written as .05.

Quartiles Divisions of a distribution representing quarters; the first quartile representing the 25th percentile, the second quartile the 50th percentile (or median), and the third quartile the 75th percentile.

Quota Sampling A method used for selecting a representative sample that is based on drawing subjects in proportion to their existing percentages in the population. If, for example, 55% of a certain population is composed of women, then 55% of the sample must also be made up of women.

Random Sample Sample selected in such a way that every element or individual in the entire population has an equal chance of being chosen. When samples are selected randomly, then sampling error should also be random and the samples representative of the population.

Range (*R*) A measure of variability that describes the entire width of the distribution. The range is the difference between the two most extreme scores in a distribution and is thus equal to the highest value minus the lowest value.

Ratio Data Data (measurements) that provide information regarding the order of numbers, the difference between numbers, and also an *absolute* zero point. It permits comparisons, such as A being three times B, or one-half of B.

Regression Line The single straight line that lies closest to all the points in a scatter plot. The regression line can be used for making correlational predictions when three important pieces of information are known: (1) the extent to which the scatter points deviate from the line, (2) the slope of the line, and (3) the point of intercept.

$$Y = bX + a$$

Representative Sample A sample that reflects the characteristics of the entire population. Random sampling is assumed to result in representative samples.

Rulon Reliability Formula A method for estimating reliability through the use of the standard error of measurement.

$$r_{tt} = 1 - \frac{SEM^2}{SD^2}$$

Sample A group of any number of observations selected from a population, as long as it is less than the total population.

Sampling Distributions Distributions made up of measures taken on successive random samples. Such measures are called statistics, and when all samples in an entire population are measured, the resulting sampling distributions are assumed to approximate normality. (*See* Central Limit Theorem.) Two important sampling distributions are the distribution of means and the distribution of differences.

Sampling Error The expected difference between a measure of the sample and a measure of the population ($M - \mu$). Under conditions of random sampling, the probability of obtaining a sample mean greater than the population mean is identical to the probability of obtaining a sample mean less than the population mean ($P = .50$).

Scatter Plot A graphic format in which each point represents a pair of scores, the score on X as well as the score on Y. The array of points in a scatter plot typically forms an elliptical shape (a result of the central tendency usually present in both the X and Y distributions).

Secondary Variance Variability of the dependent variable that is not under experimental control but is instead a result of confounding variables. Holding secondary variance to a minimum helps to ensure a high degree of internal validity.

Secular Trend Analysis A method, using correlational techniques, for predicting linear trends across *time*. Historical data are used to forecast future results, based on the not-always-wise assumption that the past trend will continue.

Significance A statistical term used to indicate that the results of a study are not simply a matter of chance. Researchers talk about significant differences and significant correlations, the assumption being that chance has been ruled out (on a probability basis) as the explanation of these phenomena.

Skewed Distribution An unbalanced distribution in which there are a few extreme scores in *one direction*. In a skewed distribution, the best measure of central tendency is the median.

Spearman, Charles E. (1863–1945) English psychologist and test expert who worked in the area of measuring intelligence and identifying the factors that make up intelligence. In pursuing his correlational studies on intellectual factors, he produced a correlation technique for the analysis of ordinal data called the Spearman rho, or the r_S.

Spearman r_S Correlation coefficient devised by Charles E. Spearman for use with rank-ordered (ordinal) data. Sometimes called the Spearman ρ (rho), the coefficient is found as follows:

$$r_S = 1 - \frac{6\Sigma d^2}{N(N^2 - 1)}$$

Standard Deviation A measure of variability that indicates how far *all* scores in a distribution vary from the mean. The standard deviation has a constant relationship with the area under the normal curve. (*See* Normal Curve.) The actual or true standard deviation of any distribution is calculated as follows:

$$SD = \sqrt{\frac{\Sigma X^2}{N} - M^2}$$

The unbiased *estimate* of the standard deviation in the population is calculated with the following equation:

$$s = \sqrt{\frac{\Sigma X^2 - (\Sigma X)^2 / N}{N - 1}}$$

When the standard deviation is calculated on the basis of all scores in the entire population, it is designated as σ (lowercase Greek letter sigma).

Standard Error of Difference The standard deviation of the entire distribution of differences between pairs of successively drawn random sample means. These pairs of sample means are taken from the population until that population is exhausted. An estimate of this value can be made on the basis of the information contained in just two samples.

$$SE_D = \sqrt{SE^2_{M_1} + SE^2_{M_2} - 2r_{1\cdot2}SE_{M_1}SE_{M_2}}$$

When sample selections are independent of one another, the correlation term ($2r_{1\cdot2}SE_{M_1}SE_{M_2}$) is equal to zero and is therefore not used. The estimated standard error of difference for independent samples is thus as follows:

$$SE_D = \sqrt{SE^2_{M_1} + SE^2_{M_2}}$$

Standard Error of Estimate–SE_{est} In a scatter plot, where the value of *Y* is being predicted, the standard error of estimate is an estimate of the variability of all the obtained data points around the linear regression line of *Y* on *X*. It can be used in assessing the accuracy of a predicted *Y* value that has been obtained from the regression equation. The higher the correlation between *X* and *Y*, the lower is the resulting value of the standard error of estimate and the more accurate is the predicted *Y* value. When $r = 0$, the standard error of estimate is equal to the standard deviation of the *Y* distribution, and when $r = \pm 1.00$, then the standard error of estimate equals zero (since when $r = \pm 1.00$, all the data points fall directly on the regression line).

Standard Error of Measurement An estimate of test score variability. It is used to predict the standard deviation of all the possible scores a given individual might obtain around that same person's theoretically true score. The standard error of measurement, thus, combines information about the test's reliability, as well as the variability occurring around the individual scores. It allows the researcher to predict what score a person might get if given the test a second time.

Standard Error of the Mean The standard deviation of the entire distribution of random sample means successively selected from a single population until that population is exhausted. An estimate of the standard error of the mean can be made on the basis of the information contained in a single random sample, that is, the variability within the sample and the size of the sample. When using the actual standard deviation of the sample scores, the equation becomes

$$SE_M = \frac{SD}{\sqrt{N-1}}$$

When using the estimated population standard deviation, the equation becomes

$$SE_M = \frac{s}{\sqrt{N}}$$

Standard Error of Multiple Estimate A technique for assessing the accuracy of a prediction that has been generated from the multiple regression equation. For the three-variable situation, the standard error of multiple estimate is as follows:

$$SE_{\text{M est}} = SD_y\sqrt{1 - R^2_{y\cdot 1,2}}$$

Statistic Any measure that is obtained from a sample as opposed to the entire population. The range (or the standard deviation or the mean) of a set of sample scores is a statistic.

Stepwise Regression This is a type of multiple regression in which IVs can be added one at a time in what is called a stepwise process. The procedure then orders the variables on the basis of their predictive power. This is called a forward stepwise regression. There is also a backward process where the researcher enters all the variables and lets the procedure eliminate those that contribute least to the overall predictive power and order those that contribute the most.

Stratified (or **Quota**) **Sampling** Selecting a sample that directly reflects the population characteristics. If it is known that 45% of the population is composed of males, and if it is assumed that gender is a relevant research variable, then the sample must contain 45% male subjects.

Sum of Squares (SS) An important concept for ANOVA; the sum of squares equals the sum of the squared deviations of all scores around the mean.

$$\text{SS} = \Sigma x^2 = \Sigma X^2 - \frac{(\Sigma X)^2}{N}$$

When the sum of squares is divided by its appropriate degrees of freedom, the resulting value is called the mean square, or variance.

***t* Ratio** Statistical test used to establish whether significant (nonchance) differences can be detected between *two* means. With two samples, it is the ratio of the difference between the sample means to an estimate of the standard error of difference. With one sample, it is the ratio of the difference between the sample mean and population mean to an estimate of the standard error of the mean. Two samples:

$$t = \frac{M_1 - M_2}{SE_D}$$

One sample:

$$t = \frac{M - \mu}{SE_M}$$

***T* Score** A converted standard score such that the mean equals 50 and the standard deviation equals 10. *T* scores, thus, range from a low of 20 to a high of 80.

Test Reliability r_{tt} Test measures that produce consistent results and afford the researcher dependable measures. The higher the proportion of true score to observed or total score, the higher is the reliability. Three methods for determining test reliability are (1) test-retest, (2) alternate-form, and (3) split-half. Correlation techniques can be used for assessing reliability for all three techniques. For the test-retest method, the correlation is computed by comparing the test scores on one administration of the test with the test scores on a second administration of the same test. For the alternate-form method, the correlation is used to compare scores on one form of the test with scores on a second version of the same test, called an alternate or parallel form. For the split-half technique, the test is deliberately divided into equal halves, usually on the basis of the odd- versus the even-numbered items. The correlation is then run by comparing scores on one-half of the test with scores on the second half.

Test Validity r_{tc} Test validity is based on whether the test is in fact measuring what it has been designed to measure. Two types of validity techniques are (1) concurrent validity, where test scores being validated are compared with test scores achieved on a different, already established test (which had been designed to measure the same construct), and (2) predictive validity, where the test scores are correlated against some independent measure of future performance.

Tukey's HSD (honestly significant difference) A multiple comparison technique developed by J. W. Tukey for establishing whether the differences among various sample means are significant. The test is performed *after* the ANOVA when the *F* ratio is significant. It is thus a post-hoc test.

Two-Tail (Nondirectional) Test The use of both tails of the distribution for testing the null hypothesis. For example, with the *t* test, if the direction of the difference is not specified in the alternative hypothesis, then the critical value of the *t* statistic, for a given number of degrees of freedom, is taken from both tails of its distribution.

Type 1 Error The error made when the researcher incorrectly rejects the null hypothesis (because in fact it should have been accepted). The alpha level sets the *probability* of this error occurring.

Type 2 Error The error made when the researcher incorrectly accepts the null hypothesis (because in fact it should have been rejected). The beta level sets the probability of this error occurring.

Unimodal Distribution A distribution of scores in which only one mode (most frequently occurring score) is present.

Variability Measures Measures that give information regarding individual differences, or how persons or events vary in their measured scores. The three most important measures of variability are the range, the standard deviation, and the variance (which is the standard deviation squared).

Variable Anything that varies *and can be measured.* In experimental research, the two most important variables to be identified are the independent variable and the dependent variable. The independent variable is a stimulus, is actively manipulated by the experimenter, and is the causal half of the cause-and-effect relationship. The dependent variable is a measure of the subject's response and is the effect half of the cause-and-effect relationship.

Variance A measure of variability that indicates how far all of the scores in a distribution vary from the *mean.* Variance is equal to the square of the standard deviation.

Wilcoxon *T* Test A test on ordinal data of the hypothesis of difference between two sample groups when the selections are correlated (as in the matched-subjects design). The Wilcoxon *T* is the ordinal counterpart of the paired *t*.

Wilks Lambda A statistic used to determine if there are significant differences among the means following univariate or multivariate analyses. Lambda equals 1.00 when all means are equal. The *smaller* the Lambda the higher the likelihood of showing significance among the means. SPSS gives a univariate Tests-of-Between-Subjects table that identifies the significance levels of each dependent variable, values that can only be interpreted if the Lambda is significant.

Within-Subjects *F* Ratio Statistical test of the hypothesis of difference among several sample means, where sample selection is not independent. It is used when samples are correlated, as in repeated-measures designs, and the data are in at least interval form.

Yates Correction A correction for continuity usually applied to a 2 × 2 chi square analysis (df = 1) whenever any or all of the expected frequencies are less than 10. The absolute difference between f_o and f_e is reduced by .50, resulting in a slightly lower chi square value.

***z* Distribution** The standard normal distribution, or as it is sometimes called, the unit-normal distribution, where the mean is equal to zero and the standard deviation equal to 1.00. It is the distribution that is shown in Table A, page 546.

***z* Score (Standard Score)** A number that results from the transformation of a raw score into units of standard deviation. The *z* score specifies how far above or below the mean a given raw score is in these standard deviation units. Any raw score above the mean converts to a positive *z* score, while scores below the mean convert to negative *z* scores. The *z* score is also referred to as a standard score. The normal deviate, *z*, has a mean equal to zero and a standard deviation equal to 1.00.

$$z = \frac{X - M}{SD} \quad \text{or} \quad \frac{X - \mu}{\sigma}$$

***z* Test** A method of hypothesis testing that can be used when the parameter values are normally distributed and the mean and standard deviation of the distribution are already known.

References

Abrams, S. (1989). The complete polygraph handbook. Lexington, MA: Lexington Books.

Adamson, D. (1995). *Blaise Pascal: Mathematician, physicist and thinker about God.* London: Basingstroke.

Agresti, A. (1990). *Categorical data analysis.* New York: John Wiley.

Aiken, L. R. (1988). *Psychological testing and assessment* (6th ed.). Boston: Allyn & Bacon.

Aiken, L. R. (1998). *Tests and examinations: Measuring abilities and performance.* New York: Wiley.

Anderson, H., & Pipkin, T. (1989). *Games you can't lose.* New York: Pocket Books (a division of Simon & Schuster).

Asok, C. (1980). A note on the comparison between sample mean and mean based on distinct units in sampling with replacement. *The American Statistician, 34,* 159.

Babbage, C. (1864). *Passages from the life of a philosopher.* London: Longmans and Green. Reprinted by Augustus M. Kelly, New York, 1969.

Babbage, C. (1825). Observations on the application of machinery to the computation of mathematical tables. *Memoirs of the Astronomical Society, 1,* 311–314.

Balanda, K. P., & MacGillivray, H. L. (1988). Kurtosis: A critical review. *The American Statistician, 42,* 11–119.

Barker, R. G., Dembo, T., & Lewin, K. (1941). Frustration and regression. *University of Iowa Studies in Child Welfare,* 18.

Barnes, F. (1995). Can you trust those polls? *Reader's Digest,* July, 49–54.

Barron's. (1998). A house is not a home. Dec. 21., MW13.

Baum, J. (1986). *The calculating passion of Ada Byron.* New York: Archon Press.

Becker, G. (2000). Magnitude of transient error. *Psychological Methods, 5,* 315–329.

Berger, B. (1994). *Beating Murphy's law: The amazing science of risk.* New York: Dell Publishing.

Bernstein, P. (1997). Chanelling Pascal. *Worth Magazine, 6,* 29.

Blissmer, R. H. (1988). *Introducing computers.* New York: John Wiley & Sons.

Bloom, B. S. (1964). *Stability and change in human characteristics.* New York: Wiley.

Blumstein, P., & Schwartz, P. (1983). *American couples.* New York: William Morrow.

Boettcher, J. V. (2001). The spirit of invention. *Syllabus, 14,* 11, 10–12.

Book, A. S., Knap, M. A., & Holden, R. R. (2001). Criterion validity of the Holden psychological screening inventory social symptomology scale in a prison sample. *Psychological Assessment, 13,* 249–253.

Borgatta, E. F., & Bohrnstedt, G. W. (1980). Level of measurement once over again. *Sociological Methods and Research, 9,* 147–160.

Brown, D. C. (1994). Measurement on the hill. *The Score, 17,* 8–9.

Brown, W. K., Miller, T. P., Jenkins, R. L., & Rhodes, W. A. (1991). The human costs of "giving the kid another chance." *International Journal of Offender Therapy and Comparative Criminology, 35,* 296–302.

Bruner, J. S., & Goodman, C. C. (1947). Value and need as organizing factors in perception. *Journal of Abnormal and Social Psychology, 42,* 33–44.

Bureau of Justice Statistics. (1992). *Capital punishment: Summary data from BJS data series.* Rockville, MD: Clearing House, U.S. Dept. of Justice.

Camara, W. J. (2001). Do accommodations improve or hinder psychometric qualities of assessment? *The Score, 23,* 4–6.

Camara, W. J. (2001). Why we will never know who won the presidential election and many other things. *Score, 23,* 2–5.

Camilli, G., & Hopkins, K. D. (1979). Testing for association in 2 × 2 contingency tables with very small sample sizes. *Psychological Bulletin, 86,* 1011–1014.

Campbell, D. T. (1968). Quasi-experimental design. In D. L. Gills (ed), *International encyclopedia of the social sciences.* New York: Macmillan and Free Press.

Carper, J. (1993). *Food your miracle medicine.* New York: HarperCollins.

Caruso, J. C., & Cliff, N. (2000). Increasing the reliability of Wechsler Intelligence Scale for Children—Third Edition difference scores with reliable component analysis. *Psychological Assessment, 12,* 89–96.

Ceci, S. J., & Bronfenbrenner, U. (1991). On the demise of everyday memory. *American Psychologist, 46,* 27–31.

Chalmers, B. J. (1987). *Understanding statistics.* New York: Marcel Dekker.

Cobb, G. (1991). Teaching statistics: more data, less lecturing. *Amstat News,* December, *182.*

Cochran, W. G. & Cox, G. M. (1950). *Experimental Designs.* New York: Wiley.

Cohen, J., & Cohen, P. (1983). *Applied multiple regression/correlation analysis for the behavioral sciences.* (2nd ed.). Hillsdale, NJ: Lawrence Erlbaum.

Cohen, J. (1988). *Statistical power analysis for the behavioral sciences* (2nd ed.). Hillsdale, NJ: Lawrence Erlbaum.

Conover, W. J. (1974). Some reasons for not using the Yates continuity correction on 2 × 2 contingency tables. *Journal of the American Statistical Association, 69,* 374–376.

Consumer Reports. (1995). November.

Covvey, H. D., & McAlister, N. H. (1982). *Computer choices.* Reading, MA: Addison Wesley.

Cowles, M., & Davis, C. (1982). On the origin of the .05 level of statistical significance. *American Psychologist, 37,* 553–558.

Cowles, M. (1989). *Statistics in psychology: An historical perspective.* Hillsdale, NJ: Lawrence Erlbaum.

Creager, E. (2001). Burning them up at the beach. *Detroit Free Press* b6–b7. Aug. 7.

Cronbach, L. J. (1951). Coefficient alpha and the internal structure of tests. *Psychometrika, 16,* 297–334.

Crotty, M. (1999). *The foundations of social research: Meaning and perspective in the research process.* London: Sage.

Daly, M., & Wilson, M. (2000). Not quite right. *American Psychologist, 55,* 679–680.

David, F. N. (1962). *Games, gods and gambling.* New York: Hafner.

Dolan, K. A. (1998). The world's working rich. *Forbes Magazine,* July 6, 37–42.

Dwyer, C. A. (1991). Measurement and research issues in teacher assessment. *Educational Psychologist, 26,* 3–22.

Educational Testing Service (1991). *Sex, race, ethnicity and performance on the GRE general test.* Princeton, NJ: ETS.

Edwards, A. E. (1967). *Statistical methods* (2nd ed.). New York: Holt, Rinehart and Winston.

Edwards, A. E. (1972). *Experimental design in psychological research* (4th ed.). New York: Holt, Rinehart & Winston. (See especially Ch. 14.)

Embertson, S. E., & Reise, S. P. (2000). *Item response theory for psychologists,* Mahwah, NJ: Erlbaum.

Embertson, S. E. (1998). A cognitive design system approach to generating valid tests. *Psychological Methods, 3,* 380–396.

Ennett, S. T., Tobler, N. S., Ringwall, C. L., & Flewelling, R. L. (1994). How effective is drug abuse resistance education? A meta-analysis of project DARE outcome evaluations. *American Journal of Public Health, 84,* 1394–1401.

Eron, L. D. (1982). Parent-child interaction, television violence and aggression of children. *American Psychologist, 37,* 191–211.

Everitt, B. S. (1977). *The analysis of contingency tables.* London: Chapman & Hall.

Eysenck, H. J. (1960). The concept of statistical significance and the controversy about one-tailed tests. *Psychological Review, 67,* 269–291.

Fancher, R. E. (1985). *The intelligence men.* New York: W. W. Norton.

Fienberg, S. E. (1993). An adjusted census in 1990? *Chance, 5,* 28–38.

Fisher, E. G. (2000). *Numbers R us.* Fool.com, July 10.

Fisher, R. A. (1925). *Statistical methods for research workers.* Edinburgh, Scotland: Oliver and Boyd.

Fisher, R. A. (1935). *The design of experiments.* Edinburgh, Scotland: Oliver and Boyd.

Fisher, R. A. (1938). *Statistical methods for research workers* (7th ed.). Edinburgh: Oliver and Boyd.

Fisher, R. A. (1939). Student. *Annals of Eugenics, 9,* 1–9.

Fiske, E. B. (1988). America's test mania. *The New York Times* (November 13).

Fleming, L., & Snyder, W. U. (1947). Social and personal changes following nondirective group play therapy. *American Journal of Orthopsychiatry, 17,* 101–116.

Foster, E. M., & McLanahan, S. (1996). An illustration of the use of instrumental variables: Do neighborhood conditions affect a young person's chance of finishing high school? *Psychological Methods, 1,* 249–260.

Fraenkel, J. R., & Wallen, N. E. (2000). *How to design and evaluate research in education* (4E). Boston, MA: McGraw-Hill.

Gabor, T., & Tonia, B. (1989). Probing the public's honesty: A field experiment using the "lost letter" technique. *Deviant Behavior, 10,* 387–399.

Gallup, G., & Rae, S. F. (1968). *The pulse of democracy.* New York: Greenwood Press.

Galton, F. (1869). *Hereditary genius.* London: Macmillan.

Galton, F. (1883). *Inquiries into human faculty and its development.* London: Macmillan.

Galton, F. (1888). Co-relations and their measurement. *Proceedings of the Royal Society of London, XV,* 135–145.

Galton, F. (1899). *Natural inheritance.* London: Macmillan.

Galway, L. (1993). Nuts and bolts: How much can you trust your hardware? *Chance, 5,* 73–75.

Games, P. A. (1988). Correlation and causation: An alternative view. *The Score, 11,* 9–11.

Gauss, K. F. (1801). 1966. *Disquisitiones arithmeticae.* Translation by Arthur Clarke. New Haven: Yale University Press.

Geisinger, K. F. (1992). Metamorphosis in test validation. *Educational Psychologist, 27,* 197–222.

Geraci, R. (2000). Be better than average. *Men's Health,* June, 59–68.

Gillespie, D. F. (1994). Faith in measurement. *The Behavioral Measurement Letter, 2,* 5.

Gillin, J. C. (1994). What can sleep-wake disturbances tell us about the serious mental disorders? *The Decade of the Brain, 5,* 2–4.

Glass, G. V. (1976). Primary, secondary and meta-analysis of research. *The Educational Researcher, 10,* 3–8.

Glass, G., McGaw, B., & Smith, M. L. (1981). *Meta-analysis in social research.* Beverly Hills, CA: Sage.

Goodenough, F. L. (1949). *Mental testing.* New York: Rinehart & Company.

GRE Guide. (1990). Princeton, NJ: Educational Testing Service.

Green, S. B., Salkind, N. J., & Akey, T. M. (1997). *Using spss for Windows: Analyzing and understanding data.* Upper Saddle River, NJ: Prentice Hall.

Greenwald, A. G. (1997). Validity concerns and usefulness of student ratings of instruction. *American Psychologist, 52,* 1182–1186.

Guilford, J. P. (1956). *Fundamental statistics in psychology and education* (3rd ed.). New York: McGraw-Hill.

Gurman, E. R., & Balban, M. (1990). Self evaluation of physical attractiveness as a function of self-esteem and defensiveness. *Journal of Social Behavior and Personality, 5,* 575–580.

Gutloff, K. (1999). Is high-stakes testing fair? *NEA Today, 17,* 6–7.

Hafner, K. (2001). Thirty years of e-mail. *New York Times,* Dec. 13, C4.

Haig, B. D. (2000). Explaining the use of statistical methods. *American Psychologist, 55,* 962–963.

Harrington, G. M. (1955). Smiling as a measure of teacher effectiveness. *Journal of Educational Research, 49,* 715–717.

Hays, W., & Winkler, R. (1975). *Statistics: Probability, inference and decision* (2nd ed.). New York: Holt, Rinehart & Winston.

Heffernan, P. M. (1988). New measures of spread and a simpler formula for the normal distribution. *The American Statistician, 42,* 100–102.

Helms, J. E. (1992). Why is there no study of cultural equivalence in standardized cognitive ability testing? *American Psychologist, 47,* 1083–1101.

Hite, S. (1987). *Women and love, a cultural revolution in progress.* New York: Knopf.

Hoh, J. D. (1992). Creation of an information infrastructure. *The AAHE Bulletin, 8,* 2–4.

Horwood, L. J., & Fergusson, D. M. (1998). Breast feeding and later cognitive and academic outcomes. *Pediatrics, 101,* 9–15.

Howard, G. S., Maxwell, S. E., & Fleming, K. J. (2000). The proof of the pudding: An illustration of the relative strengths of null hypothesis, meta-analysis, and Bayesian analysis. *Psychological Methods, 5,* 315–329.

Howell, D. C. (1992). *Statistical methods for psychology.* (3rd ed.). Boston: Duxbury Press.

Hoyt, C. (1941). Note on a simplified method of computing test reliability. *Educational Psychology Measurement, 1,* 93–95.

Hsu, L. M. (2002). Fail-safe ns for one versus two-tailed tests lead to different conclusions about publication bias. *Understanding Statistics, 1,* 2, 85–100.

Huff, D. (1954). *How to lie with statistics.* New York: W. W. Norton.

Jacard, J. (1998). *Interaction effects in factorial analysis of variance.* Thousand Oaks, CA: Sage.

Jaccard, J., & Becker, M. A. (1990). *Statistics for the behavioral sciences,* (2nd ed.). Belmont, CA: Wadsworth.

Jackson, B. (1980). Statistical squabble. *The Wall Street Journal,* December 9, p. 24.

Jennings, K. (1990). *The devouring fungus: Tales of the computer age.* New York: W. W. Norton.

Jensen, A. R. (1969). How much can we boost IQ and scholastic achievement? *Harvard Educational Review, 39,* 1–123.

Jensen, A. R. (1978). Genetic and behavioral effects of nonrandom mating. In R. T. Osborne, C. E. Noble, & N. Weyl (eds.), *Human variation: Biopsychology of age, race, and sex.* New York: Academic Press.

Johns, M. W., (1992). Reliability and factor analysis of the Epworth Sleepiness Scale. *Sleep, 15,* 378–381.

Johnstone, S. M. (2001). What should be capturing faculty attention? *Syllabus, 14,* No. 12, 24.

Jones, J. H. (1997). *Alfred C. Kinsey: A public/private life.* New York: W. W. Norton.

Kaplan, R. M., & Saccuzzo, D. P. (1993). *Psychological testing* (3rd ed.). Pacific Grove, CA: Brooks/Cole.

Keppel, G. (1991). *Design and analysis: A researcher's handbook* (3rd ed.). Englewood Cliffs, NJ: Prentice Hall.

Kephart, N. C. (1971). On the value of empirical data in learning disability. *Learning Disabilities, 4,* No. 7, 393–395.

Kerlinger, F. N. (1986). *Foundations of behavioral research* (3rd ed.). New York: Holt, Rinehart & Winston.

Kevles, D. J. (1984). Annals of eugenics (part 1). *The New Yorker,* October 8, p. 60.

Kimble, G. A. (1978). *How to use (and misuse) statistics.* Englewood Cliffs, NJ: Prentice-Hall.

Kinsey, A. C., Pomeroy, W. B., & Martin, C. E. (1948). *Sexual behavior in the human male.* Philadelphia: W. B. Saunders.

Kirsch, I., & Lynn, S. J. (1999). Automaticity in clinical psychology. *American Psychologist, 54,* 504–515.

Krantz, L. (1992). *What the odds are.* New York: HarperCollins

Kupfersmid, J. (1988). Improving what is published. *American Psychologist, 43,* 635–642.

Lehman, D. R., Lempert, R. O., & Nisbett, R. E. (1988). The effects of graduate training on reasoning: Formal discipline and thinking about everyday life events. *American Psychologist, 43,* 431–442.

Lemonick, M. D. (1991). Erotic electronic encounters. *Time Magazine, 138,* September 23, p. 72.

Leshowitz, B. (1989). It is time we did something about scientific illiteracy. *American Psychologist, 44,* 1159–1160.

Lewin, K. (1952). Group decision and social change. In G. E. Swanson, T. M. Newcomb, & E. L. Hartley (eds.), *Readings in social psychology.* New York: Holt.

Li, H., Rosenthal, R., & Rubin, D. B. (1996). Reliability of measurement in psychology: From Spearman-Brown to maximal reliability. *Psychological Methods, 1,* 98–107.

Lindeman, R. H., & Merenda, P. F. (1979). *Educational measurement.* Glencove, IL: Scott Forsman.

Lipsy, M. W., & Wilson, D. B. (1993). The efficacy of psychological, educational and behavioral treatment: Confirmation from meta-analysis. *American Psychologist, 48,* 1181–1209.

Literary Digest. (1936). Vol. 122, October 31, pp. 5–6, and November 14, pp. 7–8.

Loh, W. Y. (1987). Does the correlation coefficient really measure the degree of clustering around a line? *Journal of Educational Statistics, 12,* 235–239.

Maris, E. (1998). Covariance and adjustment versus gain scores revisited. *Psychological Methods, 3,* 309–327.

Marsh, H. W., & Dunkin, M. J. (1992). Student evaluations of university teaching: A multidimensional perspective. In J. A. Smart (Ed.). *Higher education handbook on theory and research.* New York: Agathon Press, 143–234.

Matarazzo, J. D. (1992). Psychological testing and assessment in the 21st century. *American Psychologist, 47,* 1007–1018.

McGraw, K. O., & Wong, S. P. (1996). Forming inferences about some intraclass correlation coefficients. *Psychological Methods, 1,* 30–46.

McClelland, G. H., (2000). Increasing statistical power without increasing sample size. *American Psychologist, 55,* 963–964.

McIntire, S. A., & Miller, L. A. (2000). *Foundations of psychological testing.* Boston, MA: McGraw-Hill.

Mertler, C. A., & Vannata, R. A. (2001). *Advanced and multivariate statistical methods.* Los Angeles, CA: Pyrczak Publishing.

Miller, G. (1990). *Substance abuse subtle screening inventory.* Spencer, IN: Spencer Evening World.

Monroe, R. J., & McVay, F. E. (1980). Gertrude Mary Cox (1900–1978). *American Statistician, 34,* 44.

Moore, D. L. (1977). *Ada, countess of lovelace.* London: John Murray.

Mortimer, E. (1959). *Blaise Pascal: The life and work of a realist.* London: Shenval Press.

Mosteller, F., & Tukey, J. W. (1977). *Data analysis and regression: A second course in statistics.* Reading, MA: Addison-Wesley, p. 318.

Nathan, P. E. (1998). Practice guidelines: Not yet ideal. *American Psychologist, 53,* 290–299.

Newsweek, Oct. 14 (1994), p. 44.

New York Times (1994) Dec. 29, p. D4.

New York Times (1996), March 16, p. 9.

New York Times (1998), Kline & Co. as reported, Sept, 24, Fri, p. d5.

Norusis, M. J. (1988). *SPSS/PC+ studentware.* Chicago: SPSS Inc.

Novaco, R. W. (1975). *Anger control: the development and evaluation of an experimental treatment.* Lexington, MA: D.C. Heath.

Oakes, M. (1986). *Statistical inference: A commentary for the social and behavioral sciences.* New York: Wiley.

Osburn, H. G. (2000). Coefficient alpha and related internal consistency reliability coefficients. *Psychological Methods, 5,* 343–355.

Overall, J. E. (1980) Power of chi square tests for 2 × 2 contingency tables with small expected frequencies. *Psychological Bulletin, 87,* 132–135.

Page, R. M. (1990). Shyness and sociability: A dangerous combination for illicit substance use in adolescent males? *Adolescence, 25,* 803–806.

Parade. (1985). April 28, p. 14.

Passell, P. (1991). Can't count on numbers. *The New York Times,* August 6, 1991, pp. A1, A14.

Pate, R. R., Trost, S. G., Levin, S., & Dowda, M. (2000). Sports participation and health related behaviors among U.S. youth. *Archives of Pediatric and Adolescent Medicine, 154,* 904–911.

Paulos, J. A. (1988). *Innumeracy.* New York: Hill and Wang.

Pearson, E. S. (1938). *Karl Pearson: An appreciation of some aspects of his life and work.* Cambridge, England: Cambridge University Press.

Pearson, K. (1880). *The new werther.* London: Kagan Press.

Pearson, K. (1894). On the dissection of asymmetrical frequency curves. *Philosophical Transactions, 185,* 1–40.

Pearson, K. (1914–1930). *Life letters and labours of Francis Galton.* 3 vols. Cambridge, England: Cambridge University Press.

Pearson, K. (1936). Old tripos days at Cambridge. *Mathematical Gazette, 20,* 27–36.

Pocket World in Figures, 2001 Edition. London. *The Economist,* p. 70.

Poe, R. (2001). *The seven myths of gun control.* Roseville, CA: Forum.

Popham, W. J., & Sirotnik, K. A. (1973). *Educational statistics* (2nd ed.). New York: Harper & Row.

Posamentier, A. S. (2002). Madam, I'm 2002—a numerically beautiful year. *New York Times,* Jan. 2, a25.

Potash, J. B., Kane, H. S., Chiu, Y., Simpson, S. G., MacKinnon, D. F., McInnis, M. G., McMahon, F. J., & DePaulo, J. R. (2000). Attempted suicide and alcoholism in bipolar disorder. *American Journal of Psychiatry, 157,* 2048–2050.

Presby, L. (2001). Seven tips for highly effective online courses. *Syllabus 14,* No. 11, 17.

PsychSoft, Inc. (1992). *Sigmund.* P. O. Box 232, North Quincy, MA. 02171.

Quick Pro+II (1987). *Automatic program writer.* Orange Park, FL: ICR Future Soft.

Rimland, B. (1982). The altruism paradox. *Psychological Reports, 51,* 521–522.

Rodgers, J. L., Cleveland, H. H., van den Oord, E., & Rowe, D. C. (2000). Resolving debate over birth order, family size and intelligence. *American Psychologist, 55,* 599–612.

Roethlisberger, F. J., & Dickson, W. J. (1939). *Management and the worker.* Cambridge, MA: Harvard University Press.

Rogosa, D. R., & Willet, J. B. (1983). Demonstrating the reliability of the difference score in the measurement of change. *Journal of Educational Measurement, 20,* 335–343.

Rokeach, M. (1960). *The open and closed mind.* New York: Basic Books.

Rosenthal, N. E., & Blehar, M. C. (eds.) (1989). *Seasonal affective disorders and phototherapy.* New York: Guilford Press.

Rosnow, R. L., & Rosenthal, R. (1993). *Beginning behavioral research: A conceptual primer.* New York: Macmillan.

Ruffolo, J. S., Javorsky, D. J., Tremont, G., Westervelt, H. J., & Stern, R. (2001). A comparison of administration procedures for the Rey-Osterrieth complex figure. *Psychological Assessment, 13,* 299-305.

Rulon, P. J. (1939). A simplified procedure for determining the reliability of a test by split-halves. *Harvard Educational Review, 9,* 99–103.

Russell, J. L., & Thalman, W. A. (1955). Personality: Does it influence teacher's marks? *Journal of Educational Research, 48,* 561–564.

Ryan, T. A., Joiner, B. L., & Ryan, B. F. (1976). *Minitab student handbook.* North Scituate, MA: Duxbury Press.

Sackett, P. R., Schmitt, N., Ellingson, J. E., & Kabin, M. B. (2001). High-stakes testing in employment, credentialing and higher education. *American Psychologist, 56,* 302–318.

SAS Institute, Inc. (1982). *SAS introductory guide.* Cary, NC: SAS Institute Inc.

Scarne, J. (1961). *Scarne's complete guide to gambling.* New York: Simon & Schuster.

Schmidt, F. L. (1996). Statistical significance testing and cumulative knowledge in psychology: Implications for training of researchers. *Psychological Methods, 1,* 115–129.

Schwarz, N. (1999). Self-reports: How the questions shape the answers. *American Psychologist, 54,* 93–105.

Science Agenda. (1990). New directions in educational testing. *Science Agenda* August–September, p. 14.

Searle, S. R. (1989). Statistical computing packages: Some words of warning. *The American Statistician, 4,* 189–190.

Shadish, W. R. (1991). Meta-analysis and the exploration of causal mediating processes. *The Score, 14,* 9.

Shaywitz, B. A., Shaywitz, S. E., Pugh, K. R., Constable, R. T., Skudlarski, P., Fulbright, R. K., Bronen, R. A., Fletcher, J. M., Shankweller, D. P., Katz, L., & Gore, J. C. (1995). Sex differences in the functional organization of the brain for language. *Nature,* February 16, *373,* 607–611.

Sherif, M., & Sherif, C. W. (1956). *An outline of social psychology.* New York: Harper.

Sherman, L. W., & Berk, R. A. (1984). The specific deterrent effects of arrest for domestic assault. *American Sociological Review, 49,* 261–272.

Sheskin, D. J. (1997). *Handbook of parametric and nonparametric statistical procedures.* New York: CRC Press.

Siegel, S., & Castellan, N.J. (1988). *Nonparametric statistics.* (2nd ed.). New York: McGraw-Hill.

Silverstein, L. B., & Auerbach, C. F. (1999), Deconstructing the essential father. *American Psychologist, 54,* 397–407.

Smith, D. L. (2000). Gambling: Recreation or road to ruin. *Liguorian, 88,* 10–13.,

Smith, D. (2001). Is too much riding on high-stakes tests? *Monitor on Psychology,* 32, 58–59.

Smith-Slep, A. M., Cascardi, M., Avery-Leaf, S., & O'Leary, K. D. (2001). Two new measures of attitudes about the acceptability of teen dating aggression. *Psychological Assessment, 13,* 306–318.

Sohn, D. (2000). Significance testing and the science. *American Psychologist, 55,* 964–965.

SomeWare in Vermont, Inc. (1988). *Ecstatic.* P. O. Box 215, Montpelier, Vermont 05602.

Spearman, C. (1904). General intelligence objectively determined and measured. *American Journal of Psychology, 15,* 201–293.

Spence, J. T., Underwood, B. J., Duncan, C. P., & Cotton, J. W. (1976). *Elementary statistics* (3rd ed.). Englewood Cliffs, NJ: Prentice-Hall.

Spencer, B. (1995). Correlations, sample size, and practical significance: A comparison of selected psychological and medical investigations. *Journal of Psychology, 129,* 469–475.

Sprinthall, R. C., & Nolan, T. E. (1991). Efficacy of presenting quantities with different pictures in solving arithmetic word problems. *Journal of Perceptual Skills, 71,* 1–5.

Sprinthall, R. C., Schmutte, G. T., & Sirois, L. (1990). *Understanding educational research.* Englewood Cliffs, NJ: Prentice-Hall.

Sprinthall, R. C., Sprinthall, N. A., & Oja, S. (1998). *Educational psychology: A developmental approach* (7th ed.). New York: McGraw-Hill.

Stallings, W. M., & Smock, H. R. (1971). The pass-fail grading option at a state university. *Journal of Educational Measurement, 8,* 153–160.

Stanley, T. J., & Danko, W. D. (1996). *The millionaire next door.* Atlanta, GA: Longstreet Press.

Steering Committee of the Physicians Health Study Research Group. (1989). Final report on the aspirin component of the ongoing physicians health study. *New England Journal of Medicine, 321,* 129–135.

Stevens, S. S. (1946). On the theory of scales of measurement. *Science,* 103, 677–680.

Stigler, S. M. (1986). *The history of statistics: The measurement of uncertainty before 1800.* Cambridge, MA: Harvard University Press.

Stinnett, S. (1990). Women in statistics: Sesquicentennial activities. *American Statistician, 44,* 74–80.

Stoline, M. R. (1981). The status of multiple comparisons: Simultaneous estimation of all pairwise comparisons in one-way ANOVA designs. *The American Statistician, 35,* 134–141.

Student. (1908). The probable error of the mean. *Biometrika, 6,* 1–25.

Tabachnick, B., & Fidell, L. S. (1996). *Using multivariate statistics.* New York: Addison-Wesley-Longman.

Time magazine. (1990). August 13.

Toole, B. A. (1992). *Ada, the enchantress of numbers.* Mill Valley, CA: Strawberry Press.

Turnbull, W. W. (1980). *Test use and validity.* Princeton, NJ: Educational Testing Service.

USA TODAY (1994) October 5, p. 5d.

Utah Law Enforcement Statistics, 1992.

Veiel, H. O. F., & Koopman, R. F. (2001). The bias in regression based indices of premorbid IQ. *Psychological Assessment, 13,* p. 356, 356–368.

Wachter, K. W., & Straf, M. L. (eds.). (1990). *The future of meta-analysis.* New York: Russell Sage Foundation.

Wake, M., Coghlan, D., & Hesketh, K. (2000). Does height influence progression through primary school grades? *Archives of Disease in Childhood, 82,* 297–301.

Walker, H. M. (1978). Karl Pearson. In W. H. Kruskal & J. M. Tanur (eds.), *International encyclopedia of statistics.* New York: The Free Press.

Waterman, B. B. (1994). Assessing children for the presence of a disability. *News Digest, 4,* 1–25. [Published in Washington, DC, by the National Information Center for Children and Youth with Disabilities (NICHCY).]

Welch, G. S. (1952). An anxiety index and an internalization ratio for the MMPI. *Journal of Consulting Psychology, 16,* 65–72.

Wilkinson, L., & The Task Force on Statistical Thinking. (1999). Statistical methods in psychology journals: Guidelines and explanations. *American Psychologist, 54,* 594–604.

Yarenko, R. M., Harari, H., Harrison, R. C., & Lynn, E. (1982). *Reference handbook of research and statistical methods in psychology.* New York: Harper & Row.

Yates, F. (1979). Gertrude, Mary Cox, 1900–1978. *Journal of the Royal Statistical Society, 142,* 516–517.

Youngman, H. (1991). *Take my life, please!* New York: William Morrow.

Zajonc, R. B. (1986). Family configuration and intelligence. *Science, 192,* 227–236.

Answers to Odd-Numbered Items (and Within-Chapter Exercises)

The answers given here are those for all the odd-numbered problems found at the end of each chapter. For all problems requiring mathematical calculations, the rounding method used is to round up one digit when the last dropped number is a five or greater. This is consistent with the method built into electronic hand calculators (on those models that can be set automatically to round off). For Chapters 2 through 6, the answers will be rounded to two places. From Chapter 7 on, we've decided to round to three places to make our answers more consistent with those found using the computer program. This means that every calculation in the formula has been rounded to three places, not just the eventual solution. For those of you who feel that two-place accuracy is sufficient, two-place answers will also be given in parentheses. These are not just rounded to two places after calculations have been carried out to three but are two-place answers created by rounding every calculation leading up to the final answer. Except for a few problems in the later chapters where you have been instructed to chain-multiply, all calculations are to be rounded, that is, round each and every time you add, subtract, multiply, divide, square, or take a square root.

Chapter 1

1. The difference between the two population sizes makes it impossible to compare the fatality numbers directly. Although the population of Americans in Vietnam during the war exceeded 500,000, the number of Americans in the United States at the time was well over 200 million. Percentage, or per-capita comparisons, would be more meaningful.
3. On the sole basis of the "evidence" from the commercial, the acid can do nothing at all to your stomach—unless your stomach is made out of a napkin.
5. The assumption may be criticized on a number of points. First, no comparison or control group is mentioned. One could therefore ask, "Twenty-seven percent fewer than what?" If it is assumed that 27% fewer means fewer than before the toothpaste was used, then some mention of other variables should be made, for example, age, sex, and brushing habits. It is known, for one thing, that teenagers have more dental caries than do adults. Was the before measure done on teenagers and the after measure taken on these same individuals when they reached their twenties? Or does the 27% fewer comparison refer to another group of individuals who used a different kind of toothpaste? If so, again, were the groups equivalent before the study began, that is, same age, sex, brushing habits, and so on?
7. The consumer should first learn the answer to the question, "Thirty-five percent more than what?" Without any basis for comparison, the advertiser's claim is meaningless. Perhaps the new detergent is 35% more effective than seawater, bacon fat, or beer.
9. The conclusions can be criticized on at least three counts. First, antisocial behavior is not always followed by a court appearance. Perhaps the seemingly high number of these particular juveniles appearing in court is because they didn't possess the skills needed to avoid arrest. Second, we would have to know the percentage of juveniles in that particular area who are assumed to be afflicted with learning disabilities. That is, if 40% of the juveniles in that area are considered to be learning disabled, then the arrest statistics are simply reflecting the population parameters. In other words, we should be asking the question, "Compared to what?" What percentage of the total population of learning-disabled juveniles in that area do not appear in court? Third, we should have some information regarding how the researchers made the diagnosis of "learning disabled."
11. First, there would have to be a comparison made with the graduation percentages for the nonathletes (which in many colleges and universities is also about 50%). Second, there should be a comparison drawn with regard to college major. In some academic majors it is reasonable to extend beyond the traditional four years. In other situations, such as an athlete simply majoring in "eligibility," graduating within the four-year span may not present a hardship. Perhaps age is an important variable in this analysis. The student athletes are typically the youngest members of the student body. Very few older students who decide later in life to complete a college degree program also try out for the varsity football or basketball team.
13. Many other factors could have caused this increase in drug charges. It may be that senior citizens are using MORE prescription drugs than eight years earlier, and that the individual drugs may even be getting cheaper. Newer drugs also may be used in lieu of expensive operations that had previously been the common mode of treatment. For example, heart by-pass patients who typically paid between 40,000 and 50,000 dollars, are now being treated with drugs that are producing results superior to surgery. Also, there may be MORE older Americans in 2000 than was the case 8 years earlier. This may be due to people living longer than previously, thus increasing the size of the senior citizen population. Finally, has the definition of senior citizen changed over that time span to include persons who eight years earlier would not have been so labelled, again increasing the size of the census? Therefore, it may be that aspirins were no more expensive in 2000 than they were in 1992.
15. We would have to know whether the record was based on ticket prices, adjusted for inflation, or on the total number of tickets sold. It turns out that this show also introduced the highest cost,

up to $100 per ticket, of any show in history up to that time.

17. Not at all. As Poe (2001) has said, civilians can pick and choose whether to get involved in a crime scene. They can often avoid situations that are confusing and threatening. Police, however, must take action every time, and thus are put in situations where guilt or innocence are not so clear-cut and in which the chance for mistakes is far higher.

19. One would have to know what percent of the drivers are men and what percent of the driving is done by men. In fact, since 74% of drivers are men, the actual numbers of fatalities is considerably higher for the male driver. In 1997 there were roughly 40,000 auto fatalities, and almost 30,000 of them involved male drivers. Men are actually twice as likely to die in an auto accident than are women. Men drive faster, more often, and are less apt to use seat belts (and won't let a women who is drunk drive home.)

Chapter 2

1. Mean = 11.500 median = 12.000 mode = 12
3. Mean = 10.620 median = 12.000 mode = 12 and 3
5. **a.** Mean = 21.250 median = 12.000 mode = 12
   b. The median, since the distribution is severely skewed.
7. **a.** Not skewed Sk = 0
   b. Sk+ (skewed to the right)
   c. Sk– (skewed to the left)
9. **a.** Mode
   b. Mean
11. Sk–
13. Mode
15. T
17. F
19. F
21. T

Chapter 3

1. Mean = 10.375 (10.38) range = 19 *SD* = 4.81
3. Although the means and ranges are identical, the standard deviations differ. Because the *SD* is smaller, company *X* is more homogeneous (indicating that the scores are more centrally located).
5. **a.** Mean = 2.71
   b. range = 2.00
   c. *SD* = .69
7. $M = 15.000$, $Mdn = 15.000$, $Mo = 15$, $R = 10$, $SD = 2.449$ (or as estimated by SPSS std. dev. = 2.582)
9. $M = 13.10$, $Mdn = 13.50$, $Mo = 14$, $R = 12$, $SD = 3.08$ (or as estimated by SPSS std. dev. = 3.247)
11. Range = 0 and the *SD* = 0
13. The larger the *SD*, the more platykurtic the distribution. The smaller the *SD*, the more leptokurtic the distribution.
15. *SD* should approximate 20.
17. Range should approximate 90.
19. T
21. T
23. T
25. T

Chapter 4

Within chapter

4a. 1. 39.44%
2. 24.86%
3. 49.51%
4. 3.98%
5. 44.74%

4b. 1. 99.49%
2. 14.69%
3. 21.48%
4. 94.63%
5. 55.57%

4c. 1. 82.90% (83rd percentile)
2. 97.50% (98th percentile)
3. 25.14% (25th percentile)
4. 4.95% (5th percentile)
5. 69.15% (69th percentile)

4d. 1. 55.96%
2. 2.28%
3. 15.15%
4. 99.56%
5. 46.41%

4e. 1. 78.19%
2. 53.47%
3. 15.34%
4. 37.99%
5. 2.69%

4f. 1. 23.89%
2. 33.15%
3. 36.86%
4. 46.16%
5. 39.97%

4g. 1. 76.11%
2. 11.90%
3. 92.07%
4. 54.78%
5. 5.59%

4h. 1. 7.78%
2. 46.81%
3. 73.57%
4. 56.75%
5. 22.66%

4i. 1. 90.04%
2. 81.15%
3. 20.62%
4. 23.28%
5. 36.91%

End of chapter

1. **a.** 25.80%
b. 39.97%
c. 16.28%
d. 3.59%

3. **a.** 10.56%
b. 95.05%
c. 9.18%
d. 57.53%

5. 62.55%

7. **a.** 79.67%
b. .62%
c. 44.01%
d. 15.58%

9. **a.** 24.20%
b. 33.72%

11. **a.** .62%
b. 6.68%
c. 30.23%
d. 13.59%

13. Mean
15. F
17. T
19. T
21. F
23. F

Chapter 5

Within chapter

5a. 1. 108.40
2. 140.95
3. 75.40
4. 98.50
5. 138.70

5b. 1. 1.65
2. –1.28
3. –0.28
4. 0.13
5. 0.67

5c. 1. 1488.30
2. 1511.70
3. 1406.40
4. 1572.90
5. 1351.50

5d. 1. 20.00
2. 28.85
3. 62.50
4. 20.27
5. 26.32

5e. 1. 53.90
2. 264.50
3. 10.05
4. 83.43
5. 1496.00

5f. 1. 39.50
2. 61.60
3. 32.10
4. 78.90
5. 55.30

5g. 1. 31.85
2. 42.65
3. 23.75
4. 26.45
5. 47.60

End of chapter

1. **a.** $15.25
b. $13.28
c. $7.96
d. $7.08

3. 75.20

5. **a.** 3.60
b. 21.60
c. 78.80
d. 57.20

7. **a.** 2.92
b. 2.92
c. 2.92

9. **a.** 51
b. 35.80
c. 70.30
d. 45
11. **a.** 531.50
b. 340.60
c. 614.50
d. 465.10
13. **a.** −.67
b. 43.30
15. F
17. T
19. T
21. 45.90
23. .52
25. 3.66
27. The SD is too high. It cannot be greater than half the range
29. The skew goes to the left (negative), not to the right.

Chapter 6

Within chapter

6a. **1.** .20
2. .98
3. .87
4. .29
5. .05

6b. **a.** = .08
b. = .05
c. = .01
d. = .12

End of chapter

1. **a.** .166 = .17
b. Odds are 5 to 1 against
3. **a.** .019 = .02
b. 51 to 1 against
c. .019 = .02 (the same as the first time)
5. **a.** .282 = .28
b. .128 = .13
7. $p = .09$
9. $p = .451$
11. $p = .215$
13. .239 = .24
15. **a.** .45
b. .55
c. 1.00
17. **a.** .09
b. .67
c. .17
19. **a.** (1/7)(1/7)(1/7) = .003. (It's a bad bet, since three oranges will appear only about 3 times in every 1000 pulls.)
b. (6/7)(6/7)(6/7) = .629 = .63. (It's a much better bet that no oranges will appear, but that's not much solace since it doesn't pay off.)
21. 1.00
23. 50%
25. T
27. F
29. T
31. T

Chapter 7

1. $p = 0.309$, or .31
3. **a.** Standard error of the mean = 2.372 (2.37)
b. $z = 2.108$ (2.11); reject H_0 at $p < .05$
5. **a.** 1.061 (1.06)
b. $z = 3.770$ (3.77); reject H_0 at $p < .01$. Since the calculated z of 3.77 is greater than the z score of 2.58 (which is known to exclude the extreme 1% of the distribution), the sample probably does NOT represent the known population.
7. 138 women
9. 178 men
11. 308 men
13. Parameter
15. Inferential
17. Statistic
19. Random
21. Bias
23. Normality
25. **a.** 0.50
b. 0.50
27. T
29. F
31. Statistic
33. Parameter
35. Statistic

Chapter 8

1. $t = 1.389$ (1.39)
   t at .05 with 24 df = 2.064; accept H_0. The population mean could be 6.00.
3. $t = -2.889$ (−2.89)
   t at .05 with 9 df = 2.262; reject H_0 at $p < .05$. The difference is significant. The sample is probably not representative of the population.
5. **a.** The point estimate of the population mean would be equal to 519 (the mean of the sample).
   b. From 501.457 to 536.543 (501.46 to 536.54)
   c. No, the assumed national mean of 451 falls outside (below) the interval.
7. **a.** *SE* of mean = 1.687 (1.69)
   b. At .95, the interval would be from 15.816 to 8.184 (15.82 to 8.18).
9. **a.** *SE* of mean = 1.203 (1.20)
   b. At .95, the interval is from 12.330 to 6.782 (12.33 to 6.79).
   c. At .99, the interval is from 13.592 to 5.520 (13.59 to 5.53).
11. $t = 2.935$ (2.94). Reject H_0. Sample is not representative.
13. *ES* = 2.122 (2.12). Very strong effect.
15. ($t(9) = -1.540$, ns). Accept H_0. Chances are good that this sample is representative of the population.
17. Normal (z)
19. Null
21. Type 1
23. 3.106
25. Confidence interval
27. Sample
29. T
31. T
33. T

Chapter 9

1. Hypothesis of difference: post facto
3. Hypothesis of difference: post facto
5. Hypothesis of association: post facto
7. IV is gender (assigned subject)
9. IV is level of income (assigned subject)
11. IV is made up of scores on the anxiety test. The prejudice measures comprise the DV. The IV is an assigned-subjects variable.
13. This study, since it is before-after with no separate control, is open to several possible confounding variables, such as (1) the Hawthorne effect (perhaps the flattery and attention paid by the researchers produced some gains); (2) growth and development—the children are, after all, six months more mature when the second test was given; and (3) reading experience outside the classroom might produce gains, with or without the special training.
15. The time span between the pre- and post-testing. Subjects may change as a result of a host of other variables, perhaps some of which are, unknown to the researcher, systematically related to both the independent variable and dependent variable. Thus, there should be a separate control group.
17. **a.** Experimental
    b. IV is whether or not subjects received the psychotherapy, and the DV is based on the judge's evaluations of the subjects' symptoms.
    c. It would be best to keep the judges on a "blind" basis to help avert any unconscious bias in symptom evaluation.
    d. Perhaps the subjects should also be equated on the basis of gender. The difficulty in this study is not that too many matching variables have been overlooked but in finding subjects who indeed match up on that rather long list of variables.
19. F
21. T
23. T
25. F
27. T
29. T
31. Nominal
33. Nominal
35. Interval

Chapter 10

1. **a.** $t = 1.803$ (1.80). Accept H_0
   b. No
   c. Post Facto
   d. IV handedness, DV picture arrangement scores.
3. **a.** $t = 2.749$ (2.77). Reject H_0 at $p < .05$
   b. Yes
   c. Experimental
   d. IV illumination, DV productivity scores.

5. **a.** $t = -1.517$ (−1.52). Accept H_0
   b. No
   c. Experimental
   d. IV therapy, DV anxiety scores.
7. **a.** $t = 8.282$ (8.33). Reject H_0 at $p < .01$
   b. Yes
   c. Experimental
   d. IV exercise, DV cholesterol levels.
   e. ES as $d =$ 3.704 (a very strong effect)
9. **a.** t = 3.436
   b. CI = 2.855 to 11.045
   This is experimental research. The researcher manipulated the IV (whether they got the treatment or not). Since the groups are independent of each other, the design is between-subjects (after-only).
11. Less likely to reject H_0.
13. At least interval.
15. Estimated standard error of difference.
17. Estimated standard error of difference.
19. H_0, the null hypothesis
21. H_a, the alternative hypothesis
23. T
25. T
27. F

Chapter 11

1. $r = .907$ (.91)
   r at .01 with 8 df = .765. Reject H_0, $P < .01$; the correlation is significant. Those with higher degrees of party loyalty tend to work more hours. We don't know, however, whether loyalty produces more work hours or whether more work hours produce loyalty, or even whether both loyalty and work hours result from some third factor.
3. **a.** $r = -.870$ (−.87)
   b. r at .01 with 13 df = .641. Reject H_0, $p < .01$; the correlation is significant.
   c. Both of these hypotheses may be valid, but neither is proved by this analysis.
5. $r = .921$ (.92)
   r at .01 with 5 df = .874. Reject H_0, $p < .01$; the correlation is significant. The evidence from this sample suggests that the two variables are related in the population.
7. $r_s = .636$ (.64). Accept H_0. Not significant
9. $r_s = -.908$ (−.91). Reject H_0, $p < .01$. Sample scores suggest an inverse relationship between reading scores and musical ability. The correlation itself, however, cannot produce the reason for this relationship.
11. $r = -.602$ (−.59); r at .05 (8) = .632. Accept H_0. The correlation is NOT significant. There is not a significant relationship between the executive's total compensation and the amount of the per-capita contributions to the fund. If one were to suspect that the income *and* contribution distributions might be skewed in the population, and well they might be, then the Spearman r_s could be calculated. Spearman $r_s = -.467$ (−.47). The Spearman r_s with $N = 10$ requires a correlation of .649 to be significant at the .05 level. Therefore, the null is accepted, again showing no dependable relationship between the two variables.
13. $Z = 3.007$ (3.29). Reject H_0 at .01 and conclude that the two correlations are significantly different and do represent separate populations. In comparing the conclusions of problems 12 and 13, notice how important sample size is to significance.
15. **a.** No, since the analysis of the data is correlational.
    b. It may be that spending more increases the enrichment of the educational system and provides for more intellectual stimulation. It also may be that higher-income parents move to states that are spending more on education (and of course bring their bright children with them).
17. The Pearson r
19. Spearman r_s
21. T
23. F
25. F
27. T
29. F
31. F

Chapter 12

1. $F = 24.850$ reject H_0 $p < .01$
3. Accept H_0 $F = 4.000$
   within rounding error, $t = 2.000$
5. **a.** $F = 1.346$
   b. Accept H_0, not significant
   c. Tukey's not appropriate
   d. This is experimental research, with active manipulation of the IV.

e. The IV is the MgPe and the DV the retention scores.
f. On the basis of these data, MgPe shows no effect on retention.

7. F for rows (exercise) = 118.814 (119.10); reject H_0, significant at $p < .01$
F for columns (temperature) = 41.616 (41.71); reject H_0, significant at $p < .01$
F for the interaction = 2.428 (2.45); accept H_0, not significant
9. **a.** The F for rows (type of training) was 108.000 and was significant at .01, since with 1 and 12 *df*, an F ratio of 9.33 was needed to reject H_0 at .01.
b. The F for columns (time) was 16.333 and was significant at .01, since with 2 and 12 *df* an F of 6.93 was needed to reject H_0 at .01.
c. The F for the interaction was 1.000 and was not significant since with 2 and 12 *df* an F of 3.88 was needed to reject at .05.
11. Total mean
13. Group mean
15. Degrees of freedom
17. Between larger than within
19. Between df = 4 and within df = 25
21. T
23. F
25. T
27. F
29. F

Chapter 13

1. Chi square = 30.000; reject H_0. Significant at $p < .01$.
3. Chi square = 1.666 (1.66); accept H_0. Results do not differ from chance.
5. Chi square = 33.586 (33.58); reject H_0. Significant at $p < .01$.
7. Chi square = 4.510 (4.50). Significant at $p < .05$.
9. Chi square = 7.504 (7.51)
a. Reject H_0: $p < .01$.
b. Post facto
c. Independent variable, sex (a subject variable). Dependent variable, union membership.
11. Chi square = 21.400 (21.39); reject H_0. Significant at $P < .01$.
13. Chi square = 4.934 (4.95); accept H_0.
15. Chi square = 22.273 (22.27); reject H_0 at $p < .01$. Significantly more subjects were judged to be assertive in the post-test.
17. Chi square = 6.618 (6.62); reject H_0 at $p < .05$. C = 0.245 (.24).
19. Chi square (with Yates) = 0.085. Accept H_0. No significant difference.
21. No, chi-square equals 0.447, and since with 2 *df* we need 5.99 to reject null we must accept the null hypothesis. Thus, there is no difference in outcome as a function of jury size.
23. Chi-Square = 22.289, which with 2 degrees of freedom is significant at $p < .01$.
25. $F_0 = F_e$
27. Before-after or matched-subjects
29. T
31. T

Chapter 14

1. **a.** 545
b. 590
c. 680
3. **a.** $r = .861$ (.86) and is significant at $p < .01$
b. 11.552 (11.55)
5. **a.** IQ of 80 predicts GPA of 2.750 (2.25)
IQ of 95 predicts GPA of 2.900 (2.70)
IQ of 105 predicts GPA of 3.000 (3.00)
IQ of 130 predicts GPA of 3.250 (3.75)
b. Post facto
c. IQ is the IV and GPA is the DV.
7. Predict an IQ of 125, the mean of the Y distribution.
9. **a.** 66.709 to 72.501
b. 68.649 to 74.441
c. 73.014 to 78.806
11. 7.830, 4.274, and .718 (7.83, 4.27, and .71)
13. F = 151.170 (131.67). Reject H_0 at $p < .01$. The multiple R is significant.
15. A partial correlation of .14 is not significant, since at .05 and 100 df, we would need an r of .195 to reject H_0.
17. **a.** Between Digit Span and Information, r = .654
b. Between Digit Span and Comprehension, r = .698
c. Between Information and Comprehension, r = .806
d. The multiple correlation for predicting Comprehension from both Digit Span and Information, R = .837
19. **a.** R = .858 (.85). F = 23.697 (21.86).
Reject H_0. $p < .01$. Yes.
b. 9.747 (9.75)

21. $F = 17.393$ (16.88). Reject H_0 at $p < .01$. The multiple R is significant.
23. Child's IQ = 112.910 (112.85)
25. Partial $r = .092$ (.08)
27. Partial $r = .375$ (.38)
29. $F = 3.795$ (3.86). Reject H_0 at $p < .05$. The multiple R is significant.
31. Slope
33. Intercept
35. Multiple R
37. .50 or 1/2
39. Negative
41. Intercept or "a" term
43. T
45. F
47. F

Chapter 15

1. **a.** $t = 2.732$ (2.74); at .05, t with 7 df = 2.365; reject H_0 at $p < .05$.
   b. Fatigue, since the muscle strength became significantly lessened.
3. $F = 12.995$ (12.95); at .01, F with 2 and 10 df = 7.56; reject H_0 at $p < .01$.
5. $F = 25.309$ (25.18); at .01, F with 2 and 12 df = 6.93; reject H_0 at $p < .01$.
7. **a.** This was experimental research that used a one-group before-after design.
   b. The IV was the SAT coaching course, and the DV was made up of their scores on the SAT.
   c. The results were as follows: First, the Pearson r of 0.845 (.84) was significant, since with 18 degrees of freedom we needed an r of .44 at the .05 level. This shows that the two sets of scores were not independent and that the paired t was appropriate. However, the paired t of –1.591 (–1.57) was not significant, since with 19 degrees of freedom, a table value of 2.093 was needed to reject H_0 at .05.
   d. Had the results been significant, the findings could have been challenged on the basis of not having an adequate control group. Students might have done better in the second test situation simply because of the passage of time (which could produce the potential for both practice and maturation).
9. Repeated-measures and matched-subjects
11. For a given number of scores, the paired t has just half the number of degrees of freedom than would be the case for the independent t.
13. Higher, assuming that the matching process works and produces a correlation
15. Dependent variable
17. Interval
19. T
21. T
23. T
25. F
27. F
29. T

Chapter 16

1. $Zu = 1.810$ (1.81); accept H_0. Difference is not significant.
3. $T = 8.500$ (8.50); accept H_0. Difference is not significant.
5. $H = 6.740$ (6.74); reject H_0. Difference is significant at $p < .05$.
7. $T = 18.500$ (18.50); accept H_0. Difference is not significant.
9. Mann–Whitney U test.
11. Friedman ANOVA by ranks
13. T
15. F
17. F
19. Null
21. Spearman's r_s
23. Mann–Whitney U test
25. Kruskal–Wallis H test

Chapter 17

1. $r_{tt} = .890$ (.89) and with 8 df is significant at $p < .01$. The test is reliable, and because of high scores subject 9 should be selected for further testing and discussion.
3. **a.** $r = .879$ (.88) and is significant at $p < .01$.
   b. As corrected by the Spearman–Brown prophecy formula, $r_{tt} = .936$ (.94)
5. **a.** 2 times longer
   b. 6 times longer
   c. 3 times longer
   d. 3½ times longer
   e. 4 times longer

7. Pearson r as point biserial = 0.909 (.91) and is significant at $p < .01$. Or point biserial = [(13.200 – 10.000)/3.521][1] = .908 (.91)
9. Standard error of measurement = 22.400 (22.00)
11. Standard error of measurement = 44.700 (45.00)
13. r_{tt} = .910 (.91)
15. KR = .858 (.86)
17. .794 (.79)
19. F = 6.015 and is significant. Reject H_0 at $p < .01$. r_{tt} = .834 (.84). The test is reliable.
21. F
23. T
25. T
27. F
29. T
31. Reliability
33. Validity
35. Validity
37. Validity

Index